CHEMICAL TABLES

for
Laboratory and Industry

MATTER ● NUMBER ● PROCESS

WOLFGANG HELBING
Director of Studies, Karlsruhe

and

ADOLF BURKART
Director of Studies, Karlsruhe

(Translated from German by R. Vardharajan and Sudhanshu Gupta)

A HALSTED PRESS BOOK

JOHN WILEY & SONS
New York   Chichester   Brisbane   Toronto

FOREWORD

Through a special way of describing and summarizing we have combined the information of the tabular work with that of a textbook. Colours have been used at certain places as methodical and didactical means. The use of technical words and comments has been made for a better understanding of the tabular and graphic matter. Therefore, this book also serves the purpose of a review course.

In the beginning of the first part 'Matter' (135 pages) an introduction of the chemical elements and their properties have been put together followed by an illustrative wave-mechanical model of the atom and a description of the chemical compounds. The synopses and tables of the inorganic compounds (matter) have been compiled in special ways based on the periodic classification of elements and according to their common properties. In the case of the organic compounds the syntheses provided the structural framework of the molecules and the kind of functional groups present as the principle of classification. Through the use of screens the tabular matter has been made easy to understand and the grouping of the various classes of matter made clear. The relocation of information through the group numbers on the respective pages has also been simplified. The inorganic chemistry part of the tables consists of information about 1300 types of matter and their properties. The organic portion has just as many compounds. The choice of the compounds included in the tabular part has been based on the matter included in the textbooks on inorganic and organic chemistry.

The second part 'Number' (45 pages) does not only contain the important values of chemical and physio-chemical quantities and constants, but also the rules of calculation and principles which are of importance for the chemical reactions and methods. Of special significance is the gas reduction table. This part is supplemented by the log tables included as appendix (36 pages) which contain apart from the 5-digit logarithm tables about 10,000 so-called decadic supplements, i.e., the logarithms of the reciprocal values of the numbers. In calculations the logarithm numbers need only to be added up. Finally, the lists also contain a large number of useful numerical figures: reciprocal values, square numbers, square roots, circumferences and areas of the circle.

The third part 'Process' (66 pages) contains principal methods for research laboratories and industry, and the numerical values used there. Graphic description of the essential apparatus and other vivid symbols are of great help for understanding the text matter that has been presented in a concise form here. Apart from the general methods qualitative as well as quantitative inorganic and organic analytical methods have also been described. In connection with the analytical tables of numerical values short notes have been inserted at places which considerably enhance the purely mathematical value of the tables. A well-compiled list of technical words with about 5,000 such key-words simplify the location of the required information about the concepts, definitions, rules, laws or methods as well as of the matter and its characteristics.

We would also like to thank all those who in the publishing house extended to us all the facility in bringing out this book. The composing and printing of the chemical formulae put the publishers to a number of difficulties.

Karlsruhe, Spring 1969

Wolfgang Helbing
Adolf Burkart

First English Edition, 1979. Authorized translation of the edition published by Georg Westermann Verlag, Braunschweig

Published in the Western Hemisphere by Halsted Press, a division of John Wiley & Sons, Inc., New York

Library of Congress Cataloging in Publication Data

Helbing, Wolfgang.
    Chemical tables for laboratory and industry.

    Translation of Chemie-Tabellen fur Labor und Betrieb.
    "A Halsted Press book."
    Includes indexes.
1. Chemistry, Technical—Tables, etc. I. Burkart, Adolf, joint author. II. Title.
TP151.H3813   540'.21'2   79-26137
ISBN 0-470-26910-3

Printed in India at Raj Bandhu Industrial Company, New Delhi

CONTENTS

MATTER

Basic Materials or Elements.

Structure of basic materials

Basic materials or elements are substances which cannot be split into more other substances through chemical methods.

All fundamental matters are built up of atoms.
Atoms have spherical shape and consist of a nucleus and a sheath.

Diameter of the sheath nearly 2 to 5.10^{-8} cm

Diameter of the nucleus nearly 1 to 3.10^{-12} cm"

Mass of the atom nearly 2 to 500.10^{-24} g

Atomic particles

Name and symbol		Mass in g	in me	in ME	Electrical charge in electrostatic units
Proton	p	$1,6725.10^{-24}$	1836	1,0076	$+ 4,80.10^{-10}$
Neutron	n	$1,6748.10^{-24}$	1838	1,0090	nil
Electron	e$^-$	$9,1089.10^{-28}$	1	$5,5.10^{-4}$	$- 4,80.10^{-10}$

The atomic nucleus consists of neutrons and protons.
In conformity to law, the electrons are distributed in the atomic shell.

The number of electrons in the shell is equal to the number of protons in the nucleus and equals the atomic number of elements in the periodic system of elements.

The chemical properties of the elements depend on the structure of the shell.

A fundamental matter or an element is a substance in which all their shells have the structure and out of which atom is built.

Till now, we know 103 such elements.

The number of neutrons in the nucleus amounts to nearly 1.2 to 2.5 times the proton number and thus does not comprehend regularity.

Till now we know about 1,350 different atomic nuclei of which 324 occur in nature. The others are produced artificially and most undergo radioactive decay in short time.

Atoms whose nuclei have the same proton number but different neutron number, possess similarly built shells and have same atomic number, that is, they stand at the same place in the periodic system and are therefore called Isotopes (Greek, isos = equal, topos = place).

Isotopes have same chemical properties and distinguish themselves only in the mass, which by admission of a neutron, increases to about 1 ME.

A fundamental matter which possesses similarly built nucleus and shells, of which atom is built, is called pure element.

In nature 20 pure elements occur (see also pp. 14 and 15).

The percentage constituent of the isotope at the structure of a natural mixed element is (practically) constant (exception Pb, see pp. 16–19). Therefore, for the mixed element a constant atomic weight as weighed average value of isotopic weight results.

Trivial variations on account of local different mixtures of isotopes result in discrepancies in atomic weights of the following mixed elements.

H	$1,00797 \pm 0,000\ 01$	O	$15,999\ 4 \pm 0,000\ 1^*$
B	$10,811 \pm 0,003$	Si	$28,086 \pm 0,001$
C	$12,01115 \pm 0,000\ 05$	S	$32,064 \pm 0,003$

*Since 1961, the fundamental unit 'atomic weight' or better the 'atom mass' 1 ME = 1/12 of mass of Carbon isotope $^{12}_{6}$C has been chosen.
In case of accuracy of measurement the usual practical determination methods for molecular weights (molecular mass) have no significance in these discrepancies.

Wave mechanical model of atom

One dimensional oscillator, e.g., vibrating chord.

Basic oscillator (fundamental tone)

1. Over oscillation (overtone) at K originates a nodal point.

2. Over oscillation (overtone) at K originates a nodal points.

The whole can superimpose at which the oscillating chord is always the minima.

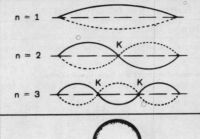

Two dimensional oscillator, e.g., drumhead basic oscillation (primary tone)

i. Over oscillation are three possible independent oscillations from one another. It results thereby a nodal line in the form of concentric circles or of diametrical nature.

ii. Over oscillations are five possible independent oscillations. The nodal lines are, thereby, in part, superimposed.

The fundamental and over oscillations can superimpose.

Through the nodal lines, the area in the region is so divided that either oscillations take place or they do not.

$n = 1 ; l = 0$

$n = 2 ; l = 0$ $n = 2 ; l = 1$ $n = 2 ; l = 1$

$n = 3 ; l = 0$ $n = 3 ; l = 1$ $n = 3 ; l = 1$

Three dimensional Oscillator

(Wave mechanic model)

By these oscillations originate nodal planes in the form of concentric spheres and sectional plane (circular area) through the oscillating sphere.

Thus the sphere in spatial region is distributed which are designated as ORBITALS.

$n = 3 ; l = 2$ $n = 3 ; l = 2$

Such figures result when a vibrating drumhead is covered with powder.

Number and Type of Nodal Planes in the Atomic Shells

Shell	n	l = No. of flat planes	n − l − 1 = No. of sphere shaped planes	Designation of electron orbitals
K	1	0	0	1s
L	2	0	1	2s
		1	0	2p
M	3	0	2	3s
		1	1	3p
		2	0	3d
N	4	0	3	4s
		1	2	4p
		2	1	4d
		3	0	4f

The limitation of shells comes as there are no further "walls" (planes) (see p. 7).

Atom Model

In every orbital two electrons can vibrate (see p. 8)	1s orbital $n = 1; l = 0$	2s orbital $n = 2, l = 0$ The inner sphere is a nodal plane	The three spatial positions of 2p orbitals $n = 2; l = 1$	

NEON	Examples for the schematic or symbol-wise writing method

1s	2s	2p	3s	3p			
⇅	⇅ ⇅ ⇅ ⇅					or $1s^2$ $2s^2$ $2p^6$	Neon
⇅	⇅ ⇅ ⇅ ⇅	⇅ ⇅ ↑ ↑				or $1s^2$ $2s^2$ $2p^6$ $3s^2$ $3p^4$	Sulphur
⇅	⇅ ⇅ ⇅ ⇅	⇅ ⇅ ⇅ ⇅				or $1s^2$ $2s^2$ $2p^6$ $3s^2$ $3p^6$	Argon

1s K-   2s 2p L-SHELL   3s 3p M-SHELL

Read: one-s-two, two-s-two, three-p-six and so on; often only the outer electrons are given

In the atomic shells, the oscillating orbitals are superimposed	The five spatial positions of 3d orbitals $n = 3, l = 2$		The nodal areas stand perpendicular to the marked plane	

Krypton	1s K-	2s 2p L-SHELL	3s 3p 3d M-SHELL	4s 4p N-SHELL	
	⇅	⇅ ⇅ ⇅ ⇅	⇅ ⇅ ⇅ ⇅		Argon
	⇅	⇅ ⇅ ⇅ ⇅	⇅ ⇅ ⇅ ⇅	⇅ $4s^2$	Calcium
	⇅	⇅ ⇅ ⇅ ⇅	⇅ ⇅ ⇅ ⇅ ⇅ ↑ ↑ ↑ ↑	⇅ $3d^6$ $4s^2$	Iron
	⇅	⇅ ⇅ ⇅ ⇅	⇅ ⇅ ⇅ ⇅ ⇅ ⇅ ⇅ ⇅ ⇅	⇅ ⇅ ⇅ ⇅	Krypton

Further examples (see pp. 26–29)

Structure of Elements and the Chemical Formula

For writing various atoms, symbols are used, which are provided with indices.

Abbreviation of names of the elements are employed as 'symbols'. The abbreviation has not more than two letters and is employed internationally. A portion of the abbreviation has been derived not only from German names but also from Latin or Greek. (See pp. 11–12)

The **Symbol** stands for	Examples: Fe
1. The substance as such	1. Ferrum (Iron)
2. **An atom** of the substance	2. Iron atom
3. One gram atom of the substance	3. 55.85 g

Mixed Elements
In case of naturally occurring mixed elements the symbol is mostly given.

Pure Elements
In case of pure elements, through rounding off the isotope weight, it is made integral to the mass number and often the atomic number is attached as index to the symbol.

Mass Number **Symbol** Atomic Number	Example $^{235}_{92}$U or U 235

Atomic Ions
To write different atomic ions the element symbols are used and the charge is written on the top right hand side as index.

To indicate single charge of atomic ion often the sign + or − is used. However, multiple charge of atomic ion is indicated by number written before the sign + or −

In case of ions in aqueous solution (solvated ions) often the signs ˙ and ˙ are employed instead of + and − respectively.

Examples are: H^+, O^{2-} or O''; Fe^{3+} or Fe

Structure and Arrangement of Atomic Shells

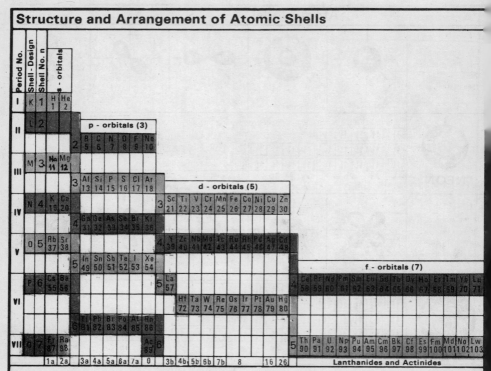

Orbital sub-groups of equal (or nearly equal) energy are arranged in equal height.
The energy increases stepwise—in these representations from above to below. First of all, the orbitals of equal energy grades.
i.e., first singlet, then doublet, are filled. Subsequently, the structure passes on to the next grade.

Fundamental Rules for the Structure and the Arrangement of Atomic Covers

The total number of electrons in an Atomic Cover is equal to the atomic number.	$Z = A.N.$
The atomic sheath is classified as shells. Instead of "Numbers" symbols are also used frequently.	$n = 1, 2, 3, \ldots$ $K, L, M, \ldots$
The total number (maximum) of possible orbitals in a shell is:	$z_{max} = 2n^2$
The total number of possible orbitals in a shell is:	$z_n = n^2$
The orbitals in orbital sub-groups are compounded. Instead of "numbers" symbols are also used.	$l = 0, 1, 2, \ldots$ $s, p, d, \ldots$
The "numbers" of orbital sub-groups are equal to:	$l = n - 1$
The number of orbital sub-group is equal to:	$z = 2l + 1$
Each orbital can take up two electrons, the spins of which must be (opposite) coupled.	Pauli - principle

Values of the first 4 Shells in an atomic cover

Shell	Designation of Sub-groups	Number of Shells	Number of Sub-groups	Maximum number of Electrons	Number of Orbitals in the Shell	in the Orbital Sub-groups
K	s	1	1	0	1 = 1	
L	s. p	2	0, 1	8	4 = 1 + 3	
M	s. p. d	3	0, 1, 2	18	9 = 1 + 3 + 5	
N	s. p. d, f	4	0, 1, 2, 3	32	16 = 1 + 3 + 5 + 7	

The fifth   Shell (O) contains these shell groups like the fourth Shell.
The sixth   Shell (P) contains only s- and p-orbitals as well as a d-Orbital.
The seventh Shell (Q) contains only an s-Orbital.

Arrangement of Elements

Principle of Arrangement in the Periodic System of Elements

1.	Successive increase in nuclear charge number or electron number gives the atomic number	A.N. 1 ... 103
2.	Arrangement according to the possible shells in the atomic cover gives the horizontal periods from:	I ... VII
3.	Arrangement according to the number of valency electrons in the outer most shell gives the vertical group from:	0 ... 8
	The Splitting of groups into main and sub-groups as well as the insertion of Lanthanides and Actinides is a result of alternately installing the electrons in the orbital sub-groups of the shell (see pp. 26–31)	1a, 1b, ... Lanthanides Actinides

Periodic system of Elements

I	1 H	2 He																														2 He
II*)	3 Li	4 Be																										5 B	6 C	7 N	8 O	9 F	10 Ne
III	11 Na	12 Mg																										13 Al	14 Si	15 P	16 S	17 Cl	18 Ar
IV	19 K	20 Ca	21 Sc															22 Ti	23 V	24 Cr	25 Mn	26 Fe	27 Co	28 Ni	29 Cu	30 Zn	31 Ga	32 Ge	33 As	34 Se	35 Br	36 Kr
V	37 Rb	38 Sr	39 Y															40 Zr	41 Nb	42 Mo	43 Tc	44 Ru	45 Rh	46 Pd	47 Ag	48 Cd	49 In	50 Sn	51 Sb	52 Te	53 I	54 Xe
VI	55 Cs	56 Ba	57 La	58 Ce	59 Pr	60 Nd	61 Pm	62 Sm	63 Eu	64 Gd	65 Tb	66 Dy	67 Ho	68 Er	69 Tm	70 Yb	71 Lu	72 Hf	73 Ta	74 W	75 Re	76 Os	77 Ir	78 Pt	79 Au	80 Hg	81 Tl	82 Pb	83 Bi	84 Po	85 At	86 Rn
VII	87 Fr	88 Ra	89 Ac	90 Th	91 Pa	92 U	93 Np	94 Pu	95 Am	96 Cm	97 Bk	98 Cf	99 Es	100 Fm	101 Md	102 No	103 Lw															
	1a	2a	3b	Lanthanides and Actinides														4b	5b	6b	7b	8			1b	2b	3a	4a	5a	6a	7a	8
	s-		d-	f- Orbitals														d- Orbitals									p- Orbitals					

Except rare gases, all atoms are qualified to either take electrons in their shell or give them out of their shell. These electrons are called **valence electrons.**
Out of a shell 1 to 8 valence electrons can be given up. Thus ionic atoms which carry 1 to 8 positive charges are formed.
Those possessing an ionic valency of +1 to +8 are called cations. (see p. 23)
In an atomic cover 1 to 4 valence electrons can be taken up. Thus ionic atoms which carry 1 to 4 negative charges are formed.
Those possessing an ionic valency of −1 to −4 are called anions. (see p. 23)
(An inert gas structure mostly results by giving up or receiving of electrons.)

Atoms, in particular of non-metallic elements, can build with the other elements to form a common structure. Besides electrons in pair always go together.
The number of possible electron pairs is equal to their cohesiveness.
In the I period upto 1 electron pair equals to 1 cohesiveness.
In the II period upto 4 electron pairs equal to 4 cohesiveness.
From the III period upto 8 electron pairs equal to 8 cohesiveness.
(Formation of a common cover also results in inert gas structure.) (see pp. 22, 26–30)

Metals are substances which can form only positive ionic atoms.
Non-metals are substances which preferably form negative ionic atoms.

A stable and permanent atomic structure can result:

1. with 2, 8, 18 or 32 electrons in the outermost shell:
   That in the above state rare gas is existing; it is designated as inert gas configuration (configuration = arrangement).
2. those which do not have completely filled shells with 18 outer most electrons.
   Examples of this state are atoms of Ni, Pd and Pt.
3. those outside the completely filled 8- or 18-shell possessing a completely filled s-orbital ($4s^2$, $5s^2$).
   Examples of this state are Sn^{+2} and other ions. (see pp. 26–29)

Properties of Elements

Group	Symbol	Name	A.N	Atomic mass	log AM	DE log	MP	BP	D/DD	$\Omega^{-1}\cdot cm^{-1}$	cal/g
O (Inert gas)	He	Helium	2	4,0026	60239	39761	-270,7	-268,94	0,17848	–	
	Ne	Neon	10	20,183	30499	69501	-248,60	-245,98	0,9004	–	
	Ar	Argon	18	39,948	60108	39892	-189,3	-186,0	1,7839	–	
	Kr	Krypton	36	83,80	92324	07676	-157,2	-152,9	3,74	–	
	Xe	Xenon	54	131,30	11826	88174	-111,9	-108,1	5,851	–	
	Rn	Radon	86	222	34635	65365	-71	-61,9	9,730	–	
	H	Hydrogen**	1	1,00797*	00346	99654	-259,14	-252,78	0,08987	–	
1a (Alkali-metals)	Li	Lithium	3	6,939	84136	15864	186	1330	0,534	$1,2\cdot10^5$	0,79
	Na	Sodium	11	22,9898	36156	63844	97,7	883	0,971	$2,3\cdot10^5$	0,295
	K	Potassium	19	39,102	59218	40782	62,3	760	0,851	$1,6\cdot10^5$	0,192
	Rb	Rubidium	37	8,547	93186	06814	39,0	700	1,532	$8,6\cdot10^4$	0,080
	Cs	Casium	55	132,905	12355	87645	28,5	670	1,90	$4,5\cdot10^4$	0,052
	Fr	Frankium	87	(223)	34830	65170				–	
2a (Alkali-earthmetals)	Be	Beryllium	4	9,0122	95521	04479	1280	2967	$\alpha=1,85$ $\beta=1,91$	$1,5\cdot10^5$	0,425
	Mg	Magnesium	12	14,312	38596	61404	650	1107	1,74	$2,2\cdot10^5$	0,246
	Ca	Calcium	20	40,08	60293	39707	850	1240	1,55	$2,2\cdot10^5$	0,145
	Sr	Strontium	38	87,62	94265	05735	757	1150	2,60	$2,3\cdot10^5$	0,068
	Ba	Barium	56	137,34	13786	86214	850	1140	3,61	$3,3\cdot10^4$	0,066
	Ra	Radium	88	226,05	35421	64579	700	1140	~6	–	–
3a (Earth-metals)	B	Boron	5	10,811*	03423	96577	2300	~2550	Cryst:2.33 amorph. 1,73	$1\cdot10^{-12}$	0,307
	Al	Aluminium	13	26,9815	43104	56896	660,1	2057	2,70	$3,8\cdot10^5$	0,214
	Ga	Gallium	31	69,72	84336	15664	29,78	1983	5,92	$1,9\cdot10^4$	0,079
	In	Indium	49	114,82	06002	93998	156,17	2000 ± 10	7,31	$1,2\cdot10^5$	0,057
	Tl	Thallium	81	204,37	31046	68954	302	1457	11,84	$5,7\cdot10^4$	0,0326
4a (Carbon group)	C	Carbon	6	12,01115*	0,7954	92046	>3550	4200	Graphite Diamond 3,51	$4,7\cdot10^3$ $3\cdot10^{-14}$	0,147
	Si	Silicon	14	28,086*	44855	55145	1420	2355	2,33	$8,3\cdot10^{-8}$	0,181
	Ge	Germanium	32	72,59	86094	13906	958,5	2700	5,35	$1,1\cdot10^{-5}$	0,074
	Sn	Stannum	50	118,69	07445	92555	231,9	2270	$\alpha=5,75$ $\beta=7,28$	$8,7\cdot10^4$	α 0,515 β 0,0542
	Pb	Plumbum	82	207,19	31641	68359	327,4	1620	11,344	$4,5\cdot10^4$	0,0306
5a (Nitrogen group)	N	Nitrogen	7	14,0067	14638	85362	-210	-195,8	1,2505	–	
	P	Phosphorus	15	30,9738	49101	50899	white:44,1 red:59b violet:593	280	w:1,64 violet:2,36 red:220 black:2,70	$5\cdot10^{-6}$	0,190
	As	Arsenic	33	74,9216	87454	12546	817 b.36at.	subl.610	2,0 - 5,73	$2,9\cdot10^4$	0,08
	Sb	Stibum	51	121,75	08550	91450	630,5	1380	6,684	$2,4\cdot10^4$	0,05
	Bi	Bismuth	83	208,980	32015	67985	271,3	~1560	9,845	$8,4\cdot10^3$	0,294
6a (Oreforming)	O	Oxygen	8	15,9994	20412	79588	-218,7	-182,97	1,42897	–	
	S	Sulphur	16	32,064*	50604	49396	$\alpha=112,8$ $\beta=119$ $\gamma=120$ 144,220	444,6 444,6 444,6 $\alpha=2,07$ $\beta=1,96$ $\gamma=1,92$		$5,3\cdot10^{-18}$ $1,3\cdot10^{-13}$	0,176 0,181
	Se	Selenium	34	78,96	89741	10259	144,220	688	$\alpha=4,47$ $\beta=4,40$ $\gamma=4,82$	$1,3\cdot10^{-7}$	
	Te	Tellurium	52	127,60	10588	89412	452	1390	6,24	$5\cdot10^3$	0,0483
	Po	Polonium	84	(210)	32222	67778			9,32		
7a (Saltforming or Halogens)	F	Fluorine	9	18,9984	27875	72125	-218/-223	-188,3	1,695	–	
	Cl	Chlorine	17	35,453	54970	45030	-101	-34,1	1,557b -34° 3,214	fl. $1\cdot10^{-16}$	
	Br	Bromine	35	79,909	90263	09737	-7,3	58,8	fl 3,14	fl $1,3\cdot10^{-15}$	0,107
	I	Iodine	53	126,9044	10349	89651	113,7	184,35	11,72 Ré	$1,3\cdot10^{-11}$	0,0523
	At	Astatine	85	(211)	32428	67572				–	
1b	Cu	Cupprum	29	63,54	80305	19695	1083	2336	8,929	$5,8\cdot10^5$	0,0921
	Ag	Argentine	47	107,870	03294	96706	960,5	1950	10,50	$6,1\cdot10^5$	0,0558
	Au	Aurum	79	196,967	29447	70553	1063	2600	19,3	$4,1\cdot10^5$	0,0312
2b	Zn	Zinc	30	65,37	81544	18456	419,4	907	7,14	$1,7\cdot10^5$	0,0925
	Cd	Cadmium	48	112,40	05080	94920	320,9	765	8,642	$1,3\cdot10^5$	0,0552
	Hg	Hydrargyrum	80	200,59	30235	69765	-38,87	356,58	13,546 b.20°	$1,1\cdot10^4$	0,03325
3b	Sc	Scandium	21	44,956	65283	34717	1400	2400	3,02	–	
	Y	Yttrium	39	88,905	94900	05100	1490	2500	4,47	–	
	La	Lanthan	57	138,91	14276	85724	920	4500/4230	6,15	–	
	Ac	Actinium	89	(227)	35603	64397	(1600)				0,0448
4b	Ti	Titanium	22	47,90	68034	31966	1800	>3000	4,50	$3,3\cdot10^5$	0,1125
	Zr	Zirconium	40	91,22	96009	03991	1857	>2900	6,52	$2,5\cdot10^4$	0,068
	Hf	Hafnium	72	178,58	25183	74817	2230	>3200	13,3	$3,0\cdot10^4$	0,033
5b	V	Ranadium	23	50,942	70714	29286		3000	6,07	$5,0\cdot10^4$	0,1153
	Nb	Niobium	41	92,906	96806	03194	2415	3700	8,56	$7,7\cdot10^4$	0,065
	Ta	Tantalum	73	180,948	25756	74244	~3000	~4100	16,6	$7,2\cdot10^4$	0,033

Properties of elements

Behaviour towards					Further properties	Isotopes**			Nomenclature	Symbol
H_2O	HCl	H_2SO_4	HNO_3	OH		a	b	c		
0.009	–	–	–	o	t_k -267.9 p_k 2.26atD$_k$ 0.0693g/cm³	4	2	–	gr. Helias = sun	He
0.012	–	–	–		„ -228.7 „ 25.9 „ 0.484 „	7	3	–	gr. Neas = new	Ne
0.056	–	–	–		„ -122 „ 48 „ 0.531 „	8	3	–	gr. Argos = inert	Ar
0.06	–	–	–		„ -63 „ 54 „ 0.78 „	25	6	–	gr. Kryptos = hidden	Kr
0.119	–	–	–		„ -16.6 „ 58.2 „ 1.155 „	29	9	–	gr. Xenon = strange	Xe
0.224	o	–			„ -104 „ 62 „ – „	16	–	3	from sodium	Rn
0.018	–	–	–	–	„ -239.9 „ 12.8 „ 0.0310	3	2	–	gr. Hydrogenium = water forming	H
d	+	+	+	o	w H_2O → LiOH; w.alk. → alcoholate	5	2	–	gr. lithos = brick	Li
d	+	+	+	o	w H_2O → NaOH; w.alk. → alcoholate	8	1	–	Natron (saltpeter)	Na
d	+	+	+	o	w H_2O → KOH; w.alk. → alcoholate	10	2	1	Arabic alkali = soda	K
d	+	+	+	o	with H_2O → RbOH;	20	1	1	L. Rubidus = red	Rb
d	+	+	+	o	with H_2O → CsOH;	22	1	–	L. Caesium = skyblue	Cs
o	o	o	o	o	properties like above	8	–	1	from France	Fr
–	+	+	+	+	corrosion resistant (oxide film)	4	1	–	from Beryl (crystal)	Be
–	+	+	+	–	corrosion resistant (oxidefilm)	3	3	–	Magnesia city	Mg
d	+	+	+	o	w H_2O → Ca(OH)$_2$	12	6		L. calx = chalk	Ca
d	+	+	+	o	w H_2O → Sr(OH)$_2$	18	4	–	Stranton city	Sr
d	+	+	+	o	w H_2O → Ba(OH)$_2$	22	7		gr. Brayo = hard	Ba
d	+	+	+	o	w H_2O → Ra(OH)$_2$; Radioact.	13	–	4	L. Radjus ray	Ra
–	–	+	+	–	crystal hardness ~ 10; semiconductor	4	2	–		B
–	+	+	–	+	corrosion resistant (oxide film)	7	1	–	from Alum (salt)	Al
–	+	+	+	+	Thermometer packing	11	2	–	L. Gallien = France	Ga
–	+	+	+	o	soft like wax	28	1	1	Indigo (blue spectral line)	In
–	(+)	+	+	o	similarity with lead	22	2	4	gr. Thallos = branch	Tl
–	–	–	–	–	Hardness of diamond = 10 Brittle n=2·4173	7	2	–	L. carbo = coal	C
–	–	–	–	+	l in HF/HNO$_3$: see crystal hardness	7	3	–	L. silex = silica, sand	Si
–	–	h+	+	–	"semiconductor"; l in HCl/HNO$_3$	16	5	–	L. Germania = Germany	Ge
–	+	+	k$^{v+}_{ox}$	h+	grey decom α 13.2° β 161° γ brittle d	29	10	–	L. Stannum = tin	Sn
–	–	h+k	+○	–	very soft and malleable	27	3	5	L. Plumbum = lead	Pb
0.015	–	–	–	–	t_k -118.8 p_k 49.7 at D$_k$ 0.430g/cm³ white P at 250-300° to red P	6	2	–	gr. Nitrogen = NaNO$_3$ forming	N
–	–	–	–	+	Ignition temperature ~ 50°	7	1	–	gr. Phosphorus = light carrier	P
–	–	–	+.	–	brittle crystal	14	1	–		As
–	–	h+k	–	+	brittle crystal; l HCl/HNO$_3$	29	2	–	L. Stibium	Sb
–	–	h+	+	–	brittle crystal; l HCl/HNO$_3$	22	–	6	white mass	Bi
0.031	–	–	–	–	t_k -118.8 p_k 47.7atm; D$_k$ 0.430g/cm³	3	3	–	gr. Oxygenium = acid forming	O
–	–	–	+	–	α or rhomb 95.6 β monocli yellow, s. CS$_2$	8	4	–	gr. Sulphur	S
–	–	+	+	+	"Semi conductor"; photoelectric	22	6	–	gr. Selene = moon goddess	Se
–	–	+	+	+	little hardness, brittle	27	8	–	h. tellus = earth	Te
o	o	o	o	o	"strong and noble like silver"	22	–	7	from Poland	Po
d	–	–	–	–	☠ weakly yellowish gas	5	1	–	L. Fluo = bleed	F
3.1(d)	–	–	–	+	☠ t_k 144°, p_k 76.1 atm D$_k$ 0.573g/cm³ yellowish green gas	11	2	–	gr. Chloros = green	Cl
3.58$^{x)}$	–	–	–	+	☠ Brown liquid corrosive vapours	19	2	–	gr. Bromos = stench	Br
0.02$^{x)}$	–	–	–	+	s. in KI, HI and alcohol	23	1	–	gr. Iodos = violet (vapour)	I
o	o	o	o	o	marked metallic property	18	–	3	gr. Astos = unstable	At
–	(+)	h+	+	–	pale red malleable and ductile	12	2	–	L. Cuprium = copper	Cu
–	–	h+	+	–	s. KCN soln. "Noble metal"	24	2	–	L. Argentum = silver	Ag
–	–	(h+)	–	–	s. KCN soln., "Noble metal"	19	1	–	L. Aurum = gold	Au
–	+	+	+	+	corrosion resistant (hydroxide film)	15	5	–		Zn
–	+	h+	+	+	corrosion resistant	27	8	–	gr. Kadmia = earth	Cd
–	–	–	k+	–	silvery gloss liquid; ☠ vapours especially	23	7	–	gr. Hydrargyrum = active silver	Hg
d	+	+	+	+	hard and not easily accessible	13	1	–	from Scandinavia	Sc
d	+	+	+	+	"rare earths" very closely	19	1	–	Ytterby, city in Sweden	Y
d	+	+	+	+	related metals	14	1	–	gr. Lanthanos = secret	La
o	o	o	o	o	radioactive, β-radiations	10	–	2	gr. Actis = rays	Ac
–	v+	v+	v+	–	very corrosion resistant, "light metal"	9	5	–	Titanen	Ti
–	((+))	((+))	((+))	–	s. in HF and HCl/HNO$_3$ corrosion resistant	15	5	–	Arabic Zargon = golden	Zr
–	–	–	–	–	analogus to Zr, corrosion resistant	16	6	–	Hafnia = Copenhagen	Hf
–	–	+	+	–	very corrosion resistant, s. HF and HCl/HNO$_3$	10	1	1	Vanadis = Venus goddess	V
–	–	((h+))	–	–	corrosion resistant solids Insol. in HF and HCl/HNO$_3$	20	1	–	Niobe gr. goddess	Nb
–	–	–	–	(+)	steel like properties	15	1	1	Tantalos = Goddess father	Ta

*g in 100g H_2O **Isotopes; a total number; b stable and natural; c unstable and natural

11

PROPERTIES OF ELEMENTS

Group	Symbol	Name	AN	Atomic mass	log	DElog	MP	BP	D/DD	$\Omega^{-1} cm^{-1}$	cal/g
6b	Cr	Chromium	24	52.01	71609	28391	1890	2480	6.92	$7.7 \cdot 10^4$	0.11
	Mo	Molebdenum	42	95.95	98204	01796	2620	(4800)	10.21	$1.8 \cdot 10^5$	0.065
	W	Tungsten	74	183.86	26449	73551	3380	~6000	19.32	$1.8 \cdot 10$	0.034
7b	Mn	Maganese	25	54.94	73989	26011	1260	1900	7.21-7.42	$2.0 \cdot 10^5$	0.1211
	Tc	Technitium	43	(99)	99564	00436				—	—
	Re	Rhenium	75	186.22	27003	72997	3170		21.0	$5.0 \cdot 10^4$	0.035
8	Fe	Ferrous	26	55.85	74702	25298	1535	3000	7.86	$1.0 \cdot 10^4$	0.107
	Co	Cobalt	27	58.94	77041	22959	1495	2900	$\alpha = 8 \cdot 89$ $\beta = 8 \cdot 64$	$1.6 \cdot 10^5$	0.1001
	Ni	Nickel	28	58.71	78671	23129	1454.8	2900	8.90	$1.5 \cdot 10^5$	0.105
	Ru	Ruthenium	44	101.1	00475	99525	2450	2700	12.43	$1.3 \cdot 10^5$	0.0611
	Rh	Rhodium	45	102.91	01246	98754	1966	>2500	12.5	$2.1 \cdot 10^5$	0.058
	Pd	Palladium	46	106.7	02816	97184	1549.4	2200	11.97	$9.1 \cdot 10^4$	0.0538
	Os	Osmium	76	190.2	27921	72079	2700	>5300	22.48	$1.7 \cdot 10^4$	0.0311
	Ir	Iridium	77	192.2	28375	71625	2454	4400	22.421	$1.6 \cdot 10^5$	0.0323
	Pt	Platin	78	195.09	29014	70986	1773.5	4300	21.45	$1.0 \cdot 10^4$	0.0324

Group	Symbol	Name	An	Atomic mass	log	DElog	MP	BP	D	Nomenclature
	Ce	Cerium	58	140.13	14653	85347	1077	3200	6.768	Plenetoid ceres
	Pr	Praesedium	59	140.92	14897	85103	1208	3290	6.769	gr. Prasimos = green gr. dydimos = twin
	Nd	Neodynium	60	£44.27	15918	84082	1297	3450	7.007	gr. neos = new gr. dydimos = twin
	Pm	Promethium	61	(145)	16137	83863	1570	3000	....	gr. God Prometheus
	Sm	Samarium	62	150.35	17711	82289	1325	1900	7.540	Samarski, Russian city
	Eu	Europium	63	152.0	18184	81816	~1200	1700	5.166	from Europe
	Gd	Gadolinium	64	157.26	19662	80338	~1520	3000	7.868	Gadolina a chemist.
	Tb	Terbium	65	158.93	20120	79880	1638	2800	8.253	Ytterby a city in Sweden
	Dy	Dysprosium	66	162.51	21088	78912	~1670	2600	8.556	Gadolin, Finish chemist
	Ho	Holmium	67	164.94	21723	78268	~1770	2600	8.700	Holmia = Stockholm
	Er	Erbium	68	167.27	22342	77658	~1800	2900	9.058	Ytterby a city in Sweden
	Tm	Thulium	69	168.94	22773	77227	~1900	2400	9.318	Thule = island
	Yb	Ytterbium	70	173.04	23815	76185	1097	1800	6.959	Ytterby = a city in Sweden
	Lu	Lutetium	71	174.09	24302	75698	~2000	2200	9.849	Lutetia = Paris

Group	Symbol	Name	AN	Atomic mass	MP	BP	D	Half-life	period	Decay	Nomenclature
	Th	Thorium	90	232.05	1845	4500	11.2	$1.4 \cdot 10^{10}$	years	α	German god
	Pa	Protactinium	91	231	—	—	—	$3.4 \cdot 10^4$	years	α	gr. protos = first
	U	Uranium	92	238.07	~1133	—	18.7	$4.5 \cdot 10^9$	years	α	Uranus = 7th planet
	Np	Neptiunium	93	237	640	—	20.5	$2.2 \cdot 10^6$	years	α	Neptune = 8th planet
	Pu	Plutonium	94	242	640	—	19.7	$2.4 \cdot 10^4$	years	α	Pluto = 9th planet
	Am	Americium	95	243	827	610	11.7	$8.3 \cdot 10^3$	years	α	from America
	Cm	Curium	96	248	—	—	~7	$4.0 \cdot 10^5$	years	α	Curie discoverer
	Bk	Berkelium	97	249	—	—	—	$7.0 \cdot 10^3$	years	α	Berkeley (city in USA)
	Cf	Californium	98	251	—	—	—	$7.0 \cdot 10^2$	years	α	California
	Es	Einsteinium	99	254	—	—	—	38.5	hours	β	from Einstein
	Fm	Fermium	100	253	—	—	—	3	days	K	Fermi, math.
	Md	Mendelevium	101	256	—	—	—	1.5	hours	K	Mandeleef
	Nb	Nobelium	102	253	—	—	—	10	min.	α	Nobel (of Noh Pr.)
	Lw	Lawrentium	103	257	—	—	—	8	see.	α	discoverer's name

CLASSIFICATION OF METALS ACCORDING TO DENSITY

Light metals				Heavy metals									
Mg.	Be	Al	Ti	Ge	V	Cr	Zn	Sn	Fe	Cu	Ag	Pb	Au	Os
1.74	1.90	2.70	4.50	5.35	6.07	6.92	7.14	7.28	7.86	8.93	10.5	11.4	19.3	22.5

Classification of Metals from the viewpoint of their industrial applications

Ferrous metals F-Metals					Non-ferrous metals					Noble metals				
Fe	Mn	Cr	Ni	V	Cu	Zn	Sn	Pb	Cd	Au	Ag	Pt	Jr	Pd

and few others with which Fe is used in allowing. | and few others which together with copper and among themselves form alloy. | and other platinum metals with which Pt is alloyed.

Properties of Elements

Behaviour towards H₂O HCl H₂SO₄ HNO₃ –OH					Further Properties	Isotopes** a b c			Nomenclature	Symbol
–	+	v+	–	–	highly corrosion resistant	9	4	–	gr. chroma = colour	Cr
–	–	k+	hK+	–	hard, brittle, corrosion resistant	16	7	–	gr. molybdos = heavy (hard)	Mo
–	–	–	–	S+	mech. tenacity, sl. s. HF or HCl/HNO₃	18	4	1		W
((+))	v+	v+	v+	–	hard, brittle, slightly resistant	9	1	–	L magnes = Magnet	Mn
o	o	o	o	o	similarity with Re	20	–	–	gr. technetos = synthetic	Tc
–	–	–	k+	–	hard resistant	19	1	1	L rhenus = Rhein	Re
–	+	+	–	–	α $\overset{906°}{\rightleftarrows}$ γ $\overset{1400°}{\rightleftarrows}$ δ: ferromagnetic	10	4	–	L ferrum = Iron	Fe
–	+	+	+	–	corrosion resistant, ferromagnetic	15	1	–	from cobalt	Co
±	(+)	(+)	v+	–	corrosion resistant; ferromagnetic	11	5	–	heavy copper nickel = coloured copper	Ni
–	–	–	–	(+)	brittle, noble metal	15	7	–	Ruthenia = Ukraine	Ru
–	–	hK+	–	–	malleable and ductile, noble metal	21	1	–	gr. rhodon = Rose	Rh
–	–	h+	–	–	malleable and ductile, noble metal	21	6	–	from Platinoid Pallas	Pd
–	–	–	(+)	–	brittle, noble metal	21	7	–	gr. osme = Odour	Os
–	–	–	–	–	brittle, noble metal	18	2	–	L iris = Rainbow	Ir
–	–	–	–	(+)	malleable and ductile, noble metal	16	4	2	Spanish platina = small Silver	Pt

Explanations for Tables on pp. 11/12 and 12/13:
Behaviour towards . . . : Numbers denote the solubility (in case of gases g/l; otherwise g/100 ml)

o	no-information	h	hot	+	reacts with	(+)	reacts slowly
d	decay, decomposes	c	concentrated	–	does not react	((+))	reacts very slowly
ox	oxidizes	v	dilute	L	Latin	gr.	Greek

Occurrence of elements in earth's crust and in atmosphere

A) Distribution in weight percentage

Number of elements of equal occurrence

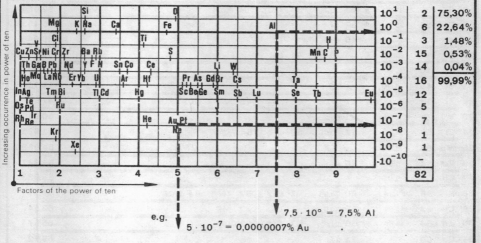

	Number of elements	
10^1	2	75,30%
10^0	6	22,64%
10^{-1}	3	1,48%
10^{-2}	15	0,53%
10^{-3}	14	0,04%
10^{-4}	16	99,99%
10^{-5}	12	
10^{-6}	5	
10^{-7}	7	
10^{-8}	1	
10^{-9}	1	
-10^{-10}	–	
	82	

Factors of the power of ten

e.g.

$7,5 \cdot 10^0 = 7,5\% \text{ Al}$

$5 \cdot 10^{-7} = 0,000\,000\,7\% \text{ Au}$

The unregistered 21 elements possess frequency of less than 10% or can only be artificially produced

B) Distribution in atom percentage

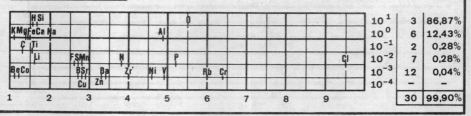

10^1	3	86,87%
10^0	6	12,43%
10^{-1}	2	0,28%
10^{-2}	7	0,28%
10^{-3}	12	0,04%
10^{-4}	–	–
	30	99,90%

** Isotopes: a total number; b stable natural; unstable natural

Stable Isotopes of Elements

Besides stable isotopes, radio-active isotopes with long life which occur in nature, are also listed here.

AN Symbol Mass No. of Isotopes and percentage constituent of the mixture. Mean atomic mass = $\bar{m}$ AM

AN	Symbol								Mean atomic mass
1	H	1 99.985	2 0.015						1.00797
2	He	3 0.000137	4 ≈ 100						4.0026
3	Li	6 7.42	7 92.58						6.939
4	Be	9 100							9.0122
5	B	10 19.8	11 80.2						10.811
6	C	12 98.893	13 1.107						12.01115
7	N	14 99.634	15 0.366						14.0067
8	O	16 99.759	17 0.0374	19 0.2039					15.9994
9	F	19 100							18.9984
10	Ne	20 90.92	21 0.257	22 8.82					20.183
11	Na	23 100							22.9898
12	Mg	24 78.70	25 10.73	26 11.17					24.312
13	Al	27 100							26.9815
14	Si	28 92.21	29 4.70	30 3.09					28.086
15	P	31 100							30.9738
16	S	32 95.00	33 0.76	34 4.22	36 0.014				32.064
17	Cl	35 75.529	37 24.4						35.453
18	Ar	36 0.337	40 99.6						39.948
19	K	39 93.10	40 0.049	41 6.81					39.102
20	Ca	40 96.97	42 0.64	43 0.3145					40.08
21	Sc	45 100							44.956
22	Ti	46 7.93	47 7.28	48 73.94	49 5.51	50 5.34			47.90
23	V	50 0.24	51 99.76						50.942
24	Cr	50 4.31	52 83.76	53 9.55	54 2.38				51.996
25	Mn	55 100							54.9380
26	Fe	54 5.82	56 91.66						55.847
27	Co	59 100							58.9332
28	Ni	58 67.88	60 26.23	61 1.19	62 3.66	64 1.08			58.71
29	Cu	63 69.09	65 30.91						63.54
30	Zn	64 48.89	66 27.81	67 4.11	68 18.57	70 0.62			65.37
31	Ga	69 60.4	71 39.6						69.72
32	Ge	70 20.52	72 27.43	73 7.76	74 36.54	76 7.76			72.59
33	As	75 100							74.9216
34	Se	74 0.87	76 9.02	77 7.58	78 23.52	80 49.82	82 9.19		78.96
35	Br	79 50.537	80 49.463						79.909
36	Kr	78 0.354	2.27	82 11.56	83 11.55	84 56.90	86 17.37		83.80
37	Rb	85 72.15	87 27.85						85.47
38	Sr	84 0.56	86 9.86	87 7.02	88 82.56				87.62
39	Y	89 100							88.905
40	Zr	90 51.46	91 11.23	92 17.11	94 17.40	96 2.80			91.22
41	Nb	93 100							92.906
42	Mo	92 15.84	94 9.04	95 15.72	96 16.53	97 9.16	98 23.78	100 9.63	95.94
43	Tc	does not occur in nature							

Stable Isotopes of Elements

Z	El.	Isotopes (mass · abundance %)	Atomic weight
44	Ru	96 · 5.51 98 · 1.87 99 · 12.72 100 · 16.62 101 · 17.07 102 · 31.61 104 · 18.58	101.07
45	Rh	103 · 100	102.905
46	Pd	102 · 0.96 104 · 10.97 105 · 22.23 106 · 27.33 108 · 26.71 110 · 11.81	106.4
47	Ag	107 · 51.37 109 · 48.65	107.870
48	Cd	106 · 1.215 108 · 0.875 110 · 12.39 · 12.75 112 · 24.07 113 · 12.26 114 · 28.86 116 · 7.58	112.40
49	In	113 · 4.28 115 · 95.67	114.82
50	Sn	112 · 0.96 114 · 0.66 115 · 0.35 116 · 14.30 117 · 7.61 118 · 24.03 119 · 8.58 120 · 35.85 122 · 4.72 124 · 5.94	118.69
51	Sb	121 · 57.35 123 · 42.75	121.75
52	Te	120 · 0.089 121 · 4.61 122 · 2.46 123 · 0.87 125 · 6.99 126 · 18.71 128 · 31.79 130 · 34.48	127.60
53	I	127 · 100	126.9044
54	Xe	124 · 0.096 126 · 0.090 128 · 1.919 129 · 26.44 130 · 4.08 131 · 21.00 132 · 26.89 134 · 10.44 136 · 8.87	131.30
55	Cs	133 · 100	132.905
56	Ba	130 · 0.101 132 · 0.097 134 · 2.42 135 · 6.59 136 · 7.81 137 · 11.32 138 · 71.66	137.34
57	La	138 · 0.089 139 · 99.911	138.91
58	Ce	136 · 0.193 138 · 0.250 140 · 88.48 142 · 11.07	140.12
59	Pr	141 · 100	140.907
60	Nd	142 · 27.21 143 · 12.17 144 · 23.9 $(2.4\cdot10^{15}\,a)$ 145 · 8.30 146 · 17.22 148 · 5.73 150 · 5.62	144.24
61	Pm	does not occur in nature	—
62	Sm	144 · 3.09 147 · 14.87 $(1.2\cdot10^{11}\,a)$ 148 · 11.24 149 · 13.83 150 · 7.44 152 · 26.72 154 · 22.71	150.35
63	Eu	151 · 47.82 153 · 52.18	151.96
64	Gd	152 · 0.20 $(1.1\cdot10^{14}\,a)$ 154 · 2.15 155 · 14.73 156 · 20.47 157 · 15.68 158 · 24.87 160 · 21.90	157.25
65	Tb	159 · 100	158.924
66	Dy	156 · 0.0524 158 · 0.0902 160 · 2.294 161 · 19.0 162 · 25.53 163 · 24.97 164 · 28.18	162.50
67	Ho	165 · 100	164.930
68	Er	162 · 0.136 164 · 1.56 166 · 33.41 167 · 22.94 168 · 27.07 170 · 14.88	167.26
69	Tm	169 · 100	168.934
70	Yb	168 · 0.135 170 · 3.03 171 · 14.31 172 · 21.82 173 · 16.13 174 · 31.84 176 · 12.72	173.04
71	Lu	175 · 97.41 176 · 2.67	174.97
72	Hf	174 · 0.18 $(2.0\cdot10^{15}\,a)$ 176 · 5.20 177 · 18.05 178 · 27.14 179 · 13.75 180 · 35.24	178.49
73	Ta	180 · 0.0123 181 · 99.9877	180.948
74	W	180 · 0.14 182 · 26.41 183 · 14.40 184 · 30.64 186 · 28.41	183.85
75	Re	185 · 37.07 187 · 62.9	186.2
76	Os	184 · 0.018 186 · 1.59 187 · 1.64 188 · 13.3 189 · 16.1 190 · 26.4 192 · 41.0	190.2
77	Ir	191 · 37.3 193 · 62.7	192.2
78	Pt	190 · 0.012 $(7.0\cdot10^{11}\,a)$ 192 · 0.78 $(\sim10^{15}\,a)$ 194 · 32.9 195 · 33.8 196 · 25.3 198 · 7.21	195.05
79	Au	197 · 100	196.967
80	Hg	196 · 0.146 198 · 10.02 199 · 16.84 200 · 23.13 201 · 13.22 202 · 29.80 204 · 6.85	200.59
81	Tl	203 · 29.5 205 · 70.5	204.37
82	Pb*)	204 · 1.37 $(1.4\cdot10^{17}\,a)$ 206 · 23.6 207 · 22.6 208 · 52.3 *) see p. 19	207.19
83	Bi	209 · 100	208.98

The following radioactive Isotopes with long life occur in nature

Z	El.	Isotopes (mass · abundance %)	Atomic weight
90	Th	232 · 100 $(1.39\cdot10^{10}\,a)$	232.038
92	U	234 · 0.0056 $(2.5\cdot10^{5}\,a)$ 235 · 0.72 $(7.1\cdot10^{8}\,a)$ 238 · 99.27 $(4.5\cdot10^{9}\,a)$	238.03

RADIOACTIVE ISOTOPES OF ELEMENTS

In the Table only those radioactive isotopes are introduced which are mentioned in radioactive protection ordinance.

The quality (and therefore the activity) is reduced to 1/1024 ~ 1‰ after 10 half life periods.

α	– partial decay	e⁻	electron conversion
β⁻	– radiations (negatrons)	m	Isomeric decay
β⁺	– decay (positrons)	NR	natural radioactivity
K, L	– shell electron capture	T	technical applications
γ	– radiations	Med	medical applications

MN Sy AN   AN = Atomic number; Sy = Element symbol; MN = Mass Number of Isotopes (rounded figures)

HLP = Half life period (a = years; d = days; h = hours; m = minutes; s = seconds)

MeV = Decay energy in million electron volts

AN Sy MN	HLP	MeV / Free limit microcurie	Uses
1 H 3	12.5 a	β⁻ 0,018	100	T¹⁾
4 Be 7	54.5 d	γ 0,479; K	100	
6 C 14	5700 a	β⁻ 0,155; no γ;	100	Med²⁾
9 F 18	1,87 h	β⁺ 0,6;	100	Med
11 Na 22	2,6 a	β⁺ 0,54; 1,8; γ 1,28;	10	
11 Na 24	14,9 h	β⁻ 1,39; γ 1,38; 2,758;	10	Med
14 Si 31	2,6 h	β⁻ 1,486; no γ;	100	
15 P 32	14,3 d	β⁻ 1,718; no γ;	10	Med
16 S 35	87,1 d	β⁻ 0,176;	10	Med
17 Cl 36	4,10⁵ a	β⁺ β⁻ 0,716; γ?;	10	Med
17 Cl 38	38,5 m	β⁻ 1,11; 2,77; 4,81; γ 1,6; 2,15;	100	
18 Ar 37	34,1 d	K; no γ;	100	
18 Ar 41	1,8 h	β⁻ 1,25; 2,55; γ 1,3;	10	
19 K 40	1,4.10⁹ a	β⁺ β⁻ 1,4; γ 1,5; K;	nb	NR
19 K 42	12,4 h	β⁻ 3,6; 2,5; γ 1,5;	10	Med
20 Ca 45	152 d	β⁻ 0,255;	1	Med
20 Ca 47	5,8 d	β⁻ 1,1; γ 1,3	1	
21 Sc 46	85 d	β⁻ 0,36; 1,2; γ 0,89;	10	
21 Sc 47	3,4 d	β⁻ 0,61; γ?;	10	
21 Sc 48	1,83 d	β⁻ 0,64; γ 0,98; 1,33; K;	10	
23 V 48	16,0 d	β⁺ 0,72; γ 1,32, 0,99, 2,22, K;	10	
24 Cr 51	26 d	γ 0,32; 0,267; e⁻; K; no β⁻;	100	
25 Mn 52m	21 m	β⁺ 2,66; γ 1,46; e⁻ 0,39;		
25 Mn 52	6,2 d	β⁺ 0,582; γ 0,734; 0,94; 1,46; K;	10	
25 Mn 54	310 d	β⁻ 1,0; γ 0,835; K;	10	
25 Mn 56	2,59 h	β⁻ 2,86; 1,05; 0,73; γ 0,854; 1,81; 2,13; 2,7; 3,0	10	
26 Fe 55	2,94 a	K;	10	Med
26 Fe 59	46 d	β⁻ 0,46; γ 1,1; 1,3;	1	Med
27 Co 57		β⁺ 0,26; γ 0,131, 0,119, 0,014; e⁻; K;	10	
27 Co 58m	9,2 h	e⁻ 0,025	10	
27 Co 58	72 d	β⁺ 0,47; γ 0,81; K;	10	
27 Co 60m	10,7 m	β⁻ 1,56; γ 1,33; e⁻ 0,059	10	
27 Co 60	5,25 a	β⁻ 0,31; γ 1,17; 1,332;	10	T³⁾ Med
28 Ni 59	7,5.10⁵ a	K;	10	
28 Ni 63	85	β⁻ 0,067; no β⁺; γ	10	
28 Ni 65	2,56 h	β⁻ 2,10; 0,60; 1,01; γ 1,49; 1,12; 0,37;	10	
29 Cu 64	12,8 h	β⁺ 0,65; β⁻ 0,57; γ 1,34; K;	10	

AN Sy MN	HLP	MeV / Free limit microcurie	Uses
30 Zn 65	250 d	β⁺ 0,32; γ 1,14; 0,201; e⁻	10	
30 Zn 69m	13,8 h	e⁻ 0,439;	10	
30 Zn 69	52 m	β⁺ 0,86; no γ	10	
31 Ga 72	14,2 h	β⁻ 0,64; 0,96; 1,48; 3,15; γ 0,84; 2,21; 2,51;	10	
32 Ge 71	11,0 d	e⁻ 0,32; K;	100	
33 As 73	76 d	e⁻ 0,052; K;	10	
33 As 74	17,5 d	β⁺ 0,92; 1,53; β⁻ 0,69; 1,36; γ 0,635; 0,596; K;	10	Med
33 As 76	27,6 h	β⁻ 3,12; 2,56; 1,4; 0,4; γ 0,581; 20; 1,76; 2,02;	10	Med
33 As 77	40 h	β⁻ 0,68;	10	
34 Se 75	128 d	γ 0,025 ... 0,402; e⁻; K;	10	
35 Br 82	35,7 h	β⁻ 0,465; γ 0,57 ... 2,0;	10	Med
36 Kr 85m	4,4 h	β⁻ 0,855; γ 0,144; e⁻ 0,305	10	
36 Kr 85	9,4 a	β⁻ 0,72; γ 0,54;	100	T⁴⁾
36 Kr 87	78 m	β⁻ 3,2;		
37 Rb 86m	19,5 m	β⁻ 1,822; 0,716; γ 1,08;	10	
37 Rb 86	1,06 m	γ 0,78; K;		
37 Rb 87	6,10¹⁰ a	β⁻ 0,275; γ?;	10	NR
38 Sr 85m	70 m	γ 0,165; e⁻ 0,0075; 0,225; 0,150; K	0,1	
38 Sr 85	65 d	γ 0,513; K; no β⁺;	0,1	
38 Sr 89	54 d	β⁻ 1,5; no γ;	1	Med
38 Sr 90	25 a	β⁻ 0,54; no γ	0,1	Med
38 Sr 91	9,7 h	β⁻ 3,2; 1,3; γ 1,3;	10	
38 Sr 92	2,7 h	β⁻	10	
39 Y 90	65 h	β⁻ 2,24; no γ;	10	Med
39 Y 91m	51 m	e⁻ 0,555; β⁻ 1,56; no γ;	0,1	
39 Y 91	57 d		1	
39 Y 92	3,5 h	β⁻ 3,5; γ 0,7;	10	
39 Y 93	10 h	β⁻ 3,1; γ 0,7;	10	
40 Zr 93	5,10⁶ a	β⁻ 0,06;	10	
40 Zr 95	65 d	β⁻ 0,39; e⁻; γ 0,73; 0,23; 0,92;	10	
40 Zr 97	17,0 d	β⁻ 2,50; γ 0,74; 1,42;	100	
41 Nb 93m	40 d		10	
41 Nb 93	stabil			
41 Nb 95m	90 h	e⁻ 0,216		
41 Nb 95	35 d	β⁻ 0,163; γ 0,771	10	
41 Nb 97	60 s	e⁻ 0,747;		
41 Nb 97	74 m	β⁻ 1,267; γ 0,665	100	

¹⁾ Lighting material ²⁾ Prehistorical and historical age determination ³⁾ Destruction free material testing ⁴⁾ Filling lighting tu...

RADIOACTIVE ISOTOPES OF ELEMENTS

MN Sy AN	HLP	MeV / Free limit microcurie	Uses
$^{99}_{42}$Mo	67 h	β⁻ 1,23; 0,445; < 0,2. / γ 0,01; 0,367; 0,740; 0,780.	10
$^{95m}_{43}$Tc	51,5 m	e⁻ 0,034;	100
$^{96}_{43}$Tc	4,35 d	γ 0,312; 1,119; 0,806; 0,77; 0,81;	10
$^{97m}_{43}$Tc	90 d	e⁻ 0,096	10
$^{97}_{43}$Tc	> 100 a	β⁺ or K	10	T
$^{99m}_{43}$Tc	5,9 h	γ 0,002; 0,140; e⁻ 0,142	100
$^{99}_{43}$Tc	5,10⁵ a	β⁻ 0,291	10
$^{97}_{44}$Ru	2,8 d	γ 0,217; e⁻; K;	10
$^{103}_{44}$Ru	39,8 d	β⁻ 0,684; 0,222; γ 0,0404; 0,494;	10
$^{105}_{44}$Ru	4,5 h	β⁻ 1,15; γ 0,726; 0,13;	10
$^{106}_{44}$Ru	1 a	β⁻ 0,044;	1
$^{103m}_{45}$Rh	57 m	e⁻ 0,040;	100
$^{103}_{45}$Rh	stable		
$^{105m}_{45}$Rh	45 s	e⁻ 1,30;	
$^{105}_{45}$Rh	36,2 m	β⁻ 0,25; 0,57; γ 0,32;	10
$^{103}_{46}$Pd	17 d	K; no e⁻; no γ;	10
$^{109m}_{46}$Pd	4,8 m	e⁻ 0,19;	
$^{109}_{46}$Pd	13,1 h	β⁻ 0,95; no γ;	10
$^{105}_{47}$Ag	40 d	γ 0,28; 0,35; 0,43; 0,063; K; no β⁻	10
$^{110m}_{47}$Ag	270 d	β⁺ 0,087; 0,53; e⁻ 0,116;	
$^{110}_{47}$Ag	24,5 s	γ 1,516; 1,389; 0,706; 0,885; 0,759; 0,656;	10
$^{111}_{47}$Ag	7,5 d	β⁻ 1,04; 0,70; 0,80; / γ 0,093; 0,243; 0,340;	10
$^{109}_{48}$Cd	330 d	γ 0,08; K;	10
$^{115m}_{48}$Cd	2,33 d	β⁻ 0,46; 1,1; γ 0,52;	10
$^{115}_{48}$Cd	43 d	β⁻ 1,41; γ 1,10;	10
$^{113m}_{49}$In	1,73 h	e⁻ 0,393;	100
$^{113}_{49}$In	stable		
$^{114m}_{49}$In	50 d	e⁻ 0,192;	10
$^{114}_{49}$In	72 s	β⁺ 0,65 β⁻ 1,98 γ 0,715 0,548 K	
$^{115m}_{49}$In	4,5 h	β⁻ 0,83; e⁻ 0,336;	10
$^{115}_{49}$In	6,10¹⁴ a	β⁻ 0,63;	nb	NR
$^{113}_{50}$Sn	118 d	γ 0,255; 0,401; K;	10
$^{125m}_{50}$Sn	9,5 m	β⁻ 1,17; 0,5; γ 1,86;	
$^{125}_{50}$Sn	9,4 d	β⁻ 2,33; 2,04; γ 0,326;	10
$^{122}_{51}$Sb	2,8 d	β⁻ 1,36; 1,44; γ 0,58; 1,24; 0,680;	10
$^{124}_{51}$Sb	60 d	β⁻ 3,2; 0,6; γ 1,7; 0,607;	10
$^{125}_{51}$Sb	2,7 a	β⁻ 0,299; 0,128; 0,616; / γ 0,465; 0,425; 0,637; 0,035;	10	T
$^{125m}_{52}$Te	58 d	e⁻ 0,110; 0,035;	10
$^{125}_{52}$Te	stable		
$^{127m}_{52}$Te	113 d	e⁻ 0,018;	10
$^{127}_{52}$Te	9,3 h	β⁻ 0,70; no γ	10
$^{129m}_{52}$Te	34 d	β⁻; e⁻ 0,106;	10
$^{129}_{52}$Te	72 m	β⁻ 1,8; γ 0,3;	10
$^{131m}_{52}$Te	1,25 d	β⁻; e⁻ 0,183;	10
$^{131}_{52}$Te	77,7 h	β⁻ 0,28; γ 0,22;	0,1
$^{126}_{53}$I	13 d	β⁺; β⁻ 0,85; 1,24; γ 0,395; 0,640; K	1
$^{129}_{53}$I	1,72,10⁷ a	β⁻ 0,13; γ 0,039;	1

MN Sy AN	HLP	MeV / Free limit microcurie	Uses
$^{131}_{53}$I	8,04 d	β⁻ 0,60; 0,32; / γ 0,364; 0,284; 0,080; 0,638;	1	Med 1)
$^{132}_{53}$I	2,4 h	β⁻ 1,5; 2,2; γ 0,6; 1,4;	10
$^{133}_{53}$I	22,4 h	β⁻ 1,4; 0,5; γ 0,53; 0,85; 1,4	10
$^{134}_{53}$I	52,5 m	β⁻ 3,9; 1,6; γ 2,3;	10
$^{135}_{53}$I	6,7 h	β⁻ 1,0; 0,47; 1,4; γ 1,3; 1,6; 2,4;	10
$^{131m}_{54}$Xe	12,0 d	e⁻ 0,163;	0,1
$^{131}_{54}$Xe	stable		
$^{133m}_{54}$Xe	2,3 d	β⁻; e⁻ 0,232;	
$^{133}_{54}$Xe	5,27 d	β⁻ 0,345; γ 0,081; 0,232;	10
$^{135m}_{54}$Xe	15,3 m	e⁻ 0,53;	
$^{135}_{54}$Xe	9,2 h	β⁻ 0,93; γ 0,247;	10
$^{131}_{55}$Cs	10 d	K; no γ;	100
$^{134m}_{55}$Cs	3,15 h	β⁻ 2,4; γ 0,7; e⁻ 0,128	100
$^{134}_{55}$Cs	2,3 a	β⁻ 0,684; 0,092; / γ 0,793; 0,601; 0,367;	10
$^{135}_{55}$Cs	2,9,10⁶ a	β⁻ 0,19; no γ;	10
$^{136}_{55}$Cs	13 d	β⁻ 0,35; 0,28; γ 0,9; 1,2;	10
$^{137}_{55}$Cs	33 a	β⁻ 0,518; 1,17; γ 0,663;	10	T 2)
$^{131}_{56}$Ba	13 d	γ 0,494; 0,122; 0,372; 0,206; K;	10
$^{140}_{56}$Ba	12,8 d	β⁻ 1,022; 0,48; γ 0,54; 0,306; 0,160;	1
$^{140}_{57}$La	40,0 h	β⁻ 1,32; 1,67; 2,26; / γ 1,65; 0,87; 2,3; 0,33;	10
$^{141}_{58}$Ce	32,5 d	β⁻ 0,41; 0,56; γ 0,141;	10
$^{143}_{58}$Ce	33 h	β⁻ 1,36; γ 0,04; 0,20; 0,87	10
$^{144}_{58}$Ce	290 d	β⁻ 0,307; 0,446; γ 0,03; 0,134	1
$^{142}_{59}$Pr	19,1 h	β⁻ 2,154; 0,636; γ 1,576; 0,135	10
$^{143}_{59}$Pr	13,8 d	β⁻ 0,93; no γ;	10
$^{144}_{60}$Nd	5,10¹⁵ a	α 1,8; no γ;	nb	NR
$^{147}_{60}$Nd	11,1 d	β⁻ 0,78; 0,35; γ 0,091; 0,520	10
$^{149}_{60}$Nd	1,8 h	β⁻ 0,95; 1,1; γ 0,03...0,65;	100
$^{147}_{61}$Pm	2,6 a	β⁻ 0,229; no γ;	10
$^{149}_{61}$Pm	4,8 h	β⁻ 1,05; γ 0,25;	10
$^{147}_{62}$Sm	6,7,10¹¹ a	α 2,1; no e⁻;	1	NR
$^{151}_{62}$Sm	70 a	β⁻ 0,075; no e⁻;	1
$^{153}_{62}$Sm	47 h	β⁻ 0,820; γ 0,069; 0,103; 0,61	
$^{152m}_{63}$Eu	9,2 h	β⁻ 0,36; 1,8; K; / γ 0,12; 0,16; 0,72; 1,0;	10
$^{152}_{63}$Eu	5,3 a	β⁻ 0,9; 1,7; γ 0,3; 0,7; K;	
$^{154}_{63}$Eu	5,4 a	β⁻ 0,3; 0,7; 1,9; γ 1,22; 1,17	1
$^{155}_{63}$Eu	1,7 a	β⁻ 0,154; 0,243; 2,23; γ 0,085; 0,099;	1
$^{153}_{64}$Gd	263 d	γ 0,106; 0,260; e⁻; K; no β⁺;	10
$^{159}_{64}$Gd	18,0 h	β⁻ 0,85; γ 0,055; 0,35;	100
$^{160}_{65}$Tb	71 d	β⁻ 0,546; 0,882; 0,71; / γ 0,086...0,297; 1,15	10
$^{165m}_{66}$Dy	1,25 m	e⁻ 0,109;	
$^{165}_{66}$Dy	2,42 h	β⁻ 0,42; 0,88; 1,25; γ 0,76; 0,095;	100
$^{166}_{66}$Dy	81 h	β⁻ 0,4; γ;	10
$^{166}_{67}$Ho	> 30 a	β⁻ 0,18; 0,28; γ 0,09...0,83;	10
$^{169}_{68}$Er	9,4 d	β⁻ 0,33; no γ;	10
$^{171}_{68}$Er	7,5 h	β⁻ 1,05; 0,65; 1,49; γ 0,305; 0,805;	100

Thyroid gland (Thyroxin and 3, 5–diiodotyrosine see p. 126)   2) Destruction free material examination

17

RADIOACTIVE ISOTOPES OF ELEMENTS

MN Sy AN	HLP	MeV Free limit microcurie	Uses
$_{69}$Tm^{170}	127 d	β⁻ 0.990; 0.886; γ 0.085;	1
$_{69}$Tm^{171}	680 d	β⁻ 0.10.	10
$_{70}$Yb^{175}	4.2 d	β⁻ 0.13. 0.5. γ 0.35.	10	T 1)
$_{71}$Lu^{177}	6.7 d	β⁻ 0.495. 0.366. 0.169. / γ 0.112. 0.206. 0.137.	10
$_{72}$Hf^{181}	45 d	β⁻ 0.410 γ 0.033 0.345 0.136 0.481	10
$_{73}$Ta^{182m}	16.4 m	β⁻ 0.2. e⁻ 0.180. K.	
$_{73}$Ta^{182}	115 d	β⁻ 0.52. 1.1. γ.	10
$_{74}$W^{181}	140 d	γ 1.83; e⁻ 0.07; 0.09; K;	10
$_{-4}$W^{185m}	1.85 m	e⁻ 0.075.	
$_{-4}$W^{185}	73 d	β⁻ 0.43. γ 0.134.	10
$_{74}$W^{187}	24.1 h	β⁻ 0.627.; 1.318. / γ 0.680 0.615 0.478 0.133 0.078	0.1
$_{75}$Re^{183m}	67 h	γ 1.75; K;	
$_{75}$Re^{183}	240 d	γ 0.081; 0.252; 1.0, K;	10
$_{75}$Re^{186}	92.8 h	β⁻ 1.07 0.93 0.3 / γ 0.137 0.746 0.627 0.123 K	10
$_{75}$Re^{187}	4.10^{12} a	β⁻ 0.043.	10	NR
$_{75}$Re^{188}	16.9 h	β⁻ 2.10 γ 0.15 0.48 0.64 0.95 1.40	10
$_{76}$Os^{185}	97 h	γ 0.65; 0.88; K;	10
$_{76}$Os^{191m}	14 h	e⁻ 0.074.	100
$_{76}$Os^{191}	15.0 d	β⁻ 0.143. γ 0.042. 0.128.	10
$_{76}$Os^{193}	32 h	β⁻ 1.10. γ 1.5. e⁻ 0.14.	10
$_{77}$Ir^{190m}	3.2 h	β⁺ 1.7; e⁻ 0.2; 0.8;	
$_{77}$Ir^{190}	12.6 d	γ 0.17; 0.55; e⁻ 0.17; 0.5; K;	10
$_{77}$Ir^{192m}	1.5 m	γ; e⁻ 0.057.	
$_{77}$Ir^{192}	74.7 d	β⁻; γ 0.137 ... 0.651.	10	T 2)
$_{77}$Ir^{194}	19.0 h	β⁻ 2.1. 0.48. γ 1.42. 0.4.	10
$_{78}$Pt^{191}	3.0 d	γ 0.57; 1.5; K;	10
$_{78}$Pt^{193m}	4.3 h	γ 0.26; 1.5; e⁻ 1.115; K;	10
$_{78}$Pt^{193}	long	L electron capture	10
$_{78}$Pt^{197m}	80 m	e⁻ 0.337.	100
$_{78}$Pt^{197}	18 h	β⁻ 0.40 0.67 γ 0.191 0.077	10
$_{79}$Au^{196m}	14 h	e⁻; K,	
$_{79}$Au^{196}	5.6 d	β⁻ 0.30 γ 0.175 0.330 0.358 K	10
$_{79}$Au^{198}	2.7 d	β⁻ 0.790. 0.957. 1.38. / γ 0.676. 0.441. 1.087.	10	Med
$_{79}$Au^{199}	3.3 d	β⁻ 0.300. 0.250. 0.460. / γ 0.23 0.156 0.207 0.024	10
$_{80}$Hg^{197m}	23 h	γ 0.276; 0.165; 0.135; e⁻ 0.165; K;	10
$_{80}$Hg^{197}	65 h	γ 0.191; 0.077; K;	10
$_{80}$Hg^{203}	46.5 d	β⁻ 0.208. γ 0.279. e⁻;	1
$_{81}$Tl^{200}	27 h	γ 0.365; 0.577; 1.210. 1.360; e⁻ C.4; K;	10
$_{81}$Tl^{201}	72 h	γ 0.21; K;	100	T
$_{81}$Tl^{202}	11.5 d	γ 0.435; e⁻ 0.35; K;	10
$_{81}$Tl^{204}	2.7 a	K; β⁻ 0.762; no γ;	10
$_{82}$Pb^{203}	47.9 h	γ 0.279; 0.47; e⁻; K;	10
$_{82}$Pb^{210}	25 a	β⁻ 0.029 0.017 γ 0.007 0.047	1	NR
$_{82}$Pb^{212}	10.6 h	β⁻ 0.355 0.590 / γ 0.238 0.115 0.176 0.250 0.300	1
$_{83}$Bi^{206}	6.2 d	γ 0.182...1.720; K; no β⁺	1

MN Sy AN	HLP	MeV Free limit microcurie	Uses
$_{83}$Bi^{207}	50 a	γ 0.565; 1.063; 1.46; 2.05—2.49; K;	1
$_{83}$Bi^{210}	4.85 d	α 5.0. β⁻ 1.030. 1.165; γ 0.08;	1	N
$_{83}$Bi^{212}	60.5 m	α 6.074; 6.113; 5.762; 5.601; / β⁻ 2.25; γ 0.040 ... 0.719;	1	N
$_{84}$Po^{210}	138 d	α 5.308; γ 0.803;	0.1	NR
$_{85}$At^{211}	7.5 h	α 5.89; K;	0.1
$_{86}$Rn^{220}	short	α 7.49;	10
$_{86}$Rn^{222}	38.0 s	α 6.51;	0.1
$_{88}$Ra^{223}	11.2 d	α 5.719. 5.607. 5.533. 5.439. / γ 0.144. 0.340.	0.1	N
$_{88}$Ra^{224}	3.64 d	α 5.66. 5.448. 5.194. γ 0.250.	1	N
$_{88}$Ra^{226}	1620 a	α 4.795 4.611 4.21 γ 0.188.	0.1	N Me
$_{88}$Ra^{228}	6.7 a	β⁻ 0.030;	0.1
$_{89}$Ac^{227}	27.7 a	α 4.94; β⁻ 0.02. γ 0.037.	0.1	N
$_{89}$Ac^{228}	6.13 h	α 4.54; β⁻ 1.5 2.0 γ 0.058 0.969	1
$_{90}$Th^{227}	18.6 d	α 5.674 ... 6.051. γ 0.043 ... 0.638;	0.1	NR
$_{90}$Th^{228}	1.90 a	α 5.423. 5.388. γ 0.063. 0.067.	0.1	NR
$_{90}$Th^{230}	8.10^{4} a	α 4.682 4.612 γ 0.066 0.15 ... 0.20	0.1	NR
$_{90}$Th^{231}	15.6	β⁻ 0.093. 0.302. 0.216. / γ 0.022. 0.085. 0.059.	10	NR
$_{90}$Th^{232}	1.39 10^{10}	α 3.98; γ 0.055;	0.1	NR
$_{90}$Th^{234}	24.10 d	β⁻ 0.192. 0.104. γ 0.090.	10	NR
$_{91}$Pa^{230}	17.7 d	α; β⁻ 1.1. γ 0.94. K;	1
$_{91}$Pa^{231}	34.3 a	α 4.66 5.042 γ 0.095 0.294 0.323;	0.1	N
$_{91}$Pa^{233}	27.4 d	β⁻ 0.530. γ 0.0289 0.4164	10
$_{92}$U^{230}	20.8 d	α 5.86;	1
$_{92}$U^{231}	70 a	α 5.31; γ 0.060;	0.1
$_{92}$U^{233}	1.62 10^{5}	α 4.80; γ 0.04; 0.08; 0.31;	1
$_{92}$U^{234}	2.48 10^{5}	α 4.76; γ 0.035;	0.1	N
$_{92}$U^{235}	7.1 10^{8}	α 4.40; 4.58;	1
$_{92}$U^{236}	2.46 10^{7}	α 4.499; γ 0.050;	1
$_{92}$U^{238}	4.498 10^{9}	α 4.180; γ 0.045;	1	N
$_{93}$Np^{237}	2.2 10^{6}	α 4.77; γ 0.065;	0.1
$_{93}$Np^{239}	2.33 d	β⁻ 0.288. 0.403. 0.676; γ 0.0249 ... 0.277;	10
$_{94}$Pu^{238}	92 a	α 5.49; γ 0.040;	0.1
$_{94}$Pu^{239}	24.1 a	α 5.147. 5.1 γ 0.035. 0.050.	0.1
$_{94}$Pu^{240}	6580 a	α 5.16;	0.1
$_{94}$Pu^{241}	14 a	α 4.91; β⁻ 0.01;	1
$_{94}$Pu^{242}	5.10^{5}	α 4.88;	0.1
$_{95}$Am^{241}	475 a	α 5.476; 5.433; γ 0.0597; 0.100; 0.041;	0.1
$_{95}$Am^{243}	10^{4} a	α 5.21;	0.1
$_{96}$Cm^{242}	162 d	α 6.08; γ 0.045;	0.1
$_{96}$Cm^{243}	35 a	α 5.777 5.732 5.935 6.003 5.672;	0.1
$_{96}$Cm^{244}	19 a	α 5.798; 5.755;	0.1
$_{96}$Cm^{245}	2.10^{4} a	α 5.36;	0.1
$_{96}$Cm^{246}	3000 a	α 5.36;	0.1
$_{97}$Bk^{249}	270 d	α 5.4; 5.08; β⁻ 0.1;	1
$_{98}$Cf^{249}	470 a	α 5.82; 5.91; 6.190;	0.1
$_{98}$Cf^{250}	12 a	α 6.024; 5.980;	0.1
$_{98}$Cf^{252}	2.2 a	α 6.112; 6.039;	0.1

1) Lighting material   2) Destruction free material examina
3) Natural thorium (isotopic mixture) 100 g for analytical and preparative chemical work permission

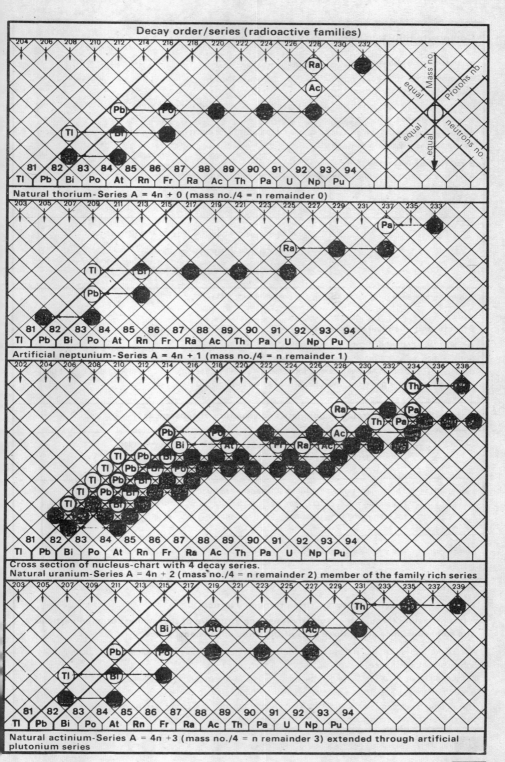

Decay order/series (radioactive families)

Mass no. / Protons no. / neutrons no. — equal / equal / equal

Natural thorium–Series A = 4n + 0 (mass no./4 = n remainder 0)

Artificial neptunium–Series A = 4n + 1 (mass no./4 = n remainder 1)

Cross section of nucleus-chart with 4 decay series.
Natural uranium–Series A = 4n + 2 (mass no./4 = n remainder 2) member of the family rich series

Natural actinium–Series A = 4n +3 (mass no./4 = n remainder 3) extended through artificial plutonium series

Composite Substances

Structure of Substances

A composite substance is a substance which can be resolved into different other substances through physical or chemical process.

Composite substances which can be split up into constituent substances merely through physical process are called mixtures.

Composite substances which can be split up into constituent substances only through chemical process are called chemical compounds.

Structure of Mixtures

A mixture consists of minimum two chemical compounds or elements.

Particle size ($\emptyset$ in cm)	Designation of mixture
$> 10^{-3}$	Heterogeneous mixture (gr. heteros = different)
10^{-4} to 10^{-6}	Colloid mixture (gr. kollos = glue)
$\sim 10^{-8}$	Homogeneous mixture (gr. homoios = equal)

The mixture is according to particle size, more or less intimate.

Structure of chemical compounds

A chemical compound consists of either molecules, or ions, or atoms.

Since atoms, ions and molecules are spatial structure, the chemical compounds also have the three dimensional structure. All symbols and formulae are simple representation of spatial conditions or processes on the paper plane and must be visualised accordingly.

Correlation between state of aggregate and structure of substances

Solids (s)	The particles of which these substances are built are locally bonded. The substance has a definite external shape and called rigid bodies.
Liquids (l)	The particles of which these substances are built are limited moveable. Only the form of surface area is definite through the attractive forces of the surrounding while the remaining liquid fills up the given vessel and assumes its form.
Gases (g)	The particles of which these substances are built are freely moveable. Gases fill uniformly any offered space.

Structure of solid bodies

The structure of solid bodies is written through a lattice.

A lattice results through the regularly repeating structure of the smallest ordered particles of a solid substance along three axes of the three dimensional space.

According to these structural particles, we differentiate between molecular lattice, ionic lattice, metallic lattice, and atomic lattice.

For writing a lattice, we imagine that the heavy points of the structural particles are bound to one another through a (stroke) line. Thus a definite geometrical figure "lattice model" results.

The binding types and the binding forces which hold together in a lattice also hold together structural particles of chemical compounds—molecules, ions, and atoms.

Bonding type*	Bonding forces	in lattice type	in structural particles
Atomic bonding see p. 22	Chemical forces	Atomic lattice	Molecules molecular ions
Ionic bonding see p. 23	Electrical attraction	Ionic lattice	Molecular ions Complex ions
Metallic bonding see p. 24	Electrical attraction	Mettalic lattice	Molecular aggregate
Hydrogen bonding see p. 32	Electrical attraction	Molecular lattice	Molecular aggregate
Vander Waal's bonding see p. 32	Physical attraction	Molecular lattice	Molecular aggregate
Inclusion bonding see p. 32	Physical and/or electrical attraction	Molecular lattice or Ionic lattice	

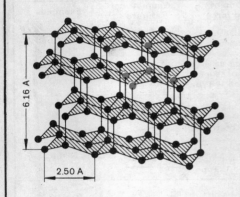

Atomic lattice of diamond

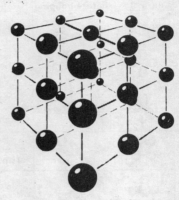

Ionic lattice of NaCl

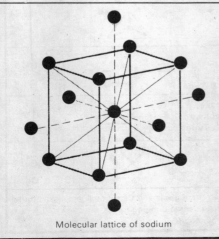

Molecular lattice of sodium

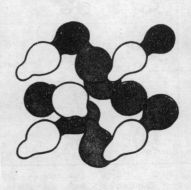

Molecular lattice of carbonmonoxide

The Atomic bonding (covalent bonds)

The atomic bonding results under the conditions when several atoms with the help of valence electrons which besides going together in pairs, build a **common shell.**

The number of concerned atoms in a compound and their geometrical structure depend on.

1. the number of valence electrons = the cohesiveness,
2. on the nature of orbital sub-groups in which the valency electrons move or pass over the excited state,
3. the distribution of valence electrons in the excited state.

The excited state results through supply of energy. Out of the doubly filled orbitals the electrons pass over to still free orbitals or other orbital sub-groups of the same shell. The existing arrangement of orbital sub-group is fully or partly dissolved whereby hybrid orbitals and other separate orbitals which serve only to build compounds, occur.

The hybrid orbitals are designated with q whereas the separate orbitals with σ, π and δ

Through the formation of hybrid and separate orbitals in the shell, directed forces result, which together with the directed forces in the shells of shared atoms in the compound stipulate the geometrical arrangement of the particles.

For the process of compound formation, the hybrid or the separate orbitals are filled with electron pair which must have opposite spins.

Examples for orbitals in the excited state (hybrid or separate orbitals)

Cohesiveness	State of orbitals normal	before bonding	Direction of bonding force	Geometrical figure	Examples
1 1	s^1 p^1	σ^1 π^1	linear in direction of partners		H in H_2 H – H Cl in Cl_2 Cl – Cl
2	p^2	π^2	linear and parallel in direction of partners **Double bond 1. Type**		O in O_2 O = O
3	p^3	π^3	in the corners of a flat pyramid **(height = 1/3 tetragon)**		N in NH_3 H – N – H H
3	$s^1 + p^2$	$\sigma + \pi^2$	linear parallel in direction of the partners **Triple bond 1. Type**		N in N_2 N ≡ N
3	$s^1 + p^2$	q^3	in the corners of a regular triangle		C^+ in $[CO_3]^-$ $\begin{bmatrix} O_{\diagdown}O \\ C \end{bmatrix}$
4	$s^1 + p^3$	q^4	in the corners of a tetrahedron		C in CH_4 H H C H H
4	$s^1 + p^3$	$q^3 + \pi^1$	in the corners of a regular triangle and parallel to its plane **Double bond 2. Type**		C in C_2H_4 H H C = C H H
4	$s^1 + p^3$	$q^2 + \pi^2$	linear, opposite and parallel to that **Triple bond 2. Type**		C in C_2H_2 H – C ≡ C – H
4	$s^1 + p^2 + d^1$	q^4	in the corners of a square		Pt in $[PtCl_4]^{2-}$ $\begin{bmatrix} Cl_{\diagdown}{}_{\diagup}Cl \\ Pt \\ Cl^{\diagup}{}^{\diagdown}Cl \end{bmatrix}^{2-}$

In case of elements that are placed before rare gases, in case of rare gases and transitional metals still further combinations are possible which give rise to three dimensional directed bonding forces.
Examples. $s^1p^3d^1$, $s^1p^3d^2$, (octahedral) $s^1p^3d^3$. (See pp. 26–30, and 31.)

Ionic bonding

The ionic bonding results under the condition that ionic atoms of opposite charges attract mutually.

The number of concerned ionic atoms and their geometrical structure besides the resulting parts or lattices depend upon:

1. the number of valence electrons given or taken up = ionic valency,

2. the size, i.e., the diameter of the concerned ionic atoms of the compound.

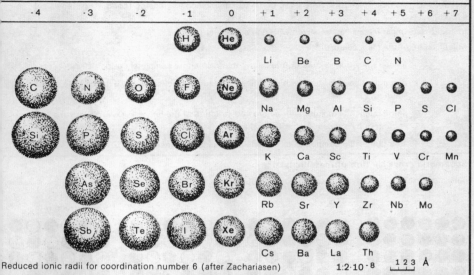

| -4 | -3 | -2 | -1 | 0 | +1 | +2 | +3 | +4 | +5 | +6 | +7 |

H   He   Li   Be   B   C   N

C   N   O   F   Ne   Na   Mg   Al   Si   P   S   Cl

Si   P   S   Cl   Ar   K   Ca   Sc   Ti   V   Cr   Mn

As   Se   Br   Kr   Rb   Sr   Y   Zr   Nb   Mo

Sb   Te   I   Xe   Cs   Ba   La   Th

Reduced ionic radii for coordination number 6 (after Zachariasen) $1:2 \cdot 10^{-8}$ 1 2 3 Å

Anion > Atom > Cation

$Anion^{-1} < Anion^{-2} < Anion^{-3} < Anion^{-4}$

$Cation^{+1} > Cation^{+2} > Cation^{+3} > Cation^{+4} > Cation^{+5} > Cation^{+6}$

The coordination number (CN) is the number of the immediate neighbouring particles which surround an ionic atom in a compound or a lattice

The coordination number is determined through the relationship: $\dfrac{\text{Diameter of cations}}{\text{Diameter of anions}} = a$

For compounds of type: AB is given by:

a	CN	Geometrical arrangement		Lattice type	
1	12	Pyramid		Hexagonal close packing	
1 - 0,732	8	Corners of a cube		Cubical packing	
0,732 - 0,414	6	Corners of octahedron		NaCl-Type <div style="text-align:right">see p. 21</div>	
0,732 - 0,414	4	Corners of a square		Complex structural particles	
0,414 - 0.22	4	Corners of tetrahedron		ZnS-Type <div style="text-align:right">see p. 21</div>	
0,22 - 0,15	3	Corners of a triangle		Complex structural particles	
< 0,15	2 1	Linear	———	Complex structural particles	

see p. 21

The metallic bonding

The metallic bonding can occur only in solid state of a substance:

The metallic bonding occurs under the condition that positively charged ionic atoms go together into a lattice. The charge composition and the cohesion in the lattice results through the sum of the given electrons. These take the place of anions of a normal ionic lattice, as these are not locally bound but distributed in spatial region in all the three directions and freely movable in space. According to the origin of valence electrons out of the orbital sub-groups s, p, d or f results different such zones, the three dimensions of which are not sharply resolved. Through these over-lapping it is possible to let the electrons to pass over to another zone through the minimum supply of energy (conductors and sub-conductors).

In liquid state, most metals can be mixed with one another in any ratio.

Solidification of the melt, can result:

> a heterogeneous mixture disintegrating the melt, either completely or partially

> a homogeneous mixture with disordered structure of the lattice = Mixed crystal or solid solution

> a homogeneous mixture with an ordered structure of the lattice = **Over structure** (by slow cooling or by annealing of **solid** bodies)

> an intermetallic compound or so-called intermetallic **Phase.**

Schematic cross section of metallic lattice:

pure metal solid solution (substitution lattice) intermetallic compound (over structure)

Between all states processes are available and these can occur side by side the solidification of melt. Therefore the structure of the solid bodies which are designated as **alloys** in technical usage, is very complicated.

The composition of a intermetallic compound (Phase) can:

> follow from stoichiometric principles,

> from the formed lattice structure (whereby only the ideal composition is given),

> as per 'HUME-ROTHERY' rule for the relationship:

$$\frac{\text{Sum of valence electrons}}{\text{Sum of ionic atoms}} = \frac{21}{14} \text{ or } \frac{21}{13} \text{ or } \frac{21}{12} \text{ follows}$$

e.g. $CuZn$ $\quad \dfrac{1+2}{1+1} = \dfrac{3}{2} = \dfrac{21}{14}$

$Cu_5 Sn$ $\quad \dfrac{5+4}{5+1} = \dfrac{9}{6} = \dfrac{21}{14}$

$Cu_5 Zn_8$ $\quad \dfrac{5+16}{5+8} = \dfrac{21}{13}$

$Cu_{31} Sn_8$ $\quad \dfrac{31+32}{31+8} = \dfrac{63}{39} = \dfrac{21}{13}$

$Cu Zn_3$ $\quad \dfrac{1+6}{1+3} = \dfrac{7}{4} = \dfrac{21}{12}$

$Cu_3 Sn$ $\quad \dfrac{3+4}{3+1} = \dfrac{7}{4} = \dfrac{21}{12}$

Metallic bonding also occurs in central atom association of compounds of transitional elements (transition metals) see pp. 31 and 47.

Mixed bonding (coordinate covalent bond)

A mixed bonding exists when an atomic and ionic bonding are shared together in a compound formation.

Thereby it gives rise to two causes:

1. To the structure of such bonds like atoms whose part of valence electrons have been given up or received only are qualified while with the other part hybrid orbitals are formed.
Such particles are called formal ionic atoms.

They possess—formal charge and simultaneous cohesiveness.
During compound formation we observe:

    Sum of formal charges = 0 neutral molecule .
    Sum of formal charges ≠ 0 molecule ion (complex)

2. The shell of an anion can be deformed through the exerting influence of electrical attractive forces.

The deformation is larger, as when the diameter of the anions is larger and the diameter of the cations is smaller, that is the cation is more charged.

Through deformation results **polar** particles.

Polar particles are particles in which the centre of gravity of the charge does not coincide with the centre of gravity of negative charge.

  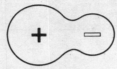 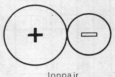

unpolar   Increasing polarity   Ionpair

The appearance of polar ions is due to excited atoms with similar cohesiveness, and it results due to their preferred bonding direction. In case of compound formation no **common shell** is built.

A bonding in which polar ions are shared, distinguishes itself from the pure ionic bonding through greater stability and in many cases it is not different from above mentioned mixed bonding.

During compound formation we observe:

When the charges of the cations equal the coordination number, and is equal to the sum of the charges of anions, an eveloped ion is built that equals a neutral molecule.

In all other cases, charged particles in the form of **complex ion** results.

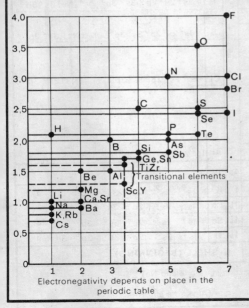

Electronegativity depends on place in the periodic table

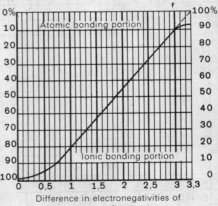

Difference in electronegativities of compound partner

The greater the electronegativity of the two elements, the greater is the strength of built ionic lattice.

(see p. 31)

In the following tables the electron distribution of the atoms of elements of maingroup is schematically given.

Electron distribution in the Atomic shell
Ionic valency—Cohesiveness—Formal charge

I		1s	(Schematic representation)												I	(Symbolic way of writing)
1	H	1													H	$1s^1$
2	He	1↓													He	$1s^2$ or (He)

II		1s	2s	2p											II	
3	Li	1↓	1												Li	$1s^2\,2s^1$ or (He)$2s^1$
4	Be	1↓	1↓												Be	$1s^2\,2s^2\,2p^1$ or (He)$2s^2$
5	B	1↓	1↓	1											B	$1s^2\,2s^2\,2p^1$ or (He)$2s^2\,2p^1$
6	C	1↓	1↓	1	1										C	$1s^2\,2s^2\,2p^2$ or (He)$2s^2\,2p^2$
7	N	1↓	1↓	1	1	1									N	$1s^2\,2s^2\,2p^3$ or (He)$2s^2\,2p^3$
8	O	1↓	1↓	1↓	1	1									O	$1s^2\,2s^2\,2p^4$ or (He)$2s^2\,2p^4$
9	F	1↓	1↓	1↓	1↓	1									F	$1s^2\,2s^2\,2p^5$ or (He)$2s^2\,2p^5$
10	Ne	1↓	1↓	1↓	1↓	1↓									Ne	$1s^2\,2s^2\,2p^6$ or (Ne)

III		1s	2s	2p			3s	3p			3d				III	
11	Na	1↓	1↓	1↓	1↓	1↓	1								Na	$1s^2\,2s^2\,2p^6\,3s^1$ or (Ne)$3s^1$
12	Mg	1↓	1↓	1↓	1↓	1↓	1↓								Mg	$1s^2\,2s^2\,2p^6\,3s^2$ or (Ne)$3s^2$
13	Al	1↓	1↓	1↓	1↓	1↓	1↓	1							Al	$1s^2\,2s^2\,2p^6\,3s^2\,3p^1$ or (Ne)$3s^2\,3p^1$
14	Si	1↓	1↓	1↓	1↓	1↓	1↓	1	1						Si	$1s^2\,2s^2\,2p^6\,3s^2\,3p^2$ or (Ne)$3s^2\,3p^2$
15	P	1↓	1↓	1↓	1↓	1↓	1↓	1	1	1					P	$1s^2\,2s^2\,2p^6\,3s^2\,3p^3$ or (Ne)$3s^2\,3p^3$
16	S	1↓	1↓	1↓	1↓	1↓	1↓	1↓	1	1					S	$1s^2\,2s^2\,2p^6\,3s^2\,3p^4$ or (Ne)$3s^2\,3p^4$
17	Cl	1↓	1↓	1↓	1↓	1↓	1↓	1↓	1↓	1					Cl	$1s^2\,2s^2\,2p^6\,3s^2\,3p^5$ or (Ne)$3s^2\,3p^5$
18	Ar	1↓	1↓	1↓	1↓	1↓	1↓	1↓	1↓	1↓					Ar	$1s^2\,2s^2\,2p^6\,3s^2\,3p^6$ or (Ne)$3s^2\,3p^6\,3d^0$

IV		1s	2s	2p			3s	3p			3d			4s	IV	
19	K	1↓	1↓	1↓	1↓	1↓	1↓	1↓	1↓	1↓				1	K	$1s^2\;2s^2\;2p^6\;3s^2\;3p^6\;3d^0\;4s^1$
20	Ca	1↓	1↓	1↓	1↓	1↓	1↓	1↓	1↓	1↓				1↓	Ca	$1s^2\;2s^2\;2p^6\;3s^2\;3p^6\;3d^0\;4s^2$
21	Sc	1↓	1↓	1↓	1↓	1↓	1↓	1↓	1↓	1↓	1			1↓	Sc	$1s^2\;2s^2\;2p^6\;3s^2\;3p^6\;3d^1\;4s^2$
22	Ti	1↓	1↓	1↓	1↓	1↓	1↓	1↓	1↓	1↓	1	1		1↓	Ti	$1s^2\;2s^2\;2p^6\;3s^2\;3p^6\;3d^2\;4s^2$

K-	L-SHELL	M-SHELL	N-SHELL (BEGINNING)

Completely filled orbital group (shells) = stabler state (He and Ne)

Completely filled orbital sub-group = stabler state (Ar, Kr, Xe and Rn)

Electron distribution in the Atomic shells
Ionic valency—Cohesiveness—Formal charge

Legend — single mark = one electron (left or right spin); paired mark (written "1↓") = electron pair with opposite spins.

Section I

I	1s	(Excited state)	C	I	(Ionised state)	IV	FC	C
x	I			H^+	PROTON (in solution $[H_3O]^+$)	+1		
				H^-	1↓	−1		

Section II

J	1s	2s	2p		C	II	1s	2s	2p		IV	FC	C
x	1↓	I			1	Li^+	1↓				+1		
x	1↓	I	I		2	Be^{2+}	1↓				+2		
x	1↓	I	I I		3	B^{3+}	1↓				+3		
x	1↓	1↓	I I		2	$C^{\ominus}$	1↓	I	I I I			+1	3
	1↓	I	I I I		4	$C^{\oplus}$	1↓	1↓	I I I			−1	3
x	1↓	1↓	I I I		3	$N^{\oplus}$	1↓	I	I I I I			+1	4
						N^{+2}	1↓	I	I I I			+2	3
						$N^{\ominus}$	1↓	1↓	1↓	I I		−1	2
x	1↓	1↓	1↓ I I		2	$O^{\oplus}$	1↓	1↓	I I I			+1	3
						$O^{\ominus}$	1↓	1↓	1↓ 1↓ I			−1	1
						O^{2-}	1↓	1↓	1↓ 1↓ 1↓		−2		
x	1↓	1↓	1↓ 1↓ I		1	F^-	1↓	1↓	1↓ 1↓ 1↓		−1		

Section III

I	K	L	3s	3p	3d		C	III	K	L	3s	3p	3d		IV	FC	C
x	2	8	I				1	Na^+	2	8					+1		
x	2	8	I I				2	Mg^{2+}	2	8					+2		
x	2	8	I I				3	Al^{3+}	2	8					+3		
x	2	8	1↓ I	I I			2	$Si^{\ominus}$	2	8	1↓	I I I				−1	3
	2	8	I I	I I			4	Si^{-2}	2	8	1↓	1↓ I I				−2	2
x	2	8	1↓ I	I I			3	$P^{\oplus}$	2	8	1↓	I I				+1	2
	2	8	I I	I I I			5	$P^{\oplus}$	2	8	I	I I I				+1	4
								$P^{\ominus}$	2	8	1↓	1↓ I I				−1	2
x	2	8	1↓ 1↓	I I			2	S^{+2}	2	8	1↓	I I				+2	2
	2	8	1↓ I	I I I			4	S^{+2}	2	8	I	I I I				+2	4
	2	8	I I	I I I I			6	S^{-2}	2	8	1↓	1↓ 1↓ 1↓			−2		
x	2	8	1↓ 1↓	1↓ I			1	Cl^{+3}	2	8	I	I I I				+3	4
	2	8	1↓ 1↓	I I I			3	Cl^{+2}	2	8	1↓	I I I				+2	3
	2	8	1↓ I	I I I I			5	$Cl^{\oplus}$	2	8	1↓	1↓ I I				+1	2
	2	8	I I	I I I I I			7	Cl^-	2	8	1↓	1↓ 1↓ 1↓			−1		
x	2	8	1↓ 1↓	1↓ I I			2		2	8	1↓	1↓ 1↓ 1↓					

Section IV

V	K	L	3s	3p	3d	4s	4p	C	IV	K	L	3s	3p	3d	4s	IV	FC	C
x	2	8	1↓ 1↓	1↓ 1↓			I	1	K^+	2	8	1↓	1↓ 1↓ 1↓			+1		
x	2	8	1↓ 1↓	1↓ 1↓			I	I	2	Ca^{2+}	2	8	1↓	1↓ 1↓ 1↓		+2		
x	2	8	1↓ 1↓	1↓ 1↓	I		I	I	3	Sc^{3+}	2	8	1↓	1↓ 1↓ 1↓		+3		
x	2	8	1↓ 1↓	1↓ 1↓	I I			1↓	2	Ti^{2+}	2	8	1↓	1↓ 1↓ 1↓	1↓	+2		
	2	8	1↓ 1↓	1↓ 1↓	I I		I	I	4	Ti^{4+}	2	8	1↓	1↓ 1↓ 1↓		+4		

Electron with left spin

Electron with right spin

Electron pair with opposite set spins

C = Cohesiveness

IV = Ionic valency

FC = Formal charge together with cohesiveness

Electron distribution in Atomic shells
Ionic valency—Cohesiveness—Formal charges

IV		1s	2s	2p	3s	3p	3d	4s	4p	4d	4f	
19	K							1				as in case of elements of
20	Ca							1↓				1 and 2 periods
21	Sc							1↓				Less stable complex
22	Ti							1↓				Stabilisation through
	Ti^{2+}											building of poly anions
	Ti^{4+}											[Poly acids and
23	V						1	1↓		•		hetero-poly acids
	V^{3+}											see p. 68]
	V^{5+}											Red edges
24	Cr						1 1 1 1					Symbol for central atom
	Cr^{3+}						1					(with charge)
25	Mn						1 1 1 1↓					for example:
	Mn^{2+}						1 1↓1↓1↓1↓1↓1↓					as central atom in
	Mn^{3+}						1↓1↓1↓1↓1↓1↓					$K_4 [Mn (CN)_6]$
26	Fe						1 1 1 1↓					$K_3 [Mn (CN)_6]$
	Fe^{2+}						1 1 1					
	Fe^{2+}											$K_4 [Fe (CN)_6]$
	Fe^{3+}						1 1 1					$K_3 [Fe (CN)_6]$
	Fe^{3+}						1 1↓1↓1↓1↓1↓1↓					
27	Co						1 1 1 1↓					
	Co^{2+}						1↓1↓1↓1↓1↓1↓	1				$[Co (NH_3)_6] Cl_2$
	Co^{3+}											$[Co (NH_3)_6] Cl_3$
28	Ni						1 1 1 1↓					
	Ni^{1}						1↓1↓1↓1↓					$K_2 [Ni (CN)_4]$
	Ni^{2}											
29	Cu						1↓1↓1↓1↓1↓	1				
	Cu						1↓1↓1↓1↓1↓					$K_3 [Cu (CN)_4]$
	Cu^{2}									1		$[Cu (NH_3)_4] Cl_2$
30	Zn						1↓1↓1↓1↓	1↓				
	Zn^{2+}											$[Zn (NH_3)_4] Cl_2$
31	Ga							1↓	1			as in case of the
32	Ge							1↓	1 1			elements of 1 and 2
33	As							1↓	1 1 1			periods
34	Se							1↓1↓	1 1			See pp. 26/27
35	Br							1↓1↓	1↓ 1			
36	Kr							1↓1↓	1↓1↓			
	Kr							1↓1↓1↓	1 1			Similarity worth for
								1↓1↓	1 1	⊥ 1		rare gas compounds
								1↓	1 1	1 1 1		of xenons. [Complex
								1	1	1 1 1 1		with d-electron bonds.]

The 4-f-electron orbital remains unoccupied

remains unoccupied

	K	L	M			N	—Shells
max:	2	8	18			32	Electrons

The transitional elements, that is elements of sub-groups especially from complexes. The Geometry of hybridisation of drop-electron is analogous to the spd electrons. (see p. 22)

28

Electron distribution in Atomic shell
Ionic valency — Cohesiveness — Formal charge

V	K	L	M	4s	4p	4d	4f	5s	5p	5d	5f
37 Rb	2	8	18	↑↓	↑↓ ↑↓ ↑↓			1			
38 Sr	2	8	18	↑↓	↑↓ ↑↓ ↑↓			↑↓			
39 Y	2	8	18	↑↓	↑↓ ↑↓ ↑↓	1		↑↓			
40 Zr	2	8	18	↑↓	↑↓ ↑↓ ↑↓	1 1		↑↓			
41 Nb	2	8	18	↑↓	↑↓ ↑↓ ↑↓	1 1 1 1		1			
42 Mo	2	8	18	↑↓	↑↓ ↑↓ ↑↓	1 1 1 1 1		1			
43 Tc	2	8	18	↑↓	↑↓ ↑↓ ↑↓	↑↓ 1 1 1		1			
44 Ru	2	8	18	↑↓	↑↓ ↑↓ ↑↓	↑↓ ↑↓ 1 1 1		1			
45 Rh	2	8	18	↑↓	↑↓ ↑↓ ↑↓	↑↓ ↑↓ ↑↓ 1 1		1			
46 Pd	2	8	18	↑↓	↑↓ ↑↓ ↑↓	↑↓ ↑↓ ↑↓ ↑↓ ↑↓					
47 Ag	2	8	18	↑↓	↑↓ ↑↓ ↑↓	↑↓ ↑↓ ↑↓ ↑↓ ↑↓		1			
48 Cd	2	8	18	↑↓	↑↓ ↑↓ ↑↓	↑↓ ↑↓ ↑↓ ↑↓ ↑↓		↑↓			
49 In	2	8	18	↑↓	↑↓ ↑↓ ↑↓	↑↓ ↑↓ ↑↓ ↑↓ ↑↓		↑↓	1		
50 Sn	2	8	18	↑↓	↑↓ ↑↓ ↑↓	↑↓ ↑↓ ↑↓ ↑↓ ↑↓		↑↓	1 1		
51 Sb	2	8	18	↑↓	↑↓ ↑↓ ↑↓	↑↓ ↑↓ ↑↓ ↑↓ ↑↓		↑↓	1 1 1		
52 Te	2	8	18	↑↓	↑↓ ↑↓ ↑↓	↑↓ ↑↓ ↑↓ ↑↓ ↑↓		↑↓	↑↓ 1 1		
53 I	2	8	18	↑↓	↑↓ ↑↓ ↑↓	↑↓ ↑↓ ↑↓ ↑↓ ↑↓		↑↓	↑↓ ↑↓ 1		
54 Xe	2	8	18	↑↓	↑↓ ↑↓ ↑↓	↑↓ ↑↓ ↑↓ ↑↓ ↑↓		↑↓	↑↓ ↑↓ ↑↓		

The 5-f-Electron orbital remain vacant

VI	K	L	M	4s	4p	4d	4f	5s	5p	5d	5f	6s
55 Cs	2	8	18	↑↓	↑↓ ↑↓ ↑↓	↑↓ ↑↓ ↑↓ ↑↓ ↑↓		↑↓	↑↓ ↑↓ ↑↓			1
56 Ba	2	8	18	↑↓	↑↓ ↑↓ ↑↓	↑↓ ↑↓ ↑↓ ↑↓ ↑↓		↑↓	↑↓ ↑↓ ↑↓			↑↓
57 La	2	8	18	↑↓	↑↓ ↑↓ ↑↓	↑↓ ↑↓ ↑↓ ↑↓ ↑↓		↑↓	↑↓ ↑↓ ↑↓	1		↑↓
58 Ce	2	8	18	↑↓	↑↓ ↑↓ ↑↓	↑↓ ↑↓ ↑↓ ↑↓ ↑↓	1 1	↑↓	↑↓ ↑↓ ↑↓			↑↓
59 Pr	2	8	18	↑↓	↑↓ ↑↓ ↑↓	↑↓ ↑↓ ↑↓ ↑↓ ↑↓	1 1 1	↑↓	↑↓ ↑↓ ↑↓			↑↓
60 Nd	2	8	18	↑↓	↑↓ ↑↓ ↑↓	↑↓ ↑↓ ↑↓ ↑↓ ↑↓	1 1 1 1	↑↓	↑↓ ↑↓ ↑↓			↑↓
61 Pm	2	8	18	↑↓	↑↓ ↑↓ ↑↓	↑↓ ↑↓ ↑↓ ↑↓ ↑↓	1 1 1 1 1	↑↓	↑↓ ↑↓ ↑↓			↑↓
62 Sm	2	8	18	↑↓	↑↓ ↑↓ ↑↓	↑↓ ↑↓ ↑↓ ↑↓ ↑↓	1 1 1 1 1 1	↑↓	↑↓ ↑↓ ↑↓			↑↓
63 Eu	2	8	18	↑↓	↑↓ ↑↓ ↑↓	↑↓ ↑↓ ↑↓ ↑↓ ↑↓	1 1 1 1 1 1 1	↑↓	↑↓ ↑↓ ↑↓			↑↓
64 Gd	2	8	18	↑↓	↑↓ ↑↓ ↑↓	↑↓ ↑↓ ↑↓ ↑↓ ↑↓	1 1 1 1 1 1 1	↑↓	↑↓ ↑↓ ↑↓	1		↑↓
65 Tb	2	8	18	↑↓	↑↓ ↑↓ ↑↓	↑↓ ↑↓ ↑↓ ↑↓ ↑↓	↑↓ ↑↓ 1 1 1 1 1	↑↓	↑↓ ↑↓ ↑↓			↑↓
66 Dy	2	8	18	↑↓	↑↓ ↑↓ ↑↓	↑↓ ↑↓ ↑↓ ↑↓ ↑↓	↑↓ ↑↓ ↑↓ 1 1 1 1	↑↓	↑↓ ↑↓ ↑↓			↑↓
67 Ho	2	8	18	↑↓	↑↓ ↑↓ ↑↓	↑↓ ↑↓ ↑↓ ↑↓ ↑↓	↑↓ ↑↓ ↑↓ ↑↓ 1 1 1	↑↓	↑↓ ↑↓ ↑↓			↑↓
68 Er	2	8	18	↑↓	↑↓ ↑↓ ↑↓	↑↓ ↑↓ ↑↓ ↑↓ ↑↓	↑↓ ↑↓ ↑↓ ↑↓ ↑↓ 1 1	↑↓	↑↓ ↑↓ ↑↓			↑↓
69 Tm	2	8	18	↑↓	↑↓ ↑↓ ↑↓	↑↓ ↑↓ ↑↓ ↑↓ ↑↓	↑↓ ↑↓ ↑↓ ↑↓ ↑↓ ↑↓ 1	↑↓	↑↓ ↑↓ ↑↓			↑↓
70 Yb	2	8	18	↑↓	↑↓ ↑↓ ↑↓	↑↓ ↑↓ ↑↓ ↑↓ ↑↓	↑↓ ↑↓ ↑↓ ↑↓ ↑↓ ↑↓ ↑↓	↑↓	↑↓ ↑↓ ↑↓			↑↓
71 Lu	2	8	18	↑↓	↑↓ ↑↓ ↑↓	↑↓ ↑↓ ↑↓ ↑↓ ↑↓	↑↓ ↑↓ ↑↓ ↑↓ ↑↓ ↑↓ ↑↓	↑↓	↑↓ ↑↓ ↑↓	1		↑↓
72 Hf	2	8	18	↑↓	↑↓ ↑↓ ↑↓	↑↓ ↑↓ ↑↓ ↑↓ ↑↓	↑↓ ↑↓ ↑↓ ↑↓ ↑↓ ↑↓ ↑↓	↑↓	↑↓ ↑↓ ↑↓	1 1		↑↓
73 Ta	2	8	18	↑↓	↑↓ ↑↓ ↑↓	↑↓ ↑↓ ↑↓ ↑↓ ↑↓	↑↓ ↑↓ ↑↓ ↑↓ ↑↓ ↑↓ ↑↓	↑↓	↑↓ ↑↓ ↑↓	1 1 1		↑↓
74 Wo	2	8	18	↑↓	↑↓ ↑↓ ↑↓	↑↓ ↑↓ ↑↓ ↑↓ ↑↓	↑↓ ↑↓ ↑↓ ↑↓ ↑↓ ↑↓ ↑↓	↑↓	↑↓ ↑↓ ↑↓	1 1 1 1		↑↓
75 Re	2	8	18	↑↓	↑↓ ↑↓ ↑↓	↑↓ ↑↓ ↑↓ ↑↓ ↑↓	↑↓ ↑↓ ↑↓ ↑↓ ↑↓ ↑↓ ↑↓	↑↓	↑↓ ↑↓ ↑↓	1 1 1 1 1		↑↓
76 Os	2	8	18	↑↓	↑↓ ↑↓ ↑↓	↑↓ ↑↓ ↑↓ ↑↓ ↑↓	↑↓ ↑↓ ↑↓ ↑↓ ↑↓ ↑↓ ↑↓	↑↓	↑↓ ↑↓ ↑↓	↑↓ 1 1 1 1		↑↓
77 Ir	2	8	18	↑↓	↑↓ ↑↓ ↑↓	↑↓ ↑↓ ↑↓ ↑↓ ↑↓	↑↓ ↑↓ ↑↓ ↑↓ ↑↓ ↑↓ ↑↓	↑↓	↑↓ ↑↓ ↑↓	↑↓ ↑↓ ↑↓ ↑↓ 1		↑↓
78 Pt	2	8	18	↑↓	↑↓ ↑↓ ↑↓	↑↓ ↑↓ ↑↓ ↑↓ ↑↓	↑↓ ↑↓ ↑↓ ↑↓ ↑↓ ↑↓ ↑↓	↑↓	↑↓ ↑↓ ↑↓	↑↓ ↑↓ ↑↓ ↑↓ 1		1

The 5-f-Electron orbital remain vacant

The 4 f-electron of 'Lanthanides' exceptionally take part in bonding.
Therefore the preferable valency (ionic valency) of "Rare earths" = 2.

#		K	L	M	N	5s	5p	5d	5f	6s	6p	6d	6f
	VI	K	L	M	N	5s	5p	5d	5f	6s	6p	6d	6f
76	Os	2	8	18	32	⇅	⇅ ⇅ ⇅	⇅ ↑ ↑ ↑ ↑		⇅			
77	Ir	2	8	18	32	⇅	⇅ ⇅ ⇅	⇅ ⇅ ⇅ ⇅ ↑		⇅			
78	Pt	2	8	18	32	⇅	⇅ ⇅ ⇅	⇅ ⇅ ⇅ ⇅ ↑		↑			
79	Au	2	8	18	32	⇅	⇅ ⇅ ⇅	⇅ ⇅ ⇅ ⇅ ↑		⇅			
80	Hg	2	8	18	32	⇅	⇅ ⇅ ⇅	⇅ ⇅ ⇅ ⇅ ⇅		⇅			
81	Tl	2	8	18	32	⇅	⇅ ⇅ ⇅	⇅ ⇅ ⇅ ⇅ ⇅		⇅	↑		
82	Pb	2	8	18	32	⇅	⇅ ⇅ ⇅	⇅ ⇅ ⇅ ⇅ ⇅		⇅	↑ ↑		
83	Bi	2	8	18	32	⇅	⇅ ⇅ ⇅	⇅ ⇅ ⇅ ⇅ ⇅		⇅	↑ ↑ ↑		
84	Po	2	8	18	32	⇅	⇅ ⇅ ⇅	⇅ ⇅ ⇅ ⇅ ⇅		⇅	⇅ ↑ ↑		
85	At	2	8	18	32	⇅	⇅ ⇅ ⇅	⇅ ⇅ ⇅ ⇅ ⇅		⇅	⇅ ⇅ ↑		
86	Rn	2	8	18	32	⇅	⇅ ⇅ ⇅	⇅ ⇅ ⇅ ⇅ ⇅		⇅	⇅ ⇅ ⇅		
	VII	K	L	M	N	5s	5p	5d	5f	6s	6p	6d	6f
87	Fr	2	8	18	32	⇅	⇅ ⇅ ⇅	⇅ ⇅ ⇅ ⇅ ⇅		⇅	⇅ ⇅ ⇅		
88	Ra	2	8	18	32	⇅	⇅ ⇅ ⇅	⇅ ⇅ ⇅ ⇅ ⇅		⇅	⇅ ⇅ ⇅		
89	Ac	2	8	18	32	⇅	⇅ ⇅ ⇅	⇅ ⇅ ⇅ ⇅ ⇅		⇅	⇅ ⇅ ⇅	↑	
90	Th	2	8	18	32	⇅	⇅ ⇅ ⇅	⇅ ⇅ ⇅ ⇅ ⇅		⇅	⇅ ⇅ ⇅	↑ ↑	
91	Pa	2	8	18	32	⇅	⇅ ⇅ ⇅	⇅ ⇅ ⇅ ⇅ ⇅	↑ ↑	⇅	⇅ ⇅ ⇅		
92	U	2	8	18	32	⇅	⇅ ⇅ ⇅	⇅ ⇅ ⇅ ⇅ ⇅	↑ ↑ ↑	⇅	⇅ ⇅ ⇅		
93	Np	2	8	18	32	⇅	⇅ ⇅ ⇅	⇅ ⇅ ⇅ ⇅ ⇅	↑ ↑ ↑ ↑	⇅	⇅ ⇅ ⇅		
94	Pu	2	8	18	32	⇅	⇅ ⇅ ⇅	⇅ ⇅ ⇅ ⇅ ⇅	↑ ↑ ↑ ↑ ↑ ↑	⇅	⇅ ⇅ ⇅		
95	Am	2	8	18	32	⇅	⇅ ⇅ ⇅	⇅ ⇅ ⇅ ⇅ ⇅	↑ ↑ ↑ ↑ ↑ ↑ ↑	⇅	⇅ ⇅ ⇅		
96	Cm	2	8	18	32	⇅	⇅ ⇅ ⇅	⇅ ⇅ ⇅ ⇅ ⇅	↑ ↑ ↑ ↑ ↑ ↑ ↑	⇅	⇅ ⇅ ⇅	↑	
97	Bk	2	8	18	32	⇅	⇅ ⇅ ⇅	⇅ ⇅ ⇅ ⇅ ⇅	⇅ ↑ ↑ ↑ ↑ ↑ ↑	⇅	⇅ ⇅ ⇅	↑	
98	Cf	2	8	18	32	⇅	⇅ ⇅ ⇅	⇅ ⇅ ⇅ ⇅ ⇅	⇅ ⇅ ↑ ↑ ↑ ↑ ↑	⇅	⇅ ⇅ ⇅	↑	
99	Es	2	8	18	32	⇅	⇅ ⇅ ⇅	⇅ ⇅ ⇅ ⇅ ⇅	⇅ ⇅ ⇅ ↑ ↑ ↑ ↑	⇅	⇅ ⇅ ⇅		
100	Fm	2	8	18	32	⇅	⇅ ⇅ ⇅	⇅ ⇅ ⇅ ⇅ ⇅	⇅ ⇅ ⇅ ⇅ ↑ ↑	⇅	⇅ ⇅ ⇅		
101	Mv	2	8	18	32	⇅	⇅ ⇅ ⇅	⇅ ⇅ ⇅ ⇅ ⇅	⇅ ⇅ ⇅ ⇅ ⇅ ↑	⇅	⇅ ⇅ ⇅		
102	No	2	8	18	32	⇅	⇅ ⇅ ⇅	⇅ ⇅ ⇅ ⇅ ⇅	⇅ ⇅ ⇅ ⇅ ⇅ ⇅ ⇅	⇅	⇅ ⇅ ⇅		
103	Lw	2	8	18	32	⇅	⇅ ⇅ ⇅	⇅ ⇅ ⇅ ⇅ ⇅	⇅ ⇅ ⇅ ⇅ ⇅ ⇅ ⇅	⇅	⇅ ⇅ ⇅	↑	
104	?	2	8	18	32	⇅	⇅ ⇅ ⇅	⇅ ⇅ ⇅ ⇅ ⇅	⇅ ⇅ ⇅ ⇅ ⇅ ⇅ ⇅	⇅	⇅ ⇅ ⇅	↑ ↑	

(right margin, group VI: "remains vacant"; group VII: "The 6 f-Electron orbital remains vacant")

The 5 f-electrons of actinides have about same energy level as the 6 d-electrons and concerned to the bonding. Therefore it is understandable of the different valencies (ionic valency) of these elements.

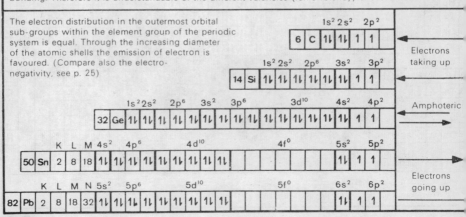

The electron distribution in the outermost orbital sub-groups within the element group of the periodic system is equal. Through the increasing diameter of the atomic shells the emission of electron is favoured. (Compare also the electronegativity, see p. 25)

			$1s^2\ 2s^2$		$2p^2$		
6 C	⇅	⇅	↑	↑			

→ Electrons taking up

$1s^2\ 2s^2 \quad 2p^6 \quad 3s^2 \quad 3p^2$

| 14 Si | ⇅ | ⇅ | ⇅ | ⇅ | ⇅ | ⇅ | ↑ | ↑ |

→

$1s^2 2s^2 \quad 2p^6 \quad 3s^2 \quad 3p^6 \quad 3d^{10} \quad 4s^2 \quad 4p^2$

32 Ge: ⇅ ⇅ ⇅ ⇅ ⇅ ⇅ ⇅ ⇅ ⇅ ⇅ ⇅ ⇅ ⇅ ⇅ ⇅ ↑ ↑

→ Amphoteric

$K\ L\ M \quad 4s^2 \quad 4p^6 \quad 4d^{10} \quad 4f^0 \quad 5s^2 \quad 5p^2$

50 Sn: 2 8 18 ⇅ ⇅ ⇅ ⇅ ⇅ ⇅ ⇅ ⇅ ⇅ ⇅ ⇅ ↑ ↑

→

$K\ L\ M\ N \quad 5s^2 \quad 5p^6 \quad 5d^{10} \quad 5f^0 \quad 6s^2 \quad 6p^2$

82 Pb: 2 8 18 32 ⇅ ⇅ ⇅ ⇅ ⇅ ⇅ ⇅ ⇅ ⇅ ⇅ ↑ ↑

→ Electrons going up

Bonding type, Structure and Properties of Elements
Summary of Correlations

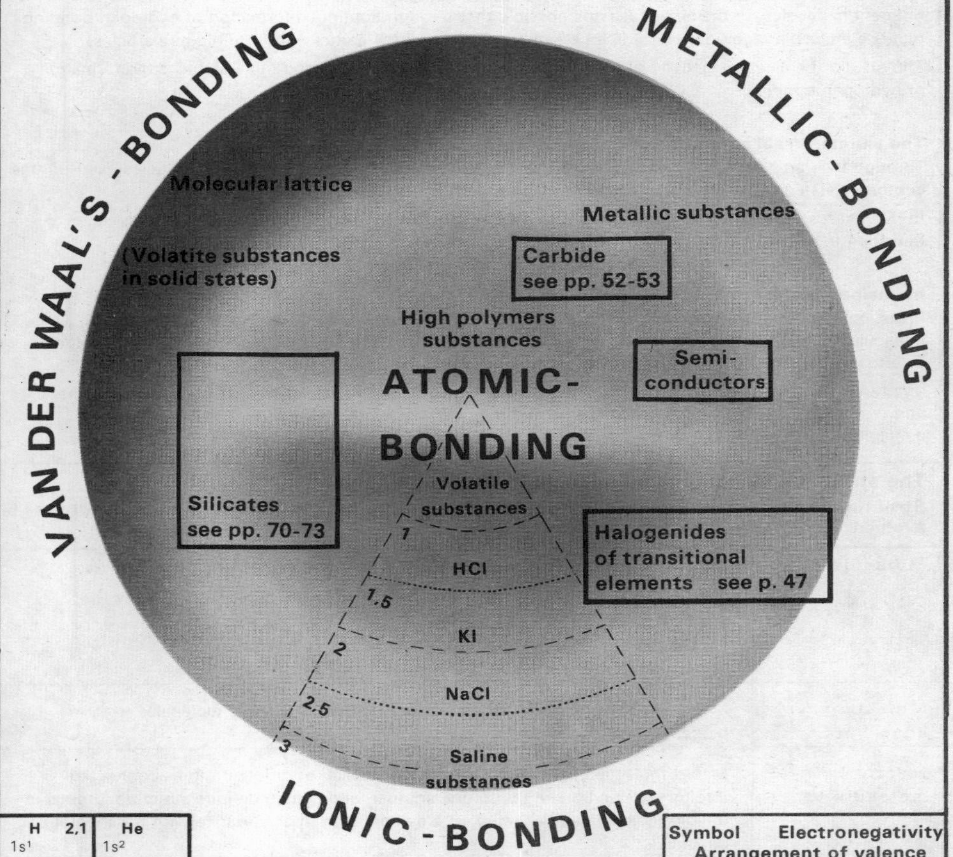

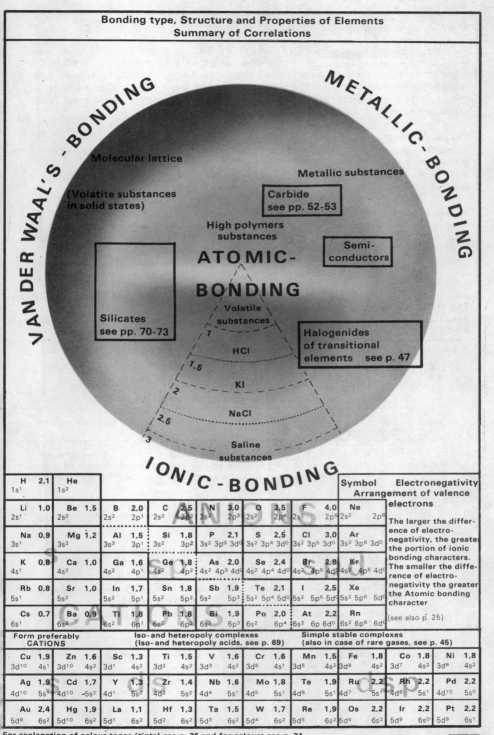

VAN DER WAAL'S - BONDING
METALLIC - BONDING
IONIC - BONDING
ATOMIC-
BONDING
Molecular lattice
(Volatite substances
in solid states)
Metallic substances
Carbide
see pp. 52-53
High polymers
substances
Semi-
conductors
Silicates
see pp. 70-73
Volatile
substances
Halogenides
of transitional
elements see p. 47
1
HCl
1,5
KI
2
NaCl
2,5
3
Saline
substances

H 2,1 $1s^1$	He $1s^2$							Symbol Electronegativity Arrangement of valence electrons
Li 1,0 $2s^1$	Be 1,5 $2s^2$	B 2,0 $2s^2$ $2p^1$	C 2,5 $2s^2$ $2p^2$	N 3,0 $2s^2$ $2p^3$	O 3,5 $2s^2$ $2p^4$	F 4,0 $2s^2$ $2p^5$	Ne $2s^2$ $2p^6$	The larger the difference of electronegativity, the greater the portion of ionic bonding characters.
Na 0,9 $3s^1$	Mg 1,2 $3s^2$	Al 1,5 $3s^3$ $3p^1$	Si 1,8 $3s^2$ $3p^2$	P 2,1 $3s^2$ $3p^3$ $3d^0$	S 2,5 $3s^2$ $3p^4$ $3d^0$	Cl 3,0 $3s^2$ $3p^5$ $3d^0$	Ar $3s^2$ $3p^6$ $3d^0$	The smaller the difference of electronegativity the greater the Atomic bonding character
K 0,8 $4s^1$	Ca 1,0 $4s^2$	Ga 1,6 $4s^2$ $4p^1$	Ge 1,8 $4s^2$ $4p^2$	As 2,0 $4s^2$ $4p^3$ $4d^0$	Se 2,4 $4s^2$ $4p^4$ $4d^0$	Br 2,8 $4s^2$ $4p^5$ $4d^0$	Kr $4s^2$ $4p^6$ $4d^0$	
Rb 0,8 $5s^1$	Sr 1,0 $5s^2$	In 1,7 $5s^2$ $5p^1$	Sn 1,8 $5s^2$ $5p^2$	Sb 1,9 $5s^2$ $5p^3$	Te 2,1 $5s^2$ $5p^4$ $5d^0$	I 2,5 $5s^2$ $5p^5$ $5d^0$	Xe $5s^2$ $5p^6$ $5d^0$	
Cs 0,7 $6s^1$	Ba 0,9 $6s^2$	Tl 1,8 $6s^2$ $6p^1$	Pb 1,8 $6s^2$ $6p^2$	Bi 1,9 $6s^2$ $6p^3$	Po 2,0 $6s^2$ $6p^4$	At 2,2 $6s^2$ 6p $6d^0$	Rn $6s^2$ $6p^6$ $6d^0$	(see also p. 25)

Form preferably CATIONS		Iso- and heteropoly complexes (Iso- and heteropoly acids, see p. 69)				Simple stable complexes (also in case of rare gases, see p. 45)			
Cu 1,9 $3d^{10}$ $4s^1$	Zn 1,6 $3d^{10}$ $4s^2$	Sc 1,3 $3d^1$ $4s^2$	Ti 1,5 $3d^2$ $4s^2$	V 1,6 $3d^3$ $4s^2$	Cr 1,6 $3d^5$ $4s^1$	Mn 1,5 $3d^5$ $4s^2$	Fe 1,8 $3d^6$ $4s^2$	Co 1,8 $3d^7$ $4s^2$	Ni 1,8 $3d^8$ $4s^2$
Ag 1,9 $4d^{10}$ $5s^1$	Cd 1,7 $4d^{10}$ $5s^2$	Y 1,3 $4d^1$ $5s^2$	Zr 1,4 $4d^2$ $5s^2$	Nb 1,6 $4d^4$ $5s^1$	Mo 1,8 $4d^5$ $5s^1$	Te 1,9 $4d^6$ $5s^1$	Ru 2,2 $4d^7$ $5s^1$	Rh 2,2 $4d^6$ $5s^1$	Pd 2,2 $4d^{10}$ $5s^0$
Au 2,4 $5d^9$ $6s^2$	Hg 1,9 $5d^{10}$ $6s^2$	La 1,1 $5d^1$ $6s^2$	Hf 1,3 $5d^2$ $6s^2$	Ta 1,5 $5d^3$ $6s^2$	W 1,7 $5d^4$ $6s^2$	Re 1,9 $5d^5$ $6s^2$	Os 2,2 $5d^6$ $6s^2$	Ir 2,2 $5d^9$ $6s^0$	Pt 2,2 $5d^9$ $6s^1$

For explanation of colour tones (tints) see p. 26 and for colours see p. 34

The hydrogen bonding

The hydrogen bonding results through the condition that a hydrogen nucleus = a proton equally belongs to two immediate neighbouring atoms. The two atoms can belong to the same molecule or also to two different molecules. In the latter case it is possible that through continued formation of hydrogen bonding builds a molecular aggregate (that is an assemblage of structural particles to form bigger particles).

The greater the electronegativity of two neighbouring atoms, the easier and more stable is the formation of hydrogen bond

The van der Waal's bond

Through the van der Waal's bonding, the primary physical attractive forces which the particles exert on one another, a state results where molecules are held together in a lattice.

In most cases these forces are very weak and only at very low temperatures a lattice structure, i.e., a transition to the solid state, results.

Inclusion bond

Inclusion compounds result through that atoms, ionic atoms, molecules or molecular ions or a big molecule, when the mutual proportions permit, can be stored in the hollow space (cavity) of the lattice. Due to the mutual effect, these inclusion compounds are more or less stable.

In a few cases, the so called crystal water is installed in such inclusion compounds in a lattice. All the ring type organic compounds form inclusion compounds (among others, also with rare gases, which directly fit in built rings system).

The structure of liquids and gaseous compounds
Structural properties of chemical compounds which can also in liquid and gaseous conditions be evolved

Molecule	Molecules are formed out of equal or different atoms held together which are bound to one another through atomic bond
	The number of atoms in a molecule amounts to mostly 2
	(In case of inorganic compounds about 10 In case of organic compounds about 100).
Molecular aggregate	Molecular aggregates are assemblages of different or same molecules to form bigger particles.
Macro-molecule	Macro-molecules are composed of either similar or different atoms connected through atomic bonding with one another, and certain definite structural groups in the big molecule are repeated on a definite plan in accordance with the theoretical principles.
	The inter linkage of the structural units can result in one or two or in all the three directions of space. Thus in case of macro-molecules often a lattice network of structure arises.
	The number of atoms amounts to about 10^4 to 10^8,
	the number of molecular structural groups to 10^3 to 10^6.
Ions	Ions are electrically charged particles atomic or molecular in size.
Cations	The electrically positive charged particles which in case of electrolytic processes move towards cathode — the negatively charged pole — are called cations.
Anions	The electrically negative charged particles which in case of electrolytic processes move towards anode — the positively charged pole — are called anion.
Ionic atoms	Ionic atoms result from atoms taking up or giving up of electrons.
Complex ions	Complex ions result through collecting of several differently charged ionic atoms whereby the sum of charges is not equal to 0.
Molecular - ions	Molecular ions result through combining formal charged atoms whereby the cohesiveness are compensated but the sum of the charges is not equal to zero.
Atoms	Rare gases exist as free atoms. Besides, in metal vapours, free atoms occur. Also during chemical reactions of very short duration free atoms result.

The structure of composite substances and their chemical formulae	
With formula:	**is discribed**
Total formula	The percentage composition (stoichiometric composition) of a substance. $\{NaCl\}$ $\{Al_2O_3\}$ $\{H_2O\}$ $\{O\}$ $\{PdH_{0,6}\}$
Molecular formula	The percentage composition and the molecular weight. O_2 HCl $(H_2O)_4$ C_2H_6 C_2H_4 C_6H_6
Structural molecular formula	The percentage composition, the molecular weight and (spatial) structure of the molecules.
Ionic formula	The percentage composition, the ionic weight and the charge. Na^+ Cl^- $[SO_4]''$ $[H_3O]^+$ O_2^{2-}
Structural ionic formula	The percentage composition, the ionic weight and the (spatial) structure of the ions.
Rigid body structural formula	The percentage composition. The unit weight/specific gravity, and the (spatial) structure of the crystals. $\left\{NaCl\frac{6}{6}\right\}_G$ $\left\{AlO\frac{4}{6}\right\}_G$ $\left\{C\frac{4}{4}\right\}_G$ $\left\{C\frac{3}{3}\right\}_N$ Diamond Graphite

Note: In case of total formula, the chain brackets can be omitted but are set to differentiate it from molecular formula in doubtful cases.

Example $[H_2O]$ as substance $-H_2O$ as vapour.

In the structural molecular formula, the hyphen signifies an electron pair in the atomic bonding. **Bond hypens may therefore be employed only in structural molecular formulae**, where actually the atomic bond is present. The use of bond hyphens in case of ionic bonds (example Na–Cl) is **wrong**. Here no electron pair is present. (See also pp. 23/24.)

In case of complex ions, where mostly mixed bond exist the coordination formula is employed in which the legends without hypens at the central atom is set. (See also p. 25.)

In case of rigid body structural formula $AB\frac{m}{n}$ signifies that every A atom of m B–atoms and every B atom of n A–atoms is surrounded in the crystal lattice.

The stoichiometric ratio of $A:B = 1:\frac{m}{n}$

The exact spatial structural nature is only reproduced in a three dimensional model. (See also p. 5.)

Properties of substances

The chemical compounds can be divided into four classes of substances:

I. Volatile II. Metallic III. Saline IV. High polymeric substances

This classification follows according to the common properties within the class of substances.

Class I. Volatile substances

1. Low melting point (MP) or freezing point (FP).
2. Low boiling point (BP) or condensation point (CP).
3. In solid state, weak crystals (hardness 1—3).
4. Also in solid and liquid state, transparent, colourless or coloured without the metallic surface gloss (lustre).
5. In aggregate state no current is conducted.

Class II. Metallic substances

1. Good conductors of electrical current. (In solid and liquid state the current is conducted without the phenomenon of decomposition.)
2. Thin layers opaque, strongly hard, metallic surface lustre. Colour silver grey. (Exceptions Au = yellow, Cu = red.)
3. Very high melting point.
4. All metals are insoluble in water. (Reaction with water is not solubility.)

Class III. Saline substances

1. The solid salts are practically non-conductor of electrical current. In liquid state the current is conducted whereby the phenomenon of decomposition occurs when a direct current is employed. The same holds good for aqueous solution also (electrolysis).
2. Thin layers are transparent, coloured, or colourless.
3. High melting or boiling point.
4. The crystals are harder than the substances of class I (hardness 2—6).
5. All saline substances are water soluble (from sparingly to unlimited solubility).

Class IV. High polymeric substances

a) Diamagnetic type substances

1. Occur only in solid state.
2. Unusually hard (hardness 6—10).
3. Extraordinarily high melting point.
4. Boiling point mostly indeterminable.
5. The electrical current is not conducted (good insulators).
6. Thin layers mostly transparent and glassy or diamond lustre.
7. In water and other solvents insoluble.

b) Plastic substances *)

1. Occur only in solid state.
2. In general only weak or medium hard (hardness 1—6).
3. Many substances are elastic (ELASTOPLAST) at room temperature, some become elastic and (THERMOPLASTIC) deformable on heating and others decompose on heating without becoming elastic and deformable at high temperatures (DUROPLAST).
4. No sharp melting point.
5. No boiling point (decompose before reaching boiling point).
6. In water and other solvents (softness) mostly insoluble or only capable of swelling.

Sometimes also a colloidal solution is formed, seldom a true solution occurs.

Compound partners		Basic structure	Bonding type
A with	B	of compound	
Non metal	Non metal	Molecule	Atomic bonding
Metal	Metal	Metal ions and electrons	Metallic bonding
Metal	Non-metal	Cations and anions	Ionic bonding
Metal	Different non-metals	Ions and complex ions	Ionic bonding
Transition member between metal and and non-metal		Macro-molecules	Atomic bonding
Non-metal	Non-metal	Macro-molecule	Atomic bonding

These tables facilitate a general survey about correlation between structure, bond type, and properties of compounds depending upon the chemical character of the elements that are concerned in compound formation.

1 H Hydrogen	see also pp. 8/9					**1 H** Hydrogen	**2 He** Helium	Rare gases	
3 Li Lithium	**4 Be** Beryllium	**5 B** Boron	**6 C** Carbon	**7 N** Nitrogen	**8 O** Oxygen	**9 F** Fluorine	**10 Ne** Neon	Non-metals	
11 Na Natrium	**12 Mg** Magnesium	**13 Al** Aluminium	**14 Si** Silicon	**15 P** Phosphor	**16 S** Sulphur	**17 Cl** Chlorine	**18 Ar** Argon	Metals	
19 K Potassium	**20 Ca** Calcium	**31 Ga** Gallium	**32 Ge** Germanium	**33 As** Arsenic	**34 Se** Selenium	**35 Br** Bromine	**36 Kr** Crypton	intermediate elements between metals and non-metals	
37 Rb Rubidium	**38 Sr** Strontium	**49 In** Indium	**50 Sn** Tin	**51 Sb** Antimony	**52 Te** Tellurium	**53 I** Iodine	**54 Xe** Xenon	Transitional elements	
55 Cs Caesium	**56 Ba** Barium	**81 Tl** Thallium	**82 Pb** Lead	**83 Bi** Bismuth	**84 Po** Polonium	**85 At** Astatine	**86 Rn** Radon		
1a	2a	3a	4a	5a	6a	7a	0		

main groups

29 Cu Copper	**30 Zn** Zinc	**21 Sc** Scandium	**22 Ti** Titanium	**23 V** Vanadium	**24 Cr** Cromium	**25 Mn** Manganese	**26 Fe** Iron	**27 Co** Cobalt	**28 Ni** Nickel
47 Ag Silver	**48 Cd** Cadmium	**39 Y** Yttrium	**40 Zr** Zirconium	**41 Nb** Niobium	**42 Mo** Molybdenum	**43 Tc** Technetium	**44 Ru** Ruthenium	**45 Rh** Rhodium	**46 Pd** Palladium
79 Au Gold	**80 Hg** Mercury	**57 La** Lanthanum	**72 Hf** Hafnium	**73 Ta** Tantalum	**74 W** Tungsten	**75 Re** Rhenium	**76 Os** Osmium	**77 Ir** Iridium	**78 Pt** Platin
1b	2b	3b	4b	5b	6b	7b		8b	

sub-groups

3 Ce rium	**59 Pr** Praseodym	**60 Nd** Neodym	**61 Pm** Promethium	**62 Sm** Samarium	**63 Eu** Europium	**64 Gd** Gadolinium	**65 Tb** Terbium	**66 Dy** Dysprosium	**67 Ho** Holmium	**68 Er** Erbium	**69 Tu** Thulium	**70 Yb** Ytterbium	**71 Lu** Lutetium

he elements No. 87 to 102 are all radioactive and in properties resemble Cs or Ba as well as the anthanides. Radium (Ra), Thorium (Th) and Uranium (U) are of greater importance.

tructure of solid bodies	Character of compound	Examples
Molecular lattice	Volatile substance	Cl_2; SO_3; H_2SO_4; H_2O; Organic compounds
Metallic lattice	Metallic substance	Zn; Mg_2Pb; Na_5Hg_2; $CuZn$; Cu_5Zn_8; $CuZn_3$;
onic lattice	Saline substance	$NaCl$; Al_2N_3; $NaOH$; Spinel;
onic lattice	Saline substance	Na_2SO_4; $CaSiO_3$; $K_3[Fe(CN)_6]$;
atomic lattice	Adamantine type** substances	C (as diamond); Si; SiO_2; SiC;
rregular or anomalous esembling lattice	Plastic substance	PVC; Teflon; Buna;

Note: * Instead of non-metals here also an intermediate member can be placed. Moreover, the metals of ub-group (transitional elements) as central atoms of complexes can appear.

    ** These substances can also have a particular character as, for example, C as graphite.

Classification of substances

Inorganic substances

The arrangement of inorganic substances in the tables from pages 39–76 and the compilations given below follow the periodic system of elements. An alphabetical order of substances in "Index of substances" at the end of the book also contains the usual trivial names of the compounds.

Compounds between two elements or binary compounds		Pages
Hydrogen compounds	Hydrides	39
Oxygen compounds	Oxides and Peroxides*	40
Nitrogen compounds	Nitrides	43
Fluorine compounds	Fluorides	44
Chlorine compounds	Chlorides	46
Bromine compounds	Bromides	48
Iodine compounds	Iodides	49
Sulphur compounds	Sulphides and Polysulphides	50
Selenium compounds	Selenides	52
Tellurium compounds	Tellurides	52
Boron compounds	Borides	52
Carbon compounds	Carbides	52
Silicon compounds	Silicides	53
Phosphorus compounds	Phosphides	53
Arsenic compounds	Arsenides	53
(Binary intermetallic compounds see in case of alloys p. 136)		

Compounds between 3 elements or ternary compounds**		
Compounds or pseudo binary compounds that dissociates into two kinds of ions as well as a true ternary compound which can decompose into three kinds of constituents		
Hydroxides		54
Oxygen acids	Oxo acids	55
Salts of oxygen acids	Oxo complexes	
Oxo complexes with non-metallic central atom		
Salts of nitric acids	Nitrates and Nitrites	58
Salts of chlorine acids	Chlorates and Perchlorates	56
Salts of bromine acids	Bromates	57
Salts of iodine acids	Iodates, periodates	57
Salts of sulphur acids	Sulphites, Sulphates, Thiosulphates and so on	57
Salts of boron acids	Borates and perborates	59
Salts of carbon acids	Carbonates	58
Salts of phosphorus acids	Hypophosphite, Phosphates, Phosphites	59
Salts of arsenic acids	Arsenites and Arsenate	59
Oxo complexes with metal central atom		60
Oxo halogenides		61
Oxo sulphides and thiohalogenides		61
Salts of N–H acids, HCN, Cynates, Cynides, thiocynates, explosives		62
Silanes, and Germane halogenides, phosphonium, halogenides, and mixed halogenides		63
Inorganic plastics		
Inorganic derivatives of NH_3		64
Oxohalogenoacids and their salts, oxonitrosyl and oxoamidoacids, thioamido complexes		64
Complexes of elements of main groups		65
Complexes of elements of sub-groups		66
Complexes with carbon monoxide	Carbonyls	68
Isopolyacids and heteropolyacids as well as thin salts		69
Salts of silicic acid	Silicates	70

* The usual notation 'oxide' is a derivative of 'oxy' and 'ide' where the letter 'y' is dropped. Instead of 'y' if 'i' is dropped, it is recommended that the consequent nomenclature is not broken after which in all binary compounds, the electronegative constituent has the ending 'ide'. [See also correct rules for the nomenclature of inorganic chemistry *Chemische Berichte* No. 7, 1959, pp. XLVII–LXXXV].

** This classification is not rigorously kept and here also compounds containing more than three elements are brought in. Of all the hydrogen salts (acid salts) those pertaining to neutral salts are registered.

Double compounds

Double compounds with same anions	Page
Double hydrides	74
Double oxides	74
Double nitrides	74
Double halides	74
Double sulphides	75
Double nitrates	75
Double sulphates	75
Double borates	76
Double carbonates	76
Double phosphates	76
Double Arsenates	76
Double Chromates	76
Double Complexes	76
Double compounds with equal cations	
Oxide salts (basic salts)	76
Hydroxide salts (basic salts)	76
Double compounds with different cations	76
Triple salts	

Explanations and abbreviations to tables on pp. 39–76 and 80–128*

1.	The number serves to facilitate the search of a substance**
2.	The straightforward and unambiguous way to designate a chemical compound is by its formula. Therefore this stands in the first place.
3.	In general, the support of the "guidelines for the nomenclature of inorganic chemistry" for rational names is employed. } Trivial names ✖ signifies a mineralogical or geological common name. } see p. 259 (Subject Index)
4.	Mol. = Molecular weight (molecular mass) corresponding to the given formula
5.	F.P. or M.P. = Freezing or melting point Press = pressure, vac = vacuum, ww = without water
6.	B.P. or C.P. = Boiling or condensation point subl = sublimes at, d or dec = decomposes, diss = dissociates
7.	D = Density V.D. = Vapour density in g/l
8.	The colour is abbreviated generally as follows: colourless = col, bl = blue, blk = black, br = brown (Sh is added with these abbreviated forms to denote bluish, blackish, brownish, etc.) Ind = Indigo. Or = Orange, grn. = green, so on.

cub	cubical crystal form	cr	crystalline
tet	tetragonal	pd	powder
rhomb	ortho rhombic	amor	amorphous
hex	hexagonal	× H_2O	water of hydration
trig	triogonal	hyg	hygroscopic
monocl	monoclinic	deliq	deliquescent
triel	triclinic	met lust	metallic lustre
oct	octahedral	✖ H	hardness (crystal)
tric	triclinic crystal form	n	refractive index

i.w.	insoluble in water		
v. sl. sw	very slightly soluble in water		1 g in 100 g water
sl. sw	slightly soluble in water	1 –	5 g in 100 g water
s.w	soluble in water	5 –	25 g in 100 g water
e.s.w.	easily soluble in water	25 –	100 g in 100 g water
v.e.s.w.	very easily soluble in water		100 g in 100 g water

Soln	solution	**exp** : Explosive	
s	soluble in water (In case of gases; sp. gr. in liq. state and solubility is . . . volume in 1 volume water)	☠ : Poisonous	
misc	miscible with	org	organic
dw	decomposes in water	solv	solvents
dil	dilute	acet	acetone
conc	concentrated	eth	ether
R	reaction	al	alcohol
n R	no reaction	Bz	benzene
alk	alkaline	ac.a	glacial acitic acid
aq. reg.	aqua regia	m al	methanol
ε	dielectric constant	pet.	petroleum ether
		CCl₄	carbon tetrachloride
		tarta	tartaric acid

** see subject index   * see p. 80
Abbreviations used follow "Handbook of Chemistry and Physics." Hogmant 'CRC' publication.

37

Nomenclature of inorganic substances			
Formula or formula part	**Name**		
	as molecule or free radical	as cation or cationic radical	as anion	as ligand in complex ions
H	H atom	Hydrogen	Hydride	Hydrido
F	F atom	Fluorine	Fluoride	Fluoro
Cl	Cl atom	Chlorine	Chloride	Chloro
Br	Br atom	Bromine	Bromide	Bromo
I	I atom	Iodine	Iodide	Iodo
ClO		Chlorosyl	Hypochlorite	Hypochlorito
ClO_2	Chlorine dioxide	Chloryl	Chlorite	Chlorito
ClO_3		Perchloryl	Chlorate	Chlorato
ClO_4			Perchlorate	
IO		Iodosyl	Hypoiodide	
O	Oxygen atom		Oxide	Oxo
O_2	Oxygen		Peroxide $[O_2]^{2-}$	Peroxo
			Hyperoxide $[O_2]^-$	Hyperoxo
OH	Hydroxyl		Hydroxide	Hydroxo
O_2H	Perhydroxyl		Hydrogenperoxide	Hydrogenperoxo
S	S atom		Sulphide	Thio
S_2	Disulphur		Disulphide	Dithio
SH	Sulphydryl		Hydrogen sulphide	Thiolo
SO	Sulphur monoxide	Thionyl		
SO_2	Sulphur dioxide	Sulphuryl	Sulphoxylate	
SO_3	Sulphur trioxide		Sulphite	Sulphito
HSO_3			Hydrogen sulphite	Hydrogen sulphite
S_2O_3			Thiosulphate	Thiosulphato
SO_4			Sulphate	Sulphato
Se	Selenium		Selenide	Seleno
Te	Tellurium		Telluride	Telluro
CrO_2		Chromyl		
UO_2		Uranyl		
N	Nitrogen atom		Nitride	Nitrido
N_3			Azide	Azido
NH			Imide	Imido
NH_2			Amide	Amido
NHOH			Hydroxylamide	Hydroxylamido
$-NH-NH_2$			Hydrazide	Hydrazido
NO	Nitric oxide	Nitrosyl		Nitrosyl
NO_2	Nitrogendioxide	Nitryl		Nitro
ONO			Nitrite	Nitrito
NO_3			Nitrate	Nitrato
N_2O_2			Hyponitrite	Hyponitrito
P	Phosphorus		Phosphide	Phosphido
PO		Phosphoryl		
PS		Phiophosphoryl		
H_2PO_2			Hypophosphite	Hypophosphito
HPO_3			Phosphite	Phosphito
PO_4			Phosphate	Phosphito
ASO_4			Arsenate	Arsenato
VO		Vanadyl		
CO	Carbon monoxide	Carbonyl		Carbonyl
CS		Thiocarbonyl		
CN		Cyan	Cyanide	Cyanido
OCN			Cyanate	Cyanato
SCN			Thiocyanate	Thiocyanato
CO_3			Carbonate (bicarbonate)	Carbonato
HCO_3			Hydrogen carbonate	Hydrogen carbonato

Compounds between two elements

Hydrogen compounds, hydrides

	1	2	3	4	5	6	7		8
II	LiH	BeH_2	Borane	CH^*	NH_3 N_2H_4	$(H_2O)n$ D_2O	$(HF)n$	a main groups	
III	NaH	MgH_2	$(AlH_3)x$	Silane	PH_3	H_2S	HCl		
IV	KH	CaH_2	Ga_2H_6	GeH_4	AsH_3	H_2Se	HBr		
V	RbH	–	–	SnH_4	SbH_3	H_2Te	HI		
VI	CsH	–	-	PbH_4	BiH_3	–	–		
IV	CuH	ZnH_2	–	$TiH_{1.7}$	–	CrH_3	–	FeH_2	– NiH_2
V	–	–	–	$ZrH_{1.9}$	NbH	–	–	–	$PdH_{0.6}$
VI	–	–	$LaH_{2.8}$	–	$TaH_{0.8}$	–	–		b sub groups

* CH = Carbon hydrogen compounds, see organic substances
HN_3 = Hydrazoic acid see p. 62/1
H_2O_2 = Hydrogen peroxide see p. 43/2

No.	Formula	Name	Mo. wt.	MP	D	Colour	Other properties
1	LiH	Lithium hydride	7.95	680	0.82	white	cubic crystals
2	NaH	Natrium hydride	24.00	d 800	1.38	white	with water decomposes → H_2
3	KH	Potassium hydride	40.11	d ~600	~1.45	white	strong reducing agents
4	CaH_2	Calcium hydride	42.10	d 600	1.7	colourless rhombic crystal

No.	B_nH_m	Name	MP	BP		SinHm	Silane	MP	BP
5	B_2H_6	Diborane	−164.5	− 92.5	12	SiH_4	Monosilane	−184.7	−112.1
6	B_4H_{10}	Tetraborane	−120.8	+ 16	13	Si_2H_6	Disilane	−129.4	− 14.8
7	B_5H_9	Pentaborane	− 46.8	+ 60	14	Si_3H_8	Trisilane	−116.9	+ 52.9
8	B_6H_{10}	Hexaborane	− 65.1	+ 94	15	Si_4H_{10}	Tetrasilane	− 91.6	+108.4
9	$B_{10}H_{14}$	Decaborane	99.3	213	16	Si_5H_{12}	Pentasilane	—	>100
10	$(AlH_3)×$	Aluminium hydride	solid	dec.	17	GeH_4	Monogermane	−165	− 90
11	Ga_2H_6	Galium hydride	solid	dec.	18	Ge_2H_6	Digermane	−109	29
					19	Ge_3H_8	Trigermane	−105.6	110.5
					20	SnH_4	Stannic hydride	—	− 52
					21	PbH_4	Lead hydride	s.l.	dec.

Boranes and silanes are very sensitive against O_2 (air) and H_2O (moisture) and react explosively.
For similarity with C–H substances see also silicones.

S.No	Formula	Name	Mol. wt.	MP	BP	D	E	Other properties
22	NH_3	Ammonia	17.03	− 77.8	− 33.5	0.7714	16.9	tₖ 132.4 Pₖ 111.5 Dₖ 0.235 ☠
23	ND_3	Heavy ammonia	20.05	− 74.0	− 30.9			
24	N_2H_4	Hydrazine	32.05	1.8	113.5	1.011	—	n = 1.470
25	PH_3	Phosphine	34.00	−133.8	− 87.4	1.530	—	s.w. 0.26 smell of carbide ☠
26	AsH_3	Arsine	77.93	−113.5	− 62.5	2.695	2.5	s.w. 0.2 ⟩ decomp. on heating ☠
27	SbH_3	Stibine	124.78	− 91	− 18	5.685	—	s.w. 0.2 ⟩ see Marsh's Test p. 197 ☠
28	BiH_3	Bismuthine	300	—	gasf		—	e.s. dec.
29	$(H_2O)n$	Water	18.016	0	100	0.9982	82	n=1.33335 =₁·1057 D max at 4.08°
30	$(D_2O)n$	Heavy water	20.028	3.82	101.42	1.105	80.5	n=1.32844 D max at 11.22°
31	H_2S	Hydrogen sulphide	34.08	− 85.5	− 60.4	1.539	—	s.w. 4.4: smell of rotten eggs ☠
32	H_2Se	Hydrogen selenide	80.98	− 64	− 42	3.67	—	s.w. 3.8: smell of rotn. blk. radish ☠
33	H_2Te	Hydrogen telluride	129.63	− 49	− 2.3		—	e.s.w: unpleasant ☠
34	$(HF)×$	Hydrogen fluoride	20.01	− 83.1	19.5	0.987	~70	miscible with water ☠
35	HCl	Hydrogen chloride	36.47	−114	− 85	1.6391	6.4	s.w. 507, pungent ☠
36	HBr	Hydrogen bromide	80.92	− 87	− 66.9	3.644	7.0	s.w. 580, dec. → Br_2 ☠
37	HI	Hydrogen iodide	127.92	− 50.9	− 35.4	5.789	—	s.w. 425, dec. → I_2

Comparison of BP of volatile hydrides of II - V periods with the rare gases (which possess equal electron number in shells). Variations in case of H_2O. HF and NH_3 stipulate the Hydrogen bonding (molecular aggregates) see p. 32

The transitional elements at moderate temperatures adsorb often large volumes of hydrogen, forming occluded hydrides (see p. 130). The hydrogen is easily given up. Pd adsorbs 900 vol. parts by H_2 whose formula conforms to $PdHo_{.6}$. For nickel hydride the above formula corresponds to the "ideal composition". The catalytic action of these metals as hydrogen collector can also be explained. Ni and Pd hydrides are especially strong reducing agents.

For the densities of aqueous solutions of NH_3 HCl and so on.
see pp. 134/135

39

Oxygen compounds—oxides and peroxides

	1	2	3	4	5	6	7
II	Li_2O	BeO	B_2O_3	CO CO_2 C_3O_2	N_2O NO N_2O_3 $N_2O_4 \rightleftharpoons 2NO_2$ N_2O_5	–	F_2O
III	Na_2O	MgO	Al_2O_3	SiO_2	P_4O_6 P_2O_5	SO_2 SO_3	Cl_2O ClO_2 Cl_2O_7
IV	K_2O	CaO	Ga_2O_3	GeO GeO_2	As_2O_3 As_2O_5	SeO_2	Br_2O BrO_2
V	Rb_2O	SrO	In_2O In_2O_3	SnO SnO_2	Sb_2O_3 Sb_2O_4 Sb_2O_5	TeO_2 TeO_3	I_2O_5
VI	Cs_2O	BaO	Tl_2O Tl_2O_3	PbO PbO_2	Bi_2O_3 Bi_2O_5	–	–

a main groups

C_3O_2 = "anhydride of malonic acid" better Dioeopropadiene M.P. 111; B.P. 7 colourless see p. 111 ☠

	1	2	3	4	5	6	7	8		
IV	Cu_2O CuO	ZnO	Sc_2O_3	Ti_2O_3 TiO_2	V_2O_2 V_2O_3 V_2O_4 V_2O_5	CrO Cr_2O_3 CrO_3	MnO Mn_2O_3 MnO_2 Mn_2O_7	FeO Fe_2O_3	CoO Co_2O_3	NiO Ni_2O_3
V	Ag_2O	CdO	Y_2O_3	ZrO_2	Nb_2O_5	MoO_2 MoO_3		RuO_4	RhO_2 Rh_2O_3	PdO
VI	Au_2O Au_2O_3	Hg_2O HgO	La_2O_3	HfO_2	Ta_2O_5	WO_2 WO_3	ReO_2 ReO_3 Re_2O_7	OsO_2 OsO_4	IrO_2	PtO PtO_2

b sub groups

No.	Formula	Name	Mol. wt.	MP	BP	D	Colour and other properties	
1	CO	Carbon monoxide	28.01	−205.1	−191.5	1.2500	cols.	t× 140° p× 35; s.w 0.033 ☠
2	CO_2	Carbon dioxide	44.01	− 57.6	s −78.5	1.9768	cols.	t× 32° p× 73, s.w 0.878*
3	N_2O	Dinitrogen monoxide	44.016	− 90.7	− 89.5	1.978	cols.	s.w. 1.048; "narcosis"
4	NO	Nitric oxide	30.008	−163.7	−151.8	1.3402	cols.	s.w. 0.0738
5	N_2O_3	Dinitrogen trioxide	76.016	−111	d ab 0°	1.447	bl	$+ H_2O \rightarrow 2HNO_2$ ☠
6	N_2O_4	Dinitrogen tetroxide	92.016	− 10.3	d 21°	1.491	cols.	$+ H_2O \rightarrow HNO_2 + HNO_3$ ☠
7	NO_2	Nitrogen dioxide	46.008	—		2.05	br.	$2NO_2 \rightarrow N_2O_4$ s. No. 6 ☠
8	N_2O_5	Dinitrogen pentoxide	108.016	30	d 45°	1.642	cols.	$+H_2O \rightarrow 2HNO_3$ **ex** ☠
9	P_4O_6	Phosphorus trioxide	219.9	22.7	173	2.135	wh.	$+H_2O \rightarrow 2H_3PO_3$ 3s. Eth; Bz ☠
10	P_2O_5	Pentoxide	141.95	p 566	s 358	2.114	wh.	$+H_2O \rightarrow 2HPO_3; + 3H_2O \rightarrow 2H_3PO_4$
11	SO_2	Sulphur dioxide	64.07	− 75.5	− 10	2.9263	cols.	s.w 80; $+ H_2O \rightarrow H_2SO_3$
12	SO_3	Sulphur trioxide	80.07	∼40	44.8	∼2	cols.	$+ H_2O \rightarrow H_2SO_4$
13	SeO_2	Selenium dioxide	110.96	p 340	s 315	3.95	wh.	$+ H_2O \rightarrow H_2SeO_3$
14	TeO_2	Tellurium dioxide	159.61	733	—	∼6	wh.	v. sl. s.w; amphoteric
15	TeO_3	Tellurium trioxide	175.61	d		∼5	yel.	i.s.w; saline type subs
16	F_2O	Difluorine oxide	54.00	−223.8	−145	1.53	cols.	v. sl. s.w; corrosive ☠
17	Cl_2O	Dichlorine monoxide	86.91	−116	3.8	3.887	brsh.	yel; s.w; 200; unpleasant **ex** ☠
18	ClO_2	Chlorine dioxide	67.46	− 79	11.8	3.013	yel.	s.w 20; pungent very **ex** ☠
19	Cl_2O_7	Dichlorine heptoxide	182.91	− 91.5	82	—	cols.	oily liq. **ex** ☠
20	I_2O_5	Iodine pentoxide	333.82	d300	—	4.799	wh.	$+ H_2O \rightarrow 2HIO_3$
21	Mn_2O_7	Manganese heptoxide	221.88	—	—		br.	oily liq.
22	Re_2O_7	Rhenium heptoxide	484.44	296	363	6.10	yel.	s.w; Al
23	RuO_4	Ruthenium tetroxide	165.10	27	d100	3.29	yel.	sl. s.w; s. Al. **ex**
24	OsO_4	Osmium tetroxide	254.2	41.8	130	4.906	pa. yel.	sl. s.w; s. Al. CCl_4

* see also pp. 55 and 130.

S. No.	Formula	Name	Mol. wt.	MP	BP	D/V.D sp. gr.	Colour and other properties
1	Li_2O	Lithium oxide	29.88	>1700	—	2.10	wh. cub; + H_2O → LiOH
2	Na_2O	Natrium oxide	61.98	920	d	2.17	wh. cub; + H_2O → NaOH
3	K_2O	Potassium oxide	94.20	d ~375	—	2.32	pa yel; cub. + H_2O → KOH
4	Rb_2O	Rabidium oxide	186.96	d 400	—	3.72	pa. gr; cub. + H_2O → RbOH
5	Cs_2O	Caesium oxide	281.82	d 380	—	4.36	or; rhomb; + H_2O → CsOH
6	MgO	Magnesium oxide	40.32	2800	—	3.65	wh; cub, ✲ i.s.w. H5.5-6 n1.736
7	CaO	Calcium oxide	56.08	~2570	2850	3.352	wh. cub. v. sl. s.w. → $Ca(OH)_2$
8	SrO	Strontiumoxide	103.63	2430	—	4.7	wh. cub. v. sl. s.w. → $Sr(OH)_2$
9	BaO	Barium oxide	153.36	1923	~2000	5.72	wh. cub. sl. s.w. → $Ba(OH)_2$
10	B_2O_3	Borontrioxide	69.64	450	~2300	1.844	wh. cub. sl. s.w. → H_3BO_3
11	In_2O_3	Indium-III oxide	277.64	d 1000	—	7.18	pagr; hex; R: acidic
12	Tl_2O	Thallium-I oxide	424.78	~300	—	9.52	blk; cub; v. sl. s.w; ~ Ag_2O. R: acidic
13	Tl_2O_3	Thallium-III oxide	456.78	717	—	10.19	br; cub; R: acidic
14	GeO	Germanium-II oxide	88.60	s 710	—		grsh. blk; R: alk
15	SnO	Stannous oxide	143.70	d 700	950	6.446	blk. tet. R: HCl, H_2SO_4
16	PbO	Lead-II oxide	223.21	890	1470	yellow 9.66 red 9.36	yel; v. sl. s.w; R: HNO_3. Alk ✲rhomb H_2
17	PbO_2	Lead-IV oxide	239.21	d 290	—	9.375	br; tet. i.s.w; R: HNO_3 + H_2O_2
18	As_2O_3	Arsenic-III oxide	197.82	s 321	—	3.86	wh; ✲ cub. H 1.5-2.5, sl. s.w ☠
19	As_2O_5	Arsenic-V oxide	229.85	d 315	—	4.09 k 5.25	wh. + $3H_2O$ → $2H_3AsO_4$. dilig. ☠
20	Sb_2O_3	Antimony-III oxide	291.52	654.8	1456	rhomb 5.7	wh; ✲ rhomb. cub; H2-2.5, i.s.w, R:HCl
21	Sb_2O_4	Antimony-III, V oxide	307.52	—	—	7.52	wh. in the Flame yel. R: alk
22	Sb_2O_5	Antimony-II oxide	323.52	d 300	—	~ 4	yelsh. wh; v. sl. s.w; R: HCl, alk
23	Bi_2O_3	Bismuth-III oxide	466.00	817	1890	9.04	yel; moncl; R: acidic
24	Bi_2O_5	Bismuth-V oxide	489.00	d 150	—	5.10	br; R: acid, KoH
25	Cu_2O	Copper-I oxide	143.08	1232	—	6.0	red ✲ cub. H 3.5-4; R: HCl, amm
26	CuO	Copper-II oxide	79.54	d 1000	—	6.45	blk; ✲ monocl. H3-4 R: acid, amm
27	Ag_2O	Silver oxide	231.76	d 300	—	7.22	br; cub; v. sl. s.w; R: HNO_3
28	Au_2O	Gold-I oxide	410.00	d 200	—	3.6	gr. viol., R: HCl
29	Au_2O_3	Gold-III oxide	442.00	d 150	—		brsh. blk; R: HCl
30	ZnO	Zinc oxide	81.38	>1800	—	5.606	wh; ✲ hex. H4-4.5 R: a, alk
31	CdO	Cadmium oxide	128.41	s 1500	—	8.15	br; cub; H3; R: a
32	Hg_2O	Mercury-I oxide	417.22	d 100	—	9.8	blksh. br.; R: ac. a
33	HgO	Mercury-II oxide	216.61	d 400	—	11.14	Yelsh. red; v.sl. s.w; ✲ monocl; H2. R: HNO_3 HCl
34	Sc_2O_3	Scandium oxide	137.92			3.86	wh; cub; R: v. sl. s.w; a-300 DM/g
35	Y_2O_3	Yttrium oxide	225.84	2410	—	4.84	wh; cub; R: a-8.50 DM/g
36	La_2O_3	Lanthanum oxide	325.84	2315	4200	6.51	wh; rhomb; v. sl. s.w; R: a 2.50 DM/g
37	V_2O_2	Vanadium-II oxide	133.90	—	—	5.76	pa. gr; cub; R: a
38	V_2O_3	Vanadium-III oxide	149.90	1970	—	4.82	blk; rhomb; R: HNO_3 HF
39	V_2O_4	Vanadium-IV oxide	165.90	1967	—	4.65	bl; tet; R: a
40	V_2O_5	Vanadium-V oxide	181.90	658	d 1750	3.36	red; rhomb; v. sl. s.w; R: alk
41	Nb_2O_5	Niobium-V oxide	265.82	~1460	d	4.47	wh; rhomb, i.w. R: H_2SO_4, HF
42	Ta_2O_5	Tantalum-V oxide	441.90	d 1470	—	8.02	wh; rhomb; R: HF
43	CrO	Chromium-II oxide	68.01	1550	—		blk. hex. R: a
44	Cr_2O_3	Chromium-III oxide	152.02	1990	d	5.21	gr; rhomb; R: a alk
45	CrO_3	Chromium-VI oxide	100.01	d 198	—	2.70	dk. red; rh; e.s.w; al, eth; R; H_2SO_4 ☠
46	MoO_3	Molybdenum-VI oxide	143.95	795	1280	4.50	wh. rhomb; v. sl. s.w; R: a
47	WO_3	Tungsten-VI oxide	231.86	1473	—	7.16	yel; monocl; R: HF, alk
48	MnO	Manganese-II oxide	70.94	1785	—	5.18	gr; ✲ cub; H5-6 n = 2.16 R: a
49	Mn_2O_3	Manganese-III oxide	157.88	d	—	4.50	blk; cub; R: a
50	MnO_2	Manganese-IV oxide	86.94	d	—	5.026	grsh. blk; ✲ tet. H1-6 R: HCl → Cl_2
51	ReO_2	Rhenium-IV oxide	218.22			11.4	blksh. br; R: HCl
52	ReO_3	Rhenium-VI oxide	234.22	d 400	—	7.43	blksh. red cub; R: conc. HNO_3
53	FeO	Iron-IV oxide	71.85	1420	—	6.04	blsh. blk. cub; R: a
54	Fe_2O_3	Iron-III oxide	159.70	~1570	—	5.25	red br; ✲ rhomb & cub; H ~6; R: a*
55	CoO	Cobalt-II oxide	74.94	1810	d.	6.43	grsh. gr; cub; R: a
56	Co_2O_3	Cobalt-III oxide	165.88	d	—	5.18	br; rhomb; R: a → Co(II) salts
57	NiO	Nickel-II oxide	74.71	1960	d	7.45	gr ✲ cub; H5.5 R: h, a, amm
58	Ni_2O_3	Nickel-III oxide	165.42			4.83	blk; R: a, amm
59	Rh_2O_3	Rhodium-III oxide	253.82	d 1100	—		gr. rhomb; R: conc. a; alk
60	RhO_2	Rhodium-IV oxide	134.91				br. R: conc. a; alk
61	PdO	Palladium-II oxide	122.40	d ~800	—	8.31	gr; tet; R: ha
62	OsO_2	Osmium-II oxide	222.20	d ~650	—	7.91	blk; tet. R: conc. a
63	IrO_2	Iridium-IV oxide	224.20	d	—	11.72	blk; tet R: conc. a
64	PtO	Platinum-II oxide	211.09	d 550	—	14.9	grsh. viol; tet R: H_2SO_3
65	PtO_2	Platinum-IV oxide	227.09	d 400	—	10.2	blsh blk; R: conc. a

*) Very highly ignited Fe_2O_3 does not react with acids   Amm = NH_4OH

S. No.	Formula	Name	Mol. wt.	MP/FP	BP/CP	D/V.D sp. gr	Colour and other properties
1	CeO_2	Cerium-IV oxide	172.13	2600	—	7.20	wh; cub; R: conc. H_2SO_4 – 40 DM/g
2	Pr_2O_3	Prasaeodymium-III oxide	329.84	d	—	6.88	grsh. yel; rhomb & cub; R: a 5 DM/g
3	PrO_2	Prasaeodymium-IV oxide	172.92			6.82	blk; cub. R: a
4	Nd_2O_3	Neodymium-III oxide	336.54			7.24	pa. bl; rhomb; & cub; R: a 4.50 DM/g
5	Sm_2O_3	Samarium-III oxide	348.70			7.43	lt. ýel; cub; R: a; 5,– DM/g
6	Eu_2O_3	Europium-III oxide	352.0			7.30	lt. rose; cub; R: a; 135,– DM/g
7	Gd_2O_3	Gadolinium-III oxide	362.52			7.407	gr. cub; R: a; 10,– DM/g
8	Tb_2O_3	Terbium-III oxide	356.86			7.90	wh. cub; R: a; 80,– DM/g
9	Dy_2O_3	Dysprosium-III oxide	373.02			7.81	wh. cub; R: a 15,– DM/g
10	Er_2O_3	Erbium-III oxide	382.54			8.64	Rose, cub; 15,– DM/g
11	Tm_2O_3	Thulium-III oxide	385.88			8.77	lt. yel; cub; 150,– DM/g
12	Yb_2O_3	Ytterbium-III oxide	394.08			9.17	wh. cub; R: a 20,– DM/g
13	ThO_2	Thorium-IV oxide	264.05	3050	4400	9.69	wh. cub; R: conc. H_2SO_4
14	UO_2	Uranium-IV oxide	270.07	2176	—	10.75	brsh. blk; ⚒ cub; H 4–6; R: HNO_3
15	U_3O_8	Uranium-IV-VI oxide	842.21	d	—	8.22	grsh. blk; rhomb; R: a
16	UO_3	Uranium-VI oxide	286.07	d	—	6.039	yelsh; red; rhomb; R: a, alk; carbonate lik
17	BeO	Beryllium oxide	25.01	2530	4120	3.025	wh. hex; R: a, alk
18	Al_2O_3	α-Aluminium oxide	101.96	2046	2700	3.99	wh. rhomb, ⚒ H9 R: conc. alk
19	Al_2O_3	β-Aluminium oxide	101.96	—	—	3.30	wh. hex
20	Al_2O_3	γ-Aluminium oxide	101.96	—	—	3.40	wh. cub;
21	Ga_2O_3	Gallium-III oxide	187.44	1741	—	5.88	wh. rhomb. R: a. alk
22	SiO_2	α-Quartz silicon dioxide	60.09	~1477	—	2.651	col. hex, ⚒ H 6–7 R: alk, HF
23	SiO_2	β-Quartz	60.09	—	—	—	col. trig
24	SiO_2	α-Tridymite	60.09	—	—	—	col. hex ⚒
25	SiO_2	β-Tridymite	60.09	1670	—	2.26	col. tet
26	SiO_2	α-Christobalite	60.09	—	—	—	col. cub ⚒
27	SiO_2	β-Christobalite	60.09	1705	—	2.32	col. tet
28	GeO_2	Germanium-IV oxide	104.60	1115	—	4.28	wh. rhomb; v. sl. s.w, R: alk
29	GeO_2	Germanium-IV oxide	104.60	—	—	6.24	tet
30	SnO_2	Tin-IV oxide	150.70	p ~1900	s 1800	6.95	wh. tet, R: H_2SO_4, ⚒ H 6–7
31	Ti_2O_3	Titanium-III oxide	143.80	~1900	—	4.49	viol; rhomb; R: H_2SO_4
32	TiO_2	Ritile, Titanium-IV oxide	79.90	~1850	<3000	4.26	wh. tet, ⚒ 5.5–6 R: H_2SO_4
33	TiO_2	Brookite	79.90	—	—	4.15	wh. rhomb ⚒
34	TiO_2	Analtase	79.90	—	—	3.94	wh. tet ⚒
35	ZrO_2	Zirconium-IV oxide	123.22	>2700	4300	5.73	wh. cub ⚒ H 6.5, R: conc. H_2SO_4, HF
36	ZrO_2	Zirconium-IV oxide	123.22	—	—	5.49	monocl
37	HfO_2	Hafnium-IV oxide	210.50	2800	—	9.68	wh. tet; & monocl
38	HfO_2	Hafnium-IV oxide	210.50	—	—	10.43	cub
39	WO_2	Tungsten-IV oxide	215.86	~1270	—	12.11	br tet, R: a, alk

18a The corundums (⚒ A_2O_3) are related:
$Al_2O_3 \cdot H_3O$ Aluminium oxide mono hydrate ⚒ Diaspore, rhomb, H 6.5–7.0, d 360 D = 3.41 and $Al_2O_3 \cdot 3H_2O$ Aluminium oxide trihydrate ⚒ Hydrargillit or gibbsite microcrystalline, H2 5–3.5, d 200, D 2.53
[$Al_2O_3 \cdot H_2O$ can also be written $AlO(OH)$ and instead of $Al_2O_3 \cdot 3H_2O$ ~ $Al(OH)_3$]

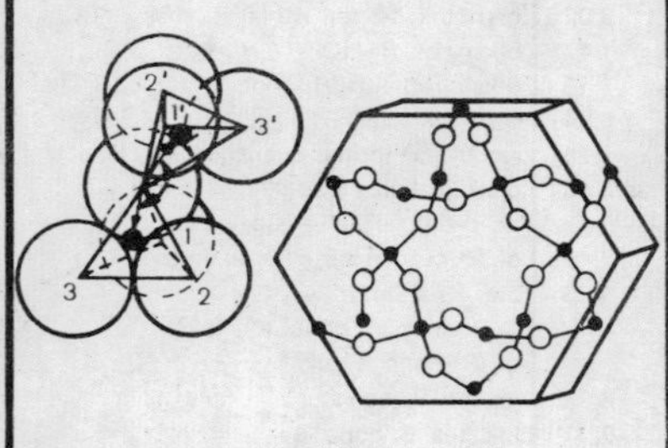

The structure of Quartz
Angle between the tetrahedron 150°
(see pp. 70–73)

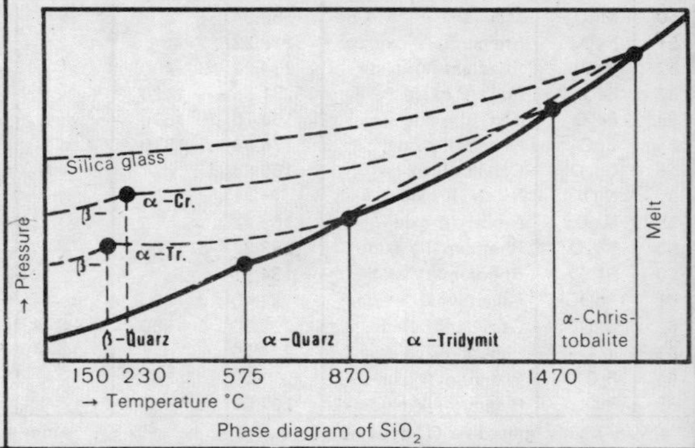

Phase diagram of SiO_2

Peroxides

No.	Formula	Name	Mol. wt.	MP	BP	D	Colour and other properties
1	O_3	Ozone	48.00	−251	−112.5	2.144	col. s w 0.49, s. CCl_4, $CHCl_3$
2	H_2O_2	Hydrogen peroxide	33.016	−0.89	151	1.4631	col. $\propto$ misc. w, n=1.406
3	Na_2O_2	Natrium peroxide	77.98	d 460	—	2.805	col. tet. R: $H_2O \rightarrow 0$
4	KO_2	Potassium peroxide	71.10	380	d	—	yel. R: $H_2O \rightarrow 0$
5	Rb_2O_4	Rubidium peroxide	234.96	412	—	3.05	dk. or; tet. R: $H_2O \rightarrow H_2O_2$
6	Cs_2O_4	Caesium peroxide	329.82	432	—	3.77	br; tet. R: $H_2O \rightarrow H_2O_2$
7	BaO_2	Barium peroxide	169.36	450	d 795	4.96	wh. tet. R: $H_2O \rightarrow 0$
8	Re_2O_8	Rhenium peroxide	500.44	145	—	~8.4	wyh.

Nitrogen compounds nitrides

	1	2	3	4	5	6	7	8
II	Li_3N	·Be_3N_2	BN	C_2N_2			F_3N	NdN U_3N_4
III	Na_3N	Mg_3N_2	AlN	Si_3N_4	P_3N_5	S_4N_2 S_4N_4	Cl_3N	
IV	K_3N	Ca_3N_2	GaN	Ge_3N_2 Ge_4N_4		Se_4N_4	Br_3N	
V							I_3N	
VI		Ba_3N_2						

a main groups

	1	2	3	4	5	6	7	8
IV	Cu_3N	Zn_3N_2		TiN	VN	CrN	Fe_2N Fe_4N	
V				ZrN	NbN			
VI					TaN			

b sub groups

S. No	Formula	Name	Mol. wt.	MP	BP	D	Colour and other properties
9	Li_3N	Lithium nitride	34.83	~840			blksh. gr; cub; d.w
10	Na_3N	Natrium nitride	83.01	d 300		1.846	wh. d.w
11	K_3N	Calium nitride	131.30	d			grsh. blk; d.w
12	Be_3N_2	Beryllium nitride	55.06	~2250	d		col. cub. d.w, d a. (conc) alk
13	Mg_3N_2	Magnesium nitride	100.98	d 1500		2.71	yelsh. gr; Cr; i.w dh. w, R: a
14	Ca_3N_2	Calcium nitride	148.26	900		2.63	br. Cr; d.w, R: a
15	Ba_3N_2	Barium nitride	440.10	1000			col. Cr;
16	Cu_3N	Copper nitride	204.63	d 300			dk. gr; d.w; R: a
17	Zn_3N_2	Zinc nitride	224.16				gr; d-w, R: HCl
18	BN	Boron nitride	24.83	~2730p	s 3000	2.20	wh. i.w. R: conc a
19	AlN	Aluminium nitride	40.99	2200p	s ~2000	3.26	gr (pa bl.) hex, d-w, R: a
20	GaN	Gallium nitride	83.73		s >800	6.10	dk. gr. hex, i.w, s. al
21	Si_3N_4	Silicon nitride	140.21	1900p		3.44	gr. wh. powd; R: HF
22	Ge_3N_2	Trigermanium dinitride	245.82		s > 650		blk. Cr;
23	Ge_3N_4	Trigermanium tetra nitride	273.83	d 450			wh. powd; i.w. a: alk
24	TiN	Titanium nitride	61.91	2950		5.43	bronze yel. cub. i.w; R a: HF aq. reg. ☠H8~9
25	ZrN	Zirconium nitride	105.23	2980		7.09	bronze br; i.w; R: HNO_3+HF, aq. reg ☠H8
26	VN	Vanadium nitride	64.96	~2050		6.13	blk; cub, i.w. R. aq. reg
27	NbN	Niobium nitride	106.92	2573		8.4	blk; cub; i.w; R: HNO_3+HF
28	TaN	Tantalum nitride	104.89	3360		16.30	bronze br-blk; i.w; R: HNO_3+HF ☠H9
29	CrN	Chromium nitride	66.02	d 1770		5.9	cub or powd; i.w; R: aq. reg
30	Fe_2N	Tetra iron dinitride	251.42	d 200		6.35	gr; i.w; R: HCl, H_2SO_4
31	Fe_4N	Tetro iron nitride	237.41			6.57?	
32	C_2N_2	Dicyanogen*)	52.04	−27.9	−20.7	2.335	col. gas. s.w-4.5; s al, eth; ☠
33	P_3N_5	Phosphorus nitride	162.98	d 800		2.51	wh; i.w. dhw
34	S_4N_2	Tetra sulphur dinitride	156.28	11	d	1.901	red liq. i.w; sl s. CS_2; so eth.
35	S_4N_4	Tetra sulphur tetra nitride	184.30	d 160		2.22	or. red; monocl. d.w; sc.s. org solv; **ex**
36	Se_4N_4	Selenium nitride	381.87	d 160			gold. yel. i.w; v.e.s. CS_2; Bz, ac, a; i. al. eth; **ex**
37	F_3N	Nitrogen trifluoride	70.01	−216.6	−120	1.537	col. gas. v.e. s.w; stable
38	Cl_3N	Trichloro nitrogen	120.38	<−40d	<−71d	1.653	yelsh. oil; i.w; s. PCl_3, CS_2, CCl_4 Bz. **ex. 95**
39	I_3N	Triiodo nitrogen	394.77	d	s vac		blk; i.w; s. $Na_2S_2O_3$, hk CNS soln. **ex**

*) Structure $N \equiv C-C \equiv N$; see also org compounds. Cynides (Nitriles): see also pseudo halogens p. 63/12

Compounds of halogens or halides

The name halogen means 'salt forming' (gr. halos = salt genao = forms). The binary compounds of halogens with the metallic elements—especially with the alkali metals—have a decided saline character.

In contrast to these, the binary compounds of halogens with the non-metallic elements have a volatile character. These specific properties depend on the whole upon:

large difference of electronegativities of halogen—metal, or

small difference of electronegativities of halogen—nonmetal

In case of many highly charged metal ions,the co-ordination number corresponds with the number of charges so that "enveloped ions" result with volatile character. This often leads to the formation of "mixed bonds" which result in stable complex particles (see p. 25) e.g. $BaCl_2$, $PbCl_2$, $PbCl_4$.

In case of halides, compounds with the ions are employed:

$[-N=N=N-]^-$	or	N_3^-	Azide ion [salts—azides]
$[-C=N]^-$	or	CN^-	Cyanide ion [salts—cyanides]
$[-O-C=N]^-$	or	OCN^-	Cyanate ion [salts—cyanates]
$[-O-N=C]^-$	or	ONC^-	Fulminate ion [salts—fulminates]
$[-S-C=N]^-$	or	SCN^-	Thiocyanate ion [salts—thiocyanates]

Comparison of properties of 'Halogens' and pseudo halogens

Halogens			page	Pseudo halogens			page
Cl_2	Cl–Cl	Chlorine	10 & 11	$(CN)_2$	(WC–CW)	Dicyanogen	62/22
Br_2	Br–Br	Bromine	10 & 11	$(SCN)_2$	(NCS–SCN)	Dithiocyanagen	62/60
I_2	I–I	Iodine	10 & 11	Halogen azides			62/6,7,8
				Halogen cyanides			62/23,24,25
HCl		Hydrogen chloride	39/49	HN_3		Hydrazoic acid	62/5
HBr		Hydrogen bromide	39/50	HCN		(Hydrogen cyanide)	62/21
HI		Hydrogen iodide	39/51	HOCN		Cyanuric acid	62/49
				HONC		Fulminic acid	62/51
				HSCN		Thiocyanuric acid	62/58

Insoluble Ag and Hg salts
Sparingly soluble Pb salts

Insoluble Ag and Hg salts
Sparingly soluble Pb salts

No.	Formula	Name	Mol. wt.	MP/FP	BP/CP	D/sp.gr V.D.	Colour and other properties
1	LiF	Lithium fluoride	25.94	844	1681	2.63	wh; cub: v. sl. s.w; n = 1.3915
2	NaF	Natrium fluoride	41.99	992	1704	2.79	wh; cub: sl. s.w; n = 1.336 ✿ H 3.5
3	NaHF₂	Natrium hydrogen fluoride	62.00	d ~ 270			wh; rhomb; sl. s.w
4	KF	Potassium fluoride	58.10	857	1503	2.48	wh; cub: e.s.w; deliq. cr;
5	KHF₂	Potassium hydrogen fluoride	78.11	239			wh; tet; e.s.w
6	NH₄F	Ammonium fluoride	37.04	s		1.015	wh; hex; e.s.w; n = 1.315
7	NH₄HF₂	Ammonium hydrogen fluoride	57.05	124		1.21	wh; rhomb; s.w; n ~ 1.390
8	BeF₂	Beryllium fluoride	47.01	800	1330	.1986	wh; tet; s.w;
9	MgF₂	Magnesium fluoride	62.32	1265	2230	3.13	wh; tet; i.w. ✿ H 5
10	CaF₂	Calcium fluoride	78.08	1392	2500	3.18	wh; cub;i.w; R: HCl, HNO_3 n=1.484 ✿ H 4
11	SrF₂	Strontium fluoride	125.63	1190	(2460)	2.44	wh; cub; i.w; n = 1.442
12	BaF₂	Barium fluoride	175.36	1320	~ 2260	4.83	wh; alk; v. sl. s.w; R: HF; n = 1.475
13	AlF₃	Aluminium fluoride	83.98	s 1260		3.18	wh; rhomb; sl. s.w; no R: a, alk;
14	TlF	Thallium fluoride	223.39	327	655	8.36	wh; rhomb; e.s.w;
15	TlF₃	Thallium-III fluoride	261.39	d 550		8.36	olive gr; d.w
16	SnF₂	Tin-II fluoride	156.70	~ 210			wh; monocl; e.s.w
17	SnF₄	Tin-IV fluoride	194.70		subl 705	rhomb;8.37	wh; cr; s.w
18	PbF₂	Lead fluoride	245.21	824	1293	cub;7.68	wh; rhomb; v. sl. s.w; R: HNO_3
19	BiF₃	Bismuth fluoride	266.00	~ 725		8.75	gr; cub; R: a;
20	CuF₂	Copper-II fluoride	137.57	d		2.93	pa. bl; monocl; $2H_2O$; v. sl. s.w; R: a
21	AgF	Silver fluoride	126.88	435	1150	5.852	yel; cub; L H_2O; v.e. s.w; R: HF
22	Ag₂F	Disilver fluoride	234.76	d 90		8.57	yel; rhomb; d.w;
23	ZnF₂	Zinc fluoride	103.38	872		4.90	wh; tet; $4H_2O$; sl. s.w;
24	CdF₂	Cadmium fluoride	150.41	1100	1748	6.64	wh; cub; sl. s.w; R: HF; n = 1.56
25	HfF₄	Hafnium fluoride	254.6			7.13	col; monocl; R: conc. a; R: HF
26	VF₃	Vanadium-III fluoride	107.95	subl >800		3.36	gr; rhomb; i.w;
27	VF₄	Vanadium-IV fluoride	126.95	d 325		2.98	yelsh br; s.w
28	CrF₃	Chromium-III fluoride	109.01	subl >1000		3.78	gr; rhomb; i.w; R: a
29	FeF₃	Iron-III fluoride	112.85	1030		3.18	gr. rhomb; v. sl. s.w;
30	CeF₃	Cerium-III fluoride	197.13	1460		6.16	wh; hex; R: a;
31	ThF₄	Thorium fluoride	380.11	> 900		6.32	wh; monocl; i.w; R: HF
32	UF₄	Uranium-IV fluoride	314.07	~1000	1417		gr; monocl; R: HNO_3

Fluoro compounds—fluorides

	1	2	3	4	5	6	7	8	
II	LiF	BeF₂	BF₃	CF₄				Nil	
III	NaF NaHF	MgF₂	AlF₃	SiF₄	PF₃ PF₅	SF₃ SF₆		SF₃	
IV	KF KHF₂ NH₄F NH₄HF₂	CaF₂		GeF₄	AsF₃ AsF₅	SeF₄ SeF₆	BrF₃ BrF₅	KrF₂ KrF₄	
V	RbF	SrF₂		SnF₂ SnF₄	SbF₃ SbF₅		IF₅ IF₇	XeF₂ XeF₄ XeF₆	a main groups
VI	CsF	BaF₂	TlF TlF₃	PbF₂	BiF₃			Rn	
IV	CuF₂	ZnF₂		TlF₄	VF₃ VF₄ VF₅	CrF₃		FeF₃	
V	AgF Ag₂F	CdF₂		ZrF₄	NbF₅	MoF₆			b sub groups
VI				HfF₄	TaF₅	WF₆	ReF₆	OsF₈	IrF₆

Side panel:

CeF₃	CeF₄
ThF₄	
UF₄	UF₆

To rare gas compounds

Relatively stable compounds

F = 140°C
F = 114°C
F = 46°C   P = 87°C

*) Fluoro compounds referred to 8

No.	Formula	Name	Mol. wt.	MP	BP	D/sp. gr.	Colour and other properties
1	BF₃	Boron trifluoride	67.82	−128.8	−101	3.07	col; d.w; s. conc. H₂SO₄, D (vap) 1.58
2	CF₄	Carbon tetrafluoride	88.011	−183.6	−128	3.94	col; sl. misc. w; D (vap) 1.96
3	SiF₄	Silicon tetrafluoride	104.09	s − 95.7	− 65p	4.68	col; cub; d.w; s. HF
4	GeF₄	Germanium tetrafluoride	148.60	s − 36.6		6.65	col; d.w
5	PF₃	Phosphorus trifluoride	87.98	−151.5	−101	3.907	col; d.w; s. al
6	PF₅	Phosphorus pentafluoride	125.98	− 93.8	− 84.5	5.80	col; d.w
7	AsF₃	Arsenic trifluoride	131.91	− 6.5	63	2.70	col; d.w; s. Bz, al, eth
8	AsF₅	Arsenic pentafluoride	169.91	− 80	− 52.9	2.33	col; s. Bz, al, eth
9	SbF₃	Antimony trifluoride	178.76	292	376	4.38	wh; rhomb; v.e. s.w
10	SbF₅	Antimony pentafluoride	216.76	7.0	149.5	2.99	wh; oily liq; misc. w
11	SF₄	Sulphur tetrafluoride	108.07	−124	− 40		col; d.w
12	SF₆	Sulphur hexafluoride	146.07	− 50.7	s − 63.8	6.602	col; s. KOH
13	SeF₄	Selenium tetrafluoride	154.96	− 13.2	~ 93	2.77	col; sl. s.w;
14	SeF₆	Selenium hexafluoride	192.96		s − 46.6	8.687	col; d.w; n = 1.895
15	BrF₃	Bromine trifluoride	136.92	8.8	127	2.49	col; d.w
16	BrF₅	Bromine pentafluoride	174.92	− 61.3	+ 41	2.47	col; d.w
17	IF₅	Iodine pentafluoride	221.91	9.6	98	3.5	col; exposed to air fumes
18	IF₇	Iodine heptafluoride	259.91	5.6	4.5	2.8	col
19	TiF₄	Titanium tetrafluoride	123.90		284	2.798	col; s. w; s. HF
20	ZrF₄	Zirconium tetrafluoride	167.22	subl.		4.433	wh; monocl; s. HF
21	VF₅	Vanadium impentafluoride	145.95	111.2		2.18	wh; sh yel; s. w; al
22	NbF₅	Niobium pentafluoride	187.91	75.5	233	3.293	wh; monocl; s.w; al
23	TaF₅	Tantalum pentafluoride	275.95	96.8	229.5	4.74	wh; tet; s.w
24	MoF₆	Molebdenium hexafluoride	209.95	17	35	2.55	wh; cr; d.w
25	WF₆	Wolfram hexafluoride	297.86	2.5	19.5	12.9	pa yel; rhomb; d.w; s. al
26	ReF₆	Rhenium hexafluoride	300.22	18.8	47.6	> 4.25	yel; s.w
27	OsF₈	Osmium octafluoride	342.20	34.4	47.3		yel; s. al
28	IrF₆	Iridium hexafluoride	307.10	44	53	6.0	yel; tet; d.w
29	CeF₄	Cerium tetrafluoride	234.14	decomp		4.77	pa yel; 1 H₂O, i.w
30	UF₆	Uranium hexafluoride	352.07	69.5	56.2	4.68	pa yel; monocl; s.w

Chloro—compounds or chlorides

a main groups

	1	2	3	4	5	6	7
II	$LiCl$	$BeCl_2$	BCl_3	CCl_4			
III	$NaCl$	$MgCl_2$	Al_2Cl_6	$SiCl_4$ Si_2Cl_6	PCl_3 PCl_5	S_2Cl_2 $SeCl_2$ SCl_4	
IV	KCl NH_4Cl	$CaCl_2$	$GaCl_2$ $GaCl_3$	$GeCl_2$ $GeCl_4$	$AsCl_3$ $AsCl_5$	Se_2Cl_2 $SeCl_4$	
V	$RbCl$	$SrCl_2$	$InCl_3$	$SnCl_2$ $SnCl_4$	$SbCl_3$ $SbCl_5$	$TeCl_2$ $TeCl_4$	ICl ICl_3
VI	$CsCl$	$BaCl_2$	$TlCl$ $TlCl_3$	$PbCl_2$ $PbCl_3$	$BiCl_3$		

b sub groups

	1	2	3	4	5	6	7	8		
IV	$CuCl$ $CuCl_2$	$ZnCl_2$	$ScCl_3$	$TiCl_2$ $TiCl_3$ $TiCl_4$	VCl_2 VCl_3 VCl_4	$CrCl_2$ $CrCl_3$	$MnCl_2$	$FeCl_2$ $FeCl_3$	$CoCl_2$	$NiCl_2$
V	$AgCl$	$CdCl_2$	YCl_3	$ZrCl_4$	$NbCl_5$	$MoCl_3$ $MoCl_4$ $MoCl_5$		$RuCl_3$	$RhCl_3$	$PdCl_2$
VI	$AuCl$ $AuCl_3$	Hg_2Cl_2 $HgCl_2$	$LaCl_3$		$TaCl_5$	WCl_4 WCl_5 WCl_6	$ReCl_3$ $ReCl_4$	$OSCl_4$	$IrCl_3$ $IrCl_4$	$PtCl_2$ $PtCl_4$

No.	Formula	Name	Mol. wt.	MP	BP	D/sp.gr.	Colour and other properties
1	BCl_3	Boron trichloride	117.19	−107.3	13	1.43	col; hex; d.w; al
2	Al_2Cl_6	Aluminium chloride	133.35	192.5	s 180	2.46	wh; rhomb monocl; s.w. s. org. solv
3	$GaCl_2$	Gallium dichloride	140.63	170.5	~535		wh; deliq. G; d.w. s. Bz
4	$GaCl_3$	Gallium trichloride	176.09	78	200	2.47	wh; deliq. G; v.e. s.w. s. pet[1]
5	$TlCl_3$	Thallium trichloride	310.76	~ 25	d		wh; hex; s.w
6	CCl_4	Carbon tetrachloride	153.84	−23	77	1.595	col; monocl cub; n=1.461, misc. al, eth
7	$SiCl_4$	Silicon tetrachloride	169.92	−67.7	56.7	1.483	col; d.w
8	Si_2Cl_6	Disilicon hexachloride	268.92	2.5	147	1.58	col; d.w
9	$GeCl_4$	Germanium tetrachloride	214.43	−51	84	1.879	col; d.w; n=1.46
10	$SnCl_4$	Stannic chloride	260.53	−33	113	2.232	wh; s.w; v.e.s. CS_2, n = 1.5112
11	$PbCl_2$	Lead tetrachloride	349.04	−15	d 105	3.18	yel; d.w; s. conc. HCl
12	PCl_3	Phosphorus trichloride	137.35	−92	74.5	1.574	col; d.w; s. eth; chl; n=1.520
13	PCl_5	Phosphorus pentachloride	208.26	149(p)	s 100	2.11	pa yel; tet. d.w; s. CCl_4
14	$AsCl_3$	Arsenic trichloride	181.28	−16.2	130.4	2.16	col; d.w; s. HCl; n = 1.598
15	$AsCl_5$	Arsenic pentachloride	252.20	~−40			col; at−25°C d → $AsCl_3$ + Cl_2
16	$SbCl_3$	Antimony trichloride	228.13	72.9	223	3.14	wh; rhomb; v.e.s.w then d; s. HCl
17	$SbCl_5$	Antimony pentachloride	299.05	4	140(p)	2.33	wh; d.w; s. conc. HCl; E = 3.22
18	S_2Cl_2	Disulphur dichloride	135.05	−76.5	137.1	1.678	brsh yel; n = 1.666. s. CS_2;
19	SCl_2	Sulphur dichloride	102.98	−78	59	1.621	dk red. n = 1.557. s. CS_2,
20	SCl_4	Sulphur tetrachloride	173.89	−30	d − 15		yelsh br; d.w;
21	Se_2Cl_2	Selenium dichloride	228.83	~−80	d 137	2.91	br; yel; n = 1.596; d.w; al; s−CS_2
22	$SeCl_4$	Selenium tetrachloride	220.79	305(p)	subl.	3.80	wh; cub; n = 1.807 d.w; i. CS_2; s. POCl
23	$TeCl_2$	Tellurium dichloride	198.52	175	327	7.05	blk gr; cr; d.w
24	$TeCl_4$	Tellurium tetrachloride	269.44	224.1	392	3.26	wh; cr; s.h.w → d.w
25	ICl	Iodine monochloride	162.37	27.2	101	$\alpha=3.182$ $\beta=3.24$	α=red cub β=brsh red rhomb ac−a s. HCl, al. eth
26	ICl_3	Iodine trichloride	233.28	101(p)	d 77	3.117	yel; rhomb; s.w; → d.w; s. al. eth
27	$TiCl_4$	Tetanium tetrachloride	189.73	−23	136	1.726	col; n = 1.607 d.w; s. HCl, al.
28	$ZrCl_4$	Zirconium tetrachloride	233.05	437(p)	s 331	2.80	wh; cub; d.w; s. HCl, al.
29	VCl_4	Vanadium tetrachloride	192.78	~−109	148.5	1.87	brsh red; s. conc. HCl, al, eth
30	$NbCl_5$	Niobium pentachloride	270.20	194	245	2.75	yel; cr needles d.w;s. hot HCl, al, eth
31	$TaCl_5$	Tantalum pentachloride	358.24	221	241.6	3.68	pa yel; cr. pris; d.w; s. al
32	$MoCl_5$	Molebdenum pentachloride	273.24	194	268	2.93	blk; rhomb; s.w. a. CS_2; org. solv
33	WCl_4	Tungsten tetrachloride	325.69	dec.		4.624	gr. br. d.w
34	WCl_5	Tungsten pentachloride	361.15	248	275.6	3.875	blk gr; d.w; s. CS_2, al, eth
35	WCl_6	Tungsten hexachloride	396.60	275	347	3.52	blk viol; rhomb; d.w;s. CS_2 al. eth, Bz

[1]pet = petroleum ether

No.	Formula	Name	Mol. wt.	MP/FP	BP/CP	D/sp. gr.	Colour and other properties
1	LiCl	Lithium chloride	42.40	614	1382	2.068	wh; cub; e.s.w; s. al; n = 1.662
2	NaCl	Sodium chloride	58.44	800	1465	2.163	wh; cub; c.s.w; n = 1544 ✂ H2.5
3	KCl	Potassium chloride	74.56	770	1407	1.984	wh; cub; e.s.w; n = 1.490 ✂H2.0
4	NH₄Cl	Ammonium chloride	53.50	s ~ 335		1.53	wh; cub; e.s.w; n = 1.642 ✂H1−2
5	RbCl	Rubidium chloride	120.94	717	1381	2.76	wh; cub; e.s.w; i. al; n = 1.493
6	CsCl	Caesium chloride	168.37	642	1300	3.97	wh; cub; v.e.s.w; s. al; n = 1.6418
7	BeCl₂	Beryllium chloride	151.99	416(ww)			wh; monocl 4H₂O; s.w. al, eth;
8	MgCl₂	Magnesium chloride	95.23	712	1418	2.325	wh; monocl; e.s.w; s. al;
9	·6H₂O	Magnesium chloride	203.33	d 120		1.56	wh; monocl; e.s.w; s.al; n ~ 15; ✂ H1 5−2
10	CaCl₂	Calcium chloride	110.99	782	>1600	2.152	wh; rhomb; e.s.w; s. al; n = 1.52
11	·6H₂O	Calcium chloride	219.09	d 29		1.68	wh; trig; e.s.w; n ~ 1.4
12	SrCl₂	Strontium chloride	158.54	872		3.052	wh; cub; e.s.w; n > 1.63
13	BaCl₂	Barium chloride	244.31	960 ww	1560	3.097	wh; monocl; e.s.w; n ~ 1.63
14	RaCl₂	Radium chloride	296.96	1000		4.91	yel; monocl; e.s.w;
15	InCl₃	Indium chloride	221.19	585		3.46	wh; hex; s.w;
16	TlCl	Thallium chloride	239.85	427	807	7.02	wh; cub; v. sl. s.w; n = 2.247
17	SnCl₂	Stannous chloride	189.61	247	603-623	3.95	wh; rhomb; e.s.w; → d.w.s. al, eth
18	PbCl₂	Lead-II chloride	278.12	498	954	5.85	wh; rhomb; v.sl.s.w; s.hw; n ~ 2.2 ✂ H2
19	BiCl₃	Bismuth chloride	315.37	230	447	4.75	wh; s. HCl
20	CuCl	Copper-I chloride	99.00	432	1490	3.53	wh; cub; sl. s.w; n = 1.93 ✂H2−2.5
21	CuCl₂	Copper-II chloride	134.45	630	d	3.045	yelsh. br. monocl; e.s.w; s. al
22	·2H₂O	Copper-II chloride	170.49	110(−2H₂O)		2.39	blsh. grn, rhomb, e.s.w; n ~ 1.7
23	AgCl	Silver chloride	143.34	455	1554	5.56	wh; cub; i.w; n = 2.071 ✂ H1−1.5
24	AuCl	Gold-I chloride	232.46	d 170		7.8	pa. yel; d.w; s. NaBr solv.
25	AuCl₃	Gold-III chloride	303.37	d 254		4.67	red; br; s.w. al;
26	ZnCl₂	Zinc chloride	136.29	~318	732	2.91	wh; rhomb; v.e.s.w; s. al, eth, n ~ 1.7
27	CdCl₂	Cadmium chloride	183.32	568	967	4.047	wh; rhomb; sl. s.w; monocl 2½H₂O; v.e.s.w;
28	Hg₂Cl₂	Mercurous chloride	472.13	525 (Dr.)	s 383.7	7.150	wh; ✂ H 1−2, n = 1.973, 2.636
29	HgCl₂	Mercuric chloride	271.52	277	304	5.53(Rö)	wh; rh; s.w; s. al, eth; n=1.725, 1.859, 1.865
30	ScCl₃	Scandium chloride	151.33	960			wh; rhomb; s.w;
31	YCl₃	Yttrium chloride	195.29	680		2.81	wh; cr; e.s.w
32	LaCl₃	Lathan chloride	245.29	860	>1000	3.842	wh; rhomb; s.w;
33	TiCl₂*	Titanous chloride	118.81			3.13	blk; rhomb; d. in w; s. al
34	TiCl₃*	Titanium-III chloride	154.27	s 432		~2.68	viol; rhomb; s.w;
35	VCl₂*	Vanadium-II chloride	121.86			3.23	pa. yel; hex; s.w; al;
36	VCl₃*	Vanadium-III chloride	157.32			3.00	pa. red; rhomb; s.w. al; eth;
37	CrCl₂	Chromium-II chloride	122.92	824	1300	2.88	wh; needles; s.w;
38	CrCl₃*	Chromium-III chloride	158.38	~1150	s	2.76	red. viol; rhomb, leaf, s.w;
39	MoCl₃*	Molebdenum-III chloride	202.32	dec.		3.578	red. br; cr; i.w; s.s. al, eth; R: HNO₃
40	MoCl₄*	Molebdenum-IV chloride	237.78	dec.			br. Cr; d.w. al, eth.
41	MnCl₂	Manganese-chloride	125.85	650	1190	2.977	pa. rose; rhomb; 4H₂O monocl; e.s.w; s.Al
42	ReCl₃*	Rhenium-III chloride	292.59	>550			red; hex; s.w
43	ReCl₄*	Rhenium-IV chloride	328.08	500			br. blk; s.h HCl
44	FeCl₂	Ferrous chloride	126.76	677	1023	2.98	wh; rhomb; e.s.w; s. al
45	FeCl₃	Ferric chloride	162.22	300	317	2.90	blk.br; rhomb; 6H₂O monocl; e.s.w; s.acet. eth
46	CoCl₂	Cobalt-II chloride	129.85	~730	1050	3.356	bl; rhomb; 6H₂O monocl; e.s.w; s. al, sbs. eth
47	NiCl₂	Nickel-II chloride	129.62	1000 (Dr.)	s 987	3.55	yel; rhomb; e.s.w; s. al
48	RuCl₃*	Ruthenium-III chloride	207.47	d >500		3.11	br. yel; trig. or hex; s.w. al;
49	RhCl₃*	Rhodium-III chloride	209.28	d 450			red; i.w. noR; conc. a; s. KCN sol.
50	PdCl₂*	Palladium-II chloride	213.35	678 (ww)	d	4.00	red. br; pris; 2H₂O; s.w
51	OsCl₄*	Osmium-IV chloride	332.03				blk; need; s.w → d in w; s. HCl
52	IrCl₃*	Iridium-III chloride	298.57	d 763		5.30	olive grn; no R; conc. a; alk;
53	IrCl₄*	Iridium-IV chloride	334.03	d			dk. red; cub; R; a;
54	PtCl₂*	Platinum-II chloride	266.00	d 581		5.87	gr. grn; rhomb; R: HCl, NH₄OH
55	PtCl₄*	Platinum-IV chloride	336.92	d 370		4.303	red. br; Cr. 5H₂O; v.e.s.w

[Mo₆Cl₈]⁸⁺ Cation

No.	Formula	Name	Mol.wt.	Colour	No.	Formula	Name	Mol.wt.	Colour
56	CeCl₃	Cerium chloride**	246.50	white	64	DyCl₃	Dysprosium chloride	268.88	yellow
57	PrCl₃	Praseodymium chloride	247.29	green	65	HoCl₃	Holmium chloride	271.31	pale yellow
58	NdCl₃	Neodymium chloride	250.64	rose	66	ErCl₃	Erbium chloride	381.74	rose
59	SmCl₂	Samarium-II chloride	221.26	Red brown	67	TmCl₃	Thulium chloride	401.42	pale yellow
60	SmCl₃	Samarium-III chloride	256.72	pale yellow	68	YbCl₃	Ytterbium chloride	387.51	green
61	EuCl₃	Europium chloride	258.4	pale rose	69	ThCl₄	Thorium chloride	373.88	white
62	GdCl₃	Gadolinium chloride	371.73	white	70	UCl₃	Uranium-III chloride	344.44	red
63	TbCl₃	Terbium chloride	265.30	white	71	UCl₄	Uranium-IV chloride	379.90	grey green

* Halides of the transitional metals with complex structure for example [Me₆X₈] X₄; [Me₆X₈] X₆; [Me₆X₈] X₈ and so on; in complex metallic bonding see p. 31.
** CeCl₃ MP = 822 D = 3.92 hex; s.w, Al; Chlorides of 'Rare Earth metals' are soluble in water.

Bromo compounds: bromides

	1	2	3	4	5	6	7		8
II	LiBr	BeBr2	BBr3						
III	NaBr	MgBr2	AlBr3		PBr3 PBr5	S2Br2		a Main groups	
IV	KBr NH4Br	CaBr2		GeBr4	AsBr3				
V	RbBr	SrBr2		SnBr2 SnBr4	SbBr5	TeBr2 TeBr4	IBr		
VI		BaBr2	TbBr TlBr3	PbBr2	BiBr3				8
IV	CuBr CuBr2	ZnBr2					FeBr2 FeBr3		NiBr2
V	AgBr	CdBr2						b Sub groups	
VI	AuBr AuBr3	Hg2Br2 HgBr2			TaBr5				PtBr2 PtBr4

No.	Formula	Name	Mol. wt.	MP/FP	BP/CP	D	Colour and other properties
1	LiBr	Lithium bromide	86.86	~551	~1312	3.464	wh; cub; v.e. s.w; n = 1.784
2	NaBr	Natrium bromide	102.90	746.8	1392	3.205	wh; cub; 2H2O monocl; v.e.s.w; n = 1.6412
3	KBr	Calium bromide	119.02	741.8	1383	2.75	wh; cub; e.s.w; sl. s. al; n = 1.559
4	NH4Br	Ammonium bromide	97.96	s		2.43	wh; cub; e.s.w; s. me. al, n = 1.7108
5	RbBr	Rubidium bromide	165.40	677	1352	3.35	wh; cub; v.e. s.w; n = 1.5539
6	BeBr2	Beryllium bromide	168.85	s 490		3.465	wh; need; s.w. al, eth;
7	MgBr2	Magnesium bromide	184.15	711	1230	3.72	wh; rhomb; v.e. s.w; s. al
8	CaBr2	Calcium bromide	199.91	760	810	3.353	wh; rhomb, v.e. s.w, s. al;
9	SrBr2	Strontium bromide	247.46	643	d	4.216	wh; rhomb; e.s.w; al;
10	BaBr2	Barium bromide	297.19	847(wW)		3.58	wh; rhomb; e.s.w, n = 1.75
11	RaBr2	Radium bromide	385.88	728		5.79	wh; monocl; e.s.w
12	TlBr	Thallium-I bromide	284.31	460	819	7.55	yelsh; cub; v. sl. s.w; s. al
13	TlBr3	Thallium-III bromide	444.14	d		pa.	yel; s.w, al;
14	SnBr2	Stannous bromide	278.53	216	620	5.117	pa. yel; rhomb; e.s.w
15	PbBr2	Lead-II bromide	367.04	370	914	6.66	wh; rhomb; v. sl. s.w; i. al
16	BiBr3	Bismuth-II bromide	448.75	218	461	5.604	yel; s. HBr
17	CuBr	Cuprous bromide	143.46	488	1345	5.05	wh; cub; R: a
18	CuBr2	Cupric bromide	223.37	498	900	4.71	blk; monocl; v. sl. s.w; s. al, eth
19	AgBr	Silver bromide	187.80	430	d 700	6.473	pa. yel; cub; i.w; n = 2.253; ✖ H2
20	AuBr	Gold-I bromide	276.92	d 115		7.9	gr. yel; d.w; s. NaBr sol
21	AuBr3	Gold-III bromide	436.75	d 160		dk.	br; s.w
22	ZnBr2	Zinc bromide	225.21	394	650	4.22	wh; rhomb; 2H2O; v.e. s.w
23	CdBr2	Cadmium bromide	272.24	566	863	5.192	wh; rhomb; e.s.w; s. al
24	Hg2Br2	Mercurous bromide	561.05	s 345		7.307	wh; tet; i.w;
25	HgBr2	Mercuric bromide	360.44	238	320	6.08	wh; rhomb; v. sl. s.w. s. al;
26	TaBr5	Tantalum-V bromide	580.53	~240	~320	4.67	yel; cr; s. al;
27	FeBr2	Ferrous bromide	215.68	684		4.636	yel; rhomb; v.e. s.w; s. al
28	FeBr3	Ferric bromide	296.60	d			br; rhomb; s.w; al, eth;
29	NiBr2	Nickel-II bromide	218.54	d 963		4.64	yel; rhomb; v.e. s.w; s. al
30	PtBr2	Platinum-III bromide	354.92	d 250		6.65	gr. br; cub; s. in bromine water
31	PtBr4	Platinum-IV bromide	514.75	d 180		5.69	br; v. sl. s.w; s. al, eth;
32	BBr3	Boron tribromide	250.57	−46	90.1	2.650	col; hex; d.w; s. CeI4, E 2.58
33	AlBr3	Aluminium bromide	266.73	97.4	257	3.20	wh; trig; v.e. s.w; s. eth;
34	GeBr4	Germanium tetrabromide	392.26	26.1	183	3.132	wh; cub; d.w. n = 1.6269
35	SnBr4	Tintetra bromide	438.36	31	203	3.340	wh; rhomb; d.w|
36	PBr3	Phosphorus tribromide	270.72	−40	173	2.852	wh; d.w; s. eth; CeI4;
37	PBr5	Phosphorus pentabromide	430.56	d 106			yel; rhomb; d.w
38	AsBr3	Arsenic tribromide	314.66	30.8	221	3.54	wh; rhomb; d.w; s. HCl
39	SbBr3	Antimony tribromide	361.51	97	288	4.148	wh; rhomb; d.w; s. h HCl
40	S2Br2	Disulphur dibromide	223.96	−46	54	2.635	red; d. w; n = 1.736
41	TeBr2	Tellurium dibromide	287.44	~210			blk; grn need; d.w; s. eth; hTart. a
42	TeBr4	Tellurium tetrabromide	447.27	~380	421	4.31	yel; pris; s.w
43	IBr	Iodine monobromide	206.83	~ 40	116	4.416	br; blk; s. al, eth; NaBr sol;

Iodo compounds—Iodides

	1	2	3	4	5	6	7	8	
II	LiI	BeI₂	BI₃						
II	NaI		AlI₃	—	PI₃				
V	KI / NH₄I	CaI₂			AsI₃				a Main groups
V	RbI	SrI₂		SnI₂ / SnI₄	SbI₃				
VI		BaI₂	TlI	PbI₂				**8**	
V	CuI	ZnI₂		TiI₄				FeI₂	b Sub groups
V	AgI	CdI₂							
VI	AuI	Hg₂I₂ / HgI₂			WI₂ / WI₄				PtI₂

No.	Formula	Name	Mol. wt.	MP	CP	D	Colour and other properties
1	LiI	Lithium iodide	133.6	450	1171	4.06	wh; cub; v.e. s.w; s. al
2	NaI	Sodium iodide	185.93	d 67.3		2.448	wh; tricl. $2H_2O$; v.e. s.w;
3	KI	Potassium iodide	166.01	681.8	1324	3.13 $\alpha=2.86$	wh; cub; v.e. s.w; s. al; n = 1.677
4	NH₄I	Ammonium iodide	144.95	s 551		$\beta=2.51$	wh; cub; v.e. s.w; s. al, n = 1.703
5	RbI	Rubidium iodide	212.39	641	1304	3.55	wh; cub; v.e. s.w; n = 1.6474
6	BeI₂	Beryllium iodide	262.83	480		4.325	wh; need, v.e. s.w; al, eth;
7	CaI₂	Calcium iodide	293.90	740	708-719	3.956	wh; rhomb; v.e.s.w; n=1.465, 1.498, 1.504
8	SrI	Strontium iodide	449.55	515		4.415	wh; rhomb $6H_2O$; v.e. s.w; s. al
9	BaI₂	Barium iodide	427.21	d 740		5.15	wh; monocl; $4H_2O$; v.e. s.w; s. al
0	TlI	Thallium-I iodide	331.30	440	823	7.29 α=yel β=red	α=rhomb β=cub i.w: n = 2.78
1	SnI₂	Stannous iodide	372.52	320	720	5.28	red; rhomb; v. sl. s.w; s. CS_2
2	PbI₂	Lead iodide	461.93	412	872	6.16	yel; rhomb; v. sl. s.w; s. conc. KI sol.
3	CuI	Copper-I iodide	190.45	588	1336	5.63	wh; cub; i. w; n = 2.346; ✿ 2.5-3
4	AgI	Silver iodide	234.79	552	1506	5.67	yel; cub; i.w; n = 2.22; no R: NH₄OH
5	AuI	Gold-I iodide	323.91	d 120		8.25	yel; s. KI sol
6	ZnI₂	Zinc iodide	319.20	446	624	4.74	wh; rhomb; v.e. s.w; s. al, eth
7	CdI₂	Cadmium iodide	366.23	387	787	5.67	wh; hex, e.s.w; v.e.s; in meal
8	Hg₂I₂	Mercurous iodide	655.04	~290	d 310	7.70	yel; tet; i.w;
9	HgI₂	Mercuric iodide	454.43	257	354	6.271	red yel; hex; i.w; s. al, eth; n = 2.748, 2.455
0	WI₂	Tungsten-II iodide	437.68			6.799	br; R: alk
1	FeI₂	Ferrous iodide	309.67	592		5.315	gr; rhomb; s.w
2	PtI₂	Platinum-II iodide	448.91	d 300		6.40	blk; i.w; R: h a, alk
3	BI₃	Boron triiodide	391.58	43	210	3.35	wh; d.w; s. CS_2 B_2 CCl_4
4	AlI₃	Aluminium iodide	407.71	191	386	3.95	wh; s.w; al, CS_2
5	SnI₄	Stannic iodide	626.34	144	341	4.46	cub; d.w; s. al, eth; B_2, CS_2, n = 2.106
6	PI₃	Phosphorus triiodide	411.71	61	d	4.18	red; hex; d.w; s. CS_2
7	AsI₃	Arsenic triiodide	455.64	141.8	~403	4.39	red; rhomb; s. al. eth; CS_2, B_2
8	SbI₃	Antimony triiodide	502.49	170	401	hex=4.85 rhomb=4.77	red yel. rhomb; d.w; s.h. HCl, HI
9	TiI₄	Titanium tetraiodide	555.54	150	>360	4.40	red br; cub; s.w;
0	WI₄	Tungsten tetraiodide	691.50	d		5.2	blk; cr; R: alk

Equilibrium constants (solubility product) sparingly soluble halides and psuedo halides

Dissociation in aqueous solution				pK_L
CuCl	⇌	Cu⁺ +	Cl⁻	6
CuBr	⇌	Cu⁺ +	Br⁻	7.4
CuI	⇌	Cu⁺ +	I⁻	11.3
CuSCN	⇌	Cu⁺ +	SCN⁻	10.8
AgCl	⇌	Ag⁺ +	Cl⁻	9.96
AgBr	⇌	Ag⁺ +	Br⁻	12.4
AgI	⇌	Ag⁺ +	I⁻	16.0
AgSCN	⇌	Ag⁺ +	SCN⁻	12

Dissociation in aqueous solution				pK_L
MgF₂	⇌	Mg²⁺ +	2F⁻	8.16
CaF₂	⇌	Ca²⁺ +	2F⁻	10.46
SrF₂	⇌	Sr²⁺ +	2F⁻	8.25
BaF₂	⇌	Ba²⁺ +	2F⁻	5.77

i.e. $K_L \times pK_L \sim 10^{-10}$ or 1.6×10^{-4} g/100 W (4×10^{19} Ag⁺)

i.e. $K_L \times pK_L \sim 4.10^{-13}$ or 1.4×10^{-5} g/100W (1.2×10^{10} Ag⁺)

i.e. $K_L \times pK_L \sim 10^{-16}$ or 2.3×10^{-7} g/100W (6×10^{7} Ag⁺)

Sulphur compounds and sulphides

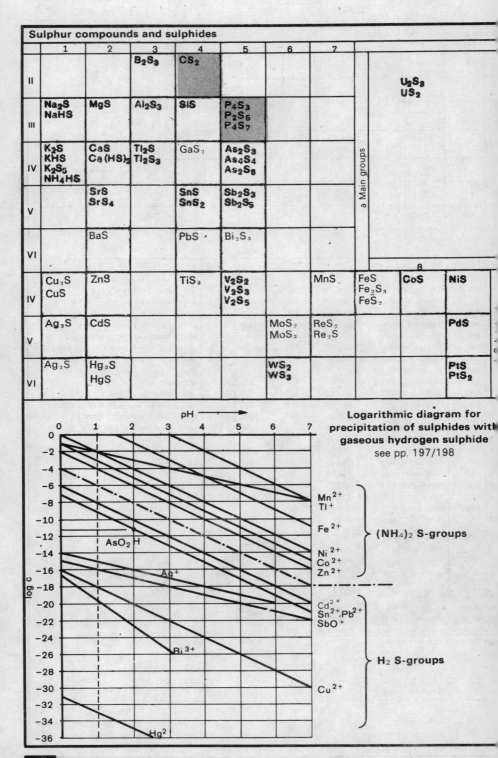

	1	2	3	4	5	6	7			
II			B_2S_3	CS_2					U_2S_3 US_2	
III	Na_2S $NaHS$	MgS	Al_2S_3	SiS	P_4S_3 P_2S_5 P_4S_7					
IV	K_2S KHS K_2S_5 NH_4HS	CaS $Ca(HS)_2$	Tl_2S Tl_2S_3	GaS_2	As_2S_3 As_4S_4 As_2S_5					
V		SrS SrS_4		SnS SnS_2	Sb_2S_3 Sb_2S_5		a Main groups			
VI		BaS		PbS	Bi_2S_3					

	1	2	3	4	5	6	7	8		
IV	Cu_2S CuS	ZnS		TiS_2	V_2S_2 V_2S_3 V_2S_5		MnS	FeS Fe_2S_3 FeS_2	CoS	NiS
V	Ag_2S	CdS				MoS_2 MoS_3	ReS_2 Re_2S			PdS
VI	Ag_2S	Hg_2S HgS			WS_2 WS_3					PtS PtS_2

Logarithmic diagram for precipitation of sulphides with gaseous hydrogen sulphide
see pp. 197/198

$pH \longrightarrow$

$\log c$

AsO_2H

Ag^+

Bi^{3+}

Hg^2

$\left.\begin{array}{l} Mn^{2+} \\ Tl^+ \\ Fe^{2+} \\ Ni^{2+} \\ Co^{2+} \\ Zn^{2+} \end{array}\right\}$ $(NH_4)_2$ S-groups

$\left.\begin{array}{l} Cd^{2+} \\ Sn^{2+}, Pb^{2+} \\ SbO^+ \\ \\ \\ Cu^{2+} \end{array}\right\}$ H_2 S-groups

No.	Formula	Name	Mol. wt	MP/FP	BP/CP	D/sp.gr.	Colour and other properties
1	Na_2S	Sodium sulphide	78.05	950		1.856	wh; cub; s.w → d.w
2	NaHS	Sodium hydrogensulphide	50.06	~350			wh; rhomb and cub; s.w. al;
3	K_2S	Potassium sulphide	110.27	~840		1.805	wh; cub; s.w → d.w; s. al
4	KHS	Potas. hydrogensulphide	72.17	455		1.71	wh; rhomb and cub; s.w; al
5	K_2S_5	Potassium pentasulphide	238.53	206			yelsh br; s.w; al
6	NH_4HS	Ammon. hydrogensulphide	51.11	120(p.)		1.17	wh; tet; n = 1.74, e.s.w; al;
7	MgS	Magnesium sulphide	56.39	d		2.80	wh; cub; n = 1.271 R: a
8	CaS	Calcium sulphide	72.15			2.56	wh; cub; n = 2.137, v. sl s.w; R: a
9	$Ca(HS)_2$	Calcium hydrogensulphide	214.32				wh; prism; v.e. s.w
10	SrS	Strontium sulphide	119.70			3.70	wh; cub; n = 2.107, R: a
11	SrS_4	Strontium tetrasulphide	323.97	25			redsh cub; $6H_2O$, s. al
12	BaS *	Barium sulphide	169.43	>2000		4.35 Ro	wh; cub; n = 2.155, R: HCl, HNO_3
13	B_2S_3	Boron sulphide	117.84	310		1.55	wh; need; d.w and al; s. PCl_3
14	Al_2S_3	Aluminium sulphide	150.16	1100	d	2.37	wh; hex; d.w, i. al
15	Tl_2S	Thallium-I sulphide	440.85	449	d	8.40	blk; rhomb; v. sl. s.w
16	Tl_2S_3	Thallium-II sulphide	504.98	~260	d		blk; d.w, R: a
17	SiS	Silicon sulphide	60.16		s940	1.853	yel; d.w
18	GeS_2	Germanium sulphide	137.73	s >600°		3.09	wh; rhomb; d.w; R: KOH; NH_4OH
19	SnS	Stannous sulphide	150.77	880	1230	5.27	brsh. blk; rhomb; i.w; R: conc. HCl
20	SnS_2	Stannic sulphide	182.83	decomp		4.49	yel; rhomb; i.w; R: a alk
21	PbS	Lead sulphide	239.28	1114	(1290)	7.58	brsh.blk; cub; n=3.912, i.w; R: HNO_3 ✻H2.5
22	As_2S_3	Arsenic sulphide	246.02	300	707	3.43	yel; monocl; i.w; R: alk; ✻ H = 1.5–2
23	As_4S_4	Arsenic-III-V sulphide	427.90	320	565	3.20	red; monocl; R: a, alk; ✻ H = 1.5 – 2
24	As_2S_5	Arsenic-V sulphide	310.15		subl.		yel; monocl; R: alk;
25	Sb_2S_3	Antimony-III sulphide	339.72	546		grey 4.65 red 4.15	rhomb; R: HCl ✻H2
26	Sb_2S_5	Antimony-V sulphide	403.85	d		4.120	or; R: alk
27	Bi_2S_3	Bismuth-III sulphide	514.18	727		6.82	br; rhomb; R: HNO_3; ✻H2
28	Cu_2S	Cuprous sulphide	159.15	1127	d	6.02	blsh.blk; hex; (rhomb) v.sl R:HCl_3✻H2.5–3
29	CuS	Cupric sulphide	95.61	d 200		4.64	blk; hex; s. KCN solv; ✻H1.5–2
30	Ag_2S	Silver sulphide	247.83	842	d	7.317	blk; rhomb; s. KCN sol; ✻ H2–2.5
31	Au_2S	Gold-I sulphide	426.07	240			dk.br; R: aq.reg.s. KCN sol; i.w. colloid
32	ZnS	Zinc sulphide	97.45	1850at 150atm	1185	Zn Bl 4.04 wt 4.01	wh; cub; ⎧ n = 2.378, 2.356 ⎫ R: a; hex ⎨ ✻ H3.5–4 ⎬
33	CdS	Cadmium sulphide	144.48	1750(D) s	980	4.82	yel; cub; or hex; R: a ✻ 3–3.5
34	Hg_2S	Mercurous sulphide	433.29	d 0			blk; R: HNO_3
35	HgS	Mercuric sulphide	232.68	s 580		8.13 red; 7.75	red; trig; ⎰ n = 2.854, 3.201 blk; cub;⎱✻2–2.5
36	TiS_2	Titanium sulphide	112.03	d 300		~3.22	yel; rhomb; R: HNO_3
37	V_2S_2	Vanadium-II sulphide	166.03	d ~1800		4.20	blk; hex; R: h. H_2SO_4, HNO_3
38	V_2S_3	Vanadium-III sulphide	198.10	d ~ 700		4.7	gr.blk; R: h. H_2SO_4, alk;
39	V_2S_5	Vanadium-V sulphide	262.23			3.00	blk; R: HNO_3; alk;
40	MoS_2	Molebdenum-IV sulphide	160.08	1185		4.8	blk; hex; R: aq reg: h. conc H_2SO_4 ✻H1–1.5
41	MoS_3	Molebdenum-V sulphide	192.15	d			dk.br; R: aq.reg: alk. sulphides
42	WS_2	Tungsten-IV sulphide	247.99	d >1150			grsh. blk; hex R: HNO_3 + HF; alk sulphides
43	WS_3	Tungsten-VI sulphide	280.06				R: HNO_3 + HF; alk sulphides
44	MnS	Manganese sulphide	87.01	1530		4.06 3.27 3.26	α=406 grn; cub; β=3.27 red; cub; R: a; ✻H3.5–4 γ=3.26 red; hex;
45	ReS_2	Rhenium-IV sulphide	250.35	d >1000			blk; trig; no R: HCl, H_2SO_4, alk
46	Re_2S_7	Rhenium-VII sulphide	410.56			4.87	blk; R: HNO_3
47	FeS	Ferric sulphide	87.92	1195		4.76	blk; hex; R: a ✻H4
48	Fe_2S_3	Ferrous-III sulphide	207.90	d		4.3	blk; R: a
49	FeS_2	Ferrous disulphide	119.98	1171		5.00 4.84	yel; ⎰cub; Pyrite ✻ H5.5–6 ⎱ R: a ⎰rhomb; Marcasite ✻ H6–6.5 ⎱
50	CoS	Cobalt-II sulphide	91.006	~1100		5.95	blk; hex; R: a
51	NiS	Nickel sulphide	90.78	d 790	d	α=5.19 β=5.60	blk; ⎰trig; v. sl. R: a; ✻ H3.5 ⎱hex;
52	PdS	Palladium sulphide	138.47	d 950		6.60	blk; tet; R: HCl
53	PtS	Platinum-II sulphide	227.16	d		10.06	silver gr; tet; i.w; no R: a, alk
54	PtS_2	Platinum-IV sulphide	259.22	d 225–50		7.66	silver gr; rhomb; R: HNO_3
55	U_2S_3	Uranium-III sulphide	572.34			8.80	blk; rhomb; R: aq.reg.
56	US_2	Uranium-IV sulphide	302.20	>1100		7.90	gr; tet; R: conc. HNO_3
	CS_2	Carbon disulphide	76.143	− 112.1	46.2	1.261	col; n = 1.629; misc.al; eth; sl. m.w;
	P_4S_3	Phosphorus trisulphide	220.10	172.5	407.5	2.03	yel; rhomb; s. CS_2; d.w
	P_2S_5	Phosphorus pentasulphide	222.28	250	515	2.09	pa. yel; cr; s. CS_2. d.w, al
	P_4S_7	Tetra phosphorus hepta sulphide	348.36	310	523	2.19	pa. yel; cr; d.w → H_2S

No.	Formula	Name	Mol. wt.	MP/FP	BP/CP	D/sp.gr	Colour and other properties
			Selenides				
1	SnSe	Tin selenide	197.66	650	—	6.179	steel gr; Cr; i.w. R: HCl, HNO₃, aq. reg
2	SnSe₂	Tin diselenide	276.62	861	—	5.133	Cr; i.w. R: h conc. alk;
3	PbSe	Lead selenide	286.17	1065	—	8.10	cub; R: HNO₃
4	As₂Se₃	Arsenic selenide	386.70	360	—	4.75	br; cr; R: alk;
5	Sb₂Se₃	Antimony selenide	480.40	611	—	—	gr; cr; v.e. s.w
6	Bi₂Se₃	Bismuth selenide	654.88	710	dec	6.82	blk; rhomb; no R: alk;
7	Cu₂Se	Copper selenide	206.04	1113	—	6.749	blk; cub; R: HCl
8	Ag₂Se	Silver selenide	294.72	880	dec	8.0	gr; cub; R: conc HNO₃, s. NH₄OH
9	ZnSe	Zinc selenide	144.34	—	—	5.42	cub; n = 2.89, R: a
10	CdSe	Cadmium selenide	191.37	>1350	—	5.81	gr. br; hex, red powd; R: a
11	CoSe	Cobalt selenide	137.90	—	—	7.65	yel; hex;
12	NiSe	Nickel selenide	137.65	—	—	8.64	wh. gr; cub; R: HNO₃, aq. reg
13	PdSe	Palladium selenide	185.66	< 960	—	—	dk. gr; i.w
			Tellurides				
14	SnTe	Tin telluride	246.31	780	—	6.48	gr; i.W, R: alk, sulphides
15	SnTe₂	Tin ditelluride	373.92	—	—	—	blk; fla; R: alk, alk. sulphides
16	PbTe	Lead telluride	334.82	917	—	8.16	wh; cub;
17	Sb₂Te₃	Antimony telluride	626.35	629	—	—	gr;
18	Bi₂Te₃	Bismuth telluride	800.83	573	—	7.7	gr; rhomb
19	Ag₂Te	Silver telluride	343.73	955	—	8.05	gr; monocl; R: HNO₃, s. KCN sol
20	ZnTe	Zinc telluride	192.99	1238	—	~6	red; cub; n = 3.56, R: a
21	CdTe	Cadmium telluride	240.02	1041	—	6.20	blk; cub; R: HNO₃
			Borides				
22	CaB₆	Calcium boride	105.00	—	—	2.33	blk; cub; R: HNO₃; conc H₂SO₄
23	SrB₆	Strontium boride	152.55	2235	—	3.773	blk; cr; R: HNO₃
24	BaB₆	Barium boride	202.28	2270	—	4.36	blk; metallic lust; R: HNO₃
25	AlB₂	Aluminium boride	48.61	—	—	5.19	hex;
26	ZrB	Zirconium boride	112.86	~3000	—	6.085	hex; H 9
27	CrB	Chromium boride	62.83	2760	—	6.17	silver gr; (O) rhomb; R: HgNa₂O₂
28	WB₂	Tungsten boride	205.55	~2900	—	12.75	hex; R: aq. reg.
29	MnB₂	Manganese boride	76.57	—	—	6–9	gr. viol; cr; d.w; R: a
30	FeB	Iron boride	66.67	—	—	7.15	gr; cr; R: HNO₃; conc H₂SO₄
31	CoB	Cobalt boride	69.76	—	—	7.25	prism; d.w; R: HNO₃
32	NiB	Nickel boride	69.51	—	—	7.39	prism; wh. gr; R: HNO₃, aq. reg.
33	LaB₆	Lanthanum boride	203.84	2210	—	2.61	red; cub; metallic lust
34	CeB₆	Cerium boride	205.05	2190	—	—	bl. cub; metallic lust.
35	ThB₆	Thorium hexaboride	297.04	2195	—	6.4	blk; cub; mettalic lust; R: HNO₃; no R: a
36	UB₂	Uranium boride	248.99	2365	—	12.7	hex;

Carbides

	1	2	3	4	5	6	7	
II	Li₂C₂	Be₂C						YC₂
III	Na₂C₂	Mg₂C₃	Al₄C₃	SiC				LaC₂ CeC₂ NdC₃
IV	K₂C₂	CaC₂						PrC₂ SmC₃
V		SrC₂						ThC₂
VI		BaC₂						UC₂
IV	CuC₂			TiC	VC	Cr₃C₂	Mn₃C	Fe₃C ... Ni₃C
V	Ag₂C₂			ZrC	NbC	MoC Mo₂C		
VI				HfC	TaC	WC W₂C		

(a Main groups / b Sub groups)

No.	Formula	Name	Mol. wt.	MP/FP	BP/CP	D/sp.gr	Colour and other properties
37	Li₂C₂	Lithium carbide	37.90	—	—	1.65	wh. sh. powd; d.w → C₂H₂
38	Na₂C₂	Natrium carbide	70.00	—	700	1.575	wh. sh. powd; d.w, a → C₂H₂
39	Be₂C	Beryllium carbide	30.04	>2100d	—	1.90	yel; hex; d.w → CH₄
40	Mg₂C₃	Magnesium carbide	84.67				d.w → allene CH₂ = C = CH₂
41	CaC₂	Calcium carbide	64.10	stable	—	2.20	wh; tet; d.w → C₂H₂ glae a; and H₂SO₄
42	SrC₂	Strontium carbide	111.65	—	—	3.2	blk; tet; d.w → C₂H₂. d.a
43	BaC₂	Barium carbide	161.38	—	—	3.75	gr. tet; d.w → C₂H₂; d.a
44	Al₄C₃	Aluminium carbide	143.91	~1400	d	2.36	yelsh. gr; hex; d.w → CH₄; d.a

Note on the structure of borides, carbides and silicates: In crystal lattice (similar as in the case of silicates) individual atoms—groups—chains—networks yield three dimensional combinations in increasing proportion. Metal: nonmetal also in some cases give saline type lattices.

No.	Formula	Name	Mol. wt.	MP/FP	BP/CP	D/sp. gr.	Colour and other properties
1	Cu_2C_2	Copper carbide	151.10	—	—	—	red, i.w, s. KCN·sol. R: a, **exp.**
2	Ag_2C_2	Silver carbide	239.78	—	—	—	wh; i.w, e.s. Al; R: a, **exp.**
3	SiC	Silicon carbide	40.07	2600	2000	3.217	col; or blk, hex cub, R: KOH ⚒ H 9.5
4	TiC	Titanium carbide	59.91	3150	4300	4.93	gr. cub; metal lust; R: HNO_3, aq. reg. ⚒ H 8–9
5	ZrC	Zirconium carbide	103.23	3540	5100	6.73	gr. cub; metal lust; R: alk; dil HF; ⚒ H 8–9
6	VC	Vanadium carbide	62.96	2810	3900	5.77	blk cub; R: HNO_3, fused KNO_3
7	NbC	Niobium carbide	104.91	3900	—	7.82	blk; no R: a; R: HNO_3 + HF
8	TaC	Tantalum carbide	192.87	3880	5500	14.65	blk cub; R: H_2SO_4 + HF
9	Cr_3C_2	Chromium carbide	180.05	1890	3810	6.68	gr; cr; R: HCl
10	MoC	Molybdenum carbide	107.96	2692	—	8.78	blk; hex; R: a, no R: alk
11	Mo_2C	Dimolybdenum carbide	203.91	2687	—	9.18	blk; hex; R: conc a; no R: alk
12	WC	Tungsten carbide	195.93	~2870	6000	15.63	blk; hex; R: HNO_3 + HF; aq. reg. ⚒ H 9
13	W_2C	Ditungsten carbide	379.85	2860	6000	17.15	blk; hex; R: aq. reg. ⚒ H 9–10
14	Mn_3C	Manganese carbide	176.80	—	—	6.98	d.w; R: a → CH_4 and H_2
15	Fe_3C	Iron carbide	179.56	1837	—	7.4	gr. cub; R: a → CH_4 and H_2
16	Ni_3C	Nickel carbide	188.08	—	—	7.957	R: a → CH_4 and H_2
17	YC_2	Yttrium carbide	112.94	—	—	4.13	yel; cr, d.w, → C_2H_2; etc.
18	LaC_2	Lanthanum carbide	162.94	—	—	5.02	yel. cr; R: H_2SO_4 → C_2H_2, etc.
19	CeC_2	Cerium carbide	164.15	—	—	5.23	red, hex; d.w, R: a → C_2H_2; etc.
20	NdC_2	Neodymium carbide	168.29	d	—	5.25	yel; R: a → C_2H_2; etc.
21	PrC_2	Praesodymium carbide	164.94	d	—	5.10	yel; cr; d.w; R: a → C_2H_2, etc.
22	SmC_2	Samarium carbide	174.45	—	—	5.86	yel; hex; d.w; R: a → C_2H_2, etc.
23	ThC_2	Thorium carbide	256.14	2773	5000	8.96	yel; tet; d.w; R: conc a → C_2H_2, etc.
24	UC_2	Uranium carbide	262.09	2260	4100	11.28	gr; cr; R: a → C_2H_2, etc.
		Silicides					
25	Mn_2Si	Dimanganese silicide	137.92	1316	—	6.20	Quadr. Prism; R: HCl, NaOH, no R: HNO_3
26	MnSi	Manganese silicide	82.99	1280	—	5.90	gr. tet; R: HF, v. sl. R: alk
27	FeSi	Iron silicide	83.91	—	—	6.1	yelsh. gr; oct, i.w, no R: aq. reg.
28	CoSi	Cobalt silicide	87.00	1395	—	6.30	rhomb;
29	$CoSi_2$	Cobalt disilicide	115.06	1277	—	5.3	dk. bl; cub;
30	Ni_2Si	Nickel silicide	145.44	1309	—	7.2	i.W; no R: alk
		Phosphides*					
31	Na_3P	Sodium phosphide	99.97	d	—	—	red; d.w → PH_3
32	Ca_3P_2	Calcium phosphide	182.20	>1600	—	2.238	red cr; d.w → PH_3; R: a
33	Cu_3P_2	Copper phosphide	252.58	d	—	6.67	gr. blk; metal lust; R: HNO_3
34	Au_2P_3	Gold phosphide	487.34	d	—	6.07	gr;
35	Zn_3P_2	Zinc phosphide	258.10	> 420	1100	4.55	dk. gr; cub; R: H_2SO_4 → PH_3
36	ZrP_2	Zirconium phosphide	153.18	—	—	4.77	gr; i.w;
37	CrP	Chromium phosphide	82.99	—	—	5.7	gr. blk; cr; R: HNO_3, HF
38	MoP	Molybdenum phosphide	126.93	—	—	6.167	gr. grn; cr; R: conc HNO_3
39	MoP_2	Molybdenum diphos.	157.99	—	—	5.35	blk, powd; R: HNO_3, conc H_2SO_4, aq. reg.
40	W_2P	Ditungsten phosphide	398.82	—	—	5.21	dk. gr. prism; R: fused Na_2CO_3 + $NaNO_3$
41	WP	Tungsten phosphide	214.10	—	—	8.5	gr. prism; R: HNO_3 + HF
42	WP_2	Tungsten diphosphide	245.88	d	—	5.8	gr. cr; R: HNO_3 + HF, aq. reg.
43	MnP	Manganese phosphide	85.91	1190	—	5.39	dk. gr; i.w, R: HNO_3
44	Mn_3P_2	Trimanganese diphos.	226.75	1095	—	5.12	dk. gr; i.W, R: dil HNO_3.
45	FeP	Iron phosphide	86.83	—	—	6.07	rhomb;
46	Fe_2P	Diiron phosphide	142.68	1290	—	6.58	bl. gr; cr; R: HNO_3 + HF; aq. reg.
47	Co_2P	Cobalt phosphide	148.90	1386	—	6.4	gr; small need; i.W; R: HNO_3
48	Ni_2P	Dinickel phosphide	148.36	1112	—	6.31	gr; cr; R: HNO_3 + HF
49	Ni_3P_2	Trinickel phosphide	238.03	—	—	5.99	dk. gr. blk; R: HNO_3
50	Ni_5P_2	Pentanickel diphosphide	355.41	1185	—	—	need or plates, cr;
		Arsenides*					
51	Ca_3As_2	Calcium arsenide	270.06	d	—	3.031	red. cr; d.w, d. alk; R: h HNO_3
52	Ba_3As_2	Barium arsenide	561.90	—	—	4.1	br; d. F_2, Cl_2, Br_2
53	Zn_3As_2	Zinc arsenide	345.96	1015	—	—	cub; i.w;
54	Cd_3As_2	Cadmium arsenide	487.05	721	—	6.21	dk. gr; cub;
55	CrAs	Chromium arsenide	126.92	—	—	6.35	gr; hex; i.w; no R: a
56	Mn_2As	Dimanganese arsenide	184.77	1400	—	—	i.w; R: aq. reg.
57	Mn_3As_2	Trimanganese arsenide	314.61	—	—	—	magnetic, R: aq.reg.
58	FeAs	Iron arsenide	130.76	1020	—	7.83	wh; v.e. s.w
59	$FeAs_2$	Iron diarsenide	205.67	990	—	7.4	silver gr; cub; i.W, R: HNO_3, no R: HCl
60	NiAs	Nickel arsenide	133.60	968	—	7.57	hex; R: aq.reg.
61	$PtAs_2$	Platinum arsenide	345.05	>800d	—	10.602	tin wh; cub; i.w; no R: a

* The properties of phosphides and arsenides of transitional elements are intermediate to those of saline and metallic substances.

Compounds between three and more elements

Complexes out of one cation and one anion

Hydroxides

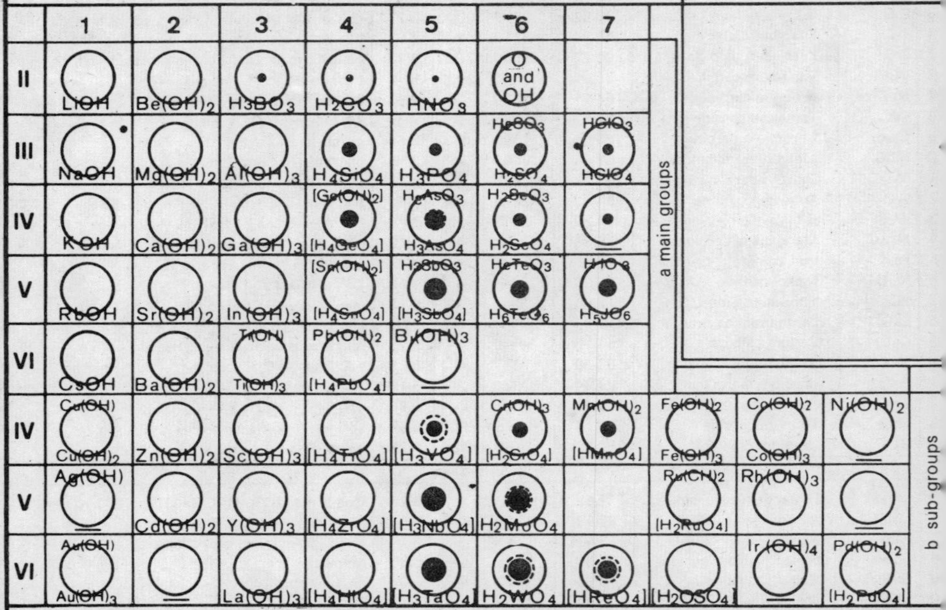

	1	2	3	4	5	6	7	
II	LiOH	Be(OH)₂	H₃BO₃	H₂CO₃	HNO₃	O and OH		a main groups
III	NaOH	Mg(OH)₂	Al(OH)₃	H₄SiO₄	H₃PO₄	H₂SO₃ / H₂SO₄	HClO₃ / HClO₄	
IV	KOH	Ca(OH)₂	Ga(OH)₃	[Ge(OH)₂] [H₄GeO₄]	H₃AsO₃ / H₃AsO₄	H₂SeO₃ / H₂SeO₄		
V	RbOH	Sr(OH)₂	In(OH)₃	[Sn(OH)₂] [H₄SnO₄]	H₃SbO₃ [H₃SbO₄]	H₆TeO₃ / H₆TeO₆	HIO₃ / H₅IO₆	
VI	CsOH	Ba(OH)₂	Tl(OH)₃	Pb(OH)₂ [H₄PbO₄]	Bi(OH)₃			

	1	2	3	4	5	6	7	8	9	10	
IV	Cu(OH) / Cu(OH)₂	Zn(OH)₂	Sc(OH)₃	[H₄TiO₄]	[H₃VO₄]	Cr(OH)₃ [H₂CrO₄]	Mn(OH)₂ [HMnO₄]	Fe(OH)₂ / Fe(OH)₃	Co(OH)₂ / Co(OH)₃	Ni(OH)₂	b sub-groups
V	Ag(OH)	Cd(OH)₂	Y(OH)₃	[H₄ZrO₄]	[H₃NbO₄]	H₂MoO₄		Ru(OH)₂ [H₂RuO₄]	Rh(OH)₃		
VI	Au(OH) / Au(OH)₃		La(OH)₃	[H₄HfO₄]	[H₃TaO₄]	H₂WO₄	[HReO₄]	[H₂OsO₄]	Ir(OH)₄	Pd(OH)₂ [H₂PdO₄]	

The black circle always represent the diameter of O^{2-} or OH^- ions and the red, blue and green circles represent the diameter of the concerned ionic atoms partially by different charges. (The higher the charge, the smaller the diameter.) ⊢—⊣ ≙ $1A = 10^{-8}$ cm

red ≙ acidic character
blue ≙ basic character
green ≙ amphoteric character

The properties of hydroxides extensively depend on the ratio of diameters of cation/anion. The sparingly soluble nature of a few hydroxides is connected with their lattice structure. (Compare also silicates pp. 70–73.)

Hydroxides

No	Formula	Name	Mol. wt.	MP/FP	BP/CP	D/sp. gr.	Colour and other properties
1	LiOH	Lithium hydroxide	23.95	461.8	d	1.43	wh; cr; 1H₂O; s.w; e.s. al
2	NaOH	Sodium hydroxide	40.00	318.4	1390	2.13	wh; rhomb; cub: deliq; e.s.w; al
3	KOH	Potassium hydroxide	56.11	360	1327	2.044	wh; cub. rhomb; deliq; v.e. e.s.w; s. al
4	NH₄OH	Ammonium hydroxide	35.05	—	—	s.S...	known only in aqueous solution
5	RbOH	Rubidium hydroxide•	102.49	300		3.203	wh; rhomb; deliq; v.e. s.w; s. al
6	CsOH	Caesium hydroxide	149.92	272.3		3.675	wh; deliq; v.e. s.w; e.s. al;
7	Be(OH)₂	Beryllium hydroxide	43.03	d300		1.909	wh; rhomb; i.w;s (NH₄)₂ CO₃sol; R: a, alk.
8	Mg(OH)₂	Magnesium hydroxide	58.34	d350		2.4	wh; rhomb; i.w; s. (NH₄)Cl sol; R: a �ęH 2.5
9	Ca(OH)₂	Calcium hydroxide	74.10	d580		2.24	wh; rhomb; v. sl. s.w; s. NH₄Cl sol; R: a
10	Sr(OH)₂	Strontium hydroxide	121.65	375		3.625	wh; deliq; v. sl. s.w; s. NH₄Cl sol; R: a
11	8H₂O	octahydrate	265.78	d100		1.90	wh; tet; deliq; v. sl. s.w; s. NH₄Cl sol. R: a
12	Ba(OH)₂	Barium hydroxide-				4.5 ow	sl. s.w; e.s. h.w; e.s. al; R: a
	8H₂O	octahydrate	315.51	d780		2.18	wh; monocl; s. 78° in w
13	Al(OH)₃	Aluminium hydroxide•	78.00	d150		2.423	wh; monocl; i.w; R: a, alk; ✦ H 2.5–3
14	AlO(OH)	Diaspore (Boehmite)	59.98	d300		≈3.4	wh; o-rhomb; i.w; R:conc. a;h alk, ✦H 6.5–
15	In(OH)₃	Indium hydroxide	165.78	d150			wh; i.w; no R: NH₄OH, R: M NaOH, a
16	Tl(OH)	Thallium-I hydroxide	221.39	d139			yel; rhomb; s.w; al; ≈ KOH, NaOH
17	Pb(OH)₂	Lead-II hydroxide	241.23	d145			wh; hex; v. sl. s.w; e.s. hw; R: a, alk
18	Bi(OH)₃	Bismuth-III hydroxide	260.02	d100	—	4.36	wh; i.w; d. hw; R: a h alk
19	BiO(OH)	Bismuth oxyhydrate		d450			br; i.w; R: a, h alk

54 • See also p. 42, Al₂O₃ ⇄ Al₂O₃·H₂O ≙ AlO(OH) ⇄ Al₂O₃·3H₂O ≙ Al(OH)₃ hydrargylite

No.	Formula	Name	Mol. wt.	MP/FP	D/sp. gr.	Colour and other properties
1	$Cu(OH)$	Copper-I hydroxide	80.55	d	3.37	yel; i.w; s. NH_4OH; R: a; Ox → $Cu(OH)_2$
2	$Cu(OH)_2$	Copper-II hydroxide	97.56	d	3.368	bl. cr; i.w; dec. h.w; s. NH_4OH, KCN sol; R: a
3	$AgOH$	Silver hydroxide				$Ag_2O + H_2O ⇌ 2Ag(OH)$ ~ 0.2 m mol/l
4	$AuOH$	Gold-I hydroxide	214.01	d 200		dk. viol; R: a, alk; s. KCN sol
5	$AuO(OH)$	Gold oxihydrate	230.01	d 250		yel. br; R: a; alk; s. KCN sol
6	$Zn(OH)_2$	Zinc hydroxide	99.40	d 125	3.05	wh; trig; rhomb; i.w; R: a, alk;
7	$Cd(OH)_2$	Cadmium hydroxide	146.43	d 300	4.79	wh; rhomb; i.w; s. NH_4OH; R: a; no R: alk.
8	$Sc(OH)_3$	Scandium hydroxide	96.12	d		wh; amor p; i.w; R: a, no R: alk.
9	$Y(OH)_3$	Yttrium hydroxide	139.94			yelsh. wh. jel; i.w; s. NH_4OH; R: alk; no R: alk;
10	$La(OH)_3$	Lanthanum hydroxide	189.94	d		wh. pond; i.w; R: a; no R: alk;
11	$Cr(OH)_2$	Chromium-II hydroxide	86.03			gr. R: a; OX → $Cr(OH)_2$
12	$Cr(OH)_3$	Chromium-III hydroxide	103.03			viol. jel; i.w; R: a, alk.
13	$Mn(OH)_2$	Manganese-I hydroxide	88.96	d	3.258	pa. rose; rhomb; i.w; s. NH_4Cl sol; no R: alk; R a ✿H 2.5
14	$MnO(OH)$	Manganese oxihydrate	87.95	d	4.2	blk. br; monocl; R: conc. H_2SO_4, HCl, ✿H 4
15	$Fe(OH)_2$	Iron-II hydroxide	89.87		3.4	wh; rhomb; i.w; R: a; no R: alk; OX → $FeO(OH)$
16	$Fe(OH)_3$	Iron-III hydroxide	106.87		3.4–3.9	red. br; i.w; R: a; no R: alk.
17	$FeO(OH)$	Iron oxihydrate	88.86		4.28	blk. br; rhomb; R: a; ✿H 5
18	$Co(OH)_2$	Cobalt-II hydroxide	92.96	d	3.597	rose red; rhomb; i.w; s. NH_4Cl sol; R: a
19	$Co(OH)_3$	Cobalt-III hydroxide	109.96	d		blk. br. powd; i.w; R: a
20	$Ni(OH)_2$	Nickel hydroxide	92.73		4.1	pa. grn. cr; i.w; s. NH_4OH; R: a.
21	$Ru(OH)_3$	Ruthenium-III hydroxide	152.12			blk. powd; v.e. s.w; R: a; no R: alk.
22	$Rh(OH)_3$	Rhodium hydroxide	153.93			yel; R: a, alk.
23	$Ir(OH)_4$	Iridium-IV hydroxide	260.23	d 350		bl. blk; R: a
24	$Pt(OH)_2$	Platinum-II hydroxide	229.11	d		blk; s. KCN, sol; R: HCl, HBr, alk;
25	$Pt(OH)_4$	Platinum-IV hydroxide	263.12	d 100		br; R: HCl, alk.

Oxygen compounds or Oxo acids

No.	Formula	Name	Mol. wt.	MP/FP	BP/CP	D/sp. gr.	Colour and other properties
26	$HClO$	Hypo chlorous acid	52.47	—	—	—	col; aq. sol. and salts very unstable
27	$HClO_2$	Chlorous acid	68.47	—	—	—	col; aq. sol. and salts very unstable
28	$HClO_3$	Chloric acid	84.47	—	d 40	1.282	col; known only in aq. sol.
29	$HClO_4$	Perchloric acid	100.47	– 112	39^{56} V	1.764	col; misc. w; conc. sol. ex.
30	$HBrO$	Hypobromous acid	96.92	—	d 40	—	col; yel; known only in aq. sol.
31	$HBrO_3$	Bromic acid	128.92		d 100		col; lt. yel, s. misc. w; dec. h.w;
32	HIO	Hypo iodous acid	143.92		d		col; lt. yel; known only in aq. sol.
33	HIO_3	Iodic acid	175.92		d 110	4.629	col; pa. yel; rhomb; v.e. s.w; s. al
34	H_5IO_6	Periodic acid (ortho)	227.94	d 140			wh; monocl; deliq. v.e. s.w; s. al, eth
35	H_2SO_3	Sulphurous acid	82.08		d		col; known only in aq. sol.
36	H_2SO_4	Sulphuric acid	98.08	10.5	d 338	1.834	col; hex; n = 1.429; ∞ misc. w
37	$H_2S_2O_7$	Pyro sulphuric acid	178.15	34.8	d	1.9	col; fumes on exposure, dec. w → H_2SO_4
38	H_2SeO_3	Selenous acid	128.98	d		3.004	col; hex; deliq. v.e. s.w; s. al ✿
39	H_2SeO_4	Selenic acid	144.98	58	d 260	2.951	col; hex; v.e. s.w; misc. H_2SO_4 ✿
40	H_6TeO_6	Telluric acid	229.66	d 300		3.07	col; monocl; s.w; ✿
41	HNO_2	Nitrous acid	47.02		d		col; known only in aq. sol.
42	HNO_3	Nitric acid	63.02	– 41.1	86	1.502	col; n = 1.397, ∞ misc. w; misc. al, (viol. sol.)
43	H_3PO_2	Hypo phosphorus acid	66.00	26.5	d	1.49	col; s.w; v.e. s.h.w, e. s. al, eth
44	H_3PO_3	Phosphorus acid	~82.00	72.6	d 200	1.651	col; deliq; cr; v.e. s.w; e.s. al
45	H_3PO_4	Ortho phosphoric acid	98.00	42.35	d 213	1.88	col; rhomb; v.e. s.w; s. al
46	$H_4P_2O_7$	Pyro phosphoric acid	177.98	61	d 300		col; need; hyg; v.e. s.w; dec. h.w. → H_3PO_4
47	$(HPO_3)×$	Metaphosphoric acid	(79.98)×		s	2.2-2.5	col; deliq; dec. w → H_3PO_4
48	H_3AsO_3	Arsenious acid	125.94		d		col; known only in aq. sol; ✿
49	H_3AsO_4	Arsenic acid (ortho)	150.94	35.5	d 160	2-2.5	wh; 1/2 H_2O; deliq; s.w; al; ✿
50	$H_3As_2O_7$	Pyro arsenic acid	265.85	d 206			col; cr; dec. w → H_2AsO_4 ✿
51	$HAsO_3$	Meta arsenic acid	123.92	d			wh; cr; dec. w → H_3AsO_4 ✿
52	H_2CO_3	Carbonic acid	62.03		d		col; known only in aq. sol;
53	H_4SiO_4	Silicic acid	96.12		d		col; only at pH 3.2 detd, → metaform (jel)
54	H_3BO_3	Boric acid	61.84	d 185		1.435	col; tricl; leaf; sl. s.w, al ✿H1
55	H_2MoO_4	Molybdic acid	161.97	d 115		3.112	wh; hex; e.s.w; s. al; s. H_2SO_4
56	H_2O	Molybdic acid monohydrate	179.98	d		3.124	yel; monocl; v.sl.s.w; 70°-H_2O → H_2MoO_4
57	H_2WO_4	Tungstic acid	249.94	d 100		5.5	yel; i.w; e.s. h.w; s. HF, i.a
58	H_2O	Tungstic acid monohydrate	267.95	d 100			wh; e.s.w; dec. → $H_2W_2O_7$

Peroxo acids

No.	Formula	Name	Mol. wt.	MP/FP	BP/CP	D/sp. gr.	Colour and other properties
59	H_2SO_5	Peroxo sulphuric acid	114.08	d 45			col; cr; dec w → O: s. H_2PO_4
60	$H_2S_2O_8$	Peroxo sulphuric acid	194.15	d 65	d		col; cr; dec, w → O: s. H_2SO_4, al, eth

Oxo complexes with non-metallic central atom

Salts of chloric acid and perchloric acid or chlorates and perchlorates

No.	Formula	Name	Mol. wt.	MP/FP	D/sp. gr.	Colour and other properties
1	$NaClO_3$	Sodium chlorate	106.44	255	2.490	wh; cub; n = 1.513; e.s.w; **ex** ☠
2	$KClO_3$	Potassium chlorate	122.56	356	2.32	wh; monocl; 400° KClO **ex** ☠
3	$(NH_4)ClO_3$	Ammonium chlorate	101.50	d	~1.8	wh; rhomb; s.w **ex** ☠
4	$Ba(ClO_3)_2H_2O$	Barium chlorate	322.29	414	3.18	wh; monocl; s.w ☠
5	$Pb(ClO_3)_2$	Lead chlorate	374.12	d	3.89	wh; monocl; v.e. s.w; s. al ☠
6	$AgClO_3$	Silver chlorate	191.34	230	4.43	wh; tetr; s.w ☠
7	$ZnClO_3 \cdot 4H_2O$	Zinc chlorate	304.36	d 60	2.15	wh; cub; v.e. s.w ☠
8	$LiClO_4 \cdot 3H_2O$	Lithium perchlorate	160.45	d 95	1.841	wh; hex; e.s.w; s. al
9	$NaClO_4$	Sodium perchlorate	122.44	d 482	2.50	wh; cub; rhomb; e.s.w; s. al; **ex**
10	$KClO_4$	Potassium perchlorate	138.56	d 610	2.52	wh; rhomb; cub; sl.s.w; v.sl. s. al **ex**
11	$(NH_4)ClO_4$	Ammonium chlorate	117.50	d 130	1.95	wh; cub; rhomb; n = 1.48; s.w; **ex**
12	$Mg(ClO_4)_2$	Magnesium perchlorate	223.23	d 251	2.60	wh; e.s.w; drying agent **ex**
13	$Ba(ClO_4)_2$	Barium perchlorate	390.32	505	2.74	wh; hex; v.e. s.w; e.s. al; **ex**
14	$AgClO_4$	Silver perchlorate	207.37	d 486	2.806	wh; cub; v.e. s.w; s. al; **ex**

Salts of bromic acid or bromates

No.	Formula	Name	Mol. wt.	MP/FP	D/sp. gr.	Colour and other properties
15	$NaBrO_3$	Sodium bromate	150.90	381	3.30	wh; cub; e.s.w; n = 1.594
16	$KBrO_3$	Potassium bromate	167.02	d 434	3.24	wh; rhomb; s.w; sl. s. al;
17	$AgBrO_3$	Silver bromate	235.80	d	5.206	wh; tetr; v.sl. s.w; s. NH_4OH
18	$Zn(BrO_3)_2$	Zinc bromate	429.31	d 100	2.57	wh; cub; n = 1.5467, sl.s.w;

Salts of iodic acid and periodic acid or iodates and periodates

No.	Formula	Name	Mol. wt.	MP/FP	D/sp. gr.	Colour and other properties
19	$NaIO_3$	Sodium iodate	197.90	d	4.277	wh; rhomb; e.s.w
20	KIO_3	Potassium iodate	214.01	560	3.990	wh; monocl; s.w; i. al
21	$(NH_4)IO_3$	Ammonium iodate	192.95	d 150	3.309	wh; rhomb; v.e. s.w
22	$Ba(IO_3)_2 \cdot H_2O$	Barium iodate	502.20	d	5.23	wh; monocl; v. sl. s.w
23	$AgIO_3$	Silver iodate	282.79	d 200	5.525	wh; rhomb; i.w; s. NH_4OH
24	$NaIO_4 \cdot 3H_2O$	Sodium periodate	267.95	d 175	3.22	wh; trig; n = 1.7745, s.w;
25	KIO_4	Potassium periodate	230.01	~582	3.618	wh; tetr; n = 1.621; 1.6479; v.sl.s.v

Salts of thiosulphuric acids and polythionic acids

No.	Formula	Name	Mol. wt.	MP/FP	D/sp. gr.	Colour and other properties
26	$Na_2S_2O_3 \cdot 5H_2O$	Sodium thiosulphate	248.19	d 48	1.685	wh; monocl; e.s.w; 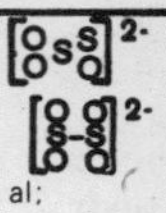
27	$K_2S_2O_3 \cdot \frac{1}{3}H_2O$	Potassium thiosulphate	196.34	d 180	2.23	wh; monocl; e.s.w;
28	$Na_2S_2O_4 \cdot 2H_2O$	Sodium dithionate	210.15	d 52		wh; cr; s.w;
29	$Na_2S_2O_6 \cdot 2H_2O$	Potassium dithionate	242.17	d 267	2.189	wh; rhomb; e.s.w;
30	$K_2S_3O_6$	Potassium trithionate	279.39	d	2.308	wh; rhomb; e.s.w; d. al;
31	$Na_2S_4O_6 \cdot 2H_2O$	Sodium tetrathionate	325.27	d		wh; e.s.w; d. al;
32	$K_2S_4O_6$	Potassium tetrathionate	302.46	d	2.296	wh; monocl; e.s.w; d. al;
33	$K_2S_5O_6 1\frac{1}{2}H_2O$	Potassium pentathionate	361.56	d	2.112	wh; rhomb; e.s.w; d. al;
34	$K_2S_6O_6$	Potassium hexathionate	366.60	d		wh; e.s.w d. al;

The structure of polythionic acids (compare fig. on p. 69 isoheteropoly acids and silicic acids)

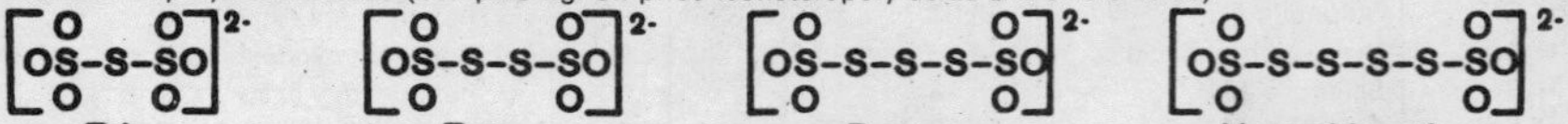

Tri- Tetra- Penta- Hexa-thionation

Salts of sulphurous acid or sulphites

No.	Formula	Name	Mol. wt.	MP/FP	D/sp. gr.	Colour and other properties
35	$Na_2SO_3 \cdot 7H_2O$	Sodium sulphite	252.16	d 150	1.561	wh; monocl; e.s.w → HSO_3^-
36	$Na_2S_2O_5$	Sodium pyrosulphite	190.11	d >150		wh; prism; e.s.w → HSO_3^-
37	K_2SO_3	Potassium sulphite	158.27	d		wh; hex; v.e. s.w → HSO_3^-
38	$KHSO_3$	Potassium hydrogensulphite	120.17			wh; e.s.w;
39	$K_2S_2O_5$	Potassium pyrosulphite	222.33	d 190	2.3	wh; monocl; e.s.w
40	$CaSO_3 \cdot 2H_2O$	Calcium sulphite	156.18	d		wh; hex; v.sl. s.w; no R: H_2SO_4
41	$BaSO_3$	Barium sulphite	217.43	d		wh; tetr; v.sl. s.w; no R: H_2SO_4

Salts of sulphuric acid or sulphites

No.	Formula	Name	Mol. wt.	MP/FP	D/sp. gr.	Colour and other properties
42	$Li_2SO_4 \cdot H_2O$	Lithium sulphate	127.96	857	2.06	wh; monocl; e.s.w
43	Na_2SO_4	Sodium sulphate	142.05	884	2.698	wh; rhomb; e.s.w; ✤ H 2.5
44	$Na_2SO_4 \cdot 10H_2O$	Glauber's salt	322.22	d32.364	1.464	wh; monocl; s.w; ✤ H 1.5
45	$NaHSO_4$	Sodium hydrogensulphate	120.07	186	2.742	wh; tricl; e.s.w
46	$Na_2S_2O_7$	Sodium pyrosulphate	222.11	400.9	2.658	wh; cr; e.s.w → SO_4^{2-}
47	K_2SO_4	Potassium sulphate	174.27	1069	2.662	wh; rhomb; s.w.
48	$KHSO_4$	Potassium hydrogensulphate	136.17	210	2.24	wh; rhomb; monocl; v.e. s.w
49	$K_2S_2O_7$	Potassium pyrosulphate	254.33	>300	2.27	wh; need; s.w → SO_4^{2-}
50	$(NH_4)_2SO_4$	Ammonium sulphate	132.15	513p	1.769	wh; rhomb; i.w → ✤ H_2
51	$(NH_4)HSO_4$	Ammonium hydrogensulphate	115.11	144	1.78	wh; rhomb; e.s.w
52	Rb_2SO_4	Rubidium sulphate	267.03	1074	3.613	wh; rhomb; e.s.w
53	Cs_2SO_4	Caesium sulphate	361.89	1019	4.243	wh; rhomb; v.e. s.w

No	Formula	Name	Mol. wt.	MP/FP	D/sp.gr.	Colour and other properties
1	BeSO$_4$·4H$_2$O	Beryllium sulphate	177.14	d 100	1.713	wh; tetr; n = 1.472; e.s.w,
2	MgSO$_4$	Magnesium sulphate	120.39	1127	2.66	wh; rhomb; e.s.w; s. eth;
3	MgSO$_4$H$_2$O	Magnesium monohydrate	138.40		2.57	wh; monocl; n = 1.523; e.s.w; ✤ H 3–3.5
4	MgSO$_4$·7H$_2$O	Magnesium heptahydrate	246.50		1.68	wh; rhomb; n=1.433; e.s.w; s. al; ✤ H 2–2.5
5	CaSO$_4$	Calcium sulphate	136.15	1450	2.96	wh; rhomb; n = 1.569; v. sl. s.w; ✤ H3–3.5
6	CaSO$_4$·2H$_2$O	Calcium dihydrate	172.18	d 128	2.32	wh; monocl; n = 1.521; v. sl. s.w; ✤ H1.5–2
7	SrSO$_4$	Stromtium sulphate	183.70	~1600	3.96	wh; rhomb; n = 1.622; i.w; ✤ H 3–3.5
8	BaSO$_4$	Barium sulphate	233.43	1350	4.50	wh; rhomb; n = 1.637; i.w; ✤ H 2–2.5
9	RaSO$_4$	Radium sulphate	322.12			wh; rhomb; i.w; co-precipitation with BaSO$_4$
10	Al$_2$(SO$_4$)$_3$	Aluminium sulphate	342.16	d 760	2.71	wh; rhomb; n = 1.47; e.s.w;
11	Al$_2$(SO$_4$)$_3$ 18H$_2$O	Aluminium octadecahydrate	666.45	d	1.62	wh; monocl; n = 1.474; e.s.w
12	Ga$_2$(SO$_4$)$_3$ 18H$_2$O	Gallium sulphate	751.92			wh; cr; e.s.w
13	In$_2$(SO$_4$)$_3$	Indium-III sulphate	517.84		3.438	wh; cr; s.w
14	Tl$_2$SO$_4$	Thallium-I sulphate	504.85	632	6.77	wh; rhomb; n = 1.860; sl. s.w; ☠
15	SnSO$_4$	Stannous sulphate	214.77	d 360		wh; need; s.w
16	PbSO$_4$	Lead sulphate	303.28	~1084	6.34	wh; rhomb; n=1.877; i.w; s. conc. H$_2$SO$_4$; ✤ H 3
17	Pb(SO$_4$)$_2$	Lead-IV sulphate	399.34			wh; cr; s. al
18	Sb$_2$(SO$_4$)$_3$	Antimony sulphate	531.70	d		wh; need; s. H$_2$SO$_4$; d.W
19	CuSO$_4$	Cupric sulphate	159.61	d 550		wh; rhomb; n = 1.739; s.w; me. al ☠
20	CuSO$_4$·5H$_2$O	„ pentahydrate	249.69	d 110	2.286	bl; tricl; n = 1.514; s.w; al; me. al; ✤ H2.5 ☠
21	Ag$_2$SO$_4$	Silver sulphate	311.83	657	5.45	wh; rhomb; v.sl. s.w;
22	ZnSO$_4$·7H$_2$O	Zinc sulphate	287.56	d 280	1.966	wh; rhomb; n = 1.457; e.s.w; me. al; ✤ H 2–2.5
23	CdSO$_4$	Cadmium sulphate	208.48	1000	4.691	wh; rhomb; s.w
24	3CdSO$_4$·8H$_2$O	Cadmium sulphate octahydrate	769.56		3.09	wh; monocl; n = 1.565; e.s.w;
25	Hg$_2$SO$_4$	Mercurous sulphate	497.29	d	7.56	wh; monocl; v. sl. s.w; ☠
26	HgSO$_4$	Mercuric sulphate	296.68	d	6.47	wh; rhomb; d.w; ☠
27	Sc$_2$(SO$_4$)$_2$	Scandium sulphate	378.12		2.579	wh; cr; e.s. w
28	Y$_2$(SO$_4$)$_2$	Yttrium sulphate	466.04	d 1000	2.612	wh; s.w;
29	La$_2$(SO$_4$)$_3$ 9H$_2$O	Lanthanum sulphate	728.18	d	2.821	wh; hex; n = 1.564; sl. s.w;
30	Ti$_2$(SO$_4$)$_3$	Titanium-III sulphate	384.0			grn. cr; i.w; s. H$_2$SO$_4$
31	Ti(SO$_4$)$_2$	Titanium-IV sulphate	240.03	d 150		wh; hyg; v.e. s.w
32	TiOSO$_4$	Titanyl sulphate	159.97			wh; s.w; d. in hot water
33	ZrSO$_4$·4H$_2$O	Zirconium sulphate	355.42	d 380		wh; rhomb; v.e. s.w
34	VSO$_4$ 7H$_2$O	Vanadium sulphate	273.13			red violet; monocl; s.w
35	VOSO$_4$	Vanadium oxysulphate-IV	163.02			bl; s.w;
36	Cr$_2$(SO$_4$)$_3$ 18H$_2$O	Chromium-III sulphate	716.15	d 100	1.86	vlt; cub; n = 1.564; v.e. s.w; s. al
37	MnSO$_4$·4H$_2$O	Manganese sulphate	223.07	d 700	2.107	pale rose; monocl; n = 1.508; e.s.w
38	FeSO$_4$	Ferrous sulphate	151.92		3.346	pale grn; rhomb; s.w
39	FeSO$_4$·7H$_2$O	„ „ hepta hydrate	278.03	d 64	1.898	grn; monocl; n = 1.471; s.w; s.a. al; ✤ H$_2$
40	Fe$_2$(SO$_4$)$_3$	Ferric sulphate	399.90	d 480	3.097	yel; rhomb; n = 1.802; s.w (dw → Fe(OH)$_3$
41	CoSO$_4$·7H$_2$O	Cobalt sulphate	281.12	d 96.8	1.948	red; monocl; n = 1.477; e.s.w; s. al
42	NiSO$_4$	Nickel sulphate	154.78		3.68	pale grn; rhomb; e.s.w
43	Rh$_2$(SO$_4$)$_3$ 12H$_2$O	Rhodium-III sulphate	720.21			pale yel; cr; s.w
44	PdSO$_4$·2H$_2$O	Palladium-II sulphate	238.50	d		brn; s.w; d. in h w
45	Ce$_2$(SO$_4$)$_3$·8H$_2$O	Cerium-III sulphate	712.59	d 630	2.886	wh; tricl; s.w;
46	Ce(SO$_4$)$_2$·4H$_2$O	Cerium-IV sulphate	404.33			yel; rhomb; s. H$_2$SO$_4$
47	Pr$_2$(SO$_4$)$_3$	Praseodymium sulphate	570.04		3.72	pale grn; s.w
48	Nd$_2$(SO$_4$)$_3$·8H$_2$O	Neodymium sulphate	720.87		2.85	rose; monocl; n = 1.541; s.w
49	Sm$_2$(SO$_4$)$_3$·8H$_2$O	Samarium sulphate	733.03	d 105	2.93	pale yel; monocl; n = 1.543; sl. s.w
50	Eu$_2$(SO$_4$)$_3$·8H$_2$O	Europium sulphate	736.3	d 375		pale rose; monocl; s. s.w
51	Gd$_2$(SO$_4$)$_3$	Gadolinium sulphate	602.72		4.139	wh; dimorph; sl. s.w
52	Tb$_2$(SO$_4$)$_3$·8H$_2$O	Terbium sulphate	790.19	d 360		wh; monocl; sl. s.w
53	Dy$_2$(SO$_4$)$_3$·8H$_2$O	Dysporsium sulphate	757.35	d 110		yel; monocl; s.w
54	Er$_2$(SO$_4$)$_3$·8H$_2$O	Erbium sulphate	766.87	d 400	3.18	rose; monocl; s.w
55	Yb$_2$(SO$_4$)$_3$	Yetterbium sulphate	634.28	d 900	3.793	wh; e.s.w; s. h w
56	Th(SO$_4$)$_2$	Thorium sulphate	424.18		4.225	wh; e.s.w
57	U(SO$_4$)$_2$·4H$_2$O	Uranium-IV sulphate	502.27	d 300		dk. grn; rhomb; ☠
58	UO$_2$SO$_4$·3H$_2$O	Uranyl sulphate	420.18	d 100	3.28	yelsh. grn; cr, s w, ☠

Salts of peroxodisulphuric acid

No	Formula	Name	Mol. wt.	MP/FP	D/sp.gr.	Colour and other properties
59	K$_2$S$_2$O$_8$	Potassium peroxodisulphate	270.33	d 100	2.447	wh; tricl; n = 1.4609, v. sl. s.w;
60	(NH$_4$)$_2$S$_2$O$_8$	Ammonium peroxodisulphate	228.21	d	1.982	wh; monocl; n = 1.498; e.s.w;

The structure of peroxo sulphuric acid

$$\begin{bmatrix} & O & O\text{-}O \\ O & S & \\ & O & O \end{bmatrix}^{2-}$$ peroxo sulphate ion

$$\begin{bmatrix} O & & & O \\ OS & O\text{-}O & SO & \\ O & & & O \end{bmatrix}^{2-}$$ peroxo disulphate ion

Salts of nitrous acid or nitrites

No.	Formula	Name	Mol. wt	MP/FP	D/sp.gr.	Colour and other properties
1	$NaNO_2$	Sodium nitrite	69.00	284	2.168	wh; rhomb; e.s.w; sl. s. al
2	KNO_2	Potassium nitrite	85.11	441	1.915	wh; monocl; v.e. s.w; i. al
3	$(NH_4)NO_2$	Ammonium nitrite	64.05	d		wh; s.w; al; OX on heating **ex**
4	$Ba(NO_2)_2$	Barium nitrite	247.39	d 217	3.173	wh; hex; e.s.w; s. al
5	$AgNO_2$	Silver nitrite	153.89	d 140	4.453	wh; rhomb; v. sl. s.w

Salts of nitric acid or nitrates

No.	Formula	Name	Mol. wt	MP/FP	D/sp.gr.	Colour and other properties
6	$LiNO_3$	Lithium nitrate	68.95	252	2.38	wh; rhomb; n = 1.737; e.s.w; s. al
7	$NaNO_3$	Sodium nitrate	85.00	312	2.2257	wh; rhomb; n = 1.587; e.s.w; ☀H1.5–2
8	KNO_3	Potassium nitrate	101.11	339	2.109	wh; rhomb; n = 1.335; e.s.w
9	$(NH_4)NO_3$	Ammonium nitrate	80.05	169.5	1.725	wh; rhomb; n=1.413; v.e. s.w; s. al Me; OX on heating **ex**
10	$RbNO_3$	Rubidium nitrate	147.49	305	3.13	wh; trig; n = 1.51; e.s.w;
11	$CsNO_3$	Caesium nitrate	194.92	417	3.685	wh; trig; n = 1.55; e.s.w
12	$Be(NO_3)_2$	Beryllium nitrate	187.07	d ~ 60		wh; v.e. s.w
13	$Mg(NO_3)_2$	Magnesium nitrate	256.43	d ~ 90	1.464	wh; monocl; e.s.w; s. al
14	$Ca(NO_3)_2 \cdot 4H_2O$	Calcium nitrate	236.16	d 42.5	1.82	wh; monocl; n = 1.465; v.e. s.w
15	$Sr(NO_3)_2$	Strontium nitrate	211.65	645	2.986	wh; cub; n = 1.567; e.s.w
16	$Ba(NO_3)_2$	Barium nitrate	261.38	592	3.24	wh; cub; n = 1.572; s.w
17	$Al(NO_3)_3 \cdot 9H_2O$	Aluminium nitrate	375.15	d 73		wh; rhomb; e.s.w; s. al; acet
18	$Tl(NO_3)$	Thallium-(I)-nitrate	266.40	207	5.56	wh; rhomb; s.w; ☠
19	$Tl(NO_3)_3$	Thallium-(III)-nitrate	444.46	d 100		wh; rhomb; d.w; ☠
20	$Pb(NO_3)_2$	Lead nitrate	331.23	d 200	4.53	wh; cub; n = 1.782; e.s.w; s. al
21	$Bi(NO_3)_3 \cdot 5H_2O$	Bismuth nitrate	485.10	d 30	2.83	wh; tricl; d.w
22	$BiO(NO_3)H_2O$	Bismuth oxinitrate	305.02	d 260	4.928	wh; hex
23	$Cu(NO_3)_2 \cdot 3H_2O$	Copper-(II)-nitrate	241.60	d 114.5	2.047	bl; v.e. s.w; s. al
24	$AgNO_3$	Silver nitrate	169.98	209	4.352	col; rhomb; v.e. s.w; s. al; me. al
25	$Zn(NO_3)_2 \cdot 6H_2O$	Zinc nitrate	297.49	d 36.4	2.065	wh; tetr; v.e. s.w; s. al
26	$Cd(NO_3)_2 \cdot 4H_2O$	Cadmium nitrate	308.49	59.5	2.455	wh; deliq; v.e. s.w
27	$Hg_2(NO_3)_2 \cdot 2H_2O$	Mercurous nitrate	561.26	d 70	4.79	wh; monocl; d.w; ☠
28	$Hg(NO_3)_2 \cdot H_2O$	Mercuric nitrate	342.64		4.3	wh; cr; s.w; ☠
29	$Sc(NO_3)_3$	Scandium nitrate	230.98	150		wh; s.w
30	$Y(NO_3)_3$	Yttrium nitrate	383.04		2.68	wh; tricl; v.e. s.w
31	$La(NO_3)_3 \cdot 6H_2O$	Lanthanum nitrate	433.04	d 40		wh; tricl; v.e. s.w
32	$Cr(NO_3)_3 \cdot 9H_2O$	Chromium-(III)-nitrate	400.18	d 37		vlt; s.w
33	$Mn(NO_3)_2 \cdot 6H_2O$	Manganese-(II)-nitrate	287.05	d 26	1.82	rose; monocl; v.e. s.w
34	$Fe(NO_3)_2 \cdot 6H_2O$	Ferrous nitrate	287.96	d 60.5		pale grn; e.s.w
35	$Fe(NO_3)_3 \cdot 9H_2O$	Ferric nitrate	404.02	d 47.2	1.684	col; monocl; e.s.w
36	$Co(NO_3)_2 \cdot 6H_2O$	Cobalt-(II)-nitrate	291.05	d 56	1.87	red; monocl; e.s.w
37	$Ni(NO_3)_2 \cdot 6H_2O$	Nickel nitrate	290.82	d 56.7	2.05	grn; monocl; e.s.w; s. al
38	$Ce(NO_3)_3 \cdot 6H_2O$	Cerium-(III)-nitrate	434.25	d 200		wh; tricl; s.w; al
39	$Ce(NO_3)_4$	Cerium-(IV)-nitrate	388.16			wh; d.w; (hydrolyses)
40	$Pr(NO_3)_3 \cdot 6H_2O$	Praseodymium nitrate	435.04	d 90		grn; s.w
41	$Sm(NO_3)_3 \cdot 6H_2O$	Samarium nitrate	444.47	d 78	2.375	pale yel; tricl; s.w
42	$Gd(NO_3)_3 \cdot 6H_2O$	Gadolinium nitrate	451.38	d 91	2.332	wh; tricl; e.s.w
43	$Tb(NO_3)_3 \cdot 6H_2O$	Terbium nitrate	453.05	d 89.3		wh; monocl; s.w
44	$Dy(NO_3)_3 \cdot 5H_2O$	Dysporsium nitrate	438.61	d 88.6		yel; tricl; s.w
45	$Th(NO_3)_4$	Thorium nitrate	480.08	d		wh; cr; s.w
46	$UO_2(NO_3)_2 \cdot 6H_2O$	Uranyl nitrate	502.18	d 59.5	2.81	yel; rhomb; v.e. s.w; al; eth, ae, a ☠

Salts of phosphorus acids—hypophosphites, phosphites; phosphates

No.	Formula	Name	Mol. wt	MP/FP	D/sp.gr.	Colour and other properties
47	$NaH_2PO_2 H_2O$	Sodium hypophosphite	106.00			wh; monocl; e.s.w; s. al
48	KH_2PO_2	Potassium hypophosphite	104.09	d		wh; hex; s.w
49	$Na_2(HPO_3) 5H_2O$	Sodium phosphite	216.05	53		wh; rhomb; v.e. s.w
50	$K_2(HPO_3)$	Potassium phosphite	158.18	d		wh; s.w;
51	$(NH_4)H(HPO_3)$	Ammo. hydrogenphosphite	99.03	~123		wh; monocl; v.e. s.w
52	Li_3PO_4	Lithium phosphate	115.80	~857	2.537	wh; rhomb; v. sl; s.w
53	$NaH_2PO_4 H_2O$	Sodiumdihydrogenphosphate	138.00	d 200	2.040	wh; rhomb; n = 1.456; e.s.w
54	$Na_2HPO_4 12H_2O$	Sodiumhydrogenphosphate	358.16	34.6	1.52	wh; monocl; n = 1.432; sl. s.w
55	$Na_3PO_4 12H_2O$	Sodiumphosphate	380.14	d 73.4	1.62	wh; rhomb; n = 1.446; s.w
56	$Na_4P_2O_7 10H_2O$	Sodiumpyrophosphate	446.07	972	1.82	wh; monocl; n = 1.450; s.w
57	$NaPO_3$	Sodium metaphosphate	101.96	619	2.476	wh; gl. mass; v.e. s.w
58	KH_2PO_4	Potassiumdihydrogenphosphate	136.09	252.6	2.338	wh; tetr; s.w
59	K_2HPO_4	Potassiumhydrogenphosphate	174.18	d ~340		wh; v.e. s.w; s. al
60	K_3PO_4	Pottassiumphosphate	212.28	1340	2.56	wh; rhomb; e.s.w
61	KPO_3	Pottassium metaphosphate	118.08	817	2.26	wh; tricl

No.	Formula	Name	Mol. wt.	FP		D/sp. gr.	Colour and other properties
1	K$_4$P$_2$O$_7$·3H$_2$O	Potassium pyrophosphate	384.40	1092		2.33	wh; s.w
2	(NH$_4$)H$_2$PO$_4$	Ammonium dihydrogen phosphate	115.03				wh; tetr; n = 1.525, e.s.w
3	(NH$_4$)$_2$HPO$_4$	Ammonium hydrogenphosphate	132.06	d		1.619	wh; monocl; n = 1.53; e.s.w; d on exposure to air
4	(NH$_4$)$_3$PO$_4$ 3H$_2$O	Ammonium phosphate	203.14				wh; s.w
5	MgHPO$_3$·3H$_2$O	Magnesium hydrogenphosphate	174.35			2.10	wh; rhomb; n = 1.514, e.s.w
6	Mg$_2$P$_2$O$_7$	Magnesium pyrophosphate	222.59	1383		2.598	wh; monocl; n = 1.602, i.w
7	Ca(H$_2$PO$_4$)$_2$ H$_2$O	Calcium dihydrogen phosphate	252.08	d	100	2.22	wh; tricl; n = 1.5292; v. sl. s.w
8	CaHPO$_4$ 2H$_2$O	Calcium dihydrogen phosphate	172.10			2.306	wh; monocl; v. sl. s.w
9	Ca$_3$(PO$_4$)$_2$	Calcium phosphate	310.19	1730		3.14	wh; rhomb; i.w
10	Ca(PO$_3$)$_2$	Calcium metaphosphate	198.02	975		2.82	wh; i.w; no R: a
11	BaHPO$_4$	Barium hydrogenphosphate	233.34			4.165	wh; rhomb; n = 1.617; i.w
12	Ba$_3$(PO$_4$)$_2$	Barium phosphate	602.03	~1730			wh; rhomb; i.w
13	AlPO$_4$	Aluminium phosphate	121.96	>1500		2.57	wh; rhomb; i.w; i. NH$_4$OH
14	Pb$_3$(PO$_4$)$_3$	Lead phosphate	811.58	~1014		6.99	wh; hex; n = 1.970; i.w
15	BiPO$_4$	Bismuth phosphate	303.98			6.323	wh; monocl; i.w
16	Ag$_3$PO$_4$	Silver phosphate	418.62	849		6.370	yel; cub; i.w; s. NH$_4$OH
17	Ag$_4$P$_2$O$_7$	Silver pyrophosphate	605.47	585			wh; i.w; s. NH$_4$OH
18	AgPO$_4$	Silver metaphosphate	186.96	~ 482		6.370	wh; cub; i.w; s. NH$_4$OH
19	Zn$_3$(PO$_4$)$_2$ 4H$_2$O	Zinc phosphate	458.15	900		3.04v 3.75t	wh; rhomb; tricl; i.w
20	Zn$_2$P$_2$O$_7$	Zinc pyrophosphate	304.71			3.756	wh; i.w
21	CrPO$_4$·6H$_2$O	Chromium-III phosphate	255.08	d	100	2.121	vlt; tricl; n = 1.568; e.s.w
22	Mn$_2$HPO$_4$·3H$_2$O	Manganese hydrogenphosphate	204.97	d	200		pale rose; rhomb; e.s.w
23	Mn$_2$P$_2$O$_7$	Manganese pyrophosphate	283.82	~1200		3.707	rose; monocl; n = 1.695; i.w
24	Fe$_3$(PO$_4$)$_2$·8H$_2$O	Ferrous-II phosphate	501.63			2.58	wh; monocl; n = 1.579; i.w; H1.5-2
25	FePO$_4$·2H$_2$O	Ferric-III phosphate	186.86	d		2.87 5.22 m	yelsh. wh; monocl; rhomb; n = 1.730; i.w; H3-4
26	CePO$_4$	Cerium-III phosphate	235.11			4.19 h	wh; monocl; hex; i.w; H5-5.5

	Salts of arsenic acids or arsenites and arsenates						
27	NaH$_2$AsO$_4$ H$_2$O	Sodium dihydrogen arsenate	181.93	d	130	2.67	wh; rhomb; monocl; n = 1.538; e.s.w
28	Na$_2$HAsO$_4$·7H$_2$O	Sodium hydrogen arsenate heptahydrate	312.01	d	57	1.871	wh; n = 1.462, s.w;
29	Na$_2$HAsO$_4$·12H$_2$O	„ decahydrate	402.09	d	22	1.72	wh; monocl; n = 1.445; s.w;
30	Na$_3$AsO$_4$·12H$_2$O	Sodium arsenate	424.07	d	85.5	1.759	wh; trig; n = 1.457; s.w;
31	KH$_2$AsO$_4$	Potassium dihydrogen arsenate	180.03	288		2.867	wh; tetr; n = 1.567; s.w;
32	Ca$_3$(AsO$_4$)$_2$ 3H$_2$O	Calcium arsenate	452.11	1450			wh; amor. powd; i.w;
33	Ag$_3$AsO$_3$	Silver arsenite	446.55	d	150		yel; i.w; s. NH$_4$OH
34	Ag$_3$AsO$_4$	Silver arsenate	462.55			6.657	brnsh. red; cub; i.w; s. NH$_4$OH

	Salts of carbonic acids or carbonates						
35	Li$_2$CO$_3$	Lithium carbonate	73.89	~ 730		2.11	wh; monocl; n = 1.428; sl.s.w; i. al
36	Na$_2$CO$_3$	Sodium carbonate	105.99	~ 850		2.533	wh; n = 1.415; e.s.w
37	Na$_2$CO$_3$·10H$_2$O	„ decahydrate	286.15	d	32	1.46	wh; monocl; n = 1.405; s.w
38	NaHCO$_3$	„ hydrogencarbonate	84.01	d	200	2.20	wh; monocl; n = 1.376; s.w
39	K$_2$CO$_3$	Potassium carbonate	138.21	897		2.43	wh; monocl; n = 1.426; i. al
40	KHCO$_3$	Potassium hydrogencarbonate	100.12	d	150	2.17	wh; monocl; n = 1.380; e.s.w
41	(NH$_4$)$_2$CO$_3$	Ammonium carbonate	114.11	d	58		wh; rhomb; e.s.w
42	(NH$_4$)HCO$_3$	Ammonium hydrogencarbonate	79.06	d	60	1.58	wh; rhomb; n = 1.423; s.w
43	Rb$_2$CO$_3$	Rubidium carbonate	230.97	837		3.468	wh; cr; v.e. s.w; sl. s. al
44	Cs$_2$CO$_3$	Casium carbonate	325.83	d	610	4.11	wh; v.e. s.w; s. al
45	MgCO$_3$	Magnesium carbonate	84.33	d	350	3.037 2.72 C	wh; rhomb; n = 1.700; i.w; s. NH$_4$Cl; H3.5-4
46	CaCO$_3$	Calcium carbonate calcite*	100.09	d	825	2.93 A	wh; rhomb; i.w; Aragonite; H3.5-4
47	SrCO$_3$	Strontium carbonate	147.64	d	1100	3.736	wh; rhomb; n = 1.516; i.w; H3.5-4
48	BaCO$_3$	Barium carbonate	197.37	d	1300	4.28	wh; rhomb; hex; n = 1.529; i.w; H3.5
49	Tl$_2$CO$_3$	Thallium-I carbonate	468.79	273		7.16	wh; monocl; s.w; melts without d
50	PbCO$_3$	Lead carbonate	267.22	d	300	6.60	wh; rhomb; n = 1.804; i.w; H 3-3.5
51	Ag$_2$CO$_3$	Silver carbonate	275.77	d	200	6.077	pale yel; i.w; s. NH$_4$OH
52	ZnCO$_3$	Zinc carbonate	125.39	d	140	4.42	wh; rhomb; n = 1.818; i.w; H4.5-5
53	CdCO$_3$	Cadmium carbonate	172.42	d	357	4.258	wh; rhomb; i.w
54	MgCO$_3$	Manganese-II carbonate	114.95			3.125	pale rose; rhomb; v.sl. s.w; H3.5-4.5
55	FeCO$_3$	Ferious-II carbonate	115.86	d		3.84	gr; rhomb; n = 1.875; i.w; H3.5-4.5
56	CoCO$_3$	Cobalt carbonate	118.95	d		4.13	pale red; rhomb; n = 1.855; i.w
57	NiCO$_3$	Nickel carbonate	118.72	d			pale grn; rhomb; i.w

	Salts of boric acid as well as perborates						
58	NaBO$_2$	Sodium metaborate	65.81	966		2.464	wh; rhomb; s.w
59	NaBO$_2$·H$_2$O$_2$ 3H$_2$O	Sodium perborate	153.88	d	63		wh; monocl; v.sl. s.w
60	Na$_2$B$_4$O$_7$·10H$_2$O	Sodium tetraborate	381.42	741		1.73	wh; monocl; 60.6° C-H$_2$O; n = 1.447; sl. s.w
61	K$_2$B$_4$O$_7$·5H$_2$O	Potassium tetraborate	325.56	815		1.74	wh; hex; e.s.w

* Calcite or calcespar occurs with ~ 1000 variations, among them is "icelands par" (double refraction separation in case of polarized light)

Oxocomplexes with metallic central atom (CA)

No.	CA	Formula	Name	Mol. wt.	FP	D/sp. gr.	Colour and other properties
1	V	$LiVO_3 \cdot 2H_2O$	Lithium metavanadate	141.92			yelsh. powd; s.w
2		$Na_3VO_4 \cdot 16H_2O$	Sodium orthovanadate	472.20	866		wh; Cr; need; e.s.w; d. in hw
3		$Na_4V_2N_7$	Sodium pyrovanadate	305.89	~ 650		hex; s.w
4		$NaVO_3$	Sodium metavanadate	121.95	630		col; monocl; s.w
5		KVO_3	Potassium metavanadate	138.05			col; cr; v.e. s.w
6		$NH_4 \cdot VO_3$	Ammonium metavanadate	116.99	d	2.326	yelsh. or col; v. sl. s.w
7		$Ba_2V_2O_7$	Barium pyrovanadate	488.62	863		wh
8	Cr	$Li_2CrO_4 \cdot 2H_2O$	Lithium chromate	165.92	d 150		gold yel; v.e. s.w
9		$Li_2Cr_2O_7 \cdot 2H_2O$	Lithium dichromate	265.93	d 130		blksh. brn; v.e. s.w
10		$Na_2CrO_4 \cdot 10H_2O$	Sodium chromate	342.16	19.92	1.483	yel; monocl; e.s.w; s. al
11		$Na_2Cr_2O_7 \cdot 2H_2O$	Sodium dichromate	298.05	320	2.52	red; monocl; v.e. s.w
12		$Na_3Cr_2O_8$	Sodium peroxychromate	249.00	d 115		or. pl; e.s.w
13		K_2CrO_4	Potassium chromate	194.20	968.3	2.732	yel; rhomb; e.s.w
14		$K_2Cr_2O_7$	Potassium dichromate	294.21	398	2.69	red; monocl; v.e. s.w
15		$K_2Cr_2O_8$	Potassium peroxy dichromate	297.30	d 170		brn; red; cub; e.s.w
16		$(NH_4)_2CrO_4$	Ammonium chromate	152.09	d	1.91	yel; monocl; s.w; s. al
17		$(NH_4)_2Cr_2O_7$	Ammonium dichromate	252.10	d	2.15	gold, yel; monocl; s.w; s. al
18		$(NH_4)_2Cr_2O_8$	Ammonium peroxy dichromate	234.13	d 40		red; brn; cub; e.s.w; 50° **ex**
19		$MgCrO_4 \cdot 7H_2O$	Magnesium chromate	266.44		1.695	yel; rhomb; v.e. s.w
20		$CaCrO_4 \cdot 2H_2O$	Calcium chromate	192.12	d 200		yel; monocl; v.sl. s.w; R. ac.a
21		$SrCrO_4$	Strontium chromate	203.64		3.895	yel; monocl; v.sl. s.w; R. ac.a
22		$BaCrO_4$	Barium chromate	253.37		4.498	yel; rhomb; i.w; no R. ac.a
23		$Ba_2Cr_2O_7$	Barium dichromate	353.38			red; monocl; e.s.w
24		$PbCrO_4$	Lead chromate	323.22	884	6.3	yel; monocl; i.w; a. no R: ac. a
25		$CuCr_2O_7 \cdot 2H_2O$	Copper dichromate	315.59	d 100	2.283	blk. cr, v.e. s.w; d in hw
26		Ag_2CrO_4	Silver chromate	331.77		5.625	red; monocl; i.w; s. NH_4OH, KCN sol
27		$Ag_2Cr_2O_7$	Silver dichromate	431.78	d	4.770	red; tricl; i.w; s. NH_4OH; KCN sol
28		$ZnCrO_4$	Zinc chromate	181.39			yelsh. prism; v.e. s.w; d in hw
29		$ZnCr_2O_7 \cdot 3H_2O$	Zinc dichromate	335.45			redsh. brn; v.e. s.w; d in hw
30		Hg_2CrO_4	Mercurous chromate	517.23	d		red; need. v.e.s.w; KCN sol
31		$HgCr_2O_7$	Mercuric chromate	316.62	d		red; rhomb; e.s.w; s. NH_4OH
32	Mo	$Na_2MoO_4 \cdot 2H_2O$	Sodium molybdate	241.98	687	3.28	wh; rhomb; e.s.w
33		$Na_2Mo_2O_7$	Sodium dimolybdate	349.89	612		wh; need; e.s.w
34		$K_2MoO_4 \times H_2O$	Potassium molybdate	...	919	2.342	wh; cr; v.e.s.w;
35		$(NH_4)_2MoO_4$	Ammonium molybdate	196.03	d	3.27	col; rhomb; s.w; d. in hw
36		$MgMoO_4$	Magnesium molybdate	184.27			col; rhomb; s.w
37		$CaMoO_4$	Calcium molybdate	200.03	561	2.36	col; v.e. s.w
38		$BaMoO_4$	Barium molybdate	297.31		~4.8	wh; tetr; v. sl. s.w
39		$PbMoO_4$	Lead molybdate	367.16	~1065	6–7	pl; i.w; R: conc. H_2SO_4 KOH ✿ H3
40	W	Li_2WO_4	Lithium tungstate	261.80	742		col; trig; e.s.w
41		$Na_2WO_4 \cdot 2H_2O$	Sodium tungstate	329.95	698	3.25	col; rhomb; e.s.w
42		$K_2WO_4 \cdot 2H_2O$	Potassium tungstate	362.14	921	3.113	col; monocl; e.s.w
43		$MgWO_4$	Magnesium tungstate	272.24		5.66	col; monocl; s.w
44		$CaWO_4$	Calcium tungstate	288.00		6.06	wh; tetr; v. sl. s.w; ✿ H4–5
45		$SrWO_4$	Strontium tungstate	335.55	d	6.187	tett; e.s.w
46		$BaWO_4$	Barium tungstate	385.28		5.04	col; tetr; e.s.w
47		$PbWO_4$	Lead tungstate	455.13	1123	8.2	col; monocl; i.w; ✿ H3
48		$CuWO_4 \cdot 2H_2O$	Cupric tungstate	347.49			grn; v.sl. s.R. ac. a., NH_4OH
49		$AgWO_4$	Silver tungstate	463.68			bright yel; cr; v.sl.s.w; NH_4OH
50		$CdWO_4$	Cadmium tungstate	360.33			yel; cr; v.sl. s.w; s. NH_4OH
51		Hg_2WO_4	Mercury-(I)-tungstate	649.14	d		yel; amor; i.w
52		$HgWO_4$	Mercury-(II)-tungstate	448.53	d		yel; d. in hw
53		$CoWO_4$	Cobalt tungstate	306.86		8.42	bl/grn; monocl; i.w
54	Mn	K_2MnO_4	Potassium manganate	197.12	d 190		grn; rhomb; d. in w
55		$KMnO_4$	Potassium permanganate	158.03	d 240		vlt; rhomb; s.w; me al; acet
56		$Mg(MnO_4)_2 \, 6H_2O$	Magnesium permanganate	370.28	d	~2.2	dk. purp. e.s.w; me al
57		$BaMnO_4$	Barium manganate	256.29		4.85	gr. grn; hex; e.s.w
58		$Ba(MnO_4)_2$	Barium permanganate	375.22			brn; vlt; cr; s.w
59		$AgMnO_4$	Silver permanganate	226.81			dk. vlt; monocl; i.w
60		$Zn(MnO_4)_2 \, 6H_2O$	Zinc permanganate	411.34	d 100	2.47	blk. vlt; e.s.w
61	Re	$NaReO_4$	Sodium perrhenate	273.31	d 300		col; hex; s.w; al
62		$Na_2Re_2O_7 \cdot H_2O$	Sodium pyrohyporhenate	594.62			yel; cr; i.w
63	Ru	$NaRuO_4 \cdot H_2O$	Sodium perruthenate	206.71	d		blk. cr; e.s.w
64		$K \, RuO_4 \cdot H_2O$	Potassium ruthenate	261.91	d 200		blk. cr; e.s.w

Oxohalogenides

No.	CA	Formula	Name	Mol. wt.	MP/FP	BP/CP	D/sp. gr	Colour and other properties
1	C	$COCl_2$	Carbon oxydichloride*	98.92	−104	8.3	1.392	col; d in w → HCl; s. Bz; ☠
2		$COBr_2$	Carbon oxydibromide	187.84		64.5	2.44	
3	Si	Si_2OCl_6	Silicon oxydichloride**	284.86	− 33	137		col; d in w; misc. Cs_2, CCl_4, acet
4	N	NOF*⁺	Nitrosyl fluoride	49.01	−134	− 56	2.176	col; d in w; s. H_2SO_4
5		NOCl	Nitrosyl chloride	65.47	− 64.5	− 5.5	2.99	yel; d in w; s. H_2SO_4
6		NOBr	Nitrosyl bromide	109.92	− 55.5	− 2		brn. d in w; R: alk
7		NO_2F	Nitril fluoride	65.01	−139	− 63.5	2.90	col; d in w
8		NO_2Cl	Nitril chloride	81.47	~ −30	5	2.57	yelsh. brn; d in w
9	P	POF_3	Phosphorus oxytrifluoride	103.98	− 68	− 39.8	4.69	col; d in w
10		$POCl_3$	Phosphorus oxytrichloride	153.35	2	105.3	1.675	col; n = 1.460, d in w
11		$POBr_3$	Phosphorus oxytribromide	286.73	56	193	2.822	col; leaves; s in org. sol.
12	Sb	SbOCl	Antimony oxychloride	173.22	d 170			wh; monocl; i.w. s. CS_2, tart a
13	Bi	BiOCl	Bismuth oxychloride	260.46			7.72	wh; tetr; R: a
14	S	SOF_2	Thionyl fluoride	86.07	−110	− 30		col; d in w; s. al
15		$SOCl_2$	Thionyl chloride	118.98	−105	79	1.655	col; n = 1.527; d in w
16		$SOBr_2$	Thionyl bromide	207.90	~ −50	138	2.68	or. yel; d in w; s. CS_2, CCl_4, B_2
17		SO_2F_2	Sulphuryl fluoride	102.07	−120	− 52	3.72	col; s. al
18		SO_2Cl_2	Sulphuryl chloride	134.98	− 54.1	69.1	1.6674	col; n = 1.444; d in w
19		S_2OCl_4	Disulphurmonoxytetrachloride	221.96		60	1.656	dk red; d in w; al
20		$S_2O_3Cl_4$	Disulphurtrioxytetrachloride	253.96	d 57			wh; rhomb; d in w; al
21		$S_2O_5Cl_2$	Pyrosulphuryl chloride	215.05	~ −38	140	1.818	col; n = 1.449; d in w
22	Se	$SeOF_2$	Selenium oxyfluoride	132.96	4.6	124	2.67	col; d in w; s. CCl_4, al
23		$SeOCl_2$	Selenium oxychloride	165.87	9	176.4	2.42	pale yel; n = 1.651; d in w
24		$SeOBr_2$	Selenium oxybromide	254.79	41.6	d ~200	3.38	yel red; cr; d in w, s in org. sol
25	Zn	$ZnOCl_2 \cdot 8H_2O$	Zirconium oxychloride	322.26	210			wh; tetr. s. w; al, 2 hot W
26		$ZnOBr_2 \times H_2O$	Zirconium oxybromide		d 120			deliq. need, v.e. s.w
27	Hf	$HfOCl_2 \cdot 8H_2O$	Hafnium oxychloride	409.64				wh; tetr; s.w
28	V	VOCl	Vanadium oxymonofluoride	102.41		127	3.64	yelsh brn; powd; i.w; R: HNO_3
29		VOBr	„ „ monochloride	146.87	d 480		4.00	vlt; oct; v.e. s.w; s. ac. a
30		VOF_2	„ „ difluoride	104.95	d		3.396	yel; e.s. ac. a
31		$VOCl_2$	„ „ dichloride	137.86			2.88	grn; deliq; d in w; R: HNO_3
32		$VOBr_2$	„ „ dibromide	226.78	d 180			brn; deliq; s.w
33		VOF_3	„ „ trifluoride	123.95	300	480	2.459	yelsh; hyg.
34		$VOCl_3$	„ „ trichloride	173.32	d		1.829	yel; v.e. s.w; s. al; eth. ac. a
35		$VOBr_3$	„ „ tribromide	306.70	d 180	~130 V	2.933	red; liq; s.w
36	Nb	$NbOCl_3$	Niobium oxychloride	215.28		s 400	10.19	col; need; d in w; R: H_2SO_4
37		$NbOBr_3$	Niobium oxybromide	348.66		s		yel; cr; d in w; R: a
38	Cr	CrO_2Cl_2	Chromyl chloride	154.92	− 96.5	117	1.911	dk red; d in w; s.eth; ac. a
39	Mo	$MoOCl_3$	Molybdenium oxytrichloride	218.32		s 100		grn; cr; d in w
40		$MoOF_4$	Molybdenium oxytetrafluoride	187.95	98	180	3.0	col; deliq. s.w; al; eth, CCl_4
41		$MoOCl_4$	Molybdenium oxytetrachloride	253.78		s		grn; cr; deliq; s.w
42		MoO_2Cl_2	Molybdenium dioxydichloride	198.86		s	3.31	yelsh; cr; s.w; al, eth
43		$Mo_2O_3Cl_5$	„ trioxypentachlorite	417.19		s		dk brn; cr; deliq; s.w
44		$Mo_2O_3Cl_6$	„ trioxyhexachloride	452.64	d			ruby red; cr; d in w; s. eth
45	W	WOF_4	Tungstenoxytetrafluoride	275.92	110	187.5		Cd. hyg; d in w; s. CS_2
46		$WOCl_4$	Tungsten oxytetrachloride	341.75	211	227.5		red. need; s. B_2 CS_2 S_2 Cl_2
47		$WOBr_4$	Tungsten oxytetrabromide	519.58	277	327		blk; deliq; d in w
48		WO_2Cl_2	Tungsten dioxydichloride	286.83	266		•	yel. pl; s.w; d in hot w; R: alk
49		WO_2Br_2	Tungsten dioxydibromide	375.75	d			red; prism
50	U	UO_2Cl_2	Uranyl chloride	340.98	d			yel; deliq; v.e. s.w; s. al, eth
51		UO_2Br_2	Uranyl bromide	429.90				yelsh; grn; need; hyg; v.e. s.w; s. ac. a
52		UO_2I_2	Uranyl iodide	523.91	d			red, deliq; s. al, eth
53	Re	$ReOF_4$	Rhenium oxytetrafluoride	278.31	39.7	62.7	4.032	col;
54		$ReOCl_4$	Rhenium oxytetrachloride	344.14	29	223		d in w
55		ReO_2F_2	Rhenium dioxydifluoride	256.31	156			col;
56		ReO_3Cl	Rhenium trioxychloride	269.77	4.5	131		col; d in w
57		ReO_3Br	Rehenium trioxybromide	314.23	39.5	163		wh;

Thiohalogens and oxosulphides

No.	CA	Formula	Name	Mol. wt.	MP/FP	BP/CP	D/sp. gr	Colour and other properties
58	C	$CSCl_2$	Carbon thiodichloride***	114.99		73.5	1.509	yelsh. red; d in H_2O
59		COS	Carbon oxysulphide	60.08	−138	− 49	2.72	s H_2O; e.s in alk
60	Zr	ZrOS	Zirconyl sulphide	139.29			4.87	yel; d in exposure to air;
61	U	UO_2S	Uranyl sulphide	302.14	d ~45			brn blk; s. H_2O

* Phosgene   ** The structure of Si_2OCl_6

*** Thiophosgene

$$\begin{array}{ccc} Cl & & Cl \\ Cl - Si - O - Si - Cl \\ Cl & & Cl \end{array}$$

Salts of hydrozoic acid, hydrogen cyanic acid, fulminic acid and thiocyanic acid

No.	Formula	Name	Mol. wt.	MP/FP		D/sp.gr.	Colour and other properties
1	$H-N=N=N$	Hydrazoic acid	43.03		$-$ 80	1.126	B.P. 37; col; ∞ misc. w; misc. al;
2	ClN_3	Chlorazide	77.48				gas. v.e. s.w; d in alk. **ex**
3	BrN_3	Bromazide	121.94		$\sim$ 45		red. s; KI sol; eth; **ex**
4	IN_2	Iodazide	168.94				yel; d in water. s. $Na_2S_2O_3$ sol; **ex**
5	NaN_3	Sodium azide	65.02	d		1.846	wh; rhomb; e.s.w; v.sl. s. al; i, eth; **ex**
6	KN_3	Potassium azide	81.12		350	2.04	wh; s.w; al; i. eth; **ex**
7	$(NH_4)N_3$	Ammonium azide	60.06	s	134	1.346	wh; pl; s.w; al; **ex**
8	$Pb(N_3)_2$	Lead azide	291.26	d	350		wh; prism; v.sl. s.w; **ex**
9	AgN_3	Silver azide	149.90	d	252		wh; rhomb; i.w; s, KCN sol; NH_4OH; **ex**
10	$Hg_2(N_3)_2$	Mercuric azide	485.27				wh; cr; v. sl. s.w; **ex**
11	$H-C\equiv N$	Hydrocyanic acid	27.03		-13.24	0.699	B.P. 25.7; col; ∞ misc. w; al; eth; ☠
12	$N\equiv C-C\equiv N$	Dicyanogen	52.04		$-$ 34	2.335	B.P. 21.2; col; s.w; (4.5); s. al; eth; ☠
13	$Cl-C\equiv N$	Cyan chloride	61.48		$-$ 6	1.186fl	B.P. 13, col; s.w; al; eth; ☠
14	$Br-C\equiv N$	Cyan bromide	105.93		52	2.015	B.P. 61.3, col; s.w; al ☠
15	$I-C\equiv N$	Cyan iodide	152.93	subl			F.P. 146.5; col; rhomb; s. al; ☠
16	$NaCN$	Sodium cyanide	49.01		562	1.857	B.P. 1487, wh; cub; s.w; n = 1.452 ☠
17	KCN	Potassium cyanide	65.12		634.5	1.52	wh; cub; e.s.w; sl. s. al; n = 1.410 ☠
18	$(NH_4)CN$	Ammonium cyanide	44.06	d	36		wh; tetr; s.w; al; ☠
19	$Ca(CN)_2$	Calcium cyanide	92.12		640		wh; rhomb; s.w; ☠
20	$Sr(CN)_2\cdot4H_2O$	Strontium cyanide	211.73	d			wh; v.e. s.w; al; ☠
21	$Ba(CN)_2$	Barium cyanide	225.43				wh; e.s.w; al; ☠
22	$Pb(CN)_2$	Lead cyanide	259.25				yel; e.s.w; s. KCN; ☠
23	$CuCN$	Cuprous cyanide	89.56		475	2.92	wh; monocl; i.w; R: a; s. KCN sol; ☠
24	$Cu(CN)_2$	Cupric cyanide	115.58	d			yel; grn; powd; i.w; s. KCN sol R: a ☠
25	$AgCN$	Silver cyanide	133.90	d	320	3.95	wh; rhomb; i.w; s. NH_4OH, KCN, sol
26	$AuCN$	Gold-I cyanide	223.02	d		7.12	yel; hex, v. sl. s.w; s. KCN sol
27	$Au(CN)_3\cdot3H_2O$	Gold-III cyanide	329.32	d	50		wh; cr; s.w; al; eth
28	$Zn(CN)_2$	Zinc cyanide	117.42	d	800		wh; cub; s. KCN sol; NH_4OH
29	$Cd(CN)_2$	Cadmium cyanide	164.65	d	$\sim$200		cr; sl. s.w; s. KCN sol; NH_4OH; ☠
30	$Hg(CN)_2$	Mercuric cyanide	252.65	d		4.00	wh; tetr; s.w; al; ☠
31	$Co(CN)_2\cdot3H_2O$	Cobalt cyanide	165.02	d	250 -280		grsh. red; powd; i.w; s. KCN sol
32	$Ni(CN)_2\cdot4H_2O$	Nickel cyanide	182.79	d	200		gr. pl; i.w; s. KCN sol R: a; ☠
33	$Pd(CN)_2$	Palladium cyanide	158.44	d			yelsh. wh; s. KCN sol; no R: a
34	$Pt(CN)_2$	Platinum cyanide	247.27				yelsh. brn; cr; s. KCN sol; no R: a
35	$HO-C\equiv N$	Cyanic acid*	43.03		$-$ 86	1.14	col; s. eth; acet. a
36	$H-N=C=O$	Isocyanic acid*	43.03				mesomeric form; known only in derivatives
37	$H-N-C=O$	Fulminic acid*	43.03			●	col; unstal, liq; **ex** ☠
38	$Na(OCN)$	Sodium cyanate	65.92			1.937	col; need; s.w; i. al; eth
39	$K(OCN)$	Potassium cyanate	81.12		d $\sim$800	2.048	wh; tetr; sl. s.w; i. al
40	$NH_4(OCN)$	Ammonium cyanate	60.06	d	60	1.342	wh; need; s.w; al; eth
41	$Pb(OCN)_2$	Lead cyanate	291.25	d			wh; need; i.w; s. in hw
42	$Hg(ONC)$	Mercuric fulminate	284.65			4.42	wh; cub; e.s.w; NH_4OH, al; **ex**
43	$Ag(ONC)$	Silver fulminate	299.80				wh; v. sl. s.w; s. NH_4OH; **ex**
44	$H-S-C\equiv N$	Thiocyanic acid	59.09	d $\to$ 10			col; gas; F.P. $-$110; v.e. s.w; al; eth
45	$H-N=C=S$	Isothiocyanic acid	59.09				mesomeric form, known only as derivatives
46	$N\equiv C-S-S-C\equiv N$	Dithiocyanogen	116.17		~-2 d		col; gas; deep yel; s. in org, sol
47	$Na(SCN)$	Sodium thiocyanate	81.07		323		wh; rhomb; v.e. s.w
48	$K(SCN)$	Potassium thiocyanate	97.18		179	1.886	wh; rhomb; d 500; v.e. s.w; s. al
49	$NH_4(SCN)$	Ammonium thiocyanate	76.12		159	1.305	wh; monocl; d 170; v.e. s.w; s. al
50	$Ca(SCN)_2\cdot3H_2O$	Calcium thiocyanate	210.30				wh; v.e. s.w; hyg.
51	$Sr(SCN)_2\cdot3H_2O$	Strontium thiocyanate	257.85	d	160		deliq. cr; v.e. s.w; s. al
52	$Tl(SCN)$	Thallium thiocyanate	262.47				col; tetr; v. sl. s.w; s. al
53	$Pb(SCN)_2$	Lead thiocyanate	323.88			3.82	wh; monocl; s. KCN sol; R: N_3H
54	$Cu(SCN)$	Cuprous thiocyanate	121.62	d	130	2.846	wh; i.w; s. NH_4OH; eth; R: a
55	$Cu(SCN)_2$	Cupric thiocyanate	179.71	d	100		blk; d in w; s. NH_4OH; R: a
56	$Ag(SCN)$	Silver thiocyanate	165.96				wh; cr; i.w; s. NH_4OH
57	$Zn(SCN)_2$	Zinc thiocyanate	181.55				wh; powd; s.w; NH_4OH; R: a
58	$Hg_2(SCN)_2$	Mercurous thiocyanate	517.38	d			wh; i.w; s. KCN sol; R: HNO_3, HCl
59	$Hg(SCN)_2$	Mercuric thiocyanate	316.78	d			wh; v. sl. s.w; s. al; eth; KCN sol; R: a
60	$Mn(SCN)_2\cdot3H_2O$	Manganese thiocyanate	225.15	d	160		deliq. cr; v.e. s.w; e.s. al
61	$Fe(SCN)_2\cdot3H_2O$	Ferrous thiocyanate	226.07	d			grn; rhomb; v.e.˙s.w; s. al; eth; acet
62	$Fe(SCN)_3$	Ferric thiocyanate	230.10	d			dk. redsh. blk; cub; v.e. s.w; s. al; eth
63	$Co(SCN)_2\cdot3H_2O$	Cobalt thiocyanate	229.16	d	105		vlt; rhomb; s.w

*Note the same total mol. formula CHON (phenomenon of isomerism)

Silane and germane – halogenides

No.	Formula	Name	Mol. wt.	MP/FP	BP/CP	Colour and other properties
1	SiH_3Cl	Silane monochloride	66.54	– 118.1	– 30.4	col; d in water → 54
2	SiH_3Br	Silane monobromide	111.00	– 94	1.9	col; d in water → 54
3	SiH_2Cl_2	Silane dichloride	100.99	– 122	8.3	col; d in water → 55
4	SiH_2Br_2	Silane dibromide	189.91	– 77	66	col; d in water → 55
5	$SiHF_3$	Silane trifluoride	86.07	~ –110	– 80.2	col; s. in toluene, d in w → 56
6	$SiHCl_3$	Silane trichloride*	135.44	– 134	33	col; s. in CS_2, CCl_4, $CHCl_3$; d in w → 56
7	$SiHBr_3$	Silane tribromide*	268.82	~ –60	109	col; d in water → 56
8	$SiHI_3$	Silane triiodide*	409.83	8	220	red; misc. CS_2, C_6H_6; d in w → 56
9	GeH_3Cl	Germane monochloride	111.08	– 52	28	col; d in water
10	GeH_3Br	Germane monobromide	155.54	– 32	52	col; d in water
11	GeH_2Cl_2	Germane dichloride	145.53	– 68	69.5	col; d in water
12	GeH_2Br_2	Germane dibromide	234.45	– 15	89	col; d in water
13	$GeHCl_3$	Germane trichloride*	179.78	– 71	75.2	col; d in water
14	$GeHBr_3$	Germane tribromide*	313.36	– 24	d	col; d in water

Phosphonium salts

No.	Formula	Name	Mol. wt.	MP/FP	BP/CP	Colour and other properties
15	PH_4Cl	Phosphonium chloride	70.47	28.46atm	subl	col; d in water
16	PH_4Br	Phosphonium bromide	114.93	—	s ~ 30	col; d in water
17	PH_4I	Phosphonium iodide	161.93	—	s 62	col; d in water

Mixed halogenides

No.	Formula	Name	Mol. wt.	MP/FP	BP/CP	Colour and other properties
18	$SiBrCl_3$	Silane bromide trichloride	214.35	~ – 60	80	col; d in water
19	$SiBr_2Cl_2$	Silane dibromide dichloride	251.81	~ – 60	104	col; d in water
20	$SiBr_3Cl$	Silane tribromide chloride	303.27	– 39	140.5	col; d in water
21	$SiICl_3$	Silane iodotrichloride	261.35	~ – 60	113.5	col; d in water
22	$GeClF_3$	Germanium monochloro trifluoride	165.08	– 66.2	– 20.3	col; s.al; d in H_2O
23	$GeCl_2F_2$	Germanium dichloro difluoride	181.51	– 51.8	– 2.8	col; s.al; d in H_2O
24	$GeCl_3F$	Germanium trichloro fluoride	191.97	– 49.8	37.5	col; s.al; d in H_2O
25	PBr_2F_3	Phosphorus dibromo trifluoride	247.81	– 20	d 15	pale.yel; attacks glass.
26	$PBrCl_4$	monobromo tetrachloride	252.72			yel; cr.
27	PBr_2Cl_3	Phosphorus dibromo trichloride	297.18	d 35	—	golden yel;
28	PBr_7Cl_2	Phosphorus heptabromo dichloride	661.31			solid; s. in PCl_3, PCl_5
29	PBr_8Cl_3	Phosphorus octabromo trichloride	776.86	25		brn; need;
30	$SeBrCl_3$	Selenium monobromotrichloride	265.25	190	—	yelsh. brn; s. CS_2
31	$SeBr_3Cl$	Selenium tribromochloride	354.17			or; s.CS_2

Mixed oxohalogenides

No.	Formula	Name	Mol. wt.	MP/FP	BP/CP	Colour and other properties
32	$POBrCl_2$	Phosphorus oxybromodichloride	197.81	13	137.6	D 2.104
33	$POBr_2Cl$	Phosphorus oxydibromochloride	242.27	30	165	D (liq at 50°) 2.45
34	$SOCIF$	Thionyl chloridefluoride	102.52	– 139.5	12.2	d in water
35	SO_2ClF	Sulphuryl chloridefluoride	118.52	– 124.7	7.1	d in water

Mixed thiohalogenides

No.	Formula	Name	Mol. wt.	MP/FP	BP/CP	Colour and other properties
36	PSF_3	Phosphorus thiofluoride	120.05	d		col; s. eth; i. CS_2, B_2
37	$PSCl_3$	Phosphorus thiochloride	169.42	– 35	125	col; s. CS_2, CCl_4; B_2
38	$PSBr_3$	Phosphorus thiobromide	302.79	38	d 175	yel. s. CS_2; eth. PCl_3
39	$PSBr_2Cl$	Phosphorus thiodibromochloride	258.34	– 60	d	pale grn; fuming
40	$PSBrCl_2$	Phosphorus thiobromodichloride	213.88	30	d 150	yel; d in water

Inorganic plastics

No.	Formula	Name	Mol. wt.	MP/FP	BP/CP	Colour and other properties
41	$(PN_2H)x$	Phosphane	60.00	—	—	wh; s.conc. H_2SO_4
42	$(PNCl_2)_3$	Phosphorus nitride dichloride	347.71	114	256.5	s. in org. solvents. i.w
43	$(PNCl_2)_4$	Phosphorus nitride dichloride	463.61	123.5	328.5	polymerizes at ~ 250°
44	$(PNCl_2)_5$	Chain structure	579.51	41	224^{13mm}	step wise up to No.46
45	$(PNCl_2)_6$		695.41	90	262^{13mm}	
46	$(PNCl_2)_7$	the $PNCl_2$	811.31	– 18	293^{13mm}	
47	$(PNCl_2)n$	"Inorganic Rubber"				(Gum elastic high polymer substance)
48	$(PNBr_2)_3$	Phosphorus nitride dibromide	614.46		—	s in org. sol; i.water;
49	$H_2Si–O–SiH_3$	Disiloxane	78.23	– 144	– 15.2	col; gas; d in air
50	$H_2Si = O$	Siloxane, Prosiloxane*	46.11	Polymerizes immediately, usually as $O = Si = O$		
51	$O = Si–O–Si = O$	Dioxodisiloxane*	106.20	to solid high polymers see also "Silicones"		

* Corresponding to

$CHCl_3$	Chlorotorm	Silico chloroform
CH_2O	Formaldehyde	Silico formaldehyde
CH_3•CH_3	Dimethylether	Silicomethylether

Inorganic derivatives of ammonia

No.	Formula	Name	Mol. wt.	MP/FP	Colour and other properties
1	NH_2Li	Lithium amide	22.96	~ 374	B.P. 430; D 1.780; d.w
2	NH_2Na	Sodium amide	39.02	206	B.P. 400; wh; d.w
3	NH_2K	Potassium amide	55.12	335	B.P. s400; wh; yel; d.w
4	NH_2OH	Hydroxyl amine	33.03	33.1	B.P. 58_{22} ToRR; wh; s. H_2O; al
5	$NH_2OH \cdot HCl$	Hydroxyl ammonium chloride	69.50	151	d; D 1.67; wh; s.H_2O, al
6	$NH_2OH \cdot NNO_3$	Hydroxyl ammonium nitrate	95.05	48	·d100, wh; e.s. H_2O; al
7	$2NH_2OH \cdot H_2SO_4$	Hydroxyl ammonium sulphate	164.15	170	d. wh; s.w; al
8	H_2N-NH_2	Hydrazine	32.05	1.4	B.P.113.5; D1.011; s.w. n_D^{23} 1.470
9	$H_2N-NH_2 \cdot H_2O$	Hydrazin hydrate	50.06	~ - 40	B.P. ~ 120. col; ∞ misc; w
10	$H_2N-NH_2 \cdot HN_3$	Hydrazonium azide	75.08	75.4	deliq; e.s.w
11	$H_2N-NH_2 HCl$	Hydrazonium monochloride	68.51	89	wh; need; e.s.w; al
12	$HCl \cdot H_2N-NH_2 \cdot HCl$	Hydrazonium dichloride	104.98	198	col; cub; e.s.w; al
13	$H_2N-NH_2 \cdot HNO_3$	Hydrazonium mononitrate	96.06	70.7	S.140, col; e.s.w; al
14	$HNO_3 \cdot H_2N-NH_2 \cdot HNO_3$	Hydrazonium dinitrate	158.08	148	
15	$H_2N-NH_2 \cdot H_3PO_4$	Hydrazonium phosphate	130.05	82	col; s.w
16	$H_2N-NH_2 \cdot H_2SO_4$	Hydrazonium monosulphate	130.13	254	col; s.w
17	$H_2SO_4 \cdot H_2N-NH_2 \cdot H_2SO_4$	Hydrazonium disulphate	162.18	85	col; e.s.w
18	HN–B–NH / HB–N–BH (Borazole ring structure)	Borazole or. "Inorganic benzene"	80.53	- 58	B.P.53; D 0.85; col; d.w
19	$H_3B-N-BH_3$	Diboramine	42.70	−66.5	
20	$H_2N-C \equiv N$	Cynamide	42.02	44	col; d; e.s.w; alk
21	$Na_2(N-C \equiv N)$	Sodium cynamide			
22	$Ca(N-C \equiv N)$	Calcium cynamide	80.11	5 ~ 1200	col; d. $H_2O \rightarrow NH_3$
23	$ClHgNH_2$	Mercuric amidochloride	252.09		D 5.7, wh; e.s.w
24	$BrHgNH_2$	Mercuric amidobromide	296.55	d	wh; d.w.s. NH_4OH
25	$IHgNH_2$	Mercuric amidoiodide	343.45		dirty wh; s. eth

Oxohalogen acids and their salts, oxonitrosyl and oxoamido acids, thio amido complexes

No.	Formula	Name	Mol. wt.	MP/FP	Colour and other properties
26	$H_2\left[\begin{smallmatrix}O & F\\O & P & O\end{smallmatrix}\right]$	Fluro phosphorus acid	100.00		D1.818, visc. liq; ∞ misc.w
27	$Na_2[PO_3F]$	Sodium fluorophosphate	143.97	~ 625	col.
28	$H\left[\begin{smallmatrix}O & P & F\\O & & F\end{smallmatrix}\right]$	Difluoro phosphorus acid	101.99	- 75	B.P. 116; D 1.583; col; s.w
29	$NH_4[PO_2F_2]$	Ammonium difluorophosphate	119.02	213	col; rhomb; s.w; al ac. a
30	$H\left[\begin{smallmatrix}O & S & F\\O & & O\end{smallmatrix}\right]$	Fluorosulphuric acid	100.07	−87.3	B.P. 165.5, D 1.743; col; d.w
31	$Li[SO_3F]$	Lithium fluorosulphate	106.01	360	wh; e.s.w; al; eth
32	$Na[SO_3F]$	Sodium fluorosulphate	122.07		Pl; Hyd; e.s.w; al, eth
33	$K[SO_3F]$	Potassium fluorosulphate	138.16	311	Prism
34	$NH_4[SO_3F]$	Ammonium fluorosulphate	117.11	244.7	col; need; s.w; e.s. al
35	$H\left[\begin{smallmatrix}O & S & Cl\\O & & O\end{smallmatrix}\right]$	Chlorosulphuric acid	116.53	- 80	B.P. 158; D 1.766; s.CS_2
36	$K\left[\begin{smallmatrix}O & Cr & Cl\\O & & O\end{smallmatrix}\right]$	Potassium chlorochromate	174.56	d	D2.497; s.w; al
37	$H\left[\begin{smallmatrix}O & S & NO\\O & & O\end{smallmatrix}\right]$	Nitrosylsulphuric acid (lead chamber crystals)	127.08	73	s.H_2SO_4; d.w;
38	$\left[\begin{smallmatrix}NO & & NO\\O & SOS & O\end{smallmatrix}\right]$	Nitrosyl sulphuric acid anhydride	236.15	217	B.P. 360; d.w
39	$H\left[\begin{smallmatrix}O & S & NH_2\\O & & O\end{smallmatrix}\right]$	Amido sulphuric acid (Amido sulphonic acid)	97.10	d200	col; s.w; e.s.eth; ac.a
40	$NH_4[SO_3NH_2]$	Ammonium amido sulphate	114.13	125	d160; big pl.
41	$\left[\begin{smallmatrix}O & S & NH_2\\O & & NH_2\end{smallmatrix}\right]$	Diamido sulphuric acid (sulphamide)	96.11	91.5	d250, col; s.w
42	$\left[\begin{smallmatrix}NH_2 & P & NH_2\\NH_2 & & S\end{smallmatrix}\right]$	Thiophosphorus acid triamide (Triamido thiophosphorus acid)	111.12	d200	yelsh; e.s.w; d. hot w; D 1.7

64

Elements existing as central atoms in complex acids and salts

	1	2	3	4	5	6	7	a main group	
II		Be	B						see also p. 31
III			Al	Si	P				
IV				Ge	As				
V				Sn	Sb				
VI				Pb					8
IV	Cu	Zn		Ti		Cr	Mn	Fe	Co	Ni
V	Ag	Cd		Zr			Mo	Ru	Rh	Pd
VI	Au	Hg					W	Os	Ir	Pt

(b sub groups)

Complexes of elements of the main groups

No.	CA	Formula	Name	M. wt.	MP/FP	Colour and other properties
1	Be	$Na_2[BeF_4]$	Sodium tetrafluoroberyllate	131.01		wh; rhomb or monocl;
2		$K_2[BeF_4]$	Potassium tetrafluoroberyllate	163.21		col; rhomb;
3	B	$H[BF_4]$	Tetrafluoroboric acid	87.83		d130; ∞ misc. H_2O
4		$Na[BF_4]$	Sodium tetrafluoroborate	109.82	384d	wh; rhomb; e.s. al
5		$K[BF_4]$	Potassium tetrafluoroborate	125.92	530	col; e.s. al; eth
6		$NH_4[BF_4]$	Ammonium tetrafluoroborate	104.68	sub.	s.w; al
7	Al	$Na_2[AlF_6]$	Sodium hexafluoroaluminate "kryolite"	209.95	1000	D = 295
8	Si	$H_2[SiF_6]$	Hexafluorosilicic acid	144.08	—	d. e.s.w
9		$H_2[SiF_6]\cdot2H_2O$	Hexafluorosilic acid dihydrate		19	col;
10		$Li_2[SiF_6]\cdot2H_2O$	Lithium hexafluoro silicate	191.97	d	wh; monocl; s. al
11		$Na_2[SiF_6]$	Sodium hexafluoro silicate	188.07	d	col; v.sl. s.w
12		$K_2[SiF_6]$	Potassium hexafluoro silicate	220.25	d	col. v.sl. s.w
13		$(NH_4)_2[SiF_6]$	Ammonium hexafluoro silicate	178.14	sub.	col; e.s.w
14		$Mg[SiF_6]$	Magnesium hexafluoro silicate	116.38		wh; prism; s.w
15		$Ca[SiF_6]$	Calcium hexafluoro silicate	18Y.17		wh; wh. s.w
16		$Sr[SiF_6]2H_2O$	Strontium hexafluoro silicate dihydrate	265.72	d	wh; s.w
17		$Ba[SiF_6]$	Barium hexafluoro silicate	279.49		rhomb; need; v.sl. s.w
18	Ge	$K_2[GeF_6]$	Potassium hexafluoro gemanate	264.79	730	B.P ~835. wh; hex
19		$(NH_4)_2[GeF_6]$	Ammonium hexafluoro germanate	222.14		col; s.w
20	Sn	$K_2[SnF_6]\cdot H_2O$	Potassium hexafluoro stannate hydrate	328.91		monocl; pr; D3.05; v.sl. s.H_2O
21		$H_2[SnCl_6]\cdot6H_2O$	Hexachlorostannic acid hexahydrate	441.55	9	D1.93. s.w
22		$K_2[SnCl_6]$	Potassium hexachloro stannate	409.63		D2.71; s.w
23		$(NH_4)_2[SnCl_6]$	Ammonium hexachlorostannate	367.52	d	D2.4. e.s.w
24		$K_2[SnBr_6]$	Potassium hexabromostannate	376.39		D = 3.78
25		$(NH_4)_2[SnBr_6]$	Ammonium hexabromostannate	634.28	d	D = 3.5; e.s.w
26		$K_2[SnS_3]$	Potassium thiostannate	293.07	100°	out of the melt
		$K_2[Sn(SH)_3(OH)_3]$	or $K_2[SnS_3]\cdot3H_2O$ - trihydrate	347.12	$-3H_2O$	dk. b. oil. s.w
27		$Na_2[SnO_3]$	Sodium stannate	212.68		out of the melt
		$Na_2[Sn(OH)_6]$	Sodium hexahydroxo stannate	266.73		D = 3.06, s.w
28		$K_2[SnO_3]$	Potassium stannate	244.90		out of the melt
		$K_2[Sn(OH)_6]$	Potassium hexahydroxo stannate	298.95		D = 3.2, e.s.w
29	Pb	$K_2[PbCl_6]$	Potassium hexachloro plumbate	498.14	d190	s. hot HCl
30		$(NH_4)_2[PbCl_6]$	Ammonium hexachloro plumbate	456.03	d120	yel; s.w; d. hw
31		$Na_2[Pb(OH)_6]$	Sodium hexahydroxo plumbate	355.25		yel; wh; d. in w; s.alk
32		$K_2[Pb(OH_6]$	Potassium hexahydroxo plumbate	387.45		col; d. in w; s.alk;
33	P	$Na[PF_6]H_2O$	Sodium hexafluoro phosphate - hydrate	185.99		D2.4; e.s.w
34		$K[PF_6]$	Potassium hexafluoro phosphate	184.1	~575	s.w;
35		$NH_4[PF_6]$	Ammonium hexafluoro phosphate	163.02	d	s.w; al; acet;
36	As	$K_3[AsS_3]$	Potassium trithioarsenate-III	288.40	d	s.w; alk;
37		$Ag_3[AsS_3]$	Silver trithioarsenate-III	494.73	>175	D = 5.5; d in HNO_3
38		$K_3[AsS_4]$	Potassium tetrathioarsenate-V	320.46	d	s.w; alk
39	Sb	$(NH_4)_2[SbF_6]$	Ammonium pentafluoro antimonate-III	252.84	s;d	e.s.w;
40		$Na[SbF_6]$	Sodium hexafluoro antimonate-V	258.76		D = 3.4; e.s.w; al, acet;
41		$Ag_3[SbS_3]$	Silver trithioantimonate-III	541.58	>175	D = 5.8; d in HNO_3
42		$Na_2[SbS_4]9H_2O$	Sodium tetrathio- antimonate-V	481.14	d	yelsh. s.w, al
43		$K_3[SbS_4]4\frac{1}{2}H_2O$	Potassium tetrathioantimonate-V			yel; s.w
44		$(NH_4)_2[SbS_4]\cdot4H_2O$	Ammonium tetrathioantimonate-V	376.21	d	yel; d. in hot, w
45		$Na[Sb(OH)_6]$	Sodium hexahydroxoantimonate-V	246.80		
46		$K[Sb(OH)_6]$	Potassium hexahydroxoantimonate-V	262.91		

*) The following are also known:
$Rb_2[SiF_6]$, $Cs_2[SiF_6]$, $Al_2[SiF_6]$, $Pb[SiF_6]$, $Cu_2[SiF_6]$ (red), $Cu[SiF_6]\cdot6H_2O$ (blue),
$Ag_2[SiF_6]$, $Zn[SiF_6]$, $Cd[SiF_6]$, $Hg_2[SiF_6]2H_2O$; $Mn[SiF_6]6H_2O$, $Fe[SiF_6]6H_2O$, $Fe_2[SiF_6]_3$, $Co[SiF_6]6H_2O$, $Ni[SiF_6]6H_2O$
The salts of heavy metals are water soluble, some are easily soluble, others are only slightly soluble or insoluble.

Complexes of the elements of subgroups

No.	CA	Formula	Name	Mol. wt.	MP/FP	Colour and other properties
1	Cu	$(NH_4)_2[CuCl_4]\cdot 2H_2O$	Ammonium tetrachlorocuprate-II	277.48	d 110	grnsh. bl. s.w
2		$Na[Cu(CN)_2]$	Sodium dicyanocuprate-I	138.57	d 100	col; e.s.w; d alk
3		$K_3[Cu(CN)_4]$	Potassium tetracyanocuprate-I	284.93	d	col; rhomb; e.s.w
4		$[Cu(NH_3)_4](OH)_2\cdot 3H_2O$	Tetrammine copperhydroxy trihydrate		d	bl; need v.e. s.w
5		$[Cu(NH_3)_4]SO_4\cdot H_2O$	Tetrammine coppersulphate monohydrate	245.75	d 150	bl; rhomb; e.s.w
6	Ag	$K[Ag(CN)_2]$	Potassium dicyanoargentate	199.20		cub; s.w
7	Au	$H[AuCl_4]\cdot 4H_2O$	Tetrachloroauric-III acid tetrahydrate	412.10	d	yelsh. need; s.w; al. eth
8		$Na[AuCl_4]\cdot 2H_2O$	Sodium tetrachloroaurate-III dihydrate	398.06	d	yel; rhomb; e.s.w; al. eth
9		$K[AuCl_4]$	Potassium tetrachloroaurate-III	378.12	d 357	yel; monocl; s.w; al
10		$K[AuCl_4]\cdot 2H_2O$	Potassium tetrachloroaurate-III dihydrate	414.16		yel; rhomb; tricl; s.w; al. eth
11		$(NH_4)[AuCl_4]$	Ammonium tetrachloroaurate-III '	357.07		yel; monocl; rhomb; e.s.w; al. eth
12		$H[AuBr_4]\cdot 5H_2O$	Tetrabromoauric-III acid pentahydrate	607.95	27	redsh. brn; e.s.w; al
13		$Na[AuBr_4]\cdot 2H_2O$	Sodium tetrabromoaurate-III dihydrate	575.89		brn; s.w
14		$K[AuBr_4]$	Potassium tetrabromoaurate-III	555.96	d	redsh; brn; e.s.w; al
15		$K[AuBr_4]\cdot 2H_2O$	Potassium tetrabromoaurate-III dihydrate	591.99		vlt. monocl; s.w; al
16		$K[AuI_4]$	Potassium tetraiodoaurate-III	743.98		blk; s.w
17		$Na[AuS]\,H_2O$	Sodium thioaurate-I	324.33	d	col; monocl; s.w
18		$K[Au(OH)_4]\cdot H_2O$	Potassium tetrahydroxoaurate-III hydrate	322.34	d	yel; need. s.w; al
19		$H[Au(CN)_4]\cdot H_2O$	Tetracyanoauric acid-III monohydrate	356.33	50	col; s.w; al; eth; tricl; d.h.w
20		$Na[Au(CN)_4]\cdot 1\frac{1}{2}H_2O$	Sodium tetracyanoaurate-III sesquihydrate	367.20	d200	col; tricl; s.w; al
21		$(NH_4)[Au(CN)_4]\cdot H_2O$	Ammonium tetracyanoaurate-III hydrate	337.33	d200	col; tricl; e.s.w; al
22		$Na[Au(CN)_2]$	Sodium dicyanoaurate-I	272.23		wh; powd; s.w
23		$K[Au(CN)_2]$	Potassium dicyanoaurate-I	288.14		col; monocl; s.w; al
24		$(NH_4)[Au(CN)_2]$	Ammonium dicyanoaurate-I	267.28	d100	col; e.s.w; al
25	Zn	$(NH_4)_2[ZnCl_4]$	Ammonium tetrachlorozincate	232.38	150	wh; rhomb; s.w.s
26		$[Zn(NH_3)_2]Cl_2$	Diamminzinc chloride	170.36	211	d. 271, col; d in w.
27	Cd	$K_2[Cd(CN)_4]$	Potassium tetracyano cadmate	294.68		col; s.w
28	Hg	$K_2[HgI_4]$	Potassium tetraiodo mercurate	786.48		yel; e.s.w
29		$Ag_2[HgI_4]$	Silver tetraiodo mercurate	924.05		yel; s. kl. alk
30		$K_2[Hg(CN)_4]$	Potassium tetracyano mercurate-II	382.87		col; s.w; al
31		$[Hg(NH_3)_2]Cl_2$	Diamine mercuric chloride	305.59	300	i.w
32		$[Hg(NH_3)_2Br_2]$	Diamine mercuric bromide	394.51	180	wh
33		$[Hg(NH_3)_2]I_2$	Diamine mercuric iodide	488.51		blsh. yel; d in w;
34		$\begin{bmatrix}HO\cdot Hg \\ HO\cdot Hg\end{bmatrix}{>}NH_2\Big]OH$	"Millon's base"	468.27		D: 4.083
35		$\begin{bmatrix}O{<}^{Hg}_{Hg}{>}NH_2\end{bmatrix}OH$	Anhydride of "Millon's base"	450.25	exp	dk. yel.
36		$\begin{bmatrix}O{<}^{Hg}_{Hg}{>}NH_2\end{bmatrix}Cl$	Chloride of no./35	468.70	d 120	blk. yel; e.s.w.
37		$\begin{bmatrix}O{<}^{Hg}_{Hg}{>}NH_2\end{bmatrix}I$	Iodide of no:/35 "Nessler's Reagent"	560.16	~128	yel. brn; i.w; **ex**
38	Ti	$K_2[TiF_6]\cdot H_2O$	Potassium hexafluoro titanate	258.11	780	at 32°–H_2O; col; monocl;
39		$(NH_4)_2[TiF_6]$	Ammonium hexafluoro titanate	197.98	d	hex; prism; s.w
40	Zr	$K_2[ZrF_6]$	Potassium hexafluoro zinconate	283.41		col; monocl; D = 3.48
41		$(NH_4)_2[ZrF_6]$	Ammonium hexafluoro zinconate	241.30		rhomb; hex; D = 1.154
42	Cr	$[Cr(H_2O)_6]Cl_3$	Hexaaquo chromium-III chloride	266.48	95	vlt; monocl; s.w. al
43		$[CrCl(H_2O)_5]Cl_2\cdot H_2O$	Chloropentaaquochromium-chloridehydrate	266.48		grn.
44		$[CrCl_2(H_2O)_4]Cl\cdot 2H_2O$	Dichlorotetra aquochromium-III chloride	266.48	83	grn; rhomb; s.w
45		$[Cr(NH_3)_6]Cl_3\cdot H_2O$	Hexaamminochromium-III chloride moho hydrate	278.59		yel; s.w.
46		$[CrCl(NH_3)_5]Cl_2$	Chloropentammino chromium-III chloride	243.54		red; octa
47	Mn	$K_3[Mn(CN)_6]$	Potassium hexacyano manganate-III	328.33		red; rhomb; s.w.
48		$H_4[Mn(CN)_6]$	Hexacyano manganese-II acid	215.07	d	e.s; al
49		$K_4[Mn(CN)_6]$	Potassium hexacyano manganate-II	421.47		bl. tetr; s.w.
50	Fe	$H_3[Fe(CN)_6]$	Hexacyano iron-III acid	214.98	d	grnsh. brn; need. s.w; al
51		$Na_3[Fe(CN)_6]$	Sodium hexacyano ferrate-III	298.97		red. s.w.
52		$K_3[Fe(CN)_6]$	Potassium hexacyano ferrate-III.	329.26	d	red; s.w.
53		$(NH_4)_3[Fe(CN)_6]$	Ammonium hexacyano ferrate-III	266.08	d	red; e.s.w.
54		$Ca_3[Fe(CN)_6]_2\cdot 12H_2O$	Calcium hexacyano ferrate-III 12 water	760.33		red. need; e.s.w.
55		$Sn_3[Fe(CN)_6]_2$	Tin hexacyano ferrate-III	780.02	d	wh; d in HCl
56		$Pb_3[Fe(CN)_6]_2\cdot 5H_2O$	Lead hexacyano ferrate-III pentahydrate	1135.65	d	red; e.s.w.
57		$Cu_3[Fe(CN)_6]_2$	Copper hexacyano ferrate-III	402.59		brnsh. red; i.w; HCl
58		$Cu_3[Fe(CN)_6]14H_2O$	Copper hexacyano ferrate-III 14 water	870.76		yelsh. grn; i.w; HCl
59		$Ag_3[Fe(CN)_6]$	Silver hexacyano ferrate-III	335.60		or; i.w.

Complexes of elements of subgroups

No.	CA	Formula	Name	Mol. wt.	FP/BP	Colour and other properties
1		$Fe_3^{II}[Fe^{III}(CN)_6]_2$*	Ferrous hexacyanoferrate-III	591.47	d	deep. bl; insoluble
2		$Co_3[Fe(CN)_6]_2$	Cobalt hexacyanoferrate-III	600.74		red, need
3		$H_4[Fe(CN)_6]$	Hexacyano ferrous acid	215.99	d	wh; need; s.w
4		$Na_4[Fe(CN)_6]\cdot10H_2O$	Sodium hexacyano ferrate-II decahydrate	484.11		bl. yel; monocl; s.w
5		$K_4[Fe(CN)_6]\cdot3H_2O$	Potassium hexacyano ferrate-II trihydrate	422.41	d	yel; monocl; s.w
6		$(NH_4)_4[Fe(CN)_6]\cdot3H_2O$	Ammonium hexacyano ferrate-II	338.17	d	yel; monocl; s.w
7		$Mg_2[Fe(CN)_6]\cdot12H_2O$	Magnesium hexacyano ferrate-II 12 water	476.79	d ~200	bl. yel; s.w
8		$Ca_2[Fe(CN)_6]\cdot11H_2O$	Calcium hexacyano ferrate-II water	490.29	d	yel. tetr; s.w
9		$Sr_2[Fe(CN)_6]\cdot15H_2O$	Strontium hexacyano ferrate-II—15 water	657.45		yel; monocl; s.w
10		$Ba_2[Fe(CN)_6]\cdot6H_2O$	Barium hexacyano ferrate-II	594.77		yel; monocl; s.w
11		$Al_4[Fe(CN)_6]_3\cdot17H_2O$	Aluminium hexacyano ferrate-II—17 water	1050.03		brn; s.w
12		$Sn[Fe(CN)_6]$	Stannic hexacyano ferrate-II	330.66		gr. wh; yel; d in HCl
13		$Sn_2[Fe(CN)_6]$	Stannous hexacyano ferrate-II	449.36		wh. gel; d in HCl
14		$Pb_2[Fe(CN)_6]3H_2O$	Lead-II hexacyano ferrate-II trihydrate	680.43	d	yel. wh; s. H_2SO_4
15		$Cu_2[Fe(CN)_6]\times H_2O$*	Copper-II hexacyano ferrate-II X water	661.49		redsh. brn; s. NH_4OH
16		$Ag_4[Fe(CN)_6]$ H_2O*	Silver hexacyano ferrate-II monohydrate			wh. s. KCN sol
17		$Zn_2[Fe(CN)_6]$ $3H_2O$	Zinc hexacyano ferrate-II trihydrate	396.77	d	wh; s. NH_4OH, HCl
18		$Cd_2[Fe(CN)_6]\times H_2O$	Cadmium hexacyano ferrate-(II) X water			s. HCl
19		$Mn_2[Fe(CN)_6]$ $7H_2O$	Manganese hexacyano ferrate-II heptahydr	447.93		gr. wh; s. HCl
20		$Fe_4[Fe(CN)_6]$*	Iron hexacyano ferrate-II	859.27	d	dk. bl; s. HCl, H_2SO_4
21		$Co_2[Fe(CN)_6]\times H_2O$	Cobalt hexacyano ferrate-II X water			gr. grn; s. KCN sol
22		$Ni_2[Fe(CN)_6]\times H_2O$	Nickel hexacyano ferrate-II X water			gr. wh; s. KCN sol; NH_4OH
23		$Na_2[Fe(CN)_5NO]$ $2H_2O$	Sodium pentacyano nitrosyl ferrate-II	297.97		red, rhomb; s.w; al
24		$K_2[Fe(CN)_5NO]\cdot2H_2O$	Potassium pentacyano nitrosyl ferrate-III dihydrate	330.17		red, monocl; hyg; s.w; al
25		$Cu[Fe(CN)_5NO]\cdot2H_2O$	Copper pentacyano nitrosyl ferrate-III dihydrate	315.52		gr. wh; s. alk
26	Co	$Na_3[Co(NO_2)_6]$	Sodium hexanitra cobaltate-III	403.98		gr. brn; s.w. s. dil. ac. a
27		$K_3[Co(NO_2)_6]\cdot H_2O$	Potassium hexanitro cobaltate-III hydrate	470.29		yel; s. hot w; s. dil. ac. a
28		$Cd_3[Co(NO_2)_6]_2$	Cadmium hexanitro cobaltate-III	1007.21	d 175	yel; s.w
29		$H_3[Co(CN)_6]\frac{1}{2}H_2O$	Hexacyano cobaltic-III acid—$\frac{1}{2}$ hydrate	227.09	d 100	col; need; s.w. al
30		$K_3[Co(CN)_6]$	Potassium hexacyano cobaltate-III	332.34		yel; monocl; s.w
31		$K_4[Co(CN)_6]$	Potassium hexacyano cobaltate-II	371·43		vlt; need; s.w
32		$[Co(NH_3)_6]Cl_2$	Hexammine cobalt-II chloride	232.05		rose red; s. NH_4OH
33		$[Co(NH_3)_6]Cl_3$	Hexammine cobalt-III chloride	267.50		or. monocl; sl. s.w
34		$[Co(NH_3)_5Cl]Cl_2$	Chloropentammine cobalt-III chloride	250.47	d	dk. red; rhomb; sl. s.w
35		$[Co(NH_3)_4(H_2O)Cl]Cl_2$	Cis-chloroaquotetrammine cobalt-III chloride	251.40	d	vlt; rhomb; sl. s.w
36		$[Co(NH_3)_5(H_2O)]Cl_3$	Aquopentammine cobalt-III chloride	268.49	d 100	red; s.w
37	Ni	$K_2[Ni(CN)_4]H_2O$	Potassium tetracyanonickolate-II hydrate	258.97		red. yel; monocl; s.w
38		$[Ni(NH_3)_6]Cl_2$	Hexammine nickel-II chloride	231.80		bl; s.w; NH_4OH
39		$[Ni(NH_3)_6]Br_2$	Hexammine nickel-II bromide	320.71		vlt; prism; e.s.w
40		$[Ni(NH_3)_6]I_2$	Hexammine nickel-II iodide	414.72	d	bl. blk; s. NH_4OH
41		$[Ni(NH_3)_6](NO_3)_2$	Hexammine nickel-II nitrate	284.90		bl. s.w
42		$[Ni(NH_3)_4(H_2O)_2](NO_3)_2$	Diaquotetrammine nickel-II nitrate	286.87		grn; s.w
43	Ru	$K_2[RuCl_6]$	Potassium hexachloro ruthenate	392.63		blk; s.w
44	Rh	$K_2[RhCl_6]$	Potassium pentachloro rhodiumate-(III)	358.39	d	red; rhomb; s.w
45		$Na_3[RhCl_6]\cdot18H_2O$	Sodium hexachloro rhodiumate-III 18 water	708.93	d	red. e.s.w
46		$K_3[RhCl_6]\cdot3H_2O$	Potassium hexachloro rhodiumate-III	486.99	d	red; d in w. e.s. al
47	Pd	$Na_2[PdCl_4]\cdot3H_2O$	Sodium tetrachloro palladinate-II trihydrate	348.57		brn. red; e.s.w; al
48		$K_2[PdCl_4]$	Potassium tetrachloro palladinate-II	326.72	d 105	red. brn; tetr; s.w
49		$(NH_4)_2[PdCl_4]$	Ammonium tetrachloro palladinate-III	284.61		olive grn; tetr; s.w
50		$K_2[PdCl_6]$	Potassium hexachloro palladinate-IV	397.63	d	red. brn; e.s.w
51		$(NH_4)_2[PdCl_6]$	Ammonium hexachloro palladinate-IV	355.22	d	red. brn; e.s.w
52		$[Pd(NH_3)_4]Cl_2\cdot H_2O$	Tetrammine palladium-II chloride—hydrate	263.76	d 120	col; tetr; e.s.w
53		$[Pd(NH_3)_4](OH)_2$	Tetrammine palladium-II hydroxide	174.78		yel; cr; e.s.w
54		$[Pd(NH_3)_4Cl_2]$	Dichloro tetrammine palladium-IV	211.68	d	yel; tetr; s. hot w
55	Os	$K_3[OsCl_6]\cdot3H_2O$	Potassium hexachloro osmate-III dihydrate	574.28		dk. red; e.s.w; s. al
56		$Na_2[OsCl_6]\cdot2H_2O$	Sodium hexachloro osmate-III dihydrate	484.97		or. red; rhomb prism; e.s.w; al
57		$K_2[OsCl_6]$	Potassium hexachloro osmate-IV	481.13		red. s.w
58		$(NH_4)[OsCl_6]$	Ammonium hexachloro osmate-IV	439.02		cub
59		$K_4[Os(CN)_6]$ $3H_2O$	Potassium hexacyano osmate-IV trihydrate	566.74	d	col; (yel); monocl; e.s.w
60	Ir	$Na_3[IrCl_6]\cdot12H_2O$	Sodium hexachloro iridium-III decahydrate	691.03		dk. grn; s.w
61		$Na_3[IrBr_6]\cdot12H_2O$	Sodium hexabromo iridium-III decahydrate	957.78	100	olive grn; tetr; s. NH_4OH
62		$K_3[IrI_6]$	Potassium hexaiodo iridiumate-III	1071.91	d	grn; e.s.w
63		$Na_2[IrCl_6]\cdot6H_2O$	Sodium hexachloro iridium-IV hexahydrate	559.93	d 600	red. blk; tetr; s.w
64		$K_2[IrCl_6]$	Potassium hexachloro iridiumate-IV	484.03	d	blk; d in water
65		$(NH_4)_2[IrCl_6]$	Ammonium hexachloro iridiumate-IV	441.02	d	dk. red; sl. s.w

* better: $Fe^{II}[Fe^{III}[Fe^{III}(CN)_6]]_2$; $[Cu(H_2O)_4][Cu[Fe(CN)_6]]$ 3 or $6H_2O$: $Ag[Ag_3[Fe(CN)_6]]$.
$Zn[Zn_3[Fe(CN)_6]_2]$; $Fe^{III}[Fe^{III}[Fe^{II}(CN)_6]]_3$

Complexes of elements of subgroups

No.	Formula	Name	Mol. wt.	MP/FP	Colour and other properties
	Pt				
1	$Li_2[PtCl_4]6H_2O$	Lithium tetrachloro platinate-II hexahydrate	529.95	d	or. red; hex; s.w. al
2	$K_2[PtCl_4]$	Potassium tetrachloro platinate-II	415.25	d	red; tetr; se. s.w
3	$(NH_4)_2[PtCl_4]$	Ammonium tetrachloro platinate-II	373.14	d	red; rhomb; s.w
4	$Ba[PtCl_4]$	Barium tetrachloro platinate	528.47		s.w; e.s. al
5	$H_2[PtCl_6]\cdot6H_2O$	Hexachloroplatinic acid-IV hexahydrate	518.08	60	red. brn; prism; e.s.al; eth
6	$Na_2[PtCl_6]$	Sodium hexachloro platinate-IV	453.97	~150	or yel; prism; s.w; al
7	$K_2[PtCl_6]$	Potassium hexachloro platinate-IV	486.16	d250	yel; v.sl. s.w
8	$(NH_4)_2[PtCl_6]$	Ammonium hexachloro platinate-IV	444.05	d	yel; v.sl. s.w
9	$Rb_2[OPtCl_6]$	Rhubidium hexachloroplatinate-IV	578.93	d	yel; v.sl. s.w
10	$Cs_2[PtCl_6]$	Caesium hexachloro platinate-IV	673.79	d	yel; v.sl. s.w
11	$Mg[PtCl_6]6H_2O$	Magnesium hexachloro platinate-IV	540.39	d	
12	$Ba[PtCl_6]6H_2O$	Barium hexachloro platinate-IV	653.43	d	or. yel; rhomb; s.w
13	$K_2[PtBr_4]$	Potassium tetrabromo platinate-II	593.09		brn; rhomb; e.s.w
14	$K_2[PtBr_4]\cdot2H_2O$	Potassium tetrabromo platinate-II dihydrate	629.12		blk; rhomb; e.s.w
15	$H_2[PtBr_6]\cdot9H_2O$	Hexabromo platinic acid-IV nona-hydrate	838.89	d ~100	red; monocl; e.s.w
16	$Na_2[PtBr_6]\cdot6H_2O$	Sodium hexabromo platinate-IV hexahydrate	828.82	d ~150	dk. red; tetr; e.s.w; al
17	$K_2[PtBr_6]$	Potassium hexabromo platinate-IV	752.92	d ~400	blk; red. brn; sl. s.w
18	$(NH_4)_2[PtBr_6]$	Ammonium hexabromo platinate-IV	710.81		red brn; v.sl. s.w
19	$Mg[PtBr_6]\cdot12H_2O$	Magnesium hexabromo platinate-IV dodecahydrate	915.24		trig.
20	$Ba[PtBr_6]\cdot10H_2O$	Barium hexabromo platinate-IV decahydrate	992.25		monocl.
21	$H_2[PtI_6]\cdot9H_2O$	Hexaiodo platinic acid-IV nona-hydrate	1120.91	~100	blk; e.s.w
22	$Na_2[PtI_2]\cdot6H_2O$	Sodium hexaiodo platinate-IV hexahydrate	1110.84		trig; s.w. al
23	$K_2[PtI_6]$	Potassium hexaiodo platinate-IV	1034.94		blk; s.w
24	$(NH_4)_2[PtI_6]$	Ammonium hexaiodo platinate-IV	992.82		cub
25	$Na_2[Pt(OH)_6]$	Sodium hexahydroxoplatinate-IV	343.27	d	red brn-yel; s.w
26	$K_2[Pt(OH)_6]$	Potassium hexahydroxo platinate-IV	375.47	d	yel. rhomb; s.w
27	$Na_2[Pt(NO_2)_4]$	Sodium tetranitro platinate-II	425.26		pl. yel; s.w
28	$K_2[Pt(NO_2)_4]$	Potassium tetranitro platinate-II	457.45	d	col; monocl
29	$K_2[Pt(CN)_4]_3H_2O$	Potassium tetracyano platinate-II trihydrate	431.54	d ~500	col. yel; rhomb; e.s.w; al
30	$(NH_4)_2[Pt(CN)_4]H_2O$	Ammonium tetracyano platinate-II monohydrate	353.40		yel; s.w
31	$Ba[Pt(CN)_4]\cdot4H_2O$	Barium tetracyanoplatinate-II tetrahydrate	508.73		yel. or grn; e.s.w
32	$[Pt(NH_3)_4]Cl_2\cdot H_2O$	Tetrammine platinum-II chloride monohydrate	352.29	250	col; tetr
33	$[Pt(NH_3)_4][PtCl_4]$	Tetrammine platinum-II tetrachloro platinate-II	600.42	d	gr.or red; tetr; e.s.w
34	$[Pt(NH_3)_2Cl_4]$	Cis-tetrachloro diammine platinum-IV	371.12	240	or. yel; rhomb or hex
35	$[Pt(NH_3)_2Cl_4]$	Trans tetra chloro diammine platinum-IV	371.12	~215	yel; octa

Complexes with carbon monoxide or carbonyls

No.	Formula	Name	Mol. wt.	MP/FP	CP	Colour and other properties
36	$Cr(CO)_6$	Chromium carbonyl	220.07	S >20	d130	col; rhomb; D = 1.77. **exp** 200
37	$MO(CO)_6$	Molybdenum carbonyl	264.01	S	d150	col. rhomb; D = 1.96
38	$W(CO)_6$	Tungsten carbonyl	351.98	S 50	d	col; rhomb; D = 2.65
39	$Mn_2(CO)_{10}$	Dimanganese decacarbonyl	389.98	155	S	gold yel; monocl
40	$Re_2(CO)_{10}$	Dirhenium decacarbonyl	652.72	177	S	col; monocl
41	$Fe(CO)_5$	Iron pentacarbonyl	195.90	− 21	105	yel; D = 1.46
42	$Fe_2(CO)_9$	Di iron nonacarbonyl	363.79	d 100		gold yel; hex
43	$Fe_3(CO)_{12}$	Tri iron dodecacarbonyl	503.67	d 140		grn; hex; prism
44	$Ru(CO)_5$	Ruthenium pentacarbonyl	241.15	− 22		col;
45	$Ru_2(CO)_9$	Diruthenium nonacarbonyl	454.29	S		or; monocl. prism
46	$Ru_3(CO)_{12}$	Triruthenium dodecacarbonyl	639.42			grn; need
47	$Os(CO)_5$	Osmium pentacarbonyl	330.25	− 15		col
48	$Os_2(CO)_9$	Diosmium nonacarbonyl	632.49	224		pale yel; hex. prism
49	$Co_2(CO)_9$	Dicobalt nonacarbonyl	369.97	51	d 52	or, cr.
50	$Co_4(CO)_{12}$	Tetra cobalt dodecacarbonyl	571.88	d 60		blk, cr.
51	$Rh_2(CO)_9$	Dirhodium nonacarbonyl	457.91	76		yelsh, red; cr
52	$Rh_4(CO)_{12}$	Tetrarhodium dodecacarbonyl	747.76	S >150		dk, red; cr
53	$Ir_2(CO)_9$	Diiridium nonacarbonyl	636.49	S 160		gr, yel; cr
54	$Ir_4(CO)_{12}$	Tetrarhodium dodecacarbonyl	1104.92	d 210		yel; trig, cr
55	$Ni(CO)_4$	Nickel tetracarbonyl	170.73	− 25	43	col; D = 1.32 v.sl. s.w

Iso and heteropolyacids and their salts

No.	Formula	Name	Mol. wt.	MP/FP	Colour and other properties
1	$Na_2[MoO_4]$	Disodium molybdate	205.94	687	wh; s.w
2	$Na_2[MoO_4]\cdot 2H_2O$	Disodium molybdate dihydrate	241.98	687	D = 3.28; wh; rhomb; s.w
3	$Na_2[Mo_2O_7]$	Disodium dimolybdate	349.89	612	wh; need; e.s.w
4	$Na_2[Mo_3O_{10}]\cdot 7H_2O$	Disodium trimolybdate heptahydrate	619.96	528	100–120°–6H₂O; need; sl. s.w
5	$Na_2[Mo_4O_{13}]\cdot 6H_2O$	Disodium tetramolybdate hexahydrate	745.89		yel; need; s.w
6	$Na_6[Mo_7O_{24}]\cdot 22H_2O$	Hexasodium heptamolybdate-22 water	1589.98	700	120–200°–H₂O; col; monocl; e.s.w
7	$(NH_4)_6[Mo_7O_{24}]\cdot 4H_2O$	Hexammonium heptamolybdate-tetrahydrate	1235.95	d	yel. s.w
8	$Na_2[Mo_8O_{25}]\cdot 17H_2O$	Disodium octamolybdate-17 water	1519.97	d	monocl; cr. e.s.w
9	$Na_2[Mo_{10}O_{31}]\cdot 21H_2O$	Disodium decamolybdate-21 water	1879.83		wh; prism; e.s.w
10	$H_3[PO_4(Mo_3O_9)_4]\cdot 14H_2O$	Dodecamolybdatophosphoric-acid-14 water	2077.63		yel; Cr; s.w; al; eth
11	$H_3[PO_4(Mo_3O_9)_4]\cdot 30H_2O$	Dodeca molybdato-phosphoric acid–30 water	2365.89		104–H₂O
12	$(NH_4)_3[PO_4(Mo_3O_9)_4]$	Triammonium dodeca-molybdatophosphate	1876.50	d	yel; prism; s.w
13	$(NH_4)_6[TeO_6(MoO_3)_6]\cdot 7H_2O$	Hexammonium hexamolybdato-tellurate heptahydrate	1321.66	d 550	col; rhomb; s.w
14	$Na_2[WO_4]$	Disodium tungstate	293.91	698	D = 4.179; wh; rhomb
15	$Na_2[WO_4]\cdot 2H_2O$	Disodium tungstate dihydrate	329.95	698	100°–2H₂O; Col; rhomb; s.w
16	$Na_6[W_7O_{24}]\cdot 16H_2O$	Hexasodium heptatungstate-16 water	2097.68	921	col; tetr. s.w
17	$K_6[W_7O_{24}]\cdot 6H_2O$	Hexapotassium heptatungstate-hexahydrate	2014.11	d	rhomb; sl. s.w
18	$(NH_4)_6[W_7O_{24}]\cdot 6H_2O$	Hexammonium heptatungstate-hexahydrate	1887.78	d	sl. s.w
19	$K_6H_2W_{12}O_{40}\cdot 18H_2O$	Hexapotassium dihydrogen-dodecatungstate–18 water	3407.92	~930	s.w
20	$Ca_3H_2W_{12}O_{40}\cdot 27H_2O$	Tricatcium dihydrogen-dodecatungstate–27 water	3455.73	d	col. tetr.
21	$Ba_3H_2W_{12}O_{40}\cdot 27H_2O$	Tribarium dihydrogen-dodecatungstate–27 water	3747.57		rhomb; d. in water; s. hot water
22	$H_5[BO_4(W_3O_{12})_4]\cdot 30H_2O$	Dodecatungstato boric acid-30 water	3403.39	~ 50	D = 3.0. col; s.w; al; eth; tetr
23	$Cd_5[BO_4(W_3O_{12})_4]_2\cdot 18H_2O$	Pentacadmium dodeca-tunstatoborate–18 water	6602.06	75	yel; e.s.w; tetr
24	$H_4[SiO_4(W_3O_{12})_4]\cdot 14H_2O$	Dodecatungstato silicic acid-14 water	3131.36	d	col; tetr; e.s.w; al; eth
25	$H_4[SiO_4(W_3O_{12})_4]\cdot 31H_2O$	Dodecatungstato silicic acid-31 water	3437.64	~ 50	yelsh; tetr; e.s.w; al; eth
26	$Na_4[SiO_4(W_3O_{12})_4]\cdot 20H_2O$	Tetrasodium dodecatungstato-silicate–24 water	3327.41	d	col; tetr; e.s.w; al
27	$K_4[SiO_4(W_3O_{12})_4]\cdot 18H_2O$	Tetra potassium dodeca-tungstato silicate–18 water	3355.77	d	col; hex; e.s.w; org. sol
28	$H_3[PO_4(W_3O_{12})_4]\cdot 14H_2O$	Dodecatungstato-phosphoric acid–14 water	3133.27		yelsh. gr; tetr; s.w; al; eth
29	$H_3[PO_4(W_3O_{12})_4]\cdot 24H_2O$	Dodecatungstato-phosphoric acid–24 water	3313.43	89	tetr. cr. s.w
30	$(NH_4)_3[PO_4(W_3O_{12})_4]$	Triammonium dodecatungstato-phosphate	2932.14		need; e.s.w

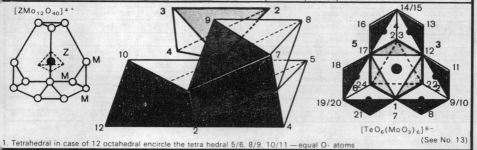

$[ZMo_{12}O_{40}]^{4+}$

$[TeO_6(MoO_3)_6]^{6-}$
(See No. 13)

1. Tetrahedral in case of 12 octahedral encircle the tetra hedral 5/6, 8/9, 10/11 — equal O- atoms

Salts of silicic acid: silicates

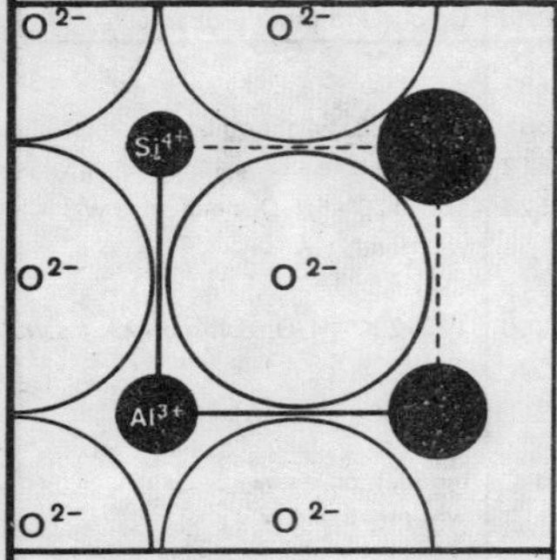

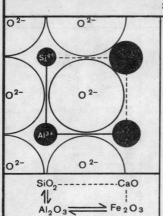

O2–
O2–
Si4+
O2–
O 2–
Al3+
O2–
O 2–

SiO_2 ---------- CaO

⇅ ⁞

Al_2O_3 ⇌ Fe_2O_3

SILICATES are the essential constituents in the mineral mixture of rocks. Together with quartz—"Silicic acid",—they form about 95% of earth's crust. (see p. 13)

Despite the complicated different chemical compositions, a simple explanation for the structure is possible.
Silicates consist of big ANIONS in a strong dense cubical packing, where charge equalization follows through the embedded central CATIONS and the peripherally stored CATIONS.

IONS before the smallest Si^{4+} ions but after the medium Al^{3+} are centrally stored, whereas the big and very big cations, e.g., alkaline earth metals and alkali metals, are accumulated.

Through the different possible combinations of the $[SiO_4]^{4-}$ tetrahedral—isolated, chained, bound, laminated, network, and three dimensional structures occur. (see pp. 71...73)

big **Anions**	very big	big	medium **Cations**	and	small (see also p. 22
	X		Y		Z
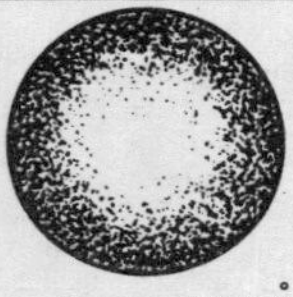	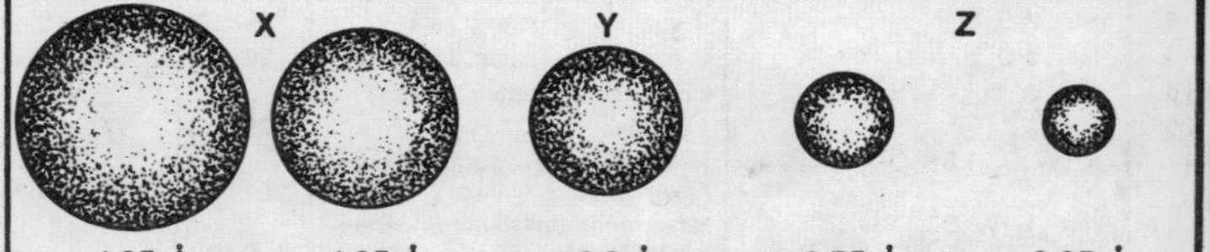				
Radius ~ 1.35 Å	~ 1,35 Å	~ 1,05 Å	~ 0,8 Å	~ 0,55 Å	~ 0,35 Å
Mostly O^{2-}	K^+	Na^+	Mg^{2+}	Al^{3+}	Si^{4+}
Fewer OH^-	Cs^+	Ca^{2+}	Fe^{2+}	Fe^{3+}	Be^{2+}
Seldom F^-	Ba^{2+}	Y^{2+}	Mn^{2+}	Cr^{3+}	P^{5+}

No.	Formula	Name	MP/FP	D/sp. gr.	Colour and other properties		
1	Li_4SiO_4	Lithium orthosilicate	1256	2.28	rhomb; col; i.w		
2	Li_2SiO_3	Lithium metasilicate	1201	2.52	rhomb; col; i.w		
3	Na_4SiO_4	Sodium orthosilicate	1018		hex; col; i.w		[1]
4	$Na_2SiO_3 \cdot 9H_2O$	Sodium metasilicate - 9 water	d(–H_2O)	2.4	rhomb; col; e.s.w		[1]
5	Na_2SiO_3	Sodium metasilicate	1088	2.61	monocl; col; s.w		[1]
6	$Na_2Si_2O_5$	Sodium disilicate	874		rhomb; col; s.w		[1]
7	K_2SiO_3	Potassium metasilicate	976		amorph; col; s.w		[2]
8	$K_2Si_2O_5$	Potassium disilicate	~1015	2.5			
9	$KHSi_2O_5$	Potassium hydrogen disilicate	515	2.42	rhomb		[2]
10	$K_2Si_4O_9 \cdot H_2O$	Potassium tetrasilicate - monohydrate	770 w H_2O	2.42	rhomb; s.w		[2]
11	$Be_2[SiO_4]$	Beryllium orthosilicate ⚒ phenakite	2200	2.98	trid; col	7.5–8	
12	$Mg_2[SiO_4]$	Magnesium orthosilicate ⚒ forsterite	1890	3.27	rhomb; wh	6–7	[3]
13	$Ca_2[SiO_4]$	α Calcium orthosilicate	2130				
		β Calcium orthosilicate	1456 →α				
14	$SrSiO_3$	Strontium silicate	1580	3.65	col; i.w		
15	$BaSiO_3 \cdot 6H_2O$	Barium metasilicate - 6 Wesser	d(–H_2O)	2.59	wh; i.w		
16	$BaSiO_3$	Barium metasilicate	1604	4.44			
17	$Zn[SiO_4]$	Zinc orthosilicate ⚒ Willemit	1509	3.9	trig; wh; i.w	5.5	
18	$ZnSiO_3$	Zinc metasilicate	1437	3.52	hex; col; i.w		
19	$Zr[SiO_4]$	Zirconium orthosilicate ⚒ zircon	~2500	3.9–4.8	tetra; brn. red	7.5	[4]
20	$Fe_2[SiO_4]$	Iron orthosilicate ⚒ fayalite	1205	4.2	yel. grn	6.5	
21	Co_2SiO_4	Cobalt orthosilicate		4.68	ruby red		
22	$Th[SiO_4]$	Thorium orthosilicate ⚒ thorite		5.3	tetr; col; ⚒ hard.	4.5–5	[5]

[1] Sodawater glass [2] Potassium water glass [3] Olevine
[4] Transparent often 4% Hf and seldom with rare earths [5] Contains He.

Salts of silicic acid: silicates

No.	Formula	Name	D/sp.gr.	H	Colour and other properties
1	$Al_2 [O\|SiO_4]$	Disthen, Kyanite	3.6–3.7	4.7 →	tricl; transparent, bl;
2	$Al_2 [O\|SiO_4]$	Andalusite	3.1–3.2	7–7.5	rhomb; gr; red; grn
3	$Al_2 [F_2\|SiO_4]$	Topas	3.5–3.6	8	col; yet; bl.
4	$(Mg,Fe)_2 [SiO_4]$	Olivine	3.3–4.2	6.5–7	Orthorhomb; grn; yel; brn
5	$Ca_2 Al_3 [(OH)\|(SiO_4)_3]$	Zoisite	3.2–3.4	6	Orthorhomb; ash. gr
6	$CaTi$	Titanite, sphene	3.4–3.6	5–5.5	monocl; blk; brn, red brn
7	$Ca_2 (Al,Fe^{III})_3 [(OH)\|(SiO_4)_3]$	Epidot, pistalite	3.3–3.5	6–7	monocl; dk. grn; bl; gr; blk. gr
8	$Ca Fe_2^{II}Fe^{III}[(OH)\|(SiO_4)_2]$	Lievrite, ilvaite	4.1	5.5–6	O-rhomb; blk. brn
9	$X_3 Y_3 [Z_3O_{12}]$	Garnet	3.4–4.6	6.5–7.5	cub;
	x=Ca, Mg, Fe^{II}, Mn.Y; Y=Fe^{III}, Al, Cr, Ti^{III}; Z=Si(sometimes upto 4% P or As. In artificial Garnets also Al.)				
10	$Mg_3 Al_2 [SiO_4]_3$	Magnesiaton garnet	~3.5		bl. red; ca; Fe; Cr
11	$Ca_3 Al_2 [SiO_4]_3$	Lime gàrnet	~3.5		col; wh; gr; pale red; Fe; Cr
12	$Ca_3 Cr_2 [SiO_4]_3$	Calcium-chromium garnet	3.4		dk. guignet's grn; gt; mass;
13	$Ca_3 Fe_2 [SiO_4]_3$	Calcium-iron garnet, andralite	~3.7		brn; Mg; al.
14	$Mn_3 Al_2 [SiO_4]_3$	Manganese garnet	~4.2		yel; red brn; Fe
15	$Fe_3 Al_2 [SiO_4]_3$	Iron garnet * "Common garnet" carbuncle	~4.2		red, brn; contains sometimes also Fe^{III}, Al, Cr, Ti^{III}
16	$Ca_2 (Fe, Mg, Mn) Al_2 BH [SiO_4]_4$	Axinite	3.2-3.3	6.5-7.0	tricl; brn; bl; ~ 5% B_2O_3
17	$(Ca, Ce, La, Na_2)_2 (Al, Fe, Mg, Mn)_3$ $[OH\|(SiO_4)_3]$	Orthite-cerium epidot } Allanite	3-4.2	5.5	blk; contains also Th; Pr, y and other rare earths.
18	$U_2 [O\|SiO_4] \cdot 2H_2O$	Soddyite*	4.6		yel
19	$Mg, U_2 [(OH)_3 SiO_4]_2 \cdot 4H_2O$	Skladowskite*	3.5		rhomb; Pr
20	$Ca U_2 [(OH)_3 SiO_4]_2 \cdot 4H_2O$	Uranotile*	3.8-3.9	2.5	tricl; O-rhomb; honey yel
21	$Pb U [O_2\|SiO_4] H_2O$	Kasolite*	6.5	4	brn, yel

$[SiO_4]^{4+-}$
tetraeder

Isolated or grouped type combinations in case of $[SiO_4]^{4-}$ tetrahedral around common (oxide) Oxygen ions.

② $[Si_2 O_7]^{6+}$ ③ $[Si_3 O_9]^{6+}$ ④ $[Si_4 O_{12}]^{8+}$ ⑥ $[Si_6 O_{18}]^{12+}$

M = Mixed forms

No.	Formula		Name	D/sp.gr.	H	Colour and other properties
22	$Zn_4 [(OH)_2\|Si_2 O_7] \cdot H_2O$	②	Hemi meorphite, Siliceous cala mine, (hydrous zinc silicate)	3.3 - 3.5	5	o-rhomb, col;
23	$(Se, Y)_2 [Si_2 O_7]$	②	Thortveitite only with y (as in 9) = thalenite	3.6	6.5	monocl; dk. grn; blk;
24	$Ca_3 [Si_3 O_9]$	③	Wollastonite	2.8 - 2.9	4.5 - 5	monocl, col, 51.7% SiO_2 ③mineral structure
25	$Mn SiO_3$		Rhodonite	3.4 - 3.7	5.5 - 6.5	tricl, 54% MnO, structurally similar to 24
26	$Ba Ti [Si_3 O_9]$	③	Benitoite	3.7	6.5	trig
27	$(Na, K)_2 (Fe^{II} Mn) Ti [Si_4 O_{12}]$	④	Neptunite	3.2	5.5	monocl, blk, gl. gloss
28	$Cu_6 [Si_6 O_{18}] \cdot 6H_2O$	⑥	Dioptas	5.3	5	trig, rhomb, blk, green, gl. gloss
29	$Be_3 Al_2 [Si_6 O_{18}]$	⑥	Beryll	2.6 - 2.9	7.5 - 8	hex, col, blk, gr, yel, red
30	$Mg_2 Al_3 [Al Si_5 O_{18}]$	⑥	Cordierite, Dichroite	2.6	7 - 7.5	o-rhomb, dk blk, gr, pa, bl
31	$Be_4 [(OH)_2\|SiO_4\|SiO_3]$	M	Bertrandite	~2.6	6 - 7	rhomb; col.
32	$Ca_{10} Al_4 (Mg, Fe)_2$ $[(OH)_4\|SiO_4\|(Si_2 O_7)_2]$	M	Vesuvian	3.3 - 3.5	6.5	tetr; pale blk; remi coloured; related to garnet
33	$XY_9(OH, F)_4 [B_3 Si_6 O_{27}]$	M	Tourmaline	3 - 3.3	7	colour according to X and Y
	X = Na or Ca Y = Al, Mg, Fe^{II}, Fe^{III}, Ti^{III}, Cr^{III} and others; Si in part through Al substitute; XY 9 adds 25 valencies.					
34	e.g. $Al_7 Na_2 Mg$		$(OH, F)_4 [B_3 Si_6 O_{27}]$			Alkali - Alumina - Tourmaline
35	e.g. $Al_5 Mg_3 Ca, Mg$		$(OH, F)_4 [B_3 Si_6 O_{27}]$			Magnesia - Alumina - Tourmaline
36	e.g. $(Al, Fe^{III}_5), Fe^{II} \cdot Ca, Fe^{II}$		$(OH, F)_4 [B_3 Si_6 O_{27}]$			Iron - Tourmaline

* Uranotile - Groups occur in Katanga.

Salts of silicic acid: silicates

No.	Formula	Name	D/sp.gr.	H	Colour and other properties
1	$Mg [SiO_3]$	Eustatite	3.1–3.4	5–6	rhomb; gr. wh; brsh. 0-5% FeO
2	$Fe [SiO_3]$	Ferrosilite	3.5		monocl; MP 1550; 54.5% FeO
3	$(Mg,Fe)[SiO_3]$	Bronzite	3.2–3.5		dk. blk; brn. gr; 5-10% FeO
4	$(Fe, Mg) [SiO_3]$	Hypersthene	3.5		blk-blk gr; 15-34% FeO
5	$Li Al [SiO_6]$	Spondumene, triphane	3.1–3.2	6.5–7	col; bl; gr;
6	$Na Fe [SiO_6]$	Agirine, aknite	3.5–3.6	6–6.5	monocl; blk; wh
7	$Ca Mg [Si_2O_6]$	Diopside	3.2–3.4	5–6	monocl; wh; gr
8	$Ca (Mg, Fe^{II}) (Al, Fe^{III})$ (Si_2O_6)	Augite (often Al instead Si)	3.3–3.6	5–6	monocl; dl. grn; blk; according to % of FeO
9	$Al IAlO_4 SiO_4$	Sillimanite fibrolite	3.2	6–7	O-rhomb; gr; see also andalusite
10	$X_2Y_5Z_8 (OH, F)_2 O_{22}$ $X = Ca, Na, K, Mn^{II}, Fe^{II}, Mg$ $Y = Mg, Fe^{II}, Fe^{III}, Al, Mn^{III}, Ti^{III}$ $Z = Si, Al$ (rarely also P^V or V^V)	Amphiboles or hornblends			
11	$Ca_2Mg_5 [Si_8O_{22}] (OH)_2$	Actinolite	2.9–3.1	5.5–6	monocl; gr; with Fe without Al, see also garnet.
12	$Ca_2Mg_5 [Si_8O_{22}] (OH)_2$	Hornblend	2.9–3.4	5–6	monocl; grn; blk; with Fe and 8-15% Al_2O_3
13	$Na_2Mg_3Al_2[Si_8O_{22}](OH)_2$	Sodium horn blend	3.0–3.1		monocl; dk. blk; bl; blk. bl; 5-10% Na_2O without FeO
14	$(Mg, Fe)_7 [Si_8O_{22}] (OH)_2$	Anthophyllite	2.9–3.2	5.5	rhomb; br. yelsh, grnsh.

$6 O^{..}$
$4 Si^{4+}$
$4O^{..} + 2OOH^-$
$6 Mg$
$4O^{..} + 2OH^-$
$4Si^{4+}$
$6O^{..}$

e.g. 15   Series of layers in **Talc** (lateral section)

$6 OH^-$
$4 AL^{3+}$
$4O^{..} + 2OH$
$4 Si^{4+}$
$6 O^{..}$

e.g. 18   Series of layers in **Kaoline** (lateral section)

No.	Formula	Name	D/sp.gr.	H	Colour and other properties
15	$Mg_6 (OH)_2 [Si_8O_{20}]$	Talc; steatite	2.7–2.8	1	monocl; col; lt. grn. wh; grnsh.
16	$Mg_6(OH)_6[Si_4 O_{11}] \cdot H_2O$	Serpentine (asbestos)	2.5–2.6	3–4	fibrous structures, wh; refer to hornblend
17	$K Ca_4 [F(Si_4O_{10})_2] \cdot 8H_2O$	Apophyllite	2.3–2.4	4.5–5	tetr; col; wh;-usually known as zaolite
18	$Al_4 (OH)_8 [Si_4O_{10}]$	Kaoline, Kaolinite	2.6	2–2.5	wh; porcelain clay
19	$Al_2 [(OH)_2 Si_4O_{10}] nH_2O$	Montmorillonite	1.7–2.7		o-rhomb; air dried n = $4H_2O$ at 400°–$4H_2O$; ~ zeolite
20	$(Ni, Mg)_6 (OH_8 [Si_4O_{10}])$	Garnierite	2.2–2.7	2–4	grn; bl. grn; 4-30% Ni
21	$Fe_2Al_4(OH)_4 [Al_4Si_4O_{20}] \cdot 2Fe(OH)_2$ Chloritoid		3.4–3.6	6–7	monocl; dk. grn; blk.
22	$XY_2(OH,F)_2[Z' Z''_3O_{10}]$ $X = K, Na, Rb, Cs, rarely Ba$ $Y = Al, Fe^{III}, Cr^{III}, Mn^{III} V^{III}, Ti^{III}, Mg^{II}, Fe^{II}, Mn^{II} Li, rarely Zr$ $Z = Al, Si, Fe^{III}, Mn^{III}, Ti^{III}, Z$ only Si	Mica		$\rightarrow$4.2–5 $\uparrow$6–3.5	Hardness in varied directions
23	$K, Li[Al(OH,F)_2][AlSi_3O_{10}]$	Lepidolite, lithium mica	2.8–2.9	2.5–4	monocl; rose wh; gr; violet
24	$K (Fe^{II}Al(OH)_3 [AlSi_3O_{10}]$	Tin waldite	2.9–3.1		monocl; gr, brn; blk; often with F instead OH
25	$K Al_2(OH,F)_2 [AlSi_3O_{10}]$	Muscovite	2.8–2.9	2–2.5	monocl; col; gr. bl. grn
26	$K (Mg, Fe^{II})_3 (OH) [(Al, Fe^{III}) Si_3O_{10}]$	Biotite**	2.8–3.2	2.5–3	monocl; dk. blk; brn; blk
27	$Ca Al_2 (OH)_2 [Al_2Si_2O_{10}]$	Margarite***	3	~4	monocl; blk; wh; gr.
28	$Mg_4 Al_2 [(OH)_8 Al_2Si_2O_{10}]$	Chlorite	2.6–3.3	~2	monocl; very few variations in mica related

** Magnesium iron mica   *** Calcium mica (lime mica)

Salts of silicic acid: silicates

Plane combinations in case of $[SiO_4]^{4-}$ Tetrahedral:

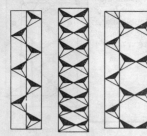 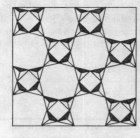 

Series or band and surface network structures in case of silicates (these tetrahedrous correspond to a SiO_4 radical)

Three dimensional combination in case of SiO_4 – tetrahedron

inverse tetrahedral layers

mirror imaged tetrahedral layers

Tridymite
Tridymite structure see pp. 42 & 85

Christobalite
Christobalite - structure see pp. 42 & 85

No.	Formula	Name	D/sp. gr.	H	Colour and other properties
1	Li[Al Si$_4$ O$_{10}$]	Petalite, castor	2.4	6.5	monocl; col.
2	Na[Al Si$_4$O$_{10}$] H$_2$O	Analcite	2.2 - 2.3	5 - 5.5	cub; col
3	Na[Al Si$_4$O$_{10}$]	Nepheline, elaolite	2.6	5.5 - 6	hex; col; 5–6% K$_2$O
4	K [Al Si$_2$O$_6$]	Leucite	2.5	5.5 - 6	tetr
5	Cs$_2$ [Al$_2$ Si$_4$O$_{12}$]·H$_2$O	Pollucite	2.9	6.5	col
6	Ca[OH|B SiO$_4$]	Datolite	2.9 - 3.0	5 - 5.5	monocl; col
7	Na[Al Si$_3$O$_8$]	Albite, soda feldspar	~2.64	6 - 6.5	trig; col; wh. gr
8	K[Al Si$_3$O$_8$]	Orthoclase, potash feldspar	2.56	6	monocl
9	K[Al Si$_3$O$_8$]	Microlin	~2.54		tricl; often grn
0	(Na, K)[Al Si$_3$O$_8$]	Anorthoclase	~2.6		tricl
1	Ca[Al$_2$Si$_2$O$_8$]	Anorthite	2.7 - 2.8	6 - 6.5	tricl; col; gr. yel
2	Ca[Al$_2$Si$_4$O$_{12}$]·4H$_2$O	Laumontite	~2.3	3 - 3.5	monocl; loses water on exposure to air
3	Na[Al Si$_3$O$_8$] isomorphous Ca[Al$_2$Si$_2$O$_8$] mixture	Plagioclase	2.6 - 2.8	6 - 6.5	tricl; few modifications with different mixing ratio
4	Na Ca$_2$Al$_5$Si$_6$O$_{20}$·6H$_2$O	Thomsonite	2.3 - 2.4	5 - 5.5	o. rhomb; dk. blk. dirty wh.
5	Na$_6$Ca$_2$[(CO$_3$)$_2$ (Al SiO$_4$)$_6$]	Cancrinite	2.4 - 2.5	5 - 6	hex; col; rose bottle grn.
6	Ca$_2$Na[Al$_5$Si$_{13}$O$_{36}$] 14H$_2$O	Desmine	2.1 - 2.2	3.5 - 4	monocl
7	Ca[Al$_2$Si$_4$O$_{12}$]·6H$_2$O	Chabasite 'analcite'	2.1	4.5	rhomb; col; bottle grn, dk. blk; col; wh
8	Ca Al$_2$[Si$_6$O$_{12}$]·5H$_2$O	Heulandite	2.2	3.5 - 4	monocl
9	(Ba,K$_2$)$_2$[Al$_4$Si$_{11}$O$_{30}$].10H$_2$O	Harmotom	~2.5	4.5	monocl; bottle grn, dk: blk

Double compounds

Double compounds in contrast to the complexes, decompose in solution or in the melt, into ions of component parts. (Limiting case, e.g. Li[AlH₄] or Pb[Pb₂O₄]—with complex binding in solid state are available.) As before, in case of all the double oxides and sulphides the cations inside the admissible radical boundary are exchangeable, a multiple of variation arises in spite of simple structures. (Examples Nos. 16, 17 and 18 see p. 75; also p. 70.)

(Examples Nos. 16, 17 and 18 see p. 75; also p. 70.)

Frequently occurring lattice structures are:

Spinel	= Sp		Zincblend	= Z
Ilmenite	= Il		Marcasite	= M
Pyrite	= P		Wurtzite	= W

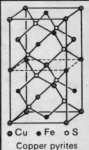

● Cu ● Fe ○ S

Copper pyrites

Double compounds with same anions

Double hydrides

No.	Formula		Name	Mol.wt	D/sp.gr.	H	Colour and other properties
1	Li Al H₄		Lithium aluminiumhydride	37.94			wh; cr; s. eth; d in w
2	Li Ga H₄		Lithium galliumhydride	80.69			wh; cr; s. eth; d in w

Double oxides

No.	Formula		Name	Mol.wt	D/sp.gr.	H	Colour and other properties
3	Be Al₂O₄	Sp	Beryllium aluminiumoxide ✿ chrysoberyll	126.95	3.76	8.5	rhomb; grn-yel; d in alk
4	Mg Al₂O₄	Sp	Magnesium aluminiumoxide ✿ Spinel	142.26	3.6	8	col; ✿ red and other colours MP 2135°C cub
5	Zn Al₂O₄	Sp	Zinc aluminiumoxide ✿ Zinc spinel, Galmite, "RINNMANN'S GREEN"	183.34	4.3	8	gl. gloss, grn.
6	Co Al₂O₄	Sp	Cobalt aluminiumoxide 'THENARD'S-BLUE"	176.90	4.37		bl. cub
7	Pb^{II} Pb$_2^{IV}$ O₄	—	Lead (II, IV) oxide	685.63	9.1		red; d in HNO₃+H₂O₂
8	Fe Cr₂O₄	Sp	Iron chromiumoxide ✿ chromium iron-steel	223.87	4.5-4.8	5.5	brn. blk
9	Mn^{II} Mn$_2^{III}$O₄	—	Manganese-II, III oxide ✿ Hansmannite	228.82	4.7	5.5	tetr; brn. red; MP=1560°C
10	Mg Fe₂O₄	Sp	Magnesium ironoxide	200.02	4.5		octa; blk; d in HCl; MP ~ 1750°C
11	Mn Fe₂O₄	Sp	Manganese ironoxide ✿ jahabsite	230.64	4.8		octa; blk;
12	Fe^{II} Fe$_2^{III}$O₄	Sp	Iron-II, III oxide ✿ Magnet iron steel	231.55	5.2	5.5	blk; ferromag; MP 1540°C
13	Ni Fe₂O₄	Sp	Nickel ironoxide ✿ trevorite	234.39			similar to No. 12
14	Co^{II} Co$_2^{III}$O₄	Sp	Cobalt-II, III oxide	240.82	6.1		blk. gr;
15	Mg TiO₃	Il	Magnesium titaniumoxide ✿	120.22	4.2	5.5	blk; metal gloss;
16	Ca TiO₃	Pe	Calcium titaniumoxide ✿ perowskite	135.98	4.0	5.5	blk;
17	Mn TiO₃	Il	Manganese titaniumoxide ✿ pyrophanite	150.84	4.5	5	trig; rhomb; bl. red; bottle grn.
18	Fe TiO	Il	Iron titaniumoxide ✿ Ilmenite	151.75	~ 5.0	5-6	trig; rhomb; blk;

Double nitrides

No.	Formula		Name	Mol.wt	D/sp.gr.	H	Colour and other properties
19	Li₃ Ga N₂		Lithium galliumnitride	118.56	d 800	3.35	gr; powd; d. in water

Double halogenides

No.	Formula	Name	Mol.wt	D/sp.gr.	H	Colour and other properties
20	NaCl·AlCl₃	Sodium aluminiumchloride	191.80	185		wh; yel; powd; s.w
21	KCl·CaCl₂	Potassium calciumchloride	185.55	754		cub; s.w
22	KCl·MgCl₂·6H₂O	Potassium magnesiumchloride 6 hydrate ✿ carnallite	277.89	265	1.60	rhomb; s.w
23	2KCl·PbCl₂	Potassiumchloride	427.23	490		yel; s.w
24	KCl·CuCl₂	Potassium cupric chloride	209.01		2.86	need; red
25	4KCl·MnCl₂	Potassium manganesechloride	424.06		2.31	trig; s.w;
26	2KCl·FeCl₃·H₂O	Potassium ironchloride	329.34		2.32	o-rhomb; red
27	(NH₄)Cl·MgCl₂·6H₂O	Ammonium magnesiumchloride-hexahydrate	256.83		1.456	rhomb; s.w
28	(NH₄)Cl·AlCl₃	Ammonium aluminiumchloride	186.84	304		cr; s.w
29	4(NH₄)Cl·CdCl₂	Ammonium cadmiumchloride	397.31		2.01	rhomb; s.w
30	(NH₄)Cl·NiCl₂·6H₂O	Ammonium nickelchloride-hexa hydrate	291.20		1.645	monocl; grn; s.w
31	2CaCl₂·MgCl₂·12H₂O	Calcium magnesiumchloride 12 water ✿ Tiechyhydrite			1.66	rhomb; cr; yel; bl; deliq.
32	2KI·CdI₂·2H₂O	Potassium cadmiumiodide	734.31		3.395	cr; powd; yel; wh; e.s.w
33	(NH₄)I·CuI·H₂O	Ammonium copperiodide	353.49			rhomb; pl. d. in water

Double sulphides

No	Formula		Name	D	H	Colour and other properties
1	$Pb_4As_2S_7$		Jordanite	6.4	3	gr; monocl
2	Cu_3AsS_4	~W	Enargite	4.4	3.5	pris. hex; metal gloss. gr. blk.
3	Ag_3AsS_3		Proustite	5.57	2.5	trig; red
4	Fe As S	~P	Arsenopyrites	5.9 - 6.2	5.5 - 6	monocl, metal lust; wh. gr.
5	Co As S	~P	Cobalt sulphides	6 - 6.4	5.5	metal lust, blk; gr
6	Ni As S	~P	Gesderfite, arsenic nickel sulphide	5.6 - 6.2	5.5	metal lust, blk. gr.
7	$Pb_5Sb_4S_{11}$		Boulangerite, antimony sulphide	5.8 - 6.2	2.5	monocl; lead gr.
8	Cu_3 Sb S_3	~Z	Tehaedrite	4.4 - 5.4	3 - 4	tetr; steel gr; cub.
9	Ag_3Sb S_3		Pyrargyrite	5.85	2.5 - 3	metal lust, grsh. blk. P. trig.
10	Ag_5 Sb S_4		Stephanite	6.2 - 6.3	2.5	o-rhomb; bl. gr., blk
11	Fe Sb S	~M	Gudmundite	6.72	4	monocl, metal gloss. wh. gr.
12	Ni Sb S	~P	Ullmanite, nickel sulphide	6.7	5	metal gloss. gr.
13	Cu Fe S_2	~Z	Copper pyrites, chalco pernite	4.1 - 4.3	3.5 - 4	tetr; gr. yel.
14	Cu_5 Fe S_4		Bornite	4.9 - 5.3	3	metal. lust; vlt. yel. cub.
15	Cu Fe_2 S_3		Cubanite, chalenersite	4.10	3.5 - 4	o-rhomb; metal lust; yel
16	$Me^{II} Me_2^{III} S_4$ ~Sp Cobalt nickel pyrites Me^{II} = Ca," Fe," Co," Ni" Me^{III} = Fe"', Co"', Ni"' e.g. Co" Co"'$_2$ S_4 = Linnert; Cu Co_2S_2 = Caroleit; (Co, Ni)$_3$ S_4 Siegnite, Ni" Ni$_2^{III}$ S_4 Polydymite			4.8 - 5.8	4.5 - 5.5	metal gloss; wh. gr. (Composition varying - mixed crystals)
17	$Me^{II}_3 R_2 S_6$ "Fahlerze" Me^{II} = Cu_2, Ag_2, Fe, Zn, Hg; R = As, Sb, Bi					e.g. No.8 (mixed structures - mixed crystals)
18	m Me^{II} S + n R_2S_3 Antimony sulphides, approx. 30 variations Me^{II} = Pb, Ag_2, Cn_2, Tl_2. Fe; R = As, Sb, Bi; m and n simple whole numbers					e.g. No.17
19	Pb Cu Sb S_3		Bonronite	5.7 - 5.9	3	o-rhomb, metal glass. gr.
20	Pb_4 Fe Sb_6 S_{14}		Jamesonite	5.63	2.5	monocl, lead gr.
21	Ag_8 Ge S_8		Argyrosonite	6.2	2.5	steel. gr.
22	Cu_6 Fe Ge S_8		Germanite	4.59	3	cub. opaque
23	Bi_2 Te_2 S		Tetradynite, "Tellurium bismuth"	7.2 - 7.9	1.5 - 2	lead grey

Double nitrides   Mol. wt.

No	Formula	Name	D	H	Colour and other properties
24	$KNO_3 \cdot Ag\ NO_3$	Potassium silver nitrate	270.99	3.219	monocl. e.s.w; MP = 125°

Double sulphates   Mol. wt.   H

No	Formula	Name	D		Colour and other properties
25	$Na_2SO_4 \cdot MgSO_4 \cdot 2.5H_2O$	Lowite ✿	307.49	2.4	wh; s.w;
26	$Na_2SO_4 \cdot MgSO_4 \cdot 4H_2O$	Astrahanite ✿	334.51	2.23	monocl; wh; s.w
27	$3Na_2SO_4 \cdot MgSO_4$	Vanthofite ✿	546.57	2.69	wh; s.w;
28	Na $SO_4 \cdot CaSO_4 \cdot 2H_2O$	Glauberite ✿	314.24	2.64	monocl; need; d in w80° (-2H_2O)
29	Na $Al(SO_4)_2 \cdot 12H_2O$	Sulphatrite ✿	458.30	1.67	cub; wh. s.w; d 61°
30	3 $K_2SO_4 \cdot Na_2SO_4$	Glaserite ✿	664.83	2.7	cub; s.w
31	$K_2SO_4 \cdot MgSO_4 \cdot 4H_2O$	Leonite ✿	366.71	2.201	monocl; col; e.s.w
32	$K_2SO_4 \cdot MgSO_4 \cdot 6H_2O$	Schonite ✿	402.74	2.15	monocl; col; d. 72°
33	$K_2SO_4 \cdot 2MgSO_4$	Langbenite ✿	415.03	2.829	tetr; MP = 927°
34	$K_2SO_4 \cdot CaSO_4 \cdot 2H_2O$	Syngenite ✿	328.42	2.60	monocl; MP = 1004° d in w
35	$KAl\ (SO_4)_2 \cdot 12H_2O$	Kalinite, alum ✿	474.40	1.75	s.w; d 92°
36	K $Cr\ (SO_4)_2 \cdot 12H_2O$	Chromalum ✿	499.42	1.83	cub; octa; red or gr
37	K $Mn\ (SO_4)_2 \cdot 12H_2O$	Potassium manganese sulphate-12 water	502.25		cub; vlt; d in w
38	K_2SO_4 2 $Mn\ SO_4$	Potassium manganese sulphate	476.23	3.02	tetr; rose; MP = 850°
39	$K_2SO_4 \cdot FeSO_4 \cdot 6H_2O$	Potassium iron - II sulphate	434.27	2.169	monocl, prism, grn
40	$K_2SO_4 \cdot Fe_2\ (SO_4)_3 \cdot 2H_2O$	Potassium iron - III sulphate - dihydrate	610.19	2.840	monocl; pale yel, grn
41	K Fe $(SO_4)_3 \cdot 12H_2O$	Potassium iron - III sulphate - 12 water	503.27	1.83	octa; col; e.s.w
42	$K_2SO_4 \cdot CoSO_4 \cdot 6H_2O$	Potassium cobalt - II sulphate - hexahydrate	437.36	2.218	monocl; prism; red; s.w
43	$K_2SO_4 \cdot NiSO_4 \cdot 6H_2O$	Potassium nickel - II sulphate	437.11	2.124	monocl; bl; s.w; d <100°
44	$(NH_4)_2SO_4 \cdot MgSO_4 \cdot 6H_2O$	Ammonium magnesium sulphate	360.63	1.723	monocl; col; s.w
45	$(NH_4)Al(SO_4)_2$	Ammonium aluminium sulphate	237.14	2.039	hex; col; s.w
46	$(NH_4)Al(SO_4)_2 \cdot 12H_2O$	Ammonium aluminium sulphate - 12 water	453.33	1.64	cub. col; e.s.w;93° s. aq reg.
47	$(NH_4)_2SO_4 \cdot ZnSO_4 \cdot 6H_2O$	Ammonium zinc sulphate	401.69	1.931	monocl; wh; s.w
48	$(NH_4)Cr(SO_4)_2 \cdot 12H_2O$	Ammonium chromium sulphate	478.38	1.72	gr. or vlt; s.w. al
49	$(NH_4)SO_4 \cdot MnSO_4 \cdot 6H_2O$	Ammonium manganese sulphate	391.24	1.83	monocl; bl. red; s.w
50	$(NH_4)_2Fe^{II}(SO_4)_2 \cdot 6H_2O$	Mohr's salt	392.16	1.864	monocl; gr; s.w; al
51	$(NH_4)Fe^{III}(SO_4)_2 \cdot 12H_2O$	Ammonium iron - III sulphate - 12 water	482.21	1.71	octa; vlt; e.s.w
52	$(NH_4)_2SO_4 \cdot COSO_4 \cdot 6H_2O$	Ammonium cobalt sulphate - hexahydrate	395.25	1.902	monocl; ruby red; s.w
53	$(NH_4)_2SO_4 \cdot NiSO_4 \cdot 6H_2O$	Ammonium nickel sulphate - hexahydrate	395.02	1.923	monocl; bl. grn; s.w

No	Formula	Name	Mol. wt.	D/sp. gr	Colour and other properties
Double borates					
1	$NaCaB_5O_9 \cdot 8H_2O$	Sodium boron calcite ✿		2.00	wh; trig. H = 1 .
Double carbonates					
2	$K_2CO_3 \cdot Na_2CO_3 \cdot 6H_2O$	Potassium sodium carbonate	230.20	1.62	monocl;
3	$KHCO_3 \cdot CoCO_3 \cdot 4H_2O$	Potassium cobalt hydrogen carbonate	291.13		redsh. need; d in w;
4	$(NH_4)_2CO_3 \cdot MgCO_3 \, 4H_2O$	Ammonium magnesium carbonate	252.49		wh; s.w;
5	$CaCo_3 \cdot MgCO_3$	Dolomite ✿		285-295	rhomb; tetr; wh; ✿ H = 3.5–4
Double phosphates					
6	$Na(NH_4)(HPO_4) \cdot 4H_2O$	Sodium ammonium hydrogen phosphate	209.09	1.554	monocl; col; s.w; 79°
7	$(NH_4)Mg(PO_4) \cdot 6H_2O$	Ammonium magnesium phosphate	245.43	1.72	rhomb; col; i.w
8	$(NH_4)Ca(PO_4) \cdot 7H_2O$	Ammonium calcium phosphate	279.21	1.561	monocl; col; i.w;
9	$(NH_4)Mn(PO_4) \cdot H_2O$	Ammonium manganese phosphate	185.97		wh; cr; i.w; s. NH_4^+ sol
10	$(NH_4)Co(PO_4) \cdot H_2O$	Ammonium cobalt phosphate	189.98		cr, prism; vlt; i.w;
Double arsenates					
11	$(NH_4)Mg(AsO_4) \cdot 6H_2O$	Ammonium magnesium arsenate	289.37	1.932	tetr; col; i.w;
12	$(NH_4)Ca(AsO_4) \cdot 6H_2O$	Ammonium calcium arsenate	305.13	1.905	monocl; col; s.w; hot NH_4^+ sol
Double chromates					
13	$K_2CrO_4 \cdot MgCrO_4 \cdot 2H_2O$	Potassium magnesium chromate	370.56	2.59	tetr;
14	$(NH_4)_2CrO_4 \cdot MgCrO_4 \cdot 6H_2O$	Ammonium magnesium chromate	400.52	1.84	monocl; eyl; e.s.w;
Double complexes					
15	$Li_3Na_3\,[AlF_6]_2$	Lithium sodium hexafluoro aluminate- ✿ "Knyolithionite"	317.75	2.8	cr; col; ✿ H = 2.5; cr, analogous to granate
16	$K_2Na[Co(NO_2)_6] \cdot H_2O$	Potassium sodium hexanitro cobaltate-III	454.19	1.633	cr; yel; i.w; s. acid
17	$K_2Na[Fe(CN)_6]$	Potassium sodium hexacyano ferrate-III	313.1		cr; or red; monocl; s.w;
Double compounds with same cations					
Oxide salts					
18	$PbCl_2 \cdot PbO \cdot H_2O$	Lead oxychloride monohydrate	519.35	6.05	monocl; prism; d 150°
19	$PbCl_2 \cdot PbO$	Lead oxychloride	501.33	7.21	rhomb; tetr; alk; d 524°
20	$PbCl_2 \cdot 2PbO$	Mendipite ✿	724.54	7.08	rhomb; yel; s. alk; MP 693°
21	$PbCl_2 \cdot 3PbO$	Tetraplumbic oxychloride	947.75		yel; sp. w
22	$PbCl_2 \, 7PbO$	"Kasseler yellow"	1840.49		yel; i.w;
23	$PbSO_4 \cdot PbO$	Lead oxysulphate	526.49	6.92	monocl; w; s.w; wh;
24	$PbCrO_4 \cdot PbO$	Lead oxychromate "chrome red"	546.43		cr; prism; red. i.w
25	$CuCrO_4 \cdot 2CuO \cdot 2H_2O$	Tricupric dioxychromate-dihydrate	374.66		yel, brn; s. NH_4OH
26	$HgF_2 \cdot HgO \cdot H_2O$	Mercuric oxydifluoride	473.24		cr, yel; d in w;
27	$HgCl_2 \cdot HgO$	Mercuric oxydichloride	704.74	~8.5	hex; red; sl. s.w;
28	$HgCl_2 \cdot 3HgO$	Tetra mercuric trioxydichloride	921.35	7.93	hex; yel; d in w;
29	$HgBr_2 \cdot 3HgO$	Tetra mercuric trioxydibromide	1010.27		cr; yel; s.w; hot;
30	$HgI_2 \cdot 3HgO$	Tetra mercuric trioxydiiodide	1104.18		yel. brn; d in w. s. HI
31	$HgSO_4 \cdot 2HgO$	Tri mercuric dioxysulphate	729.90	6.44	p. yel; e.s.w; hot.
32	$Hg(CN)_4 \cdot HgO$	Mercuric oxytetracyanide	469.26	4.437	cr; or need; wh; s.w; hot; **exp**
Hydroxide salts					
33	$MgCO_3 \cdot Mg(OH)_2 \, 3H_2O$	Dimagnesium dihydroxy carbonate	196.71	2.02	rhomb;
34	$3MgCO_3 \cdot Mg(OH)_2 \, 3H_2O$	Tetra magnesium dihydroxy carbonate	365.37	2.16	rhomb; wh; v. sl. s.w
35	$4MgCO_3 \cdot Mg(OH)_2 \, 4H_2O$	Penta magnesium dihydroxy carbonate	467.72	2.16	rhomb; wh
36	$PbCl_2 \cdot Pb(OH)_2$	Lead dihydroxy dichloride	519.35	7.24	tetr; wh; sl. s.w; d 524°
37	$2PbCO_3 \cdot Pb(OH)_2$	"White lead"	775.67	6.14	hex; wh; i.w; d 400°
38	$CuCO_3 \cdot Cu(OH)_2$	Dicupric dihydroxy carbonate ✿ Malachite	221.11	4.0	monocl; dk. grn; i.w; d 200°
39	$2CuCO_3 \cdot Cu(OH)_2$	Tricupric dihydroxy carbonate	344.66	3.88	monocl; bl; i.w; d 200°
Double salts with different anions					
40	$2(NH_4)NO_3 \cdot (NH_4)_2SO_4$	Lunasalt peter	292.24	1.68	rhomb; s.w; MP = 310
41	$CaF_2 3Ca_3(PO_4)_2$	Fluorapatite		3.16	hexa, col; ✿ H = 5
42	$CaCl_2 \cdot 3Ca_3(PO_4)_2$	Chloropatite		3.22	often muddy and coloured
Double salts with different cations and anions					
43	$KCl \cdot MgSO_4 \cdot 3H_2O$ Kainite	Potassium magnesium chlorosulphate ✿	248.99	2.4	monocl; prism; (w) ✿ H = 3
44	$LiF \cdot Al(PO_4)$ "Amblygonite"	Lithium aluminium fluorophosphate ✿	147.88	3	wh; tricl; ✿ H = 6
Triple salts					
45	$2KNO_2 \cdot Pb(NO_2)_2 \cdot Cu(NO_2)_2$ "Tripelnitrite" or $K_2PbCu(NO_2)_2$	Dipotassium lead cupric hexanitrite	624.79		very difficult to solve blk. brn–blk, cubical copper proof (~0.38)
46	$2M^I(NO_2) \cdot M^{II}(NO_2)_2 \cdot Cu(NO_2)_2$	Common formula possible for "Triple nitrite" e.g., with $M^I = (NH_4)^+$, Rb^+, Cs^+ or $M^{II} = Sr^{2+}$ (dark gr. difficulty to solve Cr.)			
47	$K_2SO_4 \cdot 2CaSO_4 \cdot MgSO_4 \, 2H_2O$ or $K_2Ca_2Mg(SO_4)_4 \cdot 2H_2O$	✿ Polyhalite Dipotassium magnesium tetrasulphate dihydrate	602.97	2.775	wh; prism; ✿ H = 3–3.5

Organic substances or carbon compounds

Classification of organic compounds	see pages
The properties of organic substances depend on the construction of the skeleton = the structure of the presence	78–104
of functional groups	105–128

PARENT SUBSTANCES | 80–104

Organic substances consisting only of carbon atoms in structural framework	80–86
1. simple and branched carbon chains	80–82
2. simple and condensed carbon rings (carbocycles)	83–86

Organic substances containing also other (hetero) atoms in structural framework	86–94
1. simple and branched hetero chains	86–89
2. simple and condensed hetero rings (heterocycles)	90–94

PARENT SUBSTANCES WITH LINKS . | 95–104

1. Carbon rings with carbon chains	95–97
2. Carbon rings linked through carbon chains	98
3. Carbon rings with hetero chains	99–100
4. Hetero rings with carbon chains	101–102
5. Hetero rings with hetero chains	102
6. Carbon rings linked through hetero chains	103
7. Carbon rings linked with hetero rings through carbon chains	104
8. Carbon rings linked with hetero rings through hetero chains	104
9. Hetero rings linked with hetero rings through carbon chains	104
10. Hetero rings linked with hetero rings through hetero chains	104

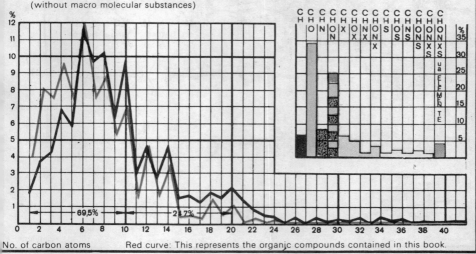

	1		2		3		4		5		6		7	
II	Li	26	Be	33	B	19	C	20	N	09	O	08	F	10
III	Na	27	Mg	34	Al	42	Si	21	P	22	S	15	Cl	11
IV	K	28	Ca	35	Ga	43	Ge	68	As	23	Se	16	Br	12
V	Rb	30	Sr	36	In	44	Sn	69	Sb	24	Te	17	I	13
VI	Cs	31	Ba	37	Tl	45	Pb	70	Bi	25	Po	18	At	14
IV	Cu	83	Zn	39	Sc	46	Ti	64	V	71	Cr	75	Mn	79
V	Ag	84	Cd	40	Y	47	Zr	65	Nb	72	Mo	76	Te	92
VI	Au	85	Hg	41	La	48	Hf	66	Ta	73	W	77	Re	93

Fe	82	Co	81	Ni	80
Ru	86	Rn	87	Pd	88
Os	89	In	90	Pt	91

Gamelin number of elements

screen colour characterizes the presence of elements in the formula of substances in the following pages.

The elements in the dark boundary sphere can also occur as hetero atoms.

Relative number of organic compounds depending on the number of C-atoms as well as on the structure of concerned elements amount to about 6000 in a compiled handbook of substances (without macro molecular substances)

69,5%   28,7%

No. of carbon atoms   Red curve: This represents the organic compounds contained in this book.

Understanding the formulae

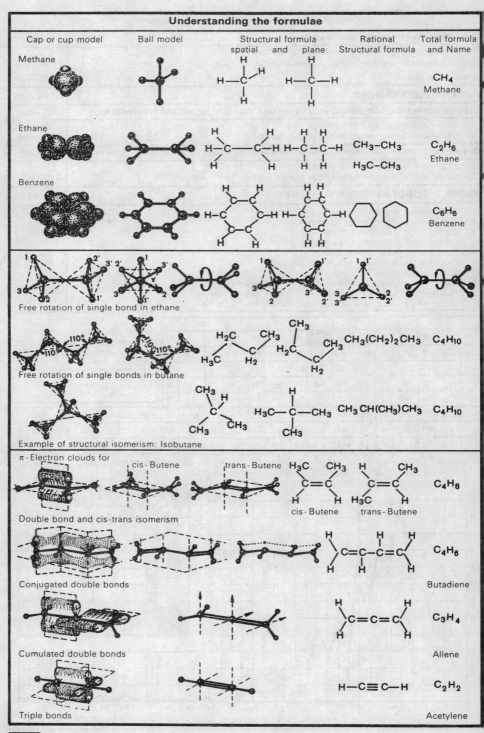

Cap or cup model	Ball model	Structural formula spatial and plane	Rational Structural formula	Total formula and Name

Methane

$$H-\underset{\underset{H}{|}}{\overset{\overset{H}{|}}{C}}-H$$

CH_4
Methane

Ethane

CH_3-CH_3
H_3C-CH_3

C_2H_6
Ethane

Benzene

C_6H_6
Benzene

Free rotation of single bond in ethane

$CH_3(CH_2)_2CH_3$ C_4H_{10}

Free rotation of single bonds in butane

Example of structural isomerism: Isobutane

$CH_3-\underset{\underset{CH_3}{|}}{\overset{\overset{H}{|}}{C}}-CH_3$ $CH_3CH(CH_3)CH_3$ C_4H_{10}

π - Electron clouds for

cis - Butene   trans - Butene

C_4H_8

cis - Butene   trans - Butene

Double bond and cis-trans isomerism

Conjugated double bonds C_4H_6 Butadiene

Cumulated double bonds C_3H_4 Allene

$H-C\equiv C-H$

Triple bonds C_2H_2 Acetylene

Bond order or binding degree

Bond order

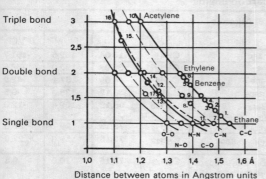

Triple bond — 3
2,5
Double bond — 2
1,5
Single bond — 1

Distance between atoms in Angstrom units

(curve labels: Acetylene, Ethylene, Benzene, Ethane)
(axis labels: 1,0 1,1 1,2 1,3 1,4 1,5 1,6 Å)
(point labels: O–O, N–N, C–N, C–C, N–O, C–O)

The bond order can be represented as a function of distance between the mid-points of two atoms (atomic nucleus) connected to one another. As the right points for the curves give the distances of normal single bond (e.g. ethane) and that of isolated double or triple bond (e.g. ethylene and acetylene) in many cases the dependence of distance result on fractional bond order (e.g. benzene and graphite). A fractional bond order signifies that the structure of the molecule of the concerning substance can be represented mostly through two or more "mesomeric" structures. With that, it explains a normal case of different properties. Often, it is through this significance a reaction outflow follow some of the mesomeric structures.
(Examples the 1, 4 addition to butadiene).

The significance of particular Points:

1. $\rightarrow$C – C =
2. = C – C =
3. $\rightarrow$C – C $\equiv$
4. = C – C $\equiv$
5. = C – C $\equiv$
6. C = C as in butadiene

7. $\rightarrow$ C – N =
8. $\rightarrow$ C – N$\leftarrow$
9. > C = N $\equiv$
10. – C $\equiv$ N
11. $\rightarrow$ C – O –
12. C – O in $-C\overset{\displaystyle O}{\underset{\displaystyle O^{\ominus}}{\diagdown}}$

13. – N = N $\equiv$ (azido groups)
14. – N = N – (azo groups)
15. = N $\equiv$ N (diazo groups)
16. = N $\equiv$ N (azido groups)
17. N – O in $-N\overset{\oplus\displaystyle O}{\underset{\displaystyle O^{\ominus}}{\diagdown}}$

Mesomeric forms of butadiene

Mesomeric forms of carboxyl and nitro groups:

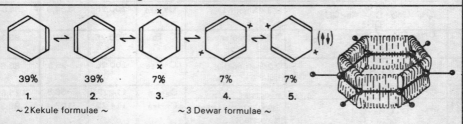

Charge clouds for 6 π-electrons in benzene

39%	39%	7%	7%	7%	
1.	2.	3.	4.	5.	

~ 2 Kekule formulae ~   ~ 3 Dewar formulae ~

For benzene five possible different mesomers, distributing the 6 π-electrons can be represented through the readily given formulae of Kekule or Dewar.*

Moreover, the above representation indicates the uniform distributing nature of 6 π electrons possible. The π-electrons have in pairs coupled spins through which the double arrow head at the end of the formula series is brought to the representation.

The percentage indicates the portion of different forms available in the whole of the existing molecules.

By fusing 2 or more benzene nucleus, the number of mesomeric forms increases.

e.g. For naphthalene   42 mesomeric forms exist.

For anthracene   429 mesomeric forms exist.

Symbol for benzene nucleus can also be represented with 6 π-electrons ⬡

Parent substances

Carbon chains

Alkanes

No	Structural formula	Name	Total (MF) Formula	M. wt.	FP/MP	BP/CP
1	CH_4	Methane	CH_4	16,04	-182,6	-161,7
2	$CH_3 - CH_3$	Ethane	C_2H_6	30,07	-172,0	-88,6
3	$CH_3 - CH_2 - CH_3$	Propane	C_3H_8	44,09	-187,1	-42,2
4	$CH_3 - CH_2 - CH_2 - CH_3$	n-Butane	C_4H_{10}	58,12	-135,0	-0,5
5	$CH_3 - CH - CH_3$ CH_3	iso-Butane (2-Methylpropane)	C_4H_{10}	58,12	-159,4	-12,0
6	$CH_3 - CH_2 - CH_2 - CH_2 - CH_3$	n-Pentane	C_5H_{12}	72,15	-129,7	36,0
7	$CH_3 - CH - CH_2 - CH_3$ CH_3	2-Methylbutane	C_5H_{12}	72,15	-159,6	27,9
8	CH_3 $CH_3 - C - CH_3$ CH_3	2.2-Dimethylpropane (Neopentane)	C_5H_{12}	72,15	16,6	9,5
9	$CH_3 - CH_2 - CH_2 - CH_2 - CH_2 - CH_3$	n-Hexane	C_6H_{14}	86,17	-94,0	68,7
10	$CH_3 - CH - CH_2 - CH_2 - CH_3$ CH_3	2-Methylpentane	C_6H_{14}	86,17	-153,7	60,3
11	$CH_3 - CH_2 - CH - CH_2 - CH_3$ CH_3	3-Methylpentane	C_6H_{14}	86,17	(-118)	63,3
12	CH_3 $CH_3 - C - CH_2 - CH_3$ CH_3	2.2-Dimethylbutane (Neohexane)	C_6H_{14}	86,17	-98,2	49,7
13	$CH_3 - CH - CH - CH_3$ H_3C CH_3	2.3-Dimethylbutane	C_6H_{14}	86,17	-128,8	58,0
14	$CH_3 - (CH_2)_5 - CH_3$	n-Heptane	C_7H_{16}	100,20	-90,6	98,4
15	$CH_3 - CH - CH_2 - CH_2 - CH_2 - CH_3$ CH_3	2-Methylhexane	C_7H_{16}	100,20	-118,2	90,0
16	$CH_3 - CH_2 - CH - CH_2 - CH_2 - CH_3$ CH_3	3-Methylhexane	C_7H_{16}	100,20	-119.0	92,0
17	CH_3 $CH_3 - C - CH_2 - CH_2 - CH_3$ CH_3	2.2-Dimethyl-pentane	C_7H_{16}	100,20	-125,0	78,9
18	$CH_3 - CH - CH - CH_2 - CH_3$ H_3C CH_3	2.3-Dimethyl pentane	C_7H_{16}	100,26		89,7
19	$CH_3 - CH - CH_2 - CH - CH_3$ H_3C CH_3	2.4-Dimethyl pentane	C_7H_{16}	100,26	-119,3	80,8
20	CH_3 $CH_3 - CH_2 - C - CH_2 - CH_3$ CH_3	3.3-Dimethyl pentane	C_7H_{16}	100,26	-134,9	86,0
21	$CH_3 - CH_2$ $CH - CH_2 - CH_3$ CH_3 CH_3	3-Ethylpentane	C_7H_{16}	100,26	-119	93,3
22	CH_3 $CH_3 - C - CH - CH_3$ H_3C CH_3	2.2.3-Trimethyl butane	C_7H_{16}	100,26	-25,0	80,8
23	$CH_3 - (CH_2)_6 - CH_3$	n-octane	C_8H_{18}	114,22	-56,8	125,6
24	CH_3 $CH_3 - C - CH_2 - CH - CH_3$ CH_3 CH_3	2.2.4 Trimethyl pentane iso-octane	C_8H_{18}	114,22	-107,4	99,3
25	H_3C CH_3 $CH_3 - C - C - CH_3$ H_3C CH_3	2.2.3.3 Tetramethyl butane (Neooctane)	C_8H_{18}	114,22	103...4	106...7

Explanations and abbreviations to the Tables on pages 80–128 (see also p. 37)

For classification of organic compounds (see pages 77, 86, 90 and 105).
The search for a compound can be followed from the structural formula of the parent compound or the presence of functional groups. The index of molecular formula on pages 267–272 offer possibilities for looking the compounds. The name index on pages 259–266 can be of further help.

For the abbreviations see p. 37. Supplement to these are:

immisc	=	immiscible with...	misc	=	miscible with...
v.sl. misc	=	very slightly miscible with...	∞ misc	=	miscible in all proportions...
sl. misc	=	slightly miscible with...	d in w	=	decomposes in water...

Carbon chains

Alkane

No	Formula	Name	M.F	M. wt.	MP/FP	BP/CP	D/sp. gr.
1	$CH_3(CH_2)_3CH_3$	n - Pentane	C_5H_{12}	72.15	-129.7	36.1	0.6264
2	$CH_3(CH_2)_4CH_3$	n - Hexane	C_6H_{14}	86.17	-94.0	68.7	0.6594
3	$CH_3(CH_2)_5CH_3$	n - Heptane	C_7H_{16}	100.20	-90.5	98.4	0.6837
4	$CH_3(CH_2)_6CH_3$	n - Octane	C_8H_{18}	114.22	-56.8	125.6	0.7028
5	$CH_3(CH_2)_7CH_3$	n - Monane	C_9H_{20}	128.25	-53.7	150.7	0.7179
6	$CH_3(CH_2)_8CH_3$	n - Decane	$C_{10}H_{22}$	142.28	-29.7	174.0	0.7298
7	$CH_3(CH_2)_9CH_3$	n - Undecane	$C_{11}H_{24}$	156.31	-25.6	195.8	0.7404
8	$CH_3(CH_2)_{10}CH_3$	n - Dodecane	$C_{12}H_{26}$	170.33	-9.6	216.3	0.7493
9	$CH_3(CH_2)_{11}CH_3$	n - Tridecane	$C_{13}H_{28}$	184.36	-6	(230)	0.7568
10	$CH_3(CH_2)_{12}CH_3$	n - Tetradecane	$C_{14}H_{30}$	198.39	5.5	251	0.7636
11	$CH_3(CH_2)_{13}CH_3$	n - Pentadecane	$C_{15}H_{32}$	212.41	10	268	0.7688
12	$CH_3(CH_2)_{14}CH_3$	n - Hexadecane	$C_{16}H_{34}$	226.44	18.1	280	0.7749
13	$CH_3(CH_2)_{15}CH_3$	n - Heptadecane	$C_{17}H_{36}$	240.46	22.0	303	0.7767
14	$CH_3(CH_2)_{16}CH_3$	n - Octadecane	$C_{18}H_{38}$	254.48	28.0	308	0.7767
15	$CH_3(CH_2)_{17}CH_3$	n - Monadecane	$C_{19}H_{40}$	268.50	32	330	0.7776
16	$CH_3(CH_2)_{18}CH_3$	n - Eicosane	$C_{20}H_{42}$	282.53	36.4	205 V	0.7777
17	$CH_3(CH_2)_{19}CH_3$	n - Heneicosane	$C_{21}H_{44}$	296.56	40.4	215 V	0.7782
18	$CH_3(CH_2)_{20}CH_3$	n - Docosane	$C_{22}H_{46}$	310.59	44.4	224 V	0.7778
19	$CH_3(CH_2)_{21}CH_3$	n - Tricosane	$C_{23}H_{48}$	324.62	47.4	234 V	0.7797
20	$CH_3(CH_2)_{22}CH_3$	n - Tetracosane	$C_{24}H_{50}$	338.64	51.1	243 V	0.7786
21	$CH_3(CH_2)_{23}CH_3$	n - Pentacosane	$C_{25}H_{52}$	352.67	53.3		
22	$CH_3(CH_2)_{28}CH_3$	n - Triacontane	$C_{30}H_{62}$	422.83	66		
23	$CH_3(CH_2)_{33}CH_3$	n - Pentatriacontane	$C_{35}H_{72}$	492.97	74.6		0.7814
24	$CH_3(CH_2)_{38}CH_3$	n - Tetracontane	$C_{40}H_{82}$	563.10	81		
25	$CH_3(CH_2)_{48}CH_3$	n - Pentacontane	$C_{50}H_{102}$	703.38	92		0.7940
26	$CH_3(CH_2)_{58}CH_3$	n - Hexacontane	$C_{60}H_{122}$	843.64	99		
27	$CH_3(CH_2)_{60}CH_3$	n - Dohexacontane	$C_{62}H_{126}$	871.69	101		
28	$CH_3(CH_2)_{62}CH_3$	n - Tetrahexacontane	$C_{64}H_{130}$	899.65	102		
29	$CH_3(CH_2)_{68}CH_3$	n - Heptacontane	$C_{70}H_{142}$	983.92	105		

Uniform gradation in properties of C-H compounds in a homologus series as an example of n-alkane. The boiling points of iso-alkanes lie low depending of the structure of molecules.

The adjacent columns show where the possible number of structural isomers rapidly increase with equal carbon numbers. Of these isomers only a few with lower number of C atoms have practical significance [e.g. antiknock (propellants) motor fuels.]

Number of C-Atoms	Isomers	Number of C-Atoms	Isomers
4	2	11	159
5	3	12	355
6	5	13	802
7	9	14	1858
8	18	15	4347
9	35	20	366319
10	75	30	4111846763
11	159	40	62491178805831

Carbon chain

Alkenes

No.	Structural formula	Name	M.F.	Mol. wt	M.P.	B.P.	D/sp. gr.
1	$CH_2 = CH_2$	Ethylene	C_2H_4	28,05	-169,4	-102,4	0,6100
2	$CH_2 = CH - CH_3$	Propene	C_3H_6		-185	-47,7	0,6104
3	$CH_2 = CH - CH_2 - CH_3$	Butene - 1	C_4H_8	56,10	-185,8	- 6,3	0,6255
4	H_3C CH_3 $C = C$ H cis H H CH_3 $C = C$ H_3C trans H	Butene - 2	C_4H_8	56,10		cis 1 trans 3,5	0,635
5	$CH_2 = C - CH_3$ CH_3	Isobutene 2 - Methyl	C_4H_8	56,10	-140,7	- 6,6	0,6266
6	$CH_2 = CH - CH_2 - CH_2 - CH_3$	Pentene - 1	C_5H_{10}	70,13	-138	29,968	0,6429
7	$CH_2 = C - CH_2 - CH_3$ CH_3	2 - Methyl Butene - 1	C_5H_{10}	70;13		31	0,6501
8	$CH_2 = CH - CH - CH_3$ CH_3	3 - Methyl Butene - 1	C_5H_{10}	70,13	-135	25,0	0,6340
9	$CH_2 = CH - CH_2 - CH_2 - CH_2 - CH_3$	Hexene - 1	C_6H_{12}	84,16	-138	63,5	0,6747
10	$CH_2 = CH - (CH_2)_4 - CH_3$	Heptene - 1	C_7H_{14}	98,18	-119	93,1	0,6976
11	$CH_2 = CH - (CH_2)_5 - CH_3$	Octene - 1	C_8H_{16}	112,21	(-104)	122,5	0,7159
12	$CH_2 = CH - (CH_2)_6 - CH_3$	Nonene - 1	C_9H_{18}	126,24	-81,4	146,7	0,7292
13	$CH_2 = CH - (CH_2)_7 - CH_3$	Decene - 1	$C_{10}H_{20}$	140,26	-66,3	170,7	0,7408

Dienes

No.	Structural formula	Name	M.F.	Mol. wt	M.P.	B.P.	D/sp. gr.
14	$CH_2 = CH - CH = CH_2$	Butadiene	C_4H_6	54,09	-108,92	-3	0,650 liq
15	$CH_2 = CH - CH = CH - CH_3$	Pentadiene - 1.3	C_5H_8	68,11		43	0,6830
16	$CH_2 = CH - CH_2 - CH = CH_2$	Pentadiene - 1.4	C_5H_8	68,11	-148,28	25,967	0,66076
17	$CH_2 = C - CH = CH_2$ CH_3	2 - Methyl-butadiene Isoprene	C_5H_8	68,11	-145,95	34	0,6806
18	$CH_2 = CH - CH_2 - CH_2 - CH = CH_2$	Hexadiene - 1.5	C_6H_{10}	82,14	-140,68	59,6	0,6880
19	$CH_3 - CH = CH - CH = CH - CH_3$	Hexadiene - 2.4	C_6H_{10}	82,14		82	0,7108
20	$CH_2 = C - C = CH_2$ H_3C CH_3	2.3 - Dimethyl butadiene	C_6H_{10}	82,14	-76,01	69,6	0,7446
21	$CH_3 - CH = CH - CH = CH - CH_2 - CH_3$	Heptadiene - 2.4	C_7H_{12}	96,17		107	0,733
22	$CH_3 - CH = CH - CH = CH - CH_2 - CH - CH_3$ CH_3	7 - Methyloctadiene 2.4	C_9H_{16}	124,22		144	0,752
23	$CH_2 = C - CH = CH - CH_2 - CH - CH_3$ CH_3 CH_3	2.6 - Dimethyl-heptadiene 1.3	C_9H_{16}	124,22		~145	0,765
24	$CH_3 - C = CH - CH_2 - CH_2 - C - CH = CH_2$ CH_3 CH_2	2 - Methyl-6 methyle-noctad - 2.7	$C_{10}H_{18}$	136,25		167	0,802
25	$CH_3 - C = C - CH_2 - CH - C = C - CH_3$ H CH_3 H CH_3	2.6 - Dimethyl-octadiene 2.6	$C_{10}H_{18}$	136,23		171,5...3,5	0,775
26	$CH_2 = C = CH_2$	Allene	C_3H_4	40,06	-146	-32	1,787 g/l

Alkynes

No.	Structural formula	Name	M.F.	Mol. wt	M.P.	B.P.	D/sp. gr.
27	$HC \equiv CH$	Acetylene	C_2H_2	26,04	-81,8	-83,4	0,6179
28	$HC \equiv C - CH_3$	Propyne - 1	C_3H_4	40,07	-101,5	-23,3	0,6714
29	$HC \equiv C - CH_2 - CH_3$	Butyne - 1	C_4H_6	54,09	-122,5	8,6	0,6682
30	$HC \equiv C - CH_2 - CH_2 - CH_3$	Pentyne - 1	C_5H_8	68,12	-98	39,7	0,695
31	$H_3C - C \equiv C - CH_2 - CH_3$	Pentyne - 2	C_5H_8	68,12	-101	55,5	0,7127
32	$HC \equiv C - CH - CH_3$ CH_3	3 - Methyl-butyne - 1	C_5H_8	68,12		28	0,665
33	$HC \equiv C - CH_2 - CH_2 - CH_2 - CH_3$	Hexyne - 1	C_6H_{10}	82,15	-124	71	0,7195
34	$H_3C - C \equiv C - CH_2 - CH_2 - CH_3$	Hexyne - 2	C_6H_{10}	82,15	-92	84	0,7305
35	$H_3C - CH_2 - C \equiv C - CH_2 - CH_3$	Hexyne - 3	C_6H_{10}	82,15	-51	82	0,7255
36	$HC \equiv C - C - CH_3$ CH_3 CH_3	3.3 - Dimethyl-butyne - 1	C_6H_{10}	82,15	-81	38	0,6686
37	$H - C \equiv C - C \equiv C - H$	Butadiyne	C_4H_2	50,06	- 36,4	10,3	2,233
38	$HC \equiv C - CH_2 - CH_2 - C \equiv CH$	Hexadiyne - 1.5	C_6H_6	78,11	-6	86	0,8049
39	$H_2C = CH - C \equiv CH$	Vinylacetylene (Butene - 1 - yne - 3)	C_4H_4	52,08		5	0,6867
40	$H_2C = CH - C \equiv C - HC = CH_2$	Divinylacetylene (Hexadiene - 1.5 - yne 3)	C_6H_6	78,11		83,5	0,7851

Carbon rings or carbocyclics

Simple rings

No.	Formula	Name			
		M.F.	B.P	D/sp. gr.	
		Mol. wt.	M.P	n$_D$	
		Other properties			
1	H$_2$C, CH$_2$, H$_2$C (cyclopropane ring)	Cyclopropan Trimethylene			
		C_3H_6	−34.4	0.6880 liq	
		42.08	−126.6		
		col: gas; i.w; e.s. acet; eth			
2	HC, CH$_2$, HC (cyclopropene ring)	Cyclopropene			
		C_3H_4			
		40.07			
		Only derivatives known			
3	H$_2$C—CH$_2$, H$_2$C—CH$_2$	Cyclobutane Tetramethylene			
		C_4H_8	13	0.7038	
		56.10	−50	1.3752	
		i.w; s. acet; eth			
4	HC—CH$_2$, HC—CH$_2$	Cyclobutene			
		C_4H_6	2	0.733 liq	
		54.09	Gas		
		s. acet			
5	Cyclopentane ring	Cyclopentane Pentamethylene			
		C_5H_{10}	49.5	0.7460	
		70.13	−93.8		
6	Cyclopentene ring	Cyclopentene			
		C_5H_8	44.242	0.77199	
		68.11	−93.3	1.42246	
		immisc. w; misc. acet; eth			
7	1,3-Cyclopentadiene ring	1,3-Cyclopentadiene			
		C_5H_6	42.5	0.8048	
		66.10			
8	Cyclohexane ring	Cyclohexane Hexahydro benzene			
		C_6H_{12}	81.4	0.7781	
		84.16	6.5	1.22900	
		i.w; misc. acet; eth			
8a		Chair form of Cyclohexane molecule			
8b		Twist form			
8c		Trough or boat form of the Cyclohexane molecule see also p. 85			
9	Cyclohexene ring	Cyclohexene 1,2,3,4 Tetrahydrobenzol			
		C_6H_{10}	83	0.8102	
		82.14	−103.7	1.44507	
		i.w; misc. acet; eth			
10	1,3-Cyclohexadiene ring	1,3-Cyclohexadiene 1,2-Dihydrobenzene			
		C_6H_8	80.5	0.8404	
		80.12	−98	1.4758	
		immisc. w; misc. acet; eth			
11	1,4-Cyclohexadiene ring	1,4 Cyclohexadiene 1,4 Dihydrobenzol			
		C_6H_8	86...7	0.8471	
		80.12	liq	1.46806	
		immisc. w; misc. acet; eth			
12	Benzene ring structures	Cyclohexatriene Benzene, Phene (see p. 79)			
		C_6H_6	80.099	0.87901	
		78.11	5.51	1.50112	
		col; rhomb; prism; or liq; v.sl. misc; w; misc. acet; eth.			
13	Cycloheptane ring	Cycloheptane Suberane			
		C_7H_{14}	118.1	0.8100	
		98.18	−12	1.4440	
		i.w. misc; acet; eth			
14	Cycloheptene ring	Cycloheptene Suberylene			
		C_7H_{12}	115	0.8228	
		96.17	oily	1.4552	
		immisc. w., misc. acet., eth.			
15	1,3-Cycloheptadiene ring	1,3-Cycloheptadiene			
		C_7H_{10}	120...		
		94.15			
		Indef. resinifies on exposure to air.			
16	Cycloheptatriene ring	Cycloheptatriene			
		C_7H_8	116		
		92.13			
		Indef. resinifies on exposure to air.			
17	Cyclooctane ring	Cyclooctane			
		C_8H_{16}	147	0.8304	
		114.23	14		
18	Cyclooctene ring	Cyclooctene			
		C_8H_{14}	145		
		112.22			
19	1,3-Cyclooctadiene ring	1,3-Cyclooctadiene			
		C_8H_{12}	48.v		
		110.20	−57		
20	1,5-Cyclooctadiene ring	1,5-Cyclooctadiene			
		C_8H_{12}	151		
		108.18	M.P-cis-70.1 (Tub)		
			M.P. Trans-62 (Chair)		
21	Cyclooctatriene ring	Cyclooctatriene			
		C_8H_{10}	147...8		
		106.17			
22	Cyclooctatetraene ring	Cyclooctatetraene			
		C_8H_8	142...3	0.9206	
		104.15	−7		
		gold yel; liq. B.P$_{14}$ 36			

Condensed rings: carbon rings or carbocyclics

No.	Structural formula	Name	M.F	Mol. wt.	MP/FP	BP/CP	D, nD, colour and other properties
1		Nepthalene	$C_{10}H_8$	128.16	80.22	217.9	1.145
					nD 1.5821; Col., monocl., v.sl.s.w., sl.s.al; e.s. eth		
2		Anthracene	$C_{14}H_{10}$	178.22	217	s305	1.25
					Col., monocl., i.w., sl. s. al; s. eth; chl; Bz		
3		Phenanthrene	$C_{14}H_{10}$	178.22	100	340.2	1.025
					Col; monocl; pl. (acet.), i.w; s. al; eth; Bz		
4		Tetracene (Nephtacen)	$C_{18}H_{12}$	228.28	335		
5		Chrycene	$C_{18}H_{12}$	228.28	254	448	Col., rhomb., pl. (Bz)
					red-viol. fluores; e.s. eth; CS₂, Bz		
6		Triphenelene	$C_{18}H_{12}$	228.28	198.5		wh., Cr. i.w., s. acet., eth.
7		Pyrene	$C_{16}H_{10}$	202.24	150	>360	lt. yel., monocl. sl. s. al., e.s. eth.
8		Pentacene	$C_{22}H_{14}$	278.33	d >300		sublimable vlt. bl.
9		1, 2, 5, 6 - Dibenz-anthracene	$C_{22}H_{14}$	278.33	267.5		sublimable Col., leaf., s. Bz.
10		3, 4 - Benzopyrene	$C_{20}H_{12}$	252.30	176-177		pa. yel., need. s. Bz, Me. al.
11		Hexacene	$C_{24}H_{16}$		d		Deep grn; (~ blk) v. sl. s. in org. sol.
12		Perylene	$C_{20}H_{12}$	252.22	273-274		gold yel., leaf. sl. s. in org. sol.
13		Coronene	$C_{24}H_{12}$	300.34	440 (in vacuum)		pa. yel., need Bz; semiconductor
14		Ovalene	$C_{32}H_{12}$	396.45	473 (in vacuum)		or. need s. 1 - Methylnaphthalene
15		Graphite*	$\{C_3^3\}N$	∞	~4500		Gr. blk., leaf., **Semiconductor**

Hexagonal arrangement of the c atoms inside the plane of graphite-layer lattices **solid lines : reference layer — dotted lines correspond to the above situated plane)**

Crystal structure of Graphite

*Graphite can also be considered as "Poly condensed Benzene"

Condensed rings—carbon rings or carbocyclics

No.	Formula	Name	M.F	Mol. wt.	F.P	B.P. colour and other properties
1		1, 2 - Dihydronaphthalene	$C_{10}H_{10}$	130.18	25	94.5_{17mm} leaf D = 0.993 s. acet., eth.
2		1, 4 - Dihydronaphthalene	$C_{10}H_{10}$	130.18	-8	89_{24mm} D = 0.9974; misc. acet; eth. $n_D^{18} = 1.58317$
3		1, 2, 3, 4 - Tetrahydro-nepthalene or Tetralin	$C_{10}H_{12}$	132.20	30	207.2
4		$\Delta^{6.9}$ - Hexahydro-nepthalene	$C_{10}H_{14}$	134.21		205.5
5		Δ^9 - Octahydro-nepthalene Δ^9 - Octanepthalene	$C_{10}H_{16}$			
6		Decahydronepthalene Decaline	$C_{10}H_{18}$	138.25	-43.3	194.6

Spatial representation of 3 - isomeric forms

I   II   III

7 | Adamantane

M.F	Mol. wt.	M.P
$C_{10}H_{16}$	136.24	269°

s. eth., pet. eth.

Numbering of C - atoms   see also p. 94

occurs in crude oil wh., Cr.

8

Inverse tetrahedron

Mirror image Tetrahedral layer

Arranged layers lying one over the other:

Diamond* structure (chair form)

Wurtzite structure (boat form)

see also p. 73

see also p. 73

*Diamond can also be considered as "Polycondensed cyclohexane"

Carbon rings or carbocyclics

Different condensed systems

No.	Formula	Name			
		M.F.	B.P.	D	
		Mol. wt.	F.P.	n_D	
		Other properties			
1		Indane, Hydrindene			
		2, 3 - Dihydroindene			
		C_9H_{11}	176.5	0.9645	
		118.17	-51.40	1.53877	
		col; i.w., misc. acet., eth.			
2		Indene			
		C_9H_8	182.4	1.006	
		116.15	-2	1.57107	
		col., i.w., misc. acet., eth., pyr. CS_2, ac. a.			
3		Fluorene			
		Diphenelenemethane			
		$C_{13}H_{10}$	295		
		166.21	116		
		Col., need. (acet.) e.s. acet., eth., Bz., CS_2			
4		Acenaphthene			
		$C_{12}H_{10}$	277.5	1.024	
		154.20	95.4	1.407	
		rhomb ; cr, (acet.), s, h. acet ; eth ; Bz, chl, tol			
5		Acenaphthylene			
		$C_{12}H_8$	265...275 (d)	0.899	
		152.18	92...3		
		yel., rhomb., pl. (acet.), e.s. acet., eth.			

No.	Formula	Name			
6		Fluoranthrene			
		Idryl			
		$C_{16}H_{10}$	256 (color)		
		202.24	110		
		Col., monocl., need. (acet.)			
		i.w., e.s. al., eth., ac. a., CS_2			
7		Cholanthrene			
		Benzo(y)aceanthrene			
		$C_{20}H_{14}$	d		
		254.31	173		
		pa., yel., leaf., i.w.			
		Carcinogenic			
8		Azulene			
		$C_{10}H_8$			
		128.17	99°		
		Bl., cr.			
9		9, 10 - Dihydroanthracene			
		Anthracene - 9, 10 - dihydride			
		$C_{14}H_{12}$	s 305	0.8976	
		180.24	180.5		
		Col., tricl. or monocl., Cr, (acet.)			
		i.w., e.s., acet., eth., Bz			
10		Triptycene View Y from above			
		$C_{20}H_{14}$			
		254.33	255		
		Col., cr.			

Hetero chains

If in a carbon chain a $-CH_2-$ (methylene group) is replaced by another atom or group of atoms (i.e. hetero), there occurs a hetero chain.

With it the changes are: bond degree, bond angle, Mol. wt., and thereby connected physical properties of the compounds as well as their chemical properties in relation with the position of these hetero atoms in the periodic systems of elements. Therefore, in such cases of installed groups or atoms in the framework are known as "Functional groups" (see p. 105). These groups are called as "Functional groups" not only due to the above reasons but also because a replacement of H-atom affects the properties of the molecules (compounds). Hetero chains can only be built from smaller units in which hetero atoms **appear** in form of "Functional groups". The hetero atom, thus, is not mere as a periphral one, but as a constituent part of the structural framework, whereby certain adaptation of properties of the hetero chain in an equally built up C-element chain follows (Compare the diagram on p. 112)

Formula	Name	Mol. wt.	Bond length	Angle	M.P.	B.P.
$CH_3-CH_2-CH_3$	Propane	44.09	1.54	110	-187.1	-42.2
CH_3-O-CH_3	2 - Oxapropane	46.07	1.42	111	-140.0	-24.9
$CH_3-NH-CH_3$	2 - Azapropane	45.08	1.47	111	- 96.0	7.4
CH_3-S-CH_3	2 - Thiapropane	62.13	1.81	105	- 83.2	38.0
$\begin{matrix}CH_3\\CH_3\end{matrix}>CH-CH_3$	2 - Methylpropane	58.12	1.54	110	-159.4	-12.0
$\begin{matrix}CH_3\\CH_3\end{matrix}>N-CH_3$	2 - Methyl - 2 - azapropane	59.11	1.47	108	-117.1	3.0

Heterogroups	Prefix	Suffix	Substance-class	Pages
$-O-$	Oxa	- ether	Ether	87
			Acetal	87
			"Ester" of ortho acids	87
$-S-$	Thia -	- sulphide	Thioether, sulphides	88
$-NH-$	Aza -	- amine	Secondary amines	88
$>N-$	Aza -	- amine	Tertiary amines	88

Only a few essential groups are given. For others see pp. 88/89

Hetero chains

Hetero chains with oxa-groups (ether, acetal, "ester" of ortho acids

No.	Formula	Name	M.F.	Mol. wt.	MP/FP	BP/CP	D	n_D
1	CH_3-O-CH_3	Dimethylether	C_2H_6O	46.07	−140	−24.9	0.661	
2	$CH_3-O-CH_2-CH_3$	Ethyl methyl ether	C_3H_8O	60.10		10.8	0.7260	1.3420
3	$CH_3-O-CH_2-CH_2-CH_3$	Methyl-n-propyl ether	$C_4H_{10}O$	74.12		39	0.7356	1.3579
4	$CH_3-CH_2-O-CH_2-CH_3$	Diethyl ether*	$C_4H_{10}O$	74.12	−116.3	34.60	0.71352	1.3526
5	$CH_3-O-CH_2-CH_2-CH_2-CH_3$	n-butyl methyl ether	$C_5H_{12}O$	88.15	−115.5	70	0.7455	1.3728
6	$CH_3-CH_2-O-CH_2-CH_2-CH_3$	Ethyl-n-propyl ether	$C_5H_{12}O$	88.15		63.6	0.7386	1.36948
7	$CH_3-O-CH_2-CH_2-CH_2-CH_3$	Methyl-n-butyl ether	$C_6H_{14}O$	102.18		100	0.7657	1.3873
8	$CH_3-CH_2-O-CH_2-CH_2-CH_2-CH_3$	Ethyl-n-butyl ether	$C_6H_{14}O$	102.18	−124	92.3	0.7505	1.3820
9	$CH_3-CH_2-CH_2-O-CH_2-CH_2-CH_3$	Di-n-propyl ether	$C_6H_{14}O$	102.18	−122	90.1	0.74698	1.38829
10	$\begin{matrix} H_3C \\ H_3C \end{matrix} > CH-O-CH_2-CH_2-CH_3$	n-iso-propyl ether	$C_6H_{14}O$	102.18		83	0.7370	1.376
11	$\begin{matrix} H_3C \\ H_3C \end{matrix} > CH-O-CH < \begin{matrix} CH_3 \\ CH_3 \end{matrix}$	Di-iso-propyl ether	$C_6H_{14}O$	102.18	−60	67.5	0.726	1.3688
		Anti knock motor fuel						
12	$CH_3-O-CH_2-(CH_2)_4-CH_3$	n-hexylmethyl ether	$C_7H_{16}O$	116.20		126	0.772	1.3972
13	$CH_3-O-CH_2-(CH_2)_3-CH_3$	Methyl-n-pentyl ether	$C_7H_{16}O$	116.20				
14	$CH_3-(CH_2)_2-CH_2-O-CH_2-(CH_2)_2-CH_3$	Di-n-butyl ether	$C_8H_{18}O$	130.23	−98	143	0.76829	1.3989
15	$CH_3-CH_2-O-CH_2-(CH_2)_4-CH_3$	Ethyl-n-hexyl ether	$C_8H_{18}O$	130.23		142	0.722	1.4008
16	$CH_3-(CH_2)_3-CH_2-O-CH_2-(CH_2)_3-CH_3$	Di-n-amyl ether	$C_{10}H_{22}O$	158.2	−69.3	187.5	0.78298	1.416
17	$\begin{matrix} H_3C \\ H_3C \end{matrix} CH-CH_2-O-CH_2-CH_2-CH \begin{matrix} CH_3 \\ CH_3 \end{matrix}$	Di-iso-amyl ether	$C_{10}H_{22}O$	158.29		172.5	0.778	1.409
		Sol. in case of Zerewitinoff method						
18	$CH_3-(CH_2)_4-CH_2-O-CH_2-(CH_2)_4-CH_3$	Di-hexyl ether	$C_{12}H_{26}O$	186.34		228	0.7936	
19	$CH_3-(CH_2)_5-CH_2-O-CH_2-(CH_2)_5-CH_3$	Di-heptyl ether	$C_{14}H_{30}O$	214.39		~260	0.8056	1.427
20	$CH_3-(CH_2)_{10}-CH_2-O-CH_2-(CH_2)_{10}-CH_3$	Di-lauryl ether	$C_{24}H_{50}O$	338.67	~30	~190		
21	$CH_2=CH-O-CH_2-CH_3$	Ethyl vinyl ether	C_4H_8O	72.11	−115.8	35.75	0.7589	1.3768
22	$CH_2=CH-CH_2-O-CH_3$	Allylmethyl ether	C_4H_8O	72.11		46		
23	$CH_2=CH-CH_2-O-CH_2-CH_3$	Allyl ethyl ether	$C_5H_{10}O$	86.14		~70	0.7651	1.3881
24	$CH_2=CH-O-CH=CH_2$	Divinyl ether	C_4H_6O	70.09		35		
25	$CH_2=CH-CH_2-O-CH_2-CH=CH_2$	Diallyl ether	$C_6H_{10}O$	98.15		94	0.826	

	Formula	Name	Mol. wt./M.F.	B.P.°C/n_D^{20}	D/other properties
26	$CH_3-O-CH_2-O-CH_3$		76.09	44	0.8560
	Dimethoxymethane, 2, 4 – oxapentane, methylal		$C_3H_8O_2$	1.35344	s.w., al., eth.
27	$CH_3-CH_2-O-CH_2-O-CH_2-CH_3$		104.15	89	0.83465
	Dimethoxymethane		$C_5H_{12}O_2$		sl. misc. w.,
28	$CH_3-(CH_2)_2-O-CH_2-O-(CH_2)_2-CH_3$		132.20	137...40	0.835
	Dipropoxymethane		$C_7H_{16}O_2$		
29	$CH_3-O-CH_2-CH_2-O-CH_3$		90.12	84.7	0.8665
	Ethyleneglycol dimethyl ether, 2, 5 – dioxahexane		$C_4H_{10}O_2$	1.37965	∞ misc. w.
30	$CH_3-O-CH_2-CH_2-O-CH_2-CH_3$		104.15	102	0.8529
	Ethyleneglycol methyl ether; 2, 5 – dioxaheptane		$C_5H_{12}O_2$	1.38677	∞ misc. w.
31	$CH_3-O-CH_2-CH_2-O-(CH_2)_2-CH_3$		118.18	124.5	0.8472
	Ethyleneglycol methyl-n-propyl ether; 2, 5 – dioxaoctane		$C_6H_{14}O_2$	1.39467	∞ misc. w.
32	$CH_3-O-(CH_2)_2-O-(CH_2)_2-O-CH_3$		134.18	162.0	0.9440
	Diethyleneglycol dimethyl ether; 2, 5, 8 – trioxanonane		$C_6H_{14}O_3$	1,4099	FP−75° ∞ misc. w
33	$CH_3-CH_2-O-(CH_2)_2-O-(CH_2)_2-O-CH_2-CH_3$		162.23	188	0.906
	Diethyleneglycol diethyl ether; 3, 6, 9 – trioxaundecane		$C_8H_{18}O_3$	1.411	∞ misc. w.,
34	$CH_3-O-(CH_2)_2-O-(CH_2)_2-O-(CH_2)_2-O-CH_3$		178.23	216	0.9871
	Triethyleneglycol dimethyl ether; 2, 5, 8, 11 – tetraoxadodecane		$C_8H_{18}O_4$	1,4233	FP −47; ∞ misc. w
35	$CH_3-O-(CH_2)-O-(CH_2)_2-O-(CH_2)_2-O-(CH_2)_2-O-CH_3$		222.28	275	1.009
	Tetraethyleneglycol dimethyl ether, 2, 5, 8, 11, 14—pentaoxa pentadecane		$C_{10}H_{22}O_5$	1.432	∞ misc. w; al.
36	$CH_3-CH < \begin{matrix} O-CH_3 \\ O-CH_3 \end{matrix}$ 1, 1−dimethoxy methane		90.12	64.5	0.8476
	2, 4−oxa−3,−methyl pentane		$C_4H_{10}O_2$		misc. w., al; eth.
37	$CH_3-CH < \begin{matrix} O-CH_2-CH_3 \\ O-CH_2-CH_3 \end{matrix}$ 1, 1−diethoxyethane; acetal		118.18	102...4	0.8461
	3,5, oxa−4 methyl heptane		$C_6H_{14}O_2$	1.38193	misc. al, eth.,
38	$CH_3-O > C < \begin{matrix} O-CH_3 \\ O-CH_3 \end{matrix}$ Trimethoxymethane $\;H$		106.12	101...2	0.974
	Orthoformic acid trimethyl ester		$C_4H_{10}O_3$		
39	$CH_3-CH_2-O > C < \begin{matrix} O-CH_2-CH_3 \\ O-CH_2-CH_3 \end{matrix}$ Triethoxymethane $\;H$		148.29	145	0.8971
	Orthoformic acid triethyl ester		$C_7H_{16}O_3$		misc. al; eth; d. in w;
40	$\begin{matrix} CH_3-CH_2-CH_2-O \\ CH_3-CH_2-CH_2-O \end{matrix} > C < \begin{matrix} O-CH_2-CH_2-CH_3 \\ H \end{matrix}$		190.28	105 V	0.8805
	Tripropoxy methane		$C_{10}H_{22}O_3$		v. sl. misc. w.,
41	$\begin{matrix} CH_3-CH_2-O \\ CH_3-CH_2-O \end{matrix} > C < \begin{matrix} O-CH_2-CH_3 \\ CH_3 \end{matrix}$ 1, 1, 1−Triethoxy ethane		162.23	142	0.8847
	Ortho acetic acid triethyl ester		$C_8H_{18}O_3$		∞ misc. al; eth.,
42	$\begin{matrix} CH_3-CH_2-O \\ CH_3-CH_2-O \end{matrix} > C < \begin{matrix} O-CH_2-CH_3 \\ O-CH_2-CH_3 \end{matrix}$ Tetraethoxy methane		192.25	159	0.9197
	Orthocarbonic acid tetraethylester		$C_9H_{20}O_4$	1.393	∞ misc. al., eth;
43	$\begin{matrix} CH_3-CH_2-CH_2-O \\ CH_3-CH_2-CH_2-O \end{matrix} > C < \begin{matrix} O-CH_2-CH_2-CH_3 \\ O-CH_2-CH_2-CH_3 \end{matrix}$ Tetrapropoxy methane		248.36	224.2	0.911
	Orthocarbonic acid tetrapropylester		$C_{13}H_{28}O_4$		

usually called as ether

Hetero chains with thia-groups—thio ethers or sulphides

No	Formula	Name	M.F.	Mol. wt.	MP/FP	BP/CP	D/sp gr	n_D and other prop.
1	CH_3-S-CH_3	Dimethylsulphide	C_2H_6S	62.13	−83.2	375...8	0.8458	col; misc. acet; eth
2	$CH_3-S-CH_2-CH_3$	Methylethylsulphide	C_3H_8S	76.15	−104.8	66	0.837	misc. acet. eth.
3	$CH_3-CH_2-S-CH_2-CH_3$	Di-ethylsulphide	$C_4H_{10}S$	90.18	−102.1	92	0.837	misc. acet. eth. sl. misc. w;
4	$CH_3-(CH_2)_2-S-(CH_2)_2-CH_3$	Di-n-propylsulphide	$C_6H_{14}S$	118.23	−101.9	141...2	0.814	misc. acet. eth.
5	$\begin{array}{l}H_3C\\H_3C\end{array}>CH-S-CH<\begin{array}{l}CH_3\\CH_3\end{array}$	Diisopropyl sulphide	$C_6H_{14}S$	118.23	liq.	120.4		misc. acet. eth,
6	$CH_3-(CH_2)_3-S-(CH_2)_3-CH_3$	Di-n-butylsulphide	$C_8H_{18}S$	146.29	−79.7	182	0.839	misc. acet. eth.
7	$\begin{array}{l}H_3C\\H_3C\end{array}>CH-(CH_2)_2-S-(CH_2)_2-CH<\begin{array}{l}CH_3\\CH_3\end{array}$ Diisoamyl sulphide		$C_{10}H_{22}S$	174.34	liq.	209...11	0.84314	col, misc. acet. eth.
8	$CH_2=CH-S-CH=CH_2$	Divinylsulphide	C_4H_8S	86.15	oil	101	0.912	misc. w. acet. eth,
9	$CH_2=CH-CH_2-S-CH_2-CH=CH_2$ Diallylsulphide		$C_6H_{10}S$	114.20	−83	138.6	0.88765	col, oily; pengent
10	CH_3-S-CH_3	Dimethyldisulphide	$C_2H_6S_2$	94.19	liq.	116...8	1.057	misc. acet. eth,
11	$CH_3-CH_2-S-S-CH_2-CH_3$	Diethyldisulphide	$C_4H_{10}S_2$	122.24	liq.	153...4	0.99267	oily misc. w, eth; acet
12	$CH_3-(CH_2)_3-S_2-(CH_2)_3-CH_3$	Dibutyldisulphide	$C_8H_{18}S_2$	178.35		100...3v		s. acet; eth,
13	$\left[\begin{array}{l}H_3C\\H_3C\end{array}>CH-(CH_2)_3-\right]_2-S_2-$	Diisoamyl disulphide	$C_{10}H_{22}S_2$	206.40	liq.	250 122...5v	0.918	i.w;
14	$CH_2=CH-CH_2-S_3-CH_2-CH=CH_2$ Diallyltrisulphide		$C_6H_{10}S_3$	178.32	liq.	140	1.085	i.w; acet, misc. eth,

Hetero chains with selenium (selena) and tellurium (tellura) groups

No	Formula	Name	M.F.	Mol. wt.	MP/FP	BP/CP	D/sp gr	n_D and other prop.
15	$CH_3-Se-CH_3$	Dimethylselenide	C_2H_6Se	109.3	liq.	58.2	1.4077	misc. acet; eth,
16	$CH_3-CH_2-Se-CH_2-CH_3$	Diethylselenide	$C_4H_{10}Se$	137.08	liq.	108	1.230	$\bar{n}=1.4768$
17	$CH_3-Te-CH_3$	Dimethyltelluride	C_2H_6Te	157.68	liq.	82		yel; misc. acet. eth,
18	$CH_3-CH_2-Te-CH_2-CH_3$	Diethyltelluride	$C_4H_{10}Te$	185.73	liq.	138	red, brn	1.599; misc. acet,

Hetero chains with aza-groups, secondary and tertiary amines

No	Formula	Name	M.F.	Mol. wt.	MP/FP	BP/CP	D/sp gr	n_D and other prop.
19	$CH_3-NH-CH_3$	Dimethylamine	C_2H_7N	45.08	−96.0	7.4	0.6804	liq misc. w; acet; eth,
20	$CH_3-NH-CH_2-CH_3$	Methylethylamine	C_3H_9N	59.11		36		
21	$CH_3-CH_2-NH-CH_2-CH_3$	Diethylamine	$C_4H_{11}N$	73.14	−50	55.5	0.7018	misc. w; acet; eth;
22	$CH_3-NH-CH<\begin{array}{l}CH_3\\CH_3\end{array}$	Methylisopropyl amine	$C_4H_{11}N$	73.14		50	0.7026	
23	$\begin{array}{l}CH_3-CH-CH_2-CH_3\\ \quad HN-CH_3\end{array}$	N-methyl-sec-butylamine (DL)	$C_5H_{13}N$	87.17		78...79		
24	$CH_3-CH_2-NH-CH_2-CH_2-CH_3$	Ethyl-n-propylamine	$C_5H_{13}N$	87.17		7	0.6804	
25	$CH_3-NH-(CH_2)_3-CH_3$	N-methyl-n-butylamine	$C_5H_{13}N$	87.17		91	0.7367	1.4018
26	$\begin{array}{l}H_3C\\H_3C\end{array}>CH-NH-CH<\begin{array}{l}CH_3\\CH_3\end{array}$	Di-isopropylamine	$C_6H_{15}N$	101.20		84	0.722	
27	$\begin{array}{l}CH_3-CH_2-CH_2-CH_3\\ \quad HN-CH_3\end{array}$	N-methyl-z-amino n-pentane	$C_6H_{15}N$	101.20		105	0.947	
28	$CH_3-(CH_2)_2-NH-(CH_2)_2-CH_3$	Di-n-propylamine	$C_6H_{15}N$	101.20		110	0.7384	1.4046
29	$CH_3-(CH_2)_3-NH-(CH_2)_3-CH_3$	Di-n-butylamine	$C_8H_{19}N$	129.25		159		
30	$\begin{array}{l}H_3C\\H_3C\end{array}>CH-CH_2-NH-CH_2-CH<\begin{array}{l}CH_3\\CH_3\end{array}$	Di-isobutyl amine	$C_8H_{19}N$	129.25		139	0.745	1.4039
31	$CH_3-(CH_2)_4-NH-(CH_2)_4-CH_3$	Di-n-amylamine	$C_{10}H_{23}N$	157.30		205		
32	$\left[\begin{array}{l}H_3C\\H_3C\end{array}>CH-(CH_2)_2\right]_2 NH$	Di-iso-amylamine	$C_{10}H_{23}N$	157.30	−44	188		1.4229
33	$CH_3(CH_2)_6-NH-(CH_2)_6-CH_3$	Di-n-heptylamine	$C_{14}H_{31}N$	213.40	80	271		
34	$\begin{array}{l}H_3C\\H_3C\end{array}>N-CH_3$	Trimethylamine	C_3H_9N	59.11	−117.1	3	0.662	
35	$\begin{array}{l}H_3C\\H_3C\end{array}>N-CH_2-CH_3$	Dimethylethylamine	$C_4H_{11}N$	73.14		37.5		
36	$\begin{array}{l}CH_3-CH_2\\CH_3-CH_2\end{array}>N-CH_2-CH_3$	Triethylamine	$C_6H_{15}N$	101.20	−114.8	89.5	0.7255	
37	$\begin{array}{l}CH_3-(CH_2)_2\\CH_3-(CH_2)_2\end{array}>N-(CH_2)_2-CH_3$	Tri-n-propyl amine	$C_9H_{21}N$	143.27	−93.5	156.5	0.753	1.4176
38	$(CH_3-CH_2-CH_2-CH_2)_3N$	Tri-n-butylamine	$C_{12}H_{27}N$	185.35	liq.	211	0.7782	
39	$(CH_3-(CH_2)_4)_3N$	Tri-n-amylamine	$C_{15}H_{33}N$	226.44		257		
40	$H_3C-NH_2-CH_3 \quad N,N'$	Dimethylhydrazine	$C_2H_8N_2$	60.10	liq.	81	0.8274	1.4029, misc. w; acet, eth
41	$H_3C-H_2C-NH_2-CH_2-CH_3 \ NN$	Diethylhydrazine	$C_4H_{12}N_2$	88.16	liq.	84...6		
42	$H_3C-HC=N-N<\begin{array}{l}CH_2-CH_3\\CH_2-CH_3\end{array}$	Acetaldehyde-diethylhydrazone	$C_6H_{14}N_2$	114.20		123...6		misc. w;
43	$\begin{array}{l}H_3C\\H_3C\end{array}>C=N-N=C<\begin{array}{l}CH_3\\CH_3\end{array}$	Dimethyl ketazine	$C_6H_{12}N_2$	112.17	liq.	131	0.8381	1.45102, col; misc. w; acet; eth,
44	$H_3C-N=N-NH-CH_3$	Dimethyltriazine	$C_2H_7N_3$	73.20	−12	92(d)		misc. w, d. exposure air

Hetero chains with other foreign atoms

Foreign atom	No.	Formula	Name	M.F.	Mol. wt.	FP/MP	BP/CP	D and other properties
P	1	$CH_3-PH-CH_3$	Dimethylphosphine	C_2H_7P	66.06	<1	25	col; i.w; misc. acet; eth. ☠
	2	$CH_3-CH_2-PH-CH_2-CH_3$	Diethylphosphine	$C_4H_{11}P$	90.11	<1	85	col; i.w; misc. acet; eth. ☠
	3	$\begin{array}{c}H_3C\\H_3C\end{array}\!>\!P-CH_3$	Trimethylphosphine	C_3H_9P	76.08	<1	42	col; i.w; misc. acet; eth.
	4	$\begin{array}{c}CH_3-CH_2\\CH_3-CH_2\end{array}\!>\!P-CH_2-CH_3$	Triethylphosphine	$C_6H_{15}P$	118.16	liq.	128	0.801, col; i.w; misc. acet; eth.
As	5	$CH_3-AsH-CH_3$	Dimethylarsine	C_2H_7As	105.99	liq.	36	1.213, col; misc. acet; eth; Bz Chl CS₂
	6	$CH_3-CH_2-AsH-CH_2-CH_3$	Diethylarsine	$C_4H_{11}As$	134.04	liq.	105	1.338, d. on exposure to air
	7	$\begin{array}{c}H_3C\\H_3C\end{array}\!>\!As-CH_3$	Trimethylarsine	C_3H_9As	120.01	liq.	52.8	1.124, col; s.w; acet; eth.
	8	$\begin{array}{c}CH_3-CH_2\\CH_3-CH_2\end{array}\!>\!As-CH_2-CH_3$	Triethylarsine	$C_6H_{15}As$	162.09	liq.	141d	1.150, col; i.w; misc. acet; eth.
	9	$\begin{array}{c}H_3C\\H_3C\end{array}\!>\!As-As\!<\!\begin{array}{c}CH_3\\CH_3\end{array}$	Cacodyl Tetramethylarsine	$C_4H_{12}As_2$	209.96	-6	170	>1; col; oil; misc. w; acet; eth. (nauseating smell) ☠
	10	$\begin{array}{c}CH_3-CH_2\\CH_3-CH_2\end{array}\!>\!As-As\!<\!\begin{array}{c}CH_2-CH_3\\CH_2-CH_3\end{array}$	Ethylcacodyl Tetraethyldiarsine	$C_8H_{20}As_2$	266.06	liq.	185.90	1.388, i.w; acet; eth.
Sb	11	$\begin{array}{c}H_3C\\H_3C\end{array}\!>\!Sb-CH_3$	Trimethylstibine Antimonytrimethyl	C_3H_9Sb	166.86	liq.	80.6	1.523, i.w; eth; acet.
	12	$\begin{array}{c}CH_3-CH_2\\CH_3-CH_2\end{array}\!>\!Sb-CH_2-CH_3$	Triethylstibini Antimonytriethyl	$C_6H_{15}Sb$	208.94	<-29	199.5	1.324, i.w; misc. acet; eth.
Bi	13	$\begin{array}{c}H_3C\\H_3C\end{array}\!>\!Bi-CH_3$	Trimethylbismuthine Bismuthtrimethyl	C_3H_9Bi	254.10	liq.	107v	1.82, i.w; misc. acet; eth.
	14	$\begin{array}{c}CH_3-CH_2\\CH_3-CH_2\end{array}\!>\!Bi-CH_2-CH_3$	Triethylbismuthine Bismuthtriethyl	$C_6H_{15}Bi$	296.19	liq.	110...	2.30, i.w; misc. acet; eth.
Si	15	$CH_3-SiH_2-CH_3$	Dimethylsilicon	C_2H_8Si	60.14	-150	-20.1	0.68 liq.
	16	$\begin{array}{c}H_3C\\H_3C\end{array}\!>\!Si\!<\!\begin{array}{c}CH_3\\CH_3\end{array}$	Tetramethylsilicon	$C_4H_{12}Si$	88.20	liq.	26.5	0.651, misc. eth; cr CH_2SO_4
	17	$\begin{array}{c}CH_3-CH_2\\CH_3-CH_2\end{array}\!>\!Si\!<\!\begin{array}{c}CH_2-CH_3\\CH_2-CH_3\end{array}$	Tetraethylsilicon	$C_8H_{20}Si$	144.30	liq.	153	0.762
	18	$\begin{array}{c}H_3C\quad CH_3\\H_3C-Si-Si-CH_3\\HC_3\quad CH_3\end{array}$	Hexamethyldisilicon	$C_6H_{18}Si_2$	146.33	12.5 ...14	112.5	
Ge	19	$\begin{array}{c}H_3C\\H_3C\end{array}\!>\!Ge\!<\!\begin{array}{c}CH_3\\CH_3\end{array}$	Tetramethyl germanium	$C_4H_{12}Ge$	132.74	-88	43.4	1006, col; misc. acet; Bz; eth.
	20	$\begin{array}{c}CH_3-CH_2\\CH_3-CH_2\end{array}\!>\!Ge\!<\!\begin{array}{c}CH_2-CH_3\\CH_2-CH_3\end{array}$	Tetraethyl germanium	$C_8H_{20}Ge$	188.24	-90	162.5 ...3.0	0.991, col; oil; misc. eth; Bz; d in w.
Sn	21	$\begin{array}{c}H_3C\\H_3C\end{array}\!>\!Sn\!<\!\begin{array}{c}CH_3\\CH_3\end{array}$	Tetramethyl tin	$C_4H_{12}Sn$	178.84	liq.	78	1.314, misc. org. sol; i.w; **exp**
	22	$\begin{array}{c}CH_3-CH_2\\CH_3-CH_2\end{array}\!>\!Sn\!<\!\begin{array}{c}CH_2-CH_3\\CH_2-CH_3\end{array}$	Tetraethyl tin	$C_8H_{20}Sn$	234.94	-112	181	1.187, misc. org. sol; i.w; **exp**
	23	$[CH_3-Sn-CH_3]x$	Dimethyltin (polymer)	$(C_2H_6Sn)*$	(148.77)*	feste	gelbe	solid, yel, cr; i.w; org. sol.
	24	$CH_3-CH_2-Sn-CH_2-CH_3$	Diethyl tin	$C_4H_{10}Sn$	176.82	<-12	150d	1.654, yel; oil; i. org. sol. **exp**
Pb	25	$\begin{array}{c}H_3C\\H_3C\end{array}\!>\!Pb\!<\!\begin{array}{c}CH_3\\CH_3\end{array}$	Tetramethyl lead Leadtetramethyl	$C_4H_{12}Pb$	267.35	-27.5	110	1.995, nD = 1.5120, i.w; misc. Bz; bet, eth; acet.
	26	$\begin{array}{c}CH_3-CH_2\\CH_3-CH_2\end{array}\!>\!Pb\!<\!\begin{array}{c}CH_2-CH_3\\CH_2-CH_3\end{array}$	Tetraethyl lead Lead tetraethyl	$C_8H_{20}Pb$	323.45	liq.	91v 200d	1.659, nD = 1.5195; col; misc. Bz; eth; acet; pet.
B	27	$\begin{array}{c}H_3C\\H_3C\end{array}\!>\!B-CH_3$	Trimethylborine Boron trimethyl	C_3H_9B	55.92	161.5	-20.2	1:9018, gas, vap; e.s.w; acet. eth.
	28	$\begin{array}{c}CH_3-CH_2\\CH_3-CH_2\end{array}\!>\!B-CH_2-CH_3$	Triethylborine Borontriethyl	$C_6H_{15}B$	98.00	-92.2	0(125Tor)	0.6961; col; i.w; misc. eth; acet.
	29	$\begin{array}{c}H_3C\\H\end{array}\!>\!B\!<\!\begin{array}{c}H\\H\end{array}\!>\!B\!<\!\begin{array}{c}H\\H\end{array}$	Methyldiborane	CH_8B_2	41.71	gas	-80(50Tor)	col; indef. d. in w.
	30	$\begin{array}{c}H_3C\\H\end{array}\!>\!B-H_2-B\!<\!\begin{array}{c}CH_3\\H\end{array}$	1, 2-dimethyl diborane	$C_2H_{10}B_2$	55.74	-125	4.9	col; indef. d. in w.
	31	$\begin{array}{c}H_3C\\H_3C\end{array}\!>\!B-H_2-B\!<\!\begin{array}{c}CH_3\\CH_3\end{array}$	1, 1, 2, 2-tetramethyl diborane	$C_4H_{14}B_2$	83.79	-72.5	68.6	col; d. in w.
Al	32	$\begin{array}{c}H_3C\\H_3C\end{array}\!>\!Al-CH_3$	Trimethyl aluminium	C_3H_9Al	72.07	0	130	nD 1.432, col; d. in w. → $Al(OH)_3$ + CH_4, misc. eth.
	33	$\begin{array}{c}CH_3-CH_2\\CH_3-CH_2\end{array}\!>\!Al-CH_2-CH_3$	Triethyl aluminium	$C_6H_{15}Al$	114.15	<-18	194	nD 1.480, col; misc. Bz; eth; d. in w → C_2H_6 + $Al(OH)_3$ **ex**
Be	34	$CH_3-Be-CH_3$	Dimethyl beryllium	C_2H_6Be	39.09	solid	s200	need; d. in w → CH_4
	35	$CH_3-CH_2-Be-CH_2-CH_3$	Diethyl beryllium	$C_4H_{10}Be$	67.14	12	110v	col; d. in w. → C_2H_6
Zn	36	$CH_3-Zn-CH_3$	Zinc dimethyl	C_2H_6Zn	95.45	-42.2	46	1.386, col; air & moisture **ex.**
	37	$CH_3-CH_2-Zn-CH_2-CH_3$	Zinc diethyl	$C_4H_{10}Zn$	123.50	liq.	118	1.182, col; air & moisture **ex.**
Cd	38	$CH_3-Cd-CH_3$	Cadmium dimethyl	C_2H_6Cd	142.48	-4.5	105.5	1.9846, oil; d. in w.
Hg	39	$CH_3-Hg-CH_3$	Mercury dimethyl	C_2H_6Hg	230.68	liq.	96	3.069, col; misc. eth; acet.
	40	$CH_3-CH_2-Hg-CH_2-CH_3$	Mercury diethyl	$C_4H_{10}Hg$	258.73	liq.	159	2.444, col; misc. eth; acet.

• $>\!B\!<\!\begin{smallmatrix}H\\H\end{smallmatrix}\!>\!B\!<$ can also be written as $>\!B-H_2-B\!<$

Simple and condensed hetero rings or cycles

Hetero rings with: Oxa-groups

If in a carbon ring a methylene group ($-CH_2-$) is replaced by another atom or group of atoms (hence hetero−), then a hetero cyclic ring occurs.

With it the changes are: bond degree, bond length; bond angle, the molecular weight and corresponding physical properties of the compound as well as the chemical properties in relation to the position of these hetero atoms in the periodic system of elements.

Hetero rings can be built from individual units which carry functional groups or can be synthesized from carbon chains with functional groups through "cyclization" reactions. The hetero atoms are then the constituent part of the structural framework whereby a definite adjustment of the properties with a similarly built up carbon ring occurs. (Compare p. 86-hetero chains).

5a	Benzene		
	C_6H_6	80.099	0.87901
	78.11	5.51	1.50112
π	misc. eth; acet; sl. misc. w.		

Examples of such similarities, see pages:
5, 11 and 12
Hetero rings containing the double bonds can also build in a similar manner
a 6π-electron system (see p. 79).
Compare here, the series:
Pyridine−Benzene−Furan−Thiophene

All have a "benzenoid" character
(substitution instead addition, etc.)

Examples for the nomenclature:

No. of members	Rings with oxygen	
	unsaturated	saturated
3	Oxirene	Oxirane
4	Oxet	Oxetane
5	Oxole (furan)	Oxolane
6	Oxine	Oxane
	Rings with nitrogen	
3	Azirine	Aziridine
4	Azet	Azetidine
5	Azole (pyrrole)	Azolidine
6	Azine (pyridine)	Perhydroazine

Most hetero rings have trivial names
(See ring index and literature)

11a	Indane Hydrindene		
	C_9H_{10}	176.5	0.9645
	118.17	−51.40	1.53877
	col; i.w; misc. acet; eth.		

12a	Indene		
	C_9H_8	182.4	1.006
	116.15	−2	1.57107
	col; i.w; misc. acet; eth; py; CCl_4, CS_2, ac. a.		

No	Formula	Name		
		M.F.	B.P/C.P	D
		Mol. wt.	F.P/M.P	nD
π^*		Other properties		
1		Ethylene exide, 1,2,-epoxyethane Oxiran		
		C_2H_4O	10.7	1.965 g/l
		44.05	−111,3	1.35988
		v.e. s.w; eth; al; col. gas.		
2		Trimethylene oxide; Oxetane		
		C_3H_6O	47.8	0.8930
		48.08	liq.	1.3897
		misc. w; penetrating odour		
3		Tetrahydrofurane, Oxolan Tetramethylenoxide		
		C_4H_8O	64...66	
		72.10	liq.	1.4040
		sl. misc. w; acet.		
4		2,5-Dihydrofuran, Dihydrooxalon Dihydrooxalan		
		C_4H_6O	67...69	0.9503
		70.09	liq.	1.44
		col.		
5		Furan, Oxole		
		C_4H_4O	32	0.9360
		68.07	liq.	1.42157
π		misc. acet.; eth, sl. misc. w.		
6		Tetrahydropyrane, Oxane Pentamethyleneoxide		
		$C_5H_{10}O$	88	0.881
		86.14	liq.	1.421
7		Δ^2-Dihydropyrane 2,3,4-Trihydroxine		
		C_5H_8O	86	0.923
		84.12		1.440
8		α-Pyran and γ-Pyran		
		C_5H_6O	—	
		82.10		
		known only in derivatives		
9		1,4-Dioxane, P-Dioxane		
		$C_4H_6O_2$	101.4	1.03361
		88.10	11.8	1.4232
		misc. w; acet; eth; and easily with all org. sol.		
10		1,3,5-Trioxane; sym-trioxane		
		$C_3H_6O_3$	s46	
		90.08	64	
		need; s.w; acet; eth.		
11		Cumarane Benzodihydrofuran		
		C_8H_8O	188...9	1.0571
		120.14		
		oil, misc. acet; eth; i.w. misc. w.		
12		Cumarone Benzofuran		
		C_8H_6O	174	1.0776
		118.13	u-20	1.5645
		misc. acet; eth; i.w.		

Simple and condensed hetero cycles

with sulphar atoms-thia groups

No	Formula	Name	M.F	BP/CP	D	Mol. wt.	MP/FP	nD	Other properties
	π^*								
1	H_2C—S, H_2C	Ethylene sulphide / Thiirane	C_2H_4S	55...56	1.0368	60.11		1.4914	misc. acet; eth; col. liq.
2	H_2C—CH_2, H_2C—S	Trimethylene sulphide / Thietane	C_3H_6S	94	1.0284	74.14	liq.	1.5059	i.w; misc. org. sol;
3	H_2 ring S	Tetramethylene sulphide / Thiophane, Thiolane	C_4H_8S	119	0.9607	88.17	liq.	1.4871	iw; misc. org. sol; strong powder
4	H_2 ring S	Dihydrothiophene / Dihydrothiol	C_4H_6S			99.17			
5	S π	Thiophene, Thiofuran / Thiolene	C_4H_4S	84.12	1.0644	84.13	-38.30	1.5287	misc. acet; eth; i.w.
6	H_2 ring S	Tetrahydrothiopyran / Thiane, pentamethylene sulphide	$C_5H_{10}S$	141	0.9849	102.20	13	1.5046	i.w; s. org. sol.
7	S ring H_2	Dihydrothiopyran / 2,3,4-trihydrothiine	C_5H_8S			100.18			
8	S H_2 S, α γ	α and γ thiopyran / 2 or 4-hydrothiine	C_5H_6S			98.163			known only in derivatives.
9	H_2 S ring S H_2	1,4-dithiane, p-dithane / Diethylenedi sulphide	$C_4H_8S_2$	200		120.22	112s		wh; need; monocl; v.e.s.w; s. acet; eth; CS_2
10	H_2 S ring S S H_2	1,3,5-trithiane, sym-trithiane	$C_3H_6S_3$	S		138.26	215...6		tetr; prism; e.s. hot w; acet; eth;
11	S H_2 bicyclic	Benzo-2,3-dihydrothiophene / 2,3-dihydrothio nephthane	C_8H_8S	234	1.1125	136.21	liq.	1.6033	Turns yel. on exposure to air
12	S π bicyclic	Benzothiophene / Benzothiol	C_8H_6S	221	1,165f	134.19	32	1.63324[36]	e.s. acet; eth; i.w.

with nitrogen atoms-aza groups

No	Formula	Name	M.F	BP/CP	D	Mol. wt.	MP/FP	nD	Other properties
	π^*								
13	H_2C—NH, H_2C	Ethylene imine / Azirine	C_2H_5N	56	0.832	43.07			
14	H_2C—CH_2, H_2C—NH	Trimethylene imine, Azetidine, propylene imine	C_3H_7N	63	0.8436	57.10			
15	H_2 ring N H	Pyrrolidine, Azolidine / Tetramethylene imine	C_4H_9N	88.5	0.8520	71.12			col; liq; misc. w; acet, eth.
16	H_2 ring N H	Pyrroline, dihydropyrrole / Dihydroazole	C_4H_7N	90	0.910	69.10			misc. w; acet; eth.
17	N H π	Pyrrole Azol	C_4H_5N	131	0.948	67.09			col; liq. misc. acet; eth; Bz; i.w
18	H_2 ring N H	Piperdine, hexahydropyridine / Hexahydroazine	$C_5H_{11}N$	106.3	0.8622	85.15	-9	1.4534	col. liq; misc. w; acet; eth.
19	H_2 ring N H	1,2,3,4-tetrahydropyridine / Tetrahydroazine, piperidein	C_5H_9N			83.14	60-61		In dimeric form monocl. cr.
20	N π	Pyridine / Azine	C_5H_5N	115.3	0.982	79.10	-42	1.50919	col. liq; misc. w; acet, eth; Bz.
21	H N ring N H	Piperazine, hexahydro-1,2-diazine / Diethylenediimine	$C_4H_{10}N_2$	145		86.14	104		col; rhomb; s.w; v.e.s. acet.
22	H N ring N N H	Hexahydro-1,3,5-triazine	$C_3H_9N_3$			59.11			Polym → hexamethylene tetramine
23	N H bicyclic	Indoline / Benzodihydropyrrole	C_8H_9N	229	1.060	119.16			liq; sl. misc. w;
24	N H π bicyclic	Indole / Benzopyrrole	C_8H_7N	253...4	1.069	117.14			leaf; s.h.w; e.s. acet; eth.

*π - aromatic system with pi-electrons.

No	Name	MF	BP/CP	D	Mol. wt.	MP/FP	nD	Other properties
1	1,3-dioxolane; Ethylene methyl dioxide	$C_3H_6O_2$	76	1.060	74.08	liq	1.3974	misc. w; smell of peppermint
2	1,3-dioxane, m-dioxane	$C_4H_8O_2$	105	1.03422	88.10		1.41652	misc. w; acet; eth
3	Chroman; 2,3-dihydro-1,4-benzopyran	$C_9H_{10}O$	95^{12}	1.064	134.17		1.544	s. in org. sol.
4	Dibenzofuran; Dipheneleneoxide	$C_{12}H_8O$	288		168.18	87		v.e.s. acet. (bl. fluores.). s. eth; Bz
5	Xanthene; Dibenzopyran	$C_{13}H_{10}O$	315		182.21	100.5		e.s.w; v.e.s. acet; Bz; chl; CS_2
6	Thianthrene; Dibenzothine	$C_{12}H_8S_2$	116.8	1.198	216.30			Col; liq; misc. acet; eth
7	Pyrazole; 1,2-Diazole	$C_3H_4N_2$	188		68.08	70	1.47027	need; (acet) e.s.w., acet., eth. s. Bz
8	2-Pyrazoline; 4,5-Dihydropyrazole	$C_3H_6N_2$	144		70.09			Col. liq; misc. w; acet; eth.
9	Imidazole, 1,3-diazole; Glyoxaline	$C_3H_4N_2$	256		68.08	90		Col. prism; e.s.w; acet; eth.
10	Pyridazine, 1,2-diazine; Diazine	$C_4H_4N_2$	208	1.107	80.09	−8	1.52311	Col. liq; misc. w; acet; eth; Bz
11	Pyrimidine, 1,3-diazine; Miazine	$C_4H_4N_2$	124		80.09	22		Cr; s.w; acet.
12	Pyrazine, 1,4-diazine; Pyazine	$C_4H_4N_2$	118	1.031 liq.	80.09	53	1.49526 liq	Col. prism; (w) v.e s.w; acet; eth; s. chl
13	1,2,3-triazole	$C_2H_3N_3$	~205	1.186	69.07	23		Cr; e.s.w; acet; eth.
14	1,2,4-triazole	$C_2H_3N_3$	260		69.07	121	1.48544	Need; (w-eth.), s.w; acet; eth.
15	1,2,3-triazine; Vic-triazine	$C_3H_3N_3$			81.08			
16	1,2,4-Triazine; asym-triazine	$C_3H_3N_3$			81.08			
17	1,3,5-Triazine; sym-Triazine	$C_3H_3N_3$			81.08	86		
18	Tetrazole	CH_2N_4	s		70.06	156		Leaf; (acet.), e.s.w; acet; sl. s. eth.
19	1,2,3,4-Tetrazine	$C_2H_2N_4$			82.07			
20	1,2,3,5-Tetrazine	$C_2H_2N_4$			82.07			
21	1,2,4,5-Tetrazine; sym-Tetrazine	$C_2H_2N_4$	s		82.07	99		Red columns; s.w, acet. eth.
22	2,3-dihydro 1,2,3,4-tetrazine	$C_2H_4N_4$			84.09			
23	1,2-dihydro 1,2,4,5-tetrazine	$C_2H_4N_4$			84.09	125...6(d)		pa. yel; prism (Bz, acet.) e.s.w; acet; eth; Bz
24	Pentazole; no free existence	N_5H			71.05			only derivatives known

Hetero rings with aza-groups

No.	Formula	Name / MF / BP/CP / D / Mol. wt. / MP/FP / nD / Other properties	No.	Formula	Name / MF / BP/CP / D / Mol. wt. / MP/FP / nD / Other properties

Column header block

No.	Formula	Name		
		MF	BP/CP	D
		Mol. wt.	MP/FP	nD
		Other properties		

1 — Quinoline 2,3-benzo pyridine

MF	BP/CP	D
C_9H_7N	237.7	1.095
129.15	−19.5	1.6245
col. liq., sl. misc. w., misc. acet., eth., CS_2		

2 — Inoquinoline 3,4-benzo pyridine

C_9H_7N	243	1.0986
129.15	23	1.62233 liq.
col., leaf., v.e. s.w.		

3 — 1,2,3,4-tetrahydroquinoline

$C_9H_{11}N$	251	1.055
133.19	20	1.5933 l liq.
pa. yel., cr. v.e. s.w., acet., eth.		

4 — cis-decahydroquinoline

$C_9H_{17}N$	205…6	0.9426
139.24	−40	
col. liq., misc. w., eth., acet.		

5 — Trans-decahydroquinoline

$C_9H_{17}N$	~210	0.9021 liq.
139.24	48	
wh. cr., s. hot w., e.s. acet., eth.		

6 — Benzimidazole, Indazole

$C_7H_6N_2$	>360	
118.13	170	
rhomb., prism., s.w., acet., eth.		

7 — Cinnoline 1,2-benzodiazine

$C_8H_6N_2$		
130.14	38…9	
Pa. yel., cr. (pet.eth.) e.s.w., acet., eth.		

8 — Pthalazine 2,3-benzodiazine

$C_8H_6N_2$		
130.14	91	

9 — Quinazoline 1,3-benzodiazine

$C_8H_6N_2$	243	
130.14	48	
Leaf. (pet. eth.) e.s.w., acet., eth.		

10 — Quinoxaline 1,4-benzodiazine

$C_8H_6N_2$	226	1.133 liq.
130.14	30.5	1.623 ll liq.
wh., cr., s.w., acet., eth., v.e.s. Bz.		

11 — Azimino benzene; 1,2,3-benzotriazole

$C_6H_5N_3$	201…4V	
119.12	100	
Need., (Bz.) s. acet., Bz., i.w.		

12 — Phenthiazine 1,2,4-benzotriazine

$C_7H_5N_3$	235…40	
13.14	74…5	
yel. or. need., (Bz), e.s. hot w., hot w., hot eth, s. acet., Bz		

13 — Carbazole, dibenzopyrrole diphenyl imine

$C_{12}H_9N$	354…5s	
167.20	245	
col. leaf (xyl) sl. s. acet., eth., Bz., i.w., v.e.s. chl. tol. CS_2		

14 — Acridine

$C_{13}H_9N$	s100	1.1005
179.21	108	
col leaf. rhom. (need.), v.e.s. acet., eth., s. Bz		

15 — 9,10,-dihydroacridine

$C_{13}H_{11}N$	s; d ~300	
181.23	169	—
col cr. (acet.) i.w., s. hot acet., s. eth.		

16 — Phenanthridine

$C_{13}H_9N$	349.5	
179.21	107	

17 — Benzo (t) quinoline 5,6-Benzoquinol

$C_{13}H_9N$	351	
179.21	93	
leaf. (w), s. hot w., e.s. acet., eth., Bz		

18 — Benzo (h) quinoline 7,8-Benzo quinol

$C_{13}H_9N$	351	
179.21	52	
monocl., cr., (eth.), e.s.w., acet., eth., s.s. Bz.		

19 — Quincludine

$C_7H_{13}N$		
111.18	158	
cr., e.s.w., acet., eth.,		

Three dimensional structure of quincludin

20 — Phenazine, dibenzo-1,4-diazine

$C_{12}H_8N_2$	>360	
180.20	171	
Pa. yel. need., v.sl. s.w. sl. s.acet., eth., Bz		

21 — 4,5-Phenanthroline o-phenanthroline

$C_{12}H_8N_2$		
168.2	without w 117	
cr. ($1H_2O$), sl. s.w., acet.		

22 — Purine Imidazo [4,5 d] pyrimidine

$C_5H_4N_4$		
120.11	217	
need. (acet.), e.s.w., s. acet., eth., s. tol		

23 — Pteridine

$C_6H_4N_4$		

24 — Cardiazole Pentamethylene tetrazole

$C_6H_{10}N_4$		
138.17	59	
cr., e.s.w., e.s. org. sol.		

Simple and condensed hetero cycles

Different hetero atoms				Adamantane type compounds		

Different hetero atoms

No.	Formula	Name		
		MF	BP/CP	D
		Mol. wt.	MP/FP	nD
		Other properties		
1		Oxazole		
		1,3-Oxazole		
		C_3H_3ON	69...70	
		69.07		
2		Isoxazole		
		1,2-oxazole		
		C_3H_3ON	95.0...5.5	1.0805
		69.07		1.4269
3		Morpholine, Tetrahydro		
		1,4-oxazine		
		C_4H_9ON	126–30	0.9998
		87.12		
		col. hyg. oil., misc. w; acet., eth., org. sol.		
4		Oxadiazole		
		1,3,4-oxadiazole		
		$C_2H_2ON_2$		
		70.06		
5		Furazon		
		1,2,5-oxadiazole		
		$C_2H_2ON_2$		
6		Phenoxazine		
		$C_{12}H_9ON$	s	
		183.20	156	
		leaf. (Bz). e.s. acet., eth.		
7		Thiazole		
		1,3-thiazole		
		C_3H_3NS	117...	1.1998
		85.12		
		misc. acet., eth., ~ pyridine smell		
8		Isothiazole		
		1,2-thiazole		
		C_3H_3NS		
		85.12		
9		Phenathiazine		
		Phenothiazine		
		$C_{12}H_9NS$	317d	
		199.26	180	
		yel. rhomb. leaf. (acet.) e.s acet., eth., s. Bz		
10		Porphin		
		$C_{20}H_{14}N_4$		
		310.34		
		dk. red. leaf. with metal gloss.		
		i.w., v.e.s. acet., e.s. eth.		
11		Chlorophyll		
		$C_{20}H_{16}N_4$		
		312.36		

Adamantane type compounds

Formula representation and numbering

adamantane molecules (see p. 85)

No.	Name		
12	2,4,10-trioxa adamantane		
	$C_7H_{14}O_3$		
	142.16	202*	
	*In sealed tubes, very volatile cryoscopic constant = 30		
13	2-thin adamantane		
	$C_9H_{18}S$		
	154.28	F 320*	
	*In sealed tubes, very volatile		
14	1,3-biaza adamantane		
	$C_8H_{16}N_2$		
	140.23		
15	1,3,5-triaza adamantane		
	$C_7H_{14}N_3$		
	140.21	F 260	
16	1,3,5,7-tetra azadamantane, Urotropine hexamethylene tetramine		
	$C_6H_{12}N_4$	s. in V	
	140.19	263	
	rhomb. cr., e.s.w., v. sl. s. acet., i. eth.		
17	1-Aza-4,6,10-trioxa adamantane trimorpholine		
	143.14	210–220	
	$C_6H_{12}O_3N$	Subl.	
	Crist., s.w., acet., eth., Bz., chl.		
18	1-phospha 2,8,9-trioxa adamantane		
	$C_6H_{11}O_3P$		
	224.09	F 207*	

Parent compounds with linkages

Carbon rings with carbon chains

No	Formula	Name		
		MF	BP/CP	D
		Mol. wt.	FP/MP	nD
		Other properties		
1	H_2C, H_2C–C–CH_3, H	1 - methylcyclopropane		
		C_4H_8	5	0.691 liq.
		56.10		
		col. gas., e.s.w., acet., eth.		
2	H_2C, H_2C–C–CH_3, CH_3	1, 1 - Dimethylcyclopropane		
		C_5H_{10}	21	0.6604
		70.13		1.366
		i.w., misc. acet., eth.		
3	H_2C–CH–CH_3, H_2C–CH_2	1 - Methylcyclobutane		
		C_5H_{10}	~40	0.6931
		70.13		1.3836
		col. liq., misc. acet., eth., immisc. w.		
4	(cyclopentane with CH_3)	Methylcyclopentane		
		C_6H_{12}	71.8	0.7488
		84.16	−142.4	1.4098
		col., i.w., misc. acet., eth., Bz		
5	(cyclopentadiene =CH_2)	Fulvene		
		Methylenecyclopentadiene		
		C_6H_6		
		yel. oil., very unstable		
6	(=C with CH_3, CH_3)	Dimethylfulvene		
		Isopropylenecyclopentadiene		
		C_8H_{10}		
		Or., coloured oil		
7	(cyclohexane with CH_3)	Methylcyclohexane		
		Hexahydrotoluene		
		C_7H_{14}	100.3	0.7864
		98.18	−126.4	1.4235
		misc. acet., eth., immisc. w.		
8	(cyclohexane CH_3, CH_3)	1, 3 - Dimethylcyclohexane		
		Hexahydro - m - xylene cis - form		
		C_8H_{16}	121	0.7735
		112.21	−85	1.4269
		misc. acet., immisc. w., trans nD 1.4254 D 0.772		
9	(cyclohexane CH_3, CH_3)	1, 4 - Dimethylcyclohexane		
		Hexahydro - p - xylene, cis - form		
		C_8H_{16}	121	0.7671
		112.21	−86	1.421
		col. liq., tran D = 0.7638, B.P. 119		
10	(cyclohexane trimethyl)	1, 3, 5 - trimethylcyclohexane		
		Hexahydromesitylene cis - form		
		C_9H_{18}	141	0.773
		126.24		1.43010
		col. liq., trans nD 1.4274, D = 0.772		
11	(cyclohexane CH_3 / CH CH_3)	p - menthane		
		1 - methyl - 4 - isopropylcyclohexane		
		$C_{10}H_{20}$	169…70	0.793
		140.26		1.437
		col. liq., misc. acet., eth., immisc. w.		
12	(cyclohexane CH_3, CH CH_3 CH_3)	m - menthane, 1 - methyl 3 iso propylcyclohexane		
		$C_{10}H_{20}$	167…8	0.8033
		140.26		1.44204
		smell like ligroin		
13	(cyclohexene CH_3)	4 - methylcyclohexene		
		C_7H_{12}	102…3	0.841
		96.17		
		liq., misc. acet., eth., i.w.		
14	(cyclohexene CH_3 / C CH_3 CH_3)	Δ^1 - menthene carvomenthene		
		$C_{10}H_{18}$	1.75	0.829
		138.25		
		col. oily liq., misc. acet., eth.		
15	(cyclohexene CH_3 / C CH_3 CH_3)	Δ^3 - menthene, (d, form)		
		$C_{10}H_{18}$	168	0.8073
		138.25		1.44813
		col. liq., misc. acet., eth., Bz.		
16	(cyclohexadiene CH_3 / C CH_3 CH_3)	Δ 1;3 or α - terpinene, (Δ1, (7), 3 or β terpinene		
		$C_{10}H_{16}$	α BP 180	D 0.846
		136.22	β BP 173	D 0.838
		col. liq., misc. acet., eth., i.w.		
17	(cyclohexene CH_3 / C= CH_3 CH_3)	Terpinene Δ1, 4 (8)		
		$C_{10}H_{16}$	185	0.855
		136.22		1.4823
		col. liq., misc. acet., eth., i.w.		
18	(cyclohexene CH_3 / C=CH_2 CH_3)	Δ1, 8, (9) limonene (d form)		
		$C_{10}H_{16}$	177	0.842
		136.22	−96.9	1.47489
		col. liq., misc. acet., eth., i.w.		
19	(cyclohexadiene CH_3 / CH CH_3 CH_3)	Δ1, 5 - phellandrene or α - phellandrene		
		$C_{10}H_{16}$	175	0.843
		136.22		
		col. liq., immisc. acet., misc. eth.		
20	(cyclohexadiene CH_2 / C CH_3 CH_3)	Δ1, (7), 2 - phellandrene or β - phellandrene		
		$C_{10}H_{16}$	171.2	0.852
		136.22		
		col. liq., immisc. w., acet., misc. eth.		
21	(cyclohexene CH_3 / C CH_2 =CH_3)	Δ1, 8, (9) - sylvestrene, (d form)		
		$C_{10}H_{16}$	177	0.863
		136.22		1.47717
		liq., immisc. w., misc. acet., eth.		
22	(numbered ring: 7 top, 6 1 2, 5 4 3, 10 8 9)	In literature usual way of numbering these structural framework.		

95

Carbon rings with carbon chains

No.	Formula	Name	MF	BP/CP	D	Mol. wt	FP/MP	nD	Other properties
1	CH3	Toluene methylbenzene phenylmethane	C7H8	110.626	0.8670	92.13	−95	1.4969	col. liq. misc. acet; eth; chl., ac.
2	CH3, CH3	o-xylene a., CS2 Bz 1,2-dimethyl benzene	C8H10	144.411	0.8802	106.16	−29	1.5054	col., liq. misc. eth., acet.
3	CH3, CH3	m-xylene 1,3-dimethyl benzene	C8H10	139.1	0.8642	106.16	−53.6	1.4972	col., liq. misc. eth., acet.
4	CH3, CH3	p-xylene 1,4-dimethyl benzene	C8H10	138.3	0.8611	106.16	13.2	1.4958	col. liq., monocl, cr., misc. eth ; acet
5	CH2—CH3	Ethyl benzene phenylethane, ethylphenyl	C8H10	136.2	0.86690	106.16	−93.9	1.49594	col. liq., misc. acet, eth ; v.sl. misc w
6	CH3, CH3, CH3	1,2,3-trimethyl benzene hemimellitene or (vic)-	C9H12	176.2	0.8944	120.19	<−15	1.5139	col. liq., misc. eth., acet.
7	CH3, CH3, CH3	1,2,4-trimethyl benzene pseudocumene (a sym)	C9H12	169.2	0.8758	120.19	−57.4	1.5049	col. liq., misc. ether, acet.
8	CH3, H3C, CH3	1,3,5-trimethyl benzene mesitylene (sym)	C9H12	164.7	0.8652	120.19	−52.7	1.4994	col. liq., misc. acet.
9	CH3, CH2—CH3	2,-methyl-2-ethyl benzene (o-ethyltoluene)	C9H12	165.2	0.8807	120.19	<−17	1.5045	col. liq., misc. acet.
10	CH3, CH2—CH3	1-methyl-3-ethyl benzene (m-ethyltoluene)	C9H12	161.3	0.8645	120.19		1.4966	col. liq., misc. acet.
11	CH3, CH2—CH3	1-methyl-4-ethyl benzene (p-ethyltoluene)	C9H12	162.1	0.8612	120.19	<−20	1.4950	col., liq., misc. acet.
12	CH2—CH2—CH3	n-propyl benzene 1-phenylpropane	C9H12	159.2	0.8620	120.19	−101.6	1.4920	col. liq., misc. acet.
13	H3C—CH—CH3	Isopropyl benzene cumene	C9H12	152.4	0.8618	120.19	−96.9	1.4915	col. liq., misc. eth., acet., Bz.
14	CH3, CH3, CH3, CH3	1,2,3,4-tetramethyl benzene prehnitene	C10H14	205.0	0.9053	134.21	−4.	1.5201	col. liq., misc. eth., acet.
15	CH3, CH3, H3C, CH3	1,2,3,5-tetramethyl benzene Isodurene	C10H14	197	0.896	134.21	−24		col. liq., misc. eth., acet.
16	H3C, CH3, H3C, CH3	1,2,4,5-tetramethyl benzene Durene	C10H14	197.9	0.8899	134.21	80	1.5125	col. monocl, leaf. s. acet. eth ; Bz, e.s. ac. a.
17	CH2—CH3, CH2—CH3	1,3-diethyl benzene ac. a. m-diethyl benzene	C10H14	181.1	0.8641	134.21	<−20	1.4953	col. liq., misc. eth., acet.
18	CH3, H3C—CH—CH3	4-isopropyltoluene p-cymene	C10H14	177.1	0.8573	134.21	−73.5	1.4909	col. misc. eth., acet., chl.
19	CH3, H3C—C—CH3 (CH3)	t-butyl benzene 2-phenyl-2-methylpropane	C10H14	169.1	0.86650	134.21	−58.1	1.4926	col. liq., misc. eth. acet.
20	CH3, CH3, H3C, CH3	Pentamethyl benzene	C11H16	230	0.847 liq.	149.28	53	1.50489 liq.	col., prism., (acet.), s. acet.
21	CH3, H3C, CH3, H3C, CH3, CH3	Hexamethyl benzene	C12H18	265		169.27	166		col. rhomb; leaf; (acet.) sl. s. acet; e.s. Bz
22	CH=CH2	Styrene, vinyl benzene phenylethylene	C8H8	146	0.9074	104.14		1.54344	col. liq., misc. acet., eth., i.w.
23	CH=CH—CH3	Propenyl benzene, Isoallyl benzene 1-phenylpropene	C9H10	175	0.914	118.17			col. liq., misc. acet., eth., i.w.
24	H3C—C=CH2	Isopropenyl benzene 2-phenylpropane	C9H10	160.5...1.5	0.9139	118.17			col. liq; misc. acet ; eth; immisc w
25	C≡CH	Ethynyl benzene phenylacetylene	C8H6	143	0.9295	102.13	−48...40	1.5524	col. liq.

Carbon rings with carbon chains

No.	Formula	Name / MF / Mol. wt / Other properties	BP / MP	D / nD
		Name		
		MF	BP	D
		Mol. wt.	MP	nD
		Other properties		
1	CH3	1 - methylnaphthalene / α - methylnaphthalene / $C_{11}H_{10}$ / 142.19 / col. liq., misc. acet., eth., i.w.	240...3 / −22	1.025 / 1.618
2	CH3	2 - methylnaphthalene / β - methylnaphthalene / $C_{11}H_{10}$ / 142.19 / col ; monocl., (acet), e.s. acet. eth.	245 / 35.1	1.029 / 1.60263 liq
3	CH3 CH3	1, 2 - dimethylnaphthalene / $C_{12}H_{12}$ / 156.22	262 / 96...7	
4	CH3 CH3	1, 3 - dimethylnaphthalene / $C_{12}H_{12}$ / 156.22	263	1.002 / 1.6078
5	CH3 CH3	1, 4 - dimethylnaphthalene / $C_{12}H_{12}$ / 156.22 / col. liq., misc. acet., eth., i.w.	262...4 / <−18	1.008 / 1.6127
6	CH3 CH3	1, 5 - dimethylnaphthalene / $C_{12}H_{12}$ / 156.22	80...80.5	
7	CH3 H3C	1, 6 - dimethylnaphthalene / $C_{12}H_{12}$ / 156.22	262...3	1.003 / 1.607
8	CH3 H3C	1, 7 - dimethylnaphthalene / $C_{12}H_{12}$ / 156.22	261...2	1.0115 / 1.60831
9	CH3 CH3	1, 8 - dimethylnaphthalene / peri - dimethylnaphthalene / $C_{12}H_{12}$ / 156.22	140V / 63	
10	CH3 CH3	2, 3 - dimethylnaphthalene / $C_{12}H_{12}$ / 156.22 / leaf. (acet.), s. acet., eth., i.w.	260...65 / 104...4.5	1.008
11	CH3 H3C	2, 6 - dimethynaphthalene / $C_{12}H_{12}$ / 156.22	261...2 / 110...1	
12	CH3 H3C	2, 7 - dimethylnaphthalene / $C_{12}H_{12}$ / 156.22	92	
13	CH3 CH3	1, 2 - dimethylazulene / $C_{12}H_{12}$ / 156.22 / bl. cr. (acet.), s. acet	58...9	
14	CH3 CH3	4, 8 - dimethylazulene / $C_{12}H_{12}$ / 156.22 / bl. Cr. (acet.), s. acet.	69...70	
15	$H_3C-CH-CH_3$ ... H CH3	Thujane / 4 - methyl - 1 - isopropyl bicyclo [0,1,3] - hexane / $C_{10}H_{18}$ / 157°(758mm) / D 0.8139 nD 1.43759		
16	$H_3C-CH-CH_3$ ... CH_2	Sabinene, 1 - isopropyl - 4 - methylene bicyclo [3,1,0] - hexane / $C_{10}H_{16}$ / 136.23 / D 0.842 nD 1.46738 / col. liq., misc. acet., eth.	162	
17	CH3 CH3 H3C	Camphane, 1,7,7,-trimethyl- bicyclo [2,2,1] heptane / $C_{10}H_{18}$ / 138.25 / hex. prism or leaf., i.w., s. hot acet., eth.	160s / 152...4	
18	CH3 CH3 =CH2	d, l-camphor., dl-2, 2, -dimethyl methylene bicyclo [2,2,1] heptane / $C_{10}H_{15}$ / 126.23 / d[α]D + 103.9° need. MP 51 / l [α]D − 52° or MP 42...52 / d. es. acet., v.e. s. eth. l:v.e. s.acet. eth.	199...60	
19	H3C H3C CH3	dl-pinene, 4, 7, 7-trimethyl bicyclo [3, 1, 1] hept. ene d, l & α-pinene / $C_{10}H_{16}$ / 136.23 / D 0.8582 nD 1.4658 / col. liq., e.s. acet., eth.. chl.	154 / −55	
20	CH3 CH3 CH3	Δ³ Carene Isodiprene / $C_{10}H_{16}$ / D 0.8561 nD 1.4750 / col. oil., (d extro rotatory- data)	170-172	
21	CH3 CH CH3 CH3	Retenene 1, methyl-7- isopropyl phenantherene / $C_{18}H_{18}$ / 234.33 / leaf., (acet), sl. s. acet., s. eth → abietic acid	390 / 100...1	
22	H CH3 H2 H2	3 - Methyl - 1, 2 - cyclo- penteno-phenantherene / $C_{18}H_{16}$ / 232.33 / dicl's hydrocarbon., →steroidal sterol		

Carbon rings linked through carbon chains

No.	Formula	Name			
		MF	BP	D	
		Mol. wt.	MP	nD	
		Other properties			
1	H₂ rings (Bicyclohexyl)	Bicyclohexyl / Dodeca hydrobiphenyl			
		$C_{12}H_{22}$	234	0.8644	
		166.30	3.65	1.4766	
		col. liq., misc. acet., eth.			
2	Cyclohexyl benzene structure	Cyclohexyl benzene / Phenylcyclohexyl			
		$C_{12}H_{16}$	237.5	0.9440	
		160.25	7	1.5329	
		oily liq., misc. acet., eth.			
3	biphenyl structure	Biphenyl, diphenyl / Phenyl benzene			
		$C_{12}H_{10}$	254...5	1.180	
		154.20	69...71	1.58822 liq	
		col., monocl., s. acet., e.s. eth.			
4	diphenylmethane structure	Diphenylmethane / Benzyl benzene			
		$C_{13}H_{12}$	261...2	1.0008	
		168.23	26...7	1.57884	
		col. rhomb., need. (acet) s. acet., eth. chl.			
5	dibenzyl structure	1,2–diphenyl ethane, Dibenzyl, sym-diphenyl ethane			
		$C_{14}H_{14}$	284	0.995	
		182.25	52.5		
		col. monocl. need (acet) s. acet.			
6	CH–CH₃ structure	1,1-diphenylethane / asym-diphenylethane	eth., CS₂		
		$C_{14}H_{14}$	272	1.006	
		182.25		1.5761	
		col. oil., misc. acet., eth.			
7	C=C structure	trans-1,2-diphenylethylene / stilbene			
		$C_{14}H_{12}$	307	0.970 liq	
		180.24	124		
		col. monocl. (acet), sl.s. acet., s. eth.			
8	C=CH₂ structure	1,1-diphenylethylene			
		$C_{14}H_{12}$	277	1.0206	
		180.24	9	1.610	
		col. liq.			
9	C≡C structure	Diphenylacetylene / Tolane			
		$C_{14}H_{10}$	300		
		178.22	62.5		
		col. monocl. leaf, (acet), e.s. hot, acet, eth, i.w.			
10	1,4-diphenylbenzene structure	1,4-diphenylbenzene Terphenyl			
		$C_{18}H_{14}$	s427	1.234	
		230.29	213		
		col. leaf(acet), v.e.s., acet, eth, ac. a, CS₂ s.h. Bz, Bl, flueres.			
11	m-terphenyl structure	1,3-diphenyl benzene / m-terphenyl			
		$C_{18}H_{14}$	363		
		230.29	86...7		
		Need. (acet), s. acet., eth. Bz.			
12	triphenylmethane structure	Triphenylmethane			
		$C_{19}H_{16}$	359.2	1.014	
		244.32	95.2	1.5839	
		col., rhomb. leaf, e.s. acet., eth., s. Bz., chl.			
13	2-benzylbiphenyl structure	2-benzylbiphenyl			
		$C_{19}H_{16}$	283...7V		
		244.32	54		
		monocl. need., s. acet., eth., Bz., CCl₄			
14	sym-triphenyl benzene structure	1,3,5-triphenyl benzene / sym-triphenyl benzene			
		$C_{24}H_{18}$			
		306.69	170	1.206	
		rhomb. pl. (eth), e.s. acet., eth., s. Bz			
15	tetraphenylmethane structure	Tetraphenylmethane			
		$C_{25}H_{20}$	431		
		320.41	285		
		col. rhomb. (Bz), i.w. acet, eth, s. hot Bz			
16	hexaphenylethane structure	Hexaphenylethane			
		$C_{38}H_{30}$			
		486.62	145...7d		
		col. cr., e.s. acet., s. chl., i.w., s. dioxane, CS₂, no. 1:16			
17	triphenylmethyl structure	Triphenylmethyl (free radical)			
		$C_{19}H_{15}$	-		
		243.31	-		
		only in solution - yel.			
18	1,1'-binaphthyl structure	1,1'-binaphthyl / α,α'-dinaphthyl			
		$C_{20}H_{14}$	240...40	(~360	
		254.31	160.5		
		col. rhomb. leaf (acet), s.h. acet.	s. eth; Bz; CS₂		
19	1,2-binaphthyl structure	1,2-binaphthyl / α,β-dinaphthyl			
		$C_{20}H_{14}$			
		254.31	76		
		bl. e.s. ligroin.			
20	2,2'-binaphthyl structure	2,2-binaphthyl / β,β'-dinaphthyl			
		$C_{20}H_{14}$	452		
		254.31	187...8		
		col. leaf., e.s. acet., eth., s. hot Bz, CS			
21	α-carotene structure	α-carotene.			
		$C_{40}H_{56}$			
		536.85	1.75		
			[α] + 364		
		i.w. v.e.s. Bz; chl, βet., CS₂			
22	β-carotene structure	β-carotene, provitamin A			
		$C_{40}H_{56}$			
		536.85	181.2		
		Red. brn., glt. cr. i.w., v.e.s. acet. ether, me al., chl. s. Bz, CS₂, pet.			

Carbon rings with hetero chains

No.	Formula	Name	MF	BP	D	Mol. wt.	MP	nD	Other properties
1	$O-CH_3$ (benzene)	Methoxy benzene, anisole Methylphenylether	C_7H_8O	155	0.9954	108.13	-37.3	1.51791	col. liq., misc. acet., eth.
2	CH_2-O-CH_3 (benzene)	Methylbenzylether α-methoxy toluene, benzylmethylether	$C_8H_{10}O$	174	0.987	122.16			liq., misc. acet., eth.
3	CH_3 / $O-CH_3$ (benzene)	2-methoxytoluene o-cresylmethylether	$C_8H_{10}O$	171.3		122.16		1.5199	liq., misc. acet., eth.
4	CH_3 / $O-CH_3$ (benzene)	3-methoxytoluene m-cresylmethylether	$C_8H_{10}O$	177.2	0.9766	122.16		1.506	liq.
5	CH_3 / $O-CH_3$ (benzene)	4-methoxytoluene p-cresylmethylether	$C_8H_{10}O$	176.5	0.9709	122.16		1.51237	
6	$O-CH_2-CH_3$ (benzene)	Ethoxy benzene, phenetole Ethylphenylether	$C_8H_{10}O$	172	0.9666	122.16	-30.2	1.5076	col. liq., misc. eth., acet.
7	$H_2C-O-CH_2-CH_3$ (benzene)	Ethylbenzylether α-ethoxy toluene, benzylethylether	$C_9H_{12}O$	185	0.949	136.19			col. liq., misc. acet., eth.
8	$O-CH=CH_2$ (benzene)	Ethenoxy benzene; Vinylphenylether Phenylvinylether	C_8H_8O	155...6		120.14			
9	$O-CH_3$ (naphthalene)	1-methoxynaphthalene methyl-1-phenylether	$C_{11}H_{10}O$	265...9	1.0964	158.19	<-10	1.6232	col. liq., misc. acet., eth., Bz
10	$O-CH_3$ (naphthalene)	2-methoxynaphthalene methylnaphthylether	$C_{11}H_{10}O$	274		158.19	72		col. leaf. (eth) e.s. acet., ether, Bz. CS₂
11	$O-CH_3$ (biphenyl)	4-methoxybiphenyl	$C_{13}H_{12}O$	274		184.23	29		prism
12	$O-CH_3$ (biphenyl)	2-methoxybiphenyl	$C_{13}H_{12}O$	99		184.23			leaf., s. hot acet.
13	$O-CH_3$ / $O-CH_3$ (benzene)	1,2-dimethoxy benzene	$C_8H_{10}O_2$			138.16			
14	$O-CH_3$ / $O-CH_3$ (benzene)	1,3-dimethoxy benzene Resorcinoldimethylether	$C_8H_{10}O_2$	216...18	1.0803	138.16	-52		col. liq., misc. acet., eth.
15	$O-CH_3$ / $O-CH_3$ (benzene)	1,4-dimethoxy benzene Hydroquinonedimethylether	$C_8H_{10}O_2$	212.6	1.053	138.16	56		col. leaf (w), e.s. acet., eth., s. Bz.
16	$O-CH_2-CH_3$ / $O-CH_2-CH_3$ (benzene)	1,2-diethoxy benzene catecholdiethylether	$C_{10}H_{14}O_2$			166.21	43...5		Cr. (pet. eth.). s. pet.
17	$O-CH_2-CH_3$ / $O-CH_2-CH_3$ (benzene)	1,3-diethoxy benzene Resorcinoldiethylether	$C_{10}H_{14}O_2$	234...5		166.21	12.4		prism, s. acet., eth.
18	$O-CH_2-CH_3$ / $O-CH_2-CH_3$ (benzene)	1,4-diethoxy benzene hydroquinonediethylether	$C_{10}H_{14}O_2$	246		166.21	71...2		leaf, e.s. acet., eth., Chl.
19	$O-CH_3$ / $O-CH_3$ / $O-CH_3$ (benzene)	1,2,3-trimethoxy benzene pyrogalloltrimethylether	$C_9H_{12}O_3$	241	1.0987 liq.	168.19	47		col. rhomb. need. (acet), e.s. acet; eth, Bz
20	$O-CH_3$ / H_3C-O / $O-CH_3$ (benzene)	1,3,5-trimethoxy benzene phloroglucinoltrimethylether	$C_9H_{12}O_3$	255.5		168.19	54...5		col. prism. (acet), e.s. acet., eth., Bz.
21	$O-CH_2-CH_3$ / H_2C-CH_3 / H_3C-CH_2 (benzene)	1,3,5-triethoxy benzene phloroglucinoltriethylether	$C_{12}H_{18}O_3$	175V		210.17	43		col. Cr., e.s. acet., eth.
22	$O-CH_3$ / $CH_2-CH=CH_2$ (benzene)	Estragole 1-methoxy-4-allyl benzene	$C_{10}H_{12}O$	215	0.9645	148.20		1.5230	oil., misc. acet., eth.
23	$O-CH_3$ / $CH=CH-CH_3$ (benzene)	Anetole 1-methoxy-4-propanyl benzene	$C_{10}H_{12}O$	235.3	0.9936	148.20	22.5	1.5624	col. leaf (acet) e.s.w., acet., eth.
24	$O-CH_3$ / $O-CH_3$ / $CH_2-CH=CH_2$ (benzene)	Eugenolmethylether 4-allylveratrole	$C_{11}H_{14}O_2$	248...9	1.055	178.22		1.5383	col. liq., misc. acet., eth. chl. Bz, CS₂
25	$O-CH_3$ / $O-CH_3$ / $CH=CH-CH_3$ (benzene)	Isoeugenolmethylether 4-propenylveratrole	$C_{11}H_{14}O_2$	262...4	1.0551	178.22		1.5720	col. liq., misc. acet., eth.

No.	Name / Other properties	MF	BP	D	Mol. wt.	MP	nD
1	N-methylcyclohexylamine — col. liq., misc. w., acet., eth.	$C_7H_{15}N$	145...7		113.20		
2	N-ethylcyclohexylamine — col. liq., misc. w. eth., acet.	$C_8H_{17}N$	164	0.868	127.23		
3	N-methylaniline / N-methylamino benzene — yel. liq., misc. w., acet., eth., cnl.	C_7H_9N	195.7	0.986	107.15	−57.0	1.57021
4	N,N-dimethylaniline / N,N-dimethylanio benzene — yel. liq., misc., w., acet., eth. and other org. sol.	$C_8H_{11}N$	192.5...3.5	0.9557	121.18	2.5	1.55819
5	N-ethylaniline / N-ethylphenylamine — col. liq., i.w., misc. acet. eth.	$C_8H_{11}N$	201	0.9193	121.18		
6	N,N-diethylaniline / N-phenyldiethylamine — yel. brn. oil., misc., acet., eth., chl. v. sl. s.w.	$C_{10}H_{15}N$	215.5	0.93507	149.23	−38.8	1.54105
7	N-methyl-N-ethylaniline — col. liq., i.w., misc. acet., eth.	$C_9H_{13}N$	201	0.9193	135.20		
8	N-methyl-o-toluidine — liq., misc. acet., eth., sl. s.w.	$C_8H_{11}N$	207	0.973	121.18		1.5649
9	N-methyl-m-toluidine — liq., immisc. w., misc. acet., eth.	$C_8H_{11}N$	206		121.18		
10	N-methyl-p-toluidine — liq., misc. acet., eth., w.	$C_8H_{11}N$	206...8		121.18		
11	N,N-dimethyl-o-toluidine — liq., misc. acet., eth., sl. s.w.	$C_9H_{13}N$	184.6	0.9286	135.20	−60.0	1.5153
12	N,N-dimethyl-m-toluidine — liq., misc. w., eth., acet.	$C_9H_{13}N$	212.5	0.941	135.20		1.5492
13	N,N-dimethyl-p-toluidine — liq., misc. w., acet., eth.	$C_9H_{13}N$	210...11	0.9287	135.20		1.536(6)
14	N,N-diethyl-o-toluidine / 1-diathylamino-2-methylbenz — prism (w.), e.s.w., s. acet., eth	$C_{11}H_{17}N$	206		163.26	72...3	
15	N,N-diethyl-p-toluidine / 1,1-diethylamino-4-methylbenzen — col. liq., misc. w., eth., acet.	$C_{11}H_{17}N$	229	0.924?	163.26		
16	N-methyl-1-naphthylamine / α-naphthylmethylamine — redsh. oil, i.w., misc; acet; eth, CS_2	$C_{11}H_{11}N$	293		157.21		
17	N-methyl-2-naphthylamine / β-naphthylmethylamine — oil.	$C_{11}H_{11}N$			157.21		
18	N-ethyl-1-naphthylamine — col. oil., i.w., misc. acet., eth.	$C_{12}H_{13}N$	305	1.060	171.23	308...10	1.647
19	N-ethyl-2-naphthylamine — col. oil., i.w., misc. eth., acet.	$C_{12}H_{13}N$	.315...6	1.057	171.23	<−15	1.654
20	N,N-dimethyl-1-naphthylam — col. viol., fluores., i.w., s. acet.,	$C_{12}H_{13}N$	274.5	1.044	171.23	60...67	
21	N,N-dimethyl-2-naphtylamine — dk. red need., i.w., s. acet., e	$C_{12}H_{13}N$	305	1.029	171.23	52...53	1.644
22	N,N-diethyl-1-naphthyl amine — brnsh. oil., i.w., misc. acet., e	$C_{14}H_{17}N$	290	1.005	199.29		1.593
23	N,N,N',N'-tetramethyl phenelenediamine — leaf. (acet), e.s.h.w., acet, eth; chl.	$C_{10}H_{16}N$	260		164.25	51	
24	1-ethyl-2-phenylhydrazine — liq., misc. eth., acet., chl.	$C_8H_{12}N_2$	240	1.004	136.19		1.571
25	Acetaldehydiphenyl hydrazone — col. need., s. pet. eth.	$C_8H_{10}N_2$	236...7V		134.18	98.101(57)	

Hetero rings with carbon chains

No.	Formula	Name			Properties
		MF	BP	D	
		Mol. wt.	MP	n_D	
		Other properties			
1		Methyloxirane, propyleneoxide 1,2-epoxypropene			C_3H_6O — 35 — 0.859; 58.08 — 1.466; col. liq., misc. w., acet., eth.
2		Methyloxirono, allyleneoxide, 1,2-epoxypropene			C_3H_4O — 63; 56.06; liq., misc. acet., w., eth.
3		Ethyloxirane, α-butyleneoxide 1,2-epoxybutane			C_4H_8O — 61...2 — 0.837; 72.10 — 1.385
4		cis. 2,3-dimethyloxirane cis. 2,3-epoxybutane			C_4H_8O — 58...59 — 0.8226; 72.10
5		trans 2,3-dimethyloxane trans 2,3-epoxybutane			C_4H_8O — 53...54 — 0.8010; 72.10; (65% trans+35% cis by wt. mix.)
6		2-methylfuran sylvan			C_5H_6O — 64 — 0.916; 82.10 — 1.434; col. liq., misc. acet., eth.
7		3-methylfuran			C_5H_6O — 65.5 — 0.923; 82.10; col. liq., misc. eth., acet.
8		2-methyltetrahydrofuran tetrahydrosylvan			$C_5H_{10}O$ — 79 — 0.855; 1.407; misc. acet., eth., chl.
9		2,5-dimethylfuran			C_6H_8O — 94 — 0.888; 96.12 — 1.4363; col. liq., misc; acet., eth., chl., ac.a., Bz.
10		2-methyl-1,3 dioxolane			C_4H_7O — 82.5 — 1.002; 88.10; col. liq., misc. w.
11		Safrole, 1-allyl-3,4-methylene dioxybenzene			$C_{10}H_{10}O_2$ — 234.5 — 1.096; 162.18 — 11 — 1.5420; col. liq. monocl. cr., misc., acet., chl.
12		Isosafrole, 3,4-methylenedioxy-1-propenylmethylbenzene			$C_{10}H_{10}O_2$ — Bp cis 242...3 trans. 248...52; 162.18 D cis 1.107 trans. 1.123; n_D cis 1.5632 trans. 1.5736
13		2-methylthiophene, thiotoluene 2-methylthiazole			C_5H_6S — 112.5 — 1.0194; 98.16 — −63.5 — 1.5203; col. liq. misc. acet., eth.
14		3-methylthiophene, β-thiotoluene 3-methylthiazole			C_5H_6S — 115.4 — 1.6216; 98.16 — −68.9 — 1.5204; col. oil., misc. acet., eth.
15		2,3-dimethylthiophene, 2,3-thioxene 2,3-dimethylthiazole			C_6H_8S — 136...7 — 0.9938; 112.18; col. liq., misc. acet., eth.
16		2,4-dimethylthiophene, 2,4-thioxene 2,4-dimethylthiazole			C_6H_8S — 138 — 0.9956; 112.18; col. liq., misc. acet., eth.
17		2,5-dimethylthiophene, 2,5-thioxene 2,5-dimethylthiazole			C_6H_8S — 137.5 — 0.9859; 112.18 — 1.51418; col. liq., misc. acet., eth.
18		2,3,5-trimethylthiophene 2,3,5-trimethylthiazole			$C_7H_{10}S$ — 160...3; 126.21; col.
19		N-methylpyrrolidine 1-methylpyrrolidine			$C_5H_{11}N$ — 81...3; 85.15; col. liq., misc. w.
20		2-methylpyrrole (α M.p.) 2-methylazole			C_5H_7N — 148 — 0.945; 81.11; col. liq., misc. acet., eth.
21		2,5-dimethylpyrrole 2,5-dimethylazole			C_6H_9N — 165 — 0.935; 95.14 — 1.50337; oil., misc. w., acet., eth.
22		Opsopyrrole 3-methyl-4-ethylpyrrole			$C_7H_{11}N$ — 74...5V — 0.9059; 43 — 1.4913
23		Haemopyrrole 2,3-dimethyl-4-ethylpyrrole			$C_8H_{13}N$ — 88V; 16...17 — 0.915; s. dil. HCl., w.-vapour
24		Cryptopyrrole 2,4-dimethyl-3-ethylpyrrole			$C_8H_{13}N$ — 84...5V — 0.913; O; prism., s. acet., eth., chl., steam
25		Phyllopyrrole 2,3,5-trimethyl-4-ethylpyrrole			$C_9H_{15}N$ — 88...90V; 66...9; leaf., e.s. acet., eth., and/w.

Hetero rings with carbon chains

No.	Formula	Name / MF / Mol. wt. / Other properties	BP / MP	D / n_D

Hetero rings with carbon chains

No.	Formula	Name	MF	BP	Mol. wt.	MP	D	n_D	Other properties
1	CH_3 / H_2 N H_2 / H_2 H_2	1-methylpiperidine	$C_6H_{13}N$	105.9	99.17		0.8207	1.4378	col. liq., misc. eth., acet.
2	H CH_3 / H_2 H / H_2 H_2	2-methylpiperidine / 2-pipercolin, α-pipercolin	$C_6H_{13}N$	119	99.17	9	0.844	1.44627	liq. → misc. w., i., dil. KOH
3	H / H_2 H_2 / H_2 CH_3 / H_2	3-methylpiperidine / 3-pipercolin, β-pipercolin	$C_6H_{13}N$	126	99.17		0.845	1.43779	liq., misc. w., acet., eth.
4	CH_3 / H_2 N CH_3 H / H_2 H_2	1,2-dimethylpiperidine / N,α-dimethylpiperidine	$C_7H_{15}N$	127.9	113.20				liq.
5	H CH_2-CH_3 / H_2 N / H_2 H_2	dl. 2-ethylpiperidine, / dl. α-ethylpiperidine	$C_7H_{15}N$	143	113.20		0.867		liq., sl. misc. w.
6	H $CH_2-CH_2-CH_3$ / H_2 N / H_2 H_2	d-coniine / d, 2-propylpiperidine	$C_8H_{17}N$	166.5	127.23	-2.5	0.845	1.45119	col., 81 [α] +13.79°, sl. misc. w. misc. acet., eth, Bz
7	N CH_3	2-picoline. α-picoline / 2-methylpyridine	C_6H_7N	128	93.12	-69.9	0.950	1.50293	col. liq., misc. w. acet., eth.
8	N / CH_3	3-picoline, β-picoline / 3-methylpyridine	C_6H_7N	143.5	93.12		0.9613	1.50432	col. liq., misc. w acet., eth.
9	N / CH_3	4-picoline, γ-picoline / 4-methylpyridine	C_6H_7N	143.1	93.12		0.9571		col. liq., misc., w., acet., eth.
10	N CH_3 / CH_3	2,4-butidine, α-γ-butidine / 2,4-dimethylpyridine	C_7H_9N	157.1	107.15		0.9493		col. liq., misc. acet., eth.
11	H_3C N CH_3	2,6-butidine, α,α-butidine / 2,6-dimethylpyridine	C_7H_9N	143	107.15		0.942		col. liq., misc. w. acet., eth.
12	N / CH_3 CH_3	3,4-butidine, $\beta\gamma$-butidine / 3,4-dimethylpyridine	C_7H_9N	163.5	167.15	4.5			col. liq., misc. acet., eth.
13	N CH_3 / CH_2 CH_3	α-Collidine / 2-Methyl-4-Ethyl pyridine	$C_8H_{11}N$	179	121.18		0.9268		col. liq., misc. w. acet., eth., Bz.
14	N / CH_2-CH_3 CH_3	β-collidine, / 4-methyl-3-ethylpyridine	$C_8H_{11}N$	195...6	121.18		0.466		col. liq., i.w., misc., acet., eth., Cl
15	H_3C N CH_3 / CH_3	γ-collidine / 2,4,6-trimethylpyridine	$C_8H_{11}N$	172	121.18		0.917		col. liq., s. misc., w., eth., acet.
16	N CH_2-CH_3 / H_3C CH_3	3,5-dimethyl-2-ethylpyridine / α-parvuline	$C_9H_{13}N$	188	135.20		0.9338		liq., misc. w., acet., eth.
17	H_3C N CH_3 / H_3C CH_3	Tetramethylpyridine / parvuline	$C_9H_{13}N$	220	135.20		0.916		
18	H_2 N H_2 / H_3C H / H CH_3	2,5-dimethylpiperazine / cis-trans	$C_6H_{12}N_2$	162	112.18	cis 114			trans-form monocl prism, trans 118. e.s.w. acet., et
19	N / H_3C N CH_3	2,5-dimethylpyrazine / ketine	$C_6H_8N_2$	155	108.14	15	0.990	1.4992	col. liq., misc., w., acet., eth.
20	N / CH_3	Skatole, 3-methylindole	C_9H_9N	266.2	131.17	95			leat, (liq) v.sl, s.w., e.s. acet., eth, Bz., chl., (lig)
21	N CH_3	Quinaldine / 2-methylquinoline	$C_{10}H_9N$	246...7	143.18		1.1013		col.liq., misc., w., acet., eth., chl
22	N / CH_3	3-methylquinoline / β-methylquinoline	$C_{10}H_9N$	250	143.18	14	1.074	1.6069	col. liq., or cr., misc., acet., eth.
23	CH_3 / N	Lepidine / 4-methylquinoline	$C_{10}H_9N$	258...263	143.19	<0	1.086		col. liq., misc., w., acet., eth.
24	N CH_3 / H_3C	2,6-dimethylquinoline	$C_{11}H_{11}N$	266.7	157.21	60			trim. cr., e.s.h.w., misc., acet., et
25	H_2 H_2 / N CH_3 H	1-methyl-1,2,3,4,-tetrahydro quinoline	$C_{10}H_{13}N$	245.5	147.21		1.021	1.4802	liq., misc., acet., eth.

Hetero rings with hetero chains

No.	Formula	Name	MF	BP	Mol. wt.	MP	D	n_D	Other properties
26	H_2 H_2 CH_3 / H_2 H CH_2 / O C O H_2	2-ethoxymethyltetrahydrofuran / tetrahydrofurfurylethylether	$C_7H_{14}O_2$	152...4	130.18	liq.	0.9386		col.
27	H H / H $C-O-CH_2-CH_3$ / H O H	2-ethoxymethylfuran / ethylfurfurylether	$C_7H_{10}O_2$	150	126.15	liq.	0.9844		col., i.w., misc., acet., eth.

Carbon rings linked through hetero chains

No	Formula	Name					
		MF	Mol. wt.	MP/FP	BP/CP	D	Other properties
1	⟨⟩-O-⟨⟩	Phenoxybenzene, Phenylether, Diphenylether					
		$C_{12}H_{10}O$	170.20	28	259	1.0728	col., monocl., e.s.w., s. acet., eth., Bz.
2	⟨⟩-CH₂-O-CH₂-⟨⟩	Benzylether, Dibenzylether					
		$C_{14}H_{14}O$	198.25	4...5	295...8	1.0428	col. oil., e.s. hot, acet., s. eth.
3	⟨⟩-O-CH₂-CH₂-O-⟨⟩	1, 2-diphenoxyethane, Glycoldiphenylether					
		$C_{14}H_{14}O_2$	214.15	98.5			col. leaf., e.s.w. acet., eth., chl.
4	⟨⟩-O-CH₂-CH₂-CH₂-O-⟨⟩	1, 3-diphenoxypropane, Trimethyleneglycoldiphenylether					
		$C_{15}H_{16}O_2$	228.28	61	338....40		leaf., i.w., s. acet., eth.
5	⟨⟩-S-⟨⟩	Phenylsulphide, Phenylthiobenzene					col. liq., $n_D = 1.635.$,
		$C_{12}H_{10}S$	186.26	< −40	296	1.1185	misc. w. acet., eth., Bz., CS_2
6	⟨⟩-S-S-⟨⟩	Diphenyldisulphide, Phenyldithiobenzene, Phenyldisulphide					
		$C_{12}H_{10}S_2$	218.32	61	310d		need., s. acet., eth., Bz., CS_2
7	⟨⟩-CH₂-S-CH₂-⟨⟩	Benzylsulphide, Dibenzylsulphide					
		$C_{14}H_{14}S$	214.31	49		1.0712	col. rhomb. leaf., s. acet., eth.
8	⟨⟩-CH₂-S-S-CH₂-⟨⟩	Benzyldisulphide, Dibenzyldisulphide					s. eth., Bz., ma
		$C_{14}H_{14}S_2$	246.37	71...2			leaf. (acet.), s.h. acet., v.e.s.w.
9	⟨⟩-N(H)-⟨⟩	Diphenylamine, N-phenylaniline					col. monocl. leaf., v.e.s.w.,
		$C_{12}H_{11}N$	169.22	53	302	1.159	e.s. acet., eth., Bz., ligroin, ma.
10	⟨⟩-CH₂-N(H)-⟨⟩	N-phenylbenzylamine, N-benzylaniline					col. monocl. prism., (acet.),
		$C_{13}H_{13}N$	183.24	37...8	306...7	1.0618	s. acet., eth., ma.
11	⟨⟩-CH₂-N(H)-CH₂-⟨⟩	Dibenzylamine					col. liq., $n_D = 1.57432$
		$C_{14}H_{15}N$	197.27	−26	300	1.026	e.s. acet., eth., i.w.
12	⟨⟩-N(H)-CH₂-CH₂-N(H)-⟨⟩	N, N'-diphenylethylenediamine, sym-diphenylethylenediamine					
		$C_{14}H_{16}N_2$	212.29	65			col. leaf. (acet.), v.e.s. acet., eth.
13	⟨⟩-N(H)-N(H)-⟨⟩	1, 2-diphenylhydrazine, Hydrazobenzene					col. yelsh., rhomb. plates.,
		$C_{12}H_{12}N_2$	184.23	131	d	1.158	sl. s. acet., s. eth., e.s.w.
14	⟨⟩-N=N-⟨⟩	Azobenzene					or. redsh. monocl. leaf., i.w.,
		$C_{12}H_{10}N_2$	182.22	68	297.4	1:203	sl. s. acet., s. eth.
15	⟨⟩-⟨⟩-N=N-⟨⟩-⟨⟩	4, 4-diphenylazobenzene, p, p'-azobiphenyl					or. redsh. leaf., (Bz), i.w., acet.,
		$C_{24}H_{18}N_2$	334.40	249...50			s. eth., Bz.
16	naphthyl-N=N-naphthyl	1, 1-azonaphthalene, α, α'-azonaphthalene					redsh. need. (ac. a), e.s. acet.,
		$C_{20}H_{14}N_2$	282.33	190	s		s. Bz. ac. a.
17	N=N...N=N (naphthalene)	1, 2-azonaphthalene, $\alpha\beta$-azonaphthalene					brn. leaf (ac. a), s. acet., Bz.,
		$C_{20}H_{14}N_2$	282.33	136			ac. a, conc. H_2SO_4
18	⟨⟩-C(H)=N-N(H)-⟨⟩	Benzaldehydephenylhydrazone					col. rose. monocl. Pr., s. Bz,
		$C_{13}H_{12}N_2$	196.24	156			eth., h. al.
19	⟨⟩-C(H)=N-N=C(H)-⟨⟩	Benzaldehydehydrazine, Benzolazine					long light yel. prism., e.s. acet,
		$C_{14}H_{12}N_2$	208.25	93	d		eth., chl., Bz.
20	⟨⟩-N=N-N(H)-⟨⟩	Diazoaminobenzene, 1, 3-diphenyltriazine					gold. yelsh. leaf. or prism.,
		$C_{12}H_{11}N_3$	197.23	98...9			e.s. acet., eth., Bz.
21	naphthyl-N=N-N(H)-naphthyl	1, 1-diazoaminonaphthalene, 1, 3-di-1-naphthyltriazine					
		$C_{20}H_{15}N_3$	297.35				yelsh. leaf., s. acet.
22	⟨⟩-N(CH₃)-⟨⟩	N-methyldiphenylamine					
		$C_{13}H_{13}N$	183.24	−7.6	293.4	1.048	col. liq. misc. acet., eth.
23	N(⟨⟩)₃	Triphenylamine					monocl. pr. (eth.), $n_D = 1.353$,
		$C_{18}H_{15}N$	228.28	198.5		0.774	e.s. acet., s. eth., e.s. Bz.
24		Tetraphenylhydrazine					rhomb. prism. (acet. + Chl.), v.e
		$C_{24}H_{20}N_2$	336.42	147			s. hot acet., s. Bz, ac. a., Chl., H_2SO_4
25	naphthyl-N-N-naphthyl ⇌ 2 naphthyl-N	No. 1 : 24 dissociates in solution: Diphenylnitrogen (free Radical)					
		$C_{12}H_{10}N$	168.21				In sol. blue colour

Carbon rings linked with hetero rings through carbon chains

No.	Formula	Name				No.	Formula	Name			
		MF	BP/CP	D		6		2 - phenylquinoline			
		Mol. wt.	MP/FD	n$_D$				$C_{15}H_{11}N$	363		
		Other properties						205.25	86		
								need. (acet.), e.s.w., v.e.s. acet., eth.			
1		2 - phenylpyridine				7		6 - phenylquinoline			
		$C_{11}H_9N$	270	>1				$C_{15}H_{11}N$	260/v77	1.195	
		155.19						205.25	111		
		liq., i.w., misc. acet., eth.						trim. cr. (acet. eth.), s. eth., v.e. s.w.			
2		3 - phenylpyridine				8		8 - phenylquinoline			
		$C_{11}H_9N$	270.4	>1				$C_{15}H_{11}N$	283v/187		
		155.19						205.25			
		oil., i.w., misc. acet.						visc. oil., misc. acet. eth.			
3		4 - phenylpyridine				9		1 - benzylisoquinoline			
		$C_{11}H_9N$	275					$C_{16}H_{13}N$			
		155.19	78								
		leaf. (w.), s. acet., eth., v.e. s.w.									
4		2 - benzylpyridine				10		1 - phenylpyrazoline			
		$C_{12}H_{11}N$	~280	1.067				$C_9H_{10}N_2$	273		
		169.22	139					146.19	52		
		need., i.w., s. acet., eth.						cr., i.w., s. acet.			
5		3 - benzylpyridine				11		3, 4 - dihydro - 3 - phenylquinazoline			
		$C_{12}H_{11}N$	~290	1.061				$C_{14}H_{13}N_2$		1.290	
		169.22	34					208.25	95		
		need., i.w., s. acet. eth.						hex. leaf., s. acet., eth.			

Hetero rings linked with carbon rings through hetero chains

12		Furfuralphenylhydrazone			
		$C_{11}H_{10}ON_2$	186.21	MP = 97°	i.w., s. acet., eth.
13		2 - thiophenecarbonalphenylhydrazone, 2 - thiopheneformaldehydrazone			
		$C_{11}H_{10}N_2S$	202.27	MP = 134.5°	yelsh. need., i.w., s. acet

Hetero rings linked with hetero rings through carbon chains

14		Nicotyrine, 3 - (1 - methyl - 2 - pyryl) Pyridine (dipyridine)			16		4, 4' - bipyridine 8, 8' - dipyridyl		
		$C_{10}H_{10}N_2$	280...1	1.124			$C_{10}H_8N_2$	304.8	
		158.20	108				156.18	114	
		need. (h.w.), v.e.s. s. h.w., s. acet., eth.					need. (+2H$_2$O) MP 73, v.e.s. acet. eth., Bz, v.e.s. s.w		
15		Nicotine, 3 - (1 - methyl - 2 - pyrrolidine) pyridine			17		2, 3' - biquinoline		
		$C_{10}H_{14}N_2$	247.3	1.00924			$C_{18}H_{12}N_2$	>400	
		126.23	< -80	1.52392			256.29	176...7	
		col. oil., [α] -161,55., misc. acet., eth., chl.					yelsh. leaf or need. (Bz) e.s. acet. hot Bz. chl		

Hetero rings linked with hetero rings through hetero chains

18	$CH_2-CH-CH_2-O-CH_2-CH_2-O-CH_2-CH-CH_2$ Ethyleneglycoldiglycidicether				22		Dimorpholinedisulphide		
	misc.w. $C_8H_{14}O_4$ 135 - 165° 1.1378 Ignition temp 98° 174.20 1.4491						$C_8H_{16}O_2N_2S_2$		
							246.24	123	
							sl. yelsh. cr., i.w., e.s. acet., Bz. to		
19		Difurfurylamine, α, α - di - 2 - furyldimethylamine							
		$C_{10}H_{11}O_2N$	177.20	BP 102...3	col. liq., i.w., misc. acet.				
20		5, 5' - diphenylazatetrazole							
		$C_{14}H_{10}N_{10}$							
21		1, 6 - ditetrazolhexazene - 1, 5							
		$C_2H_4N_{14}$	Analysis: C 10.7%, H 1.8%, N 87.5%						

Carbon compounds with functional groups

If an external (pheriphral) H-atom of a carbon compound is replaced (substituted) by another atom or a group of atoms, then the bond length, bond angle, the molecular weight and the corresponding physical as well as chemical properties of the compound dependent on the position of atom or of the central atom of the group in the periodic system of elements, undergo a change. However the structural framework of the compound is not changed through these substitutions. These groups are called as "functional groups" on account of the inter-relationship between the substitution and the properties of the compound (even when these "groups" consist of only one atom).

In principle, a methyl group (−CH₃) is also taken as a "functional group". By replacing a H-atom of methane (H−CH₃) we get ethane (CH₃−CH₃), and so on. Here too, see the steady change in properties of the homologus series (diagram p.81). For other changes in properties, for example, B.P., see diagram on p. 112

Functional group	Nomenclature prefix	with suffix	Class of substance or radical names	See pages
−OH	hydroxy	−ol	Alcohols	106−109
=O	oxo	−al	Aldehydes	110−111
		−one	Keytones	
Ester organic acids, and acid anhydrides result as oxa-compounds with an oxa-group of hetero chain				112−113
Lactones and lactams result as oxo or aza-compounds with oxygen or nitrogen hetero rings				114
	Carboxy (oxo and oxi function on a C-atom)	−acid	Organic acids	115−116
Several equal and different oxygen functions on some or different C-atoms (e.g. oxyacids)				117−118
Alcoholates and salts of organic acids				118
−NH₂	Amino	−amine	Amine	119−120
−NHOH	Hydroxylamino	−hydroxylamine	Hydroxylamine	121
−NH−NH₂	Hydrazino	−hydrazine	Hydrazine	121
−N₂	Diazo-(or-diazonium)	−compounds		121
=NH	Imino	−imine	Imine	121
=NOH	Hydroxylimino	−oxime	Oxime	121
=N−NH₂	Hydrazono	−hydrazone	Hydrazone	121
−NO	Nitroso			121
−NO₂	Nitro			121
−OCN	Cyanato	−cyanate	Cyanate	122
−NCO	Isocyanato	−isocyanate	Isocyanate	122
−N₃	Azido	−azide	Azide	122
C≡N	Nitrilo	−nitrile	Nitrile (cyanide)	122
Several nitrogen functions as also mixed functions with oxygen atom and nitrogen atom on the molecule (e.g. acid amides, amino acids)				122−124
−F	Fluoro-		Fluorides	125
−Cl	Chloro-		Chlorides	125
−Br	Bromo		Bromides	125
−I	Iodo-		Iodides	126
−IO	Iodoso-			126
−IO₂	Iodyl			126
Mixed halogens- and other functions (e.g.) acid chlorides				126
−SH	Thio-(mercapto)	−thiol	Thiols	127
−SO₂H		−sulphic acid	Sulphinic acids	127
−SO₃H		−sulphonic acid	Sulphonic acids	127−128
Functions with phosphorus, arsenic, esters of inorganic acids				128
Examples for nomenclature	Oxyethane	Ethanol	Ethyl alcohol (alcohol)	
	Oxoethane	Ethanal	Acetaldehyde	
	1—Oxopropane	Propanal	Propionaldehyde	
	2—Oxopropane	Propanone	Acetone	
		Propionic acid	Propionic acid	

The above series of functional groups result out of "Gmelin−Number" of elements and the (bonding) cohesiveness between the function and carbon atom.

Carbon compounds with functional groups

Oxy compounds (alcohols)

No.	Formula	Name	MF	Mol. wt	MP	BP	D	nD
1	CH₃OH	Methanol	CH₄O	32.04	−97	64.7	0.792	
2	CH₃-CH₂OH	Ethanol (alcohol)	C₂H₆O	46.07	−114	78.3	0.789	
3	CH₃-CH₂-CH₂OH	n-Propanol	C₃H₈O	60.09	−126	97.2	0.804	
4	CH₃CH(OH)-CH₃	iso-Propanol	C₃H₈O	60.09	−88.5	82.3	0.786	
5	CH₃-CH₂-CH₂-CH₂OH	n-Butanol	C₄H₁₀O	74.12	−90	117.7	0.810	
6	$\frac{H_3C}{H_3C}$>CH-CH₂OH	iso-Butanol 2-Methyl-1-propanol	C₄H₁₀O	74.12	−108	107.9	0.802	
7	CH₃-CH₂-CH(OH)-CH₃	sec-Butanol; 2-Butanol	C₄H₁₀O	74.12		99.5	0.808	
8	$\frac{H_3C}{H_3C}$>C<$\frac{CH_3}{OH}$	t-Butanol 2-Methyl-2-propanol	C₄H₁₀O	74.12	25	82.5	0.789	
9	CH₃-CH₂-CH₂-CH₂-CH₂OH	n-Amyl alcohol	C₅H₁₂O	88.15	−78.5	138.0	0.817	
10	$\frac{H_3C}{H_3C}$>CH-CH₂-CH₂OH	iso-Amyl alcohol 3-Methyl-1-butanol	C₅H₁₂O	88.15	−117	131.5	0.812	
11	CH₃-CH₂-CH-CH₂OH ꜜCH₃	2-Methyl-1-butanol	C₅H₁₂O	88.15		128.9	0.8193	1.4107 γ]$_D^{30}$-5.756
12	CH₃-(CH₂)₄-CH₂OH	n-Hexanol	C₆H₁₄O	102.17	−52	155.8	0.820	
13	CH₃-(CH₂)₅-CH₂OH	n-Heptanol	C₇H₁₆O	116.20	−34.6	176	0.8219	1.42410
14	CH₃-(CH₂)₆-CH₂OH	n-Octanol	C₈H₁₈O	130.23	−16	194	0.827	
15	CH₃-(CH₂)₇-CH₂OH	n-Nonanol	C₉H₂₀O	144.25	−5	213	0.8274	1.43347
16	CH₃-(CH₂)₈-CH₂OH	n-Decanol	C₁₀H₂₂O	158.28	+6	231	0.8292	1.43682
17	CH₃-(CH₂)₁₀-CH₂OH	n-Dodecanol (Lauryl alcohol)	C₁₂H₂₆O	186.33	22.6	225	0.8309	Leaf (acet)
18	CH₃-(CH₂)₁₂-CH₂OH	n-Tetradecanol (Myristyl alcohol)	C₁₄H₃₀O	214.38	37.62	263.2	0.8355	Leaf (acet)
19	CH₃-(CH₂)₁₄-CH₂OH	n-Hexadecanol (Cetyl alcohol)	C₁₆H₃₄O	242.44	49.3	344	0.8176	1.4283liq.
20	CH₃-(CH₂)₁₆-CH₂OH	n-Octadecanol (Stearyl alcohol)	C₁₈H₃₈O	270.49	59	210v	0.8124	Leaf (acet)
21	CH₃-(CH₂)₂₄-CH₂OH	n-Hexacosonal (Ceryl alcohol)	C₂₆H₅₄O	382.70	79.5…8	305dv		col. rhomb., leaf (eth,
22	CH₃-(CH₂)₂₆-CH₂OH	n-Octa cosan alcohol	C₂₈H₅₈O	410.75	83.2…4			
23	CH₃-(CH₂)₂₈-CH₂OH	n-Triacontanol (Myricyl alcohol)	C₃₀H₆₂O	438.79	88		0.777	col. need (eth. a
24	$_H^H$>C=C<$\frac{H}{OH}$⇌H-C-C$<^H_{\lessgtr O}$	Vinyl alcohol	C₂H₄O	44.05				only in equilibrium with acetaldehyde
				Esters and ethers are known and stable				known
25	CH₂-CH-CH₂OH	Allyl alcohol; 2-Propen-1-ol	C₃H₆O	58.08	−129	97	0.355	1.41345
26	HC≡C-CH₂OH	Propargylal alcohol	C₃H₄O	56.06	−17	114…5	0.9715	1.43064

No.	Formula	Name	MF	Mol. wt	MP	BP	D	nD
27	$\frac{H_3C}{H_3C}$>C=C<$\frac{H}{CH_2-H_2C}$>CH-CH₂OH	Citronellol (d form)						[α]$_D$+4°
			C₉H₁₈O	156.26	liq.	222	0.8565	1.45659
28	$\frac{H_3C}{H_3C}$>C=C<$\frac{H}{CH_2-H_2C}$>C=C<$\frac{H}{CH_2OH}$	cis 3,7-Dimethyl-2,6-Octadiene-1-ol cis 3-Methylene-2-Methyl-2,6-Octadiene, Nerol						
			C₁₀H₁₈O	154.25	oil	250	0.881	
29	$\frac{H_3C}{H_3C}$>C=C<$\frac{H}{CH_2-H_2C}$>C=C<$\frac{CH_2OH}{H}$	trans 3,7-Dimethyl-2,6-Octadiene-1-ol (trans 6 Methylene-7-Methyl-2,6-Octadiene-1 ol) Geraniol						
			C₁₀H₁₈O	154.25	liq.	229	0.8812	1.4798
30	$\frac{H_3C}{H_3C}$>C=C<$\frac{CH_3}{CH_2-H_2C}$>C<$\frac{CH=CH_2}{OH}$	d 3,7-Dimethyl-1,6-Octadiene-1-3-ol; a-Linalool						
			C₁₀H₁₈O	154.25	liq.	198.3	0.8622	1.4623
31	CH₃-(CH₂)₆-H₂C>C=C<$\frac{H}{CH_2-(CH_2)_6-CH_2OH}$	cis 9-Octadecene-1-ol; Oleylalcohol	C₁₈H₃₆O	268.47	oily	205.10V	0.8489 misc. acet.. eth
32	$\frac{H_3C}{H_3C}$>C<$\frac{(CH_2)_3}{H}$>C<$\frac{(CH_2)_3}{H}$>C<$\frac{(CH_2)_3}{H}$>C=C<$\frac{CH_2OH}{H}$	3,7,11,15-Tetramethyl-2-hexadecene-1-ol, Phylol	C₂₀H₄₀O	296.52	oily	203…4	0.852	1.46380
33	HOCH₂-CH₂OH	Glycol; 1,2-Ethanediol	C₂H₆O₂	197.2	1.1115			
			62.07	−17.4	1.4274			
34	HOCH₂-CH₂-CH₂OH	1,3-Propanediol	C₃H₈O₂	214d	1.0526			
			76.09	visc. liq.	1.4398			
35	H₃C-$\overset{OH}{\underset{H}{C}}$-CH₂OH	1,2-Propanediol	C₃H₈O₂	189	1.040			
			76.09	liq.				
36	H₃C-$\overset{OH}{\underset{CH_3}{C}}$-CH₂OH	2-Methyl-1,2-Propanediol	C₄H₁₀O₂	177	1.003			
			90.12	liq.				
37	HOCH₂-$\overset{CH_3}{\underset{CH_3}{C}}$-CH₂OH	2,2-Dimethyl-1,3-Propanediol	C₅H₁₂O₂	206	wh. need (Bz)			
			104.15	127	v.e.s.w			
38	H₂COH ꜜHCOH ꜜH₂COH	Glycerin, Glycerol	C₃H₈O₃	290	1.260			
			92.09	<−20	1.4729			
		oil., rhomb. cr., misc.w., acet., immisc. eth., chl.						
39	H₂COH H₂COH H₂COH ꜜHCOH HCOH HOCH ꜜHCOH HOCH HCOH ꜜH₂COH H₂COH H₂COH ꜜMESO D L	Erythritol 1,2,3,4-Tetroxybutane	C₄H₁₀O₄	331	1.451			
			122.12	119…20				
		MP D Form FP89°						
		MPDLForm 72						
		wh. cr.; e.s.w., acet., i. eth.		[α]$_D$-4.4°				
40	HOH₂C-$\overset{CH_2OH}{\underset{CH_2OH}{C}}$-CH₂OH	Pentaerithritol	C₅H₁₂O₄					
			136.15	260.5				
		cr., s.w.						

No.	Formula	Name				No.	Formula	Name		
		MF	BP/CP	D		13	OH / HO–OH	Phloroglucinol 1,3,5-trioxybenzene		
		Mol. wt.	MP	nD				$C_6H_6O_3$	s d	
		Other properties						126.11	219ww	
								rhomb., sl. s.w., e.s. acet., eth.		

1	Cyclopentanol structure	Cyclopentanol			14	1,2,3,5-Tetroxy benzene
		$C_5H_{10}O$	139…40	0.9488		$C_6H_6O_4$
		86.13		1.4530		142.11	165
		oily liq., misc. w., acet.				need. (w)., e.s.w., acet., i. Bz., chl.

2	Cyclohexanol structure	Cyclohexanol			15	1,2,4,5-Tetroxy benzene
		$C_6H_{12}O$	161.5	0.9449		$C_6H_6O_4$
		100.16	24	1.46560		142.11	220
		col. need; hyg; s.w. acet e.s. eth; Bz; CS₂				leaf. (ac.a), e.s.w., acet., eth.

3	Cis-quinitol structure	Cis-quinitol*) Cis-cyclohexane-1,4-diol			16	Hexahydroxy benzene
		$C_6H_{12}O_2$				$C_6H_6O_6$
		116.16	112s			174.11	300d
		prism; e.s.w; acet; v. sl s. eth; chl.				redsh. brn. need., (Bz), i.w., acet. eth., s. Bz.

4	trans-quinitol structure	trans-quinitol**) trans-cyclohexane-1,4-diol			17	1-Naphthol, α-Naphthol
		$C_6H_{12}O_2$				$C_{10}H_8O$	288	1.224
		116.16	143			144.16	96	1.6206
		cr., e.s.w; acet. v. sl. s. eth; chl.				yel. monocl., e.s. hot w., acet., eth., s. Bz.

5	Quercitol structure	Quercitol			18	2-naphthol, β-naphthol
		$C_6H_{12}O_5$	d	1.585		$C_{10}H_8O$	294.85	1.217
		164.16	234			144.16	122	
		col. monocl., s.w., e.s. acet., i. eth.				col. monocl. leaf., sl. s.w., s. acet. e.s. eth., s. chl.

6	Inositol structure	Inositol Cyclohexanehexol			19	1,5-dioxynaphthalene 1,5-naphthalenediol
		$C_6H_{12}O_6$	319dv	1.524		$C_{10}H_8O_2$	d
		180.10	225ww			160.16	265
		col. monocl; (w), s.w., i. acet; eth.				pr.(w), e.s.w., s. acet., eth., ac. a., i. Bz.

7	Phenol structure	Phenol, hydroxy benzene carbolic acid			20	1,8-dioxynaphthalene 1,8-naphthalenediol
		C_6H_6O	182	1.072		$C_{10}H_8O_2$
		94.11	41	1.54247		160.16	140
		col. rhomb. need; s.w., e.s. eth; chl; CS₂				leaf, or need. (w) e.s. h.w., e.s. eth; acet., s. Bz.

8	Catechol structure	Catechol 1,2-dioxybenzene			21	2,3-dioxynaphthalene 2,3-naphthalenediol
		$C_6H_6O_2$	240	1.371		$C_{10}H_8O_2$
		110.11	105			160.16	160…1
		col. rhomb. leaf. (Bz.), e.s.w; acet; eth; Bz, chl.				monocl. leaf. (w), s.h.w., e.s. eth; acet., s. Bz

9	Resorcinol structure	Resorcinol 1,3-dioxybenzene			22	2,7-dioxynaphthalene 2,7-naphthalenediol
		$C_6H_6O_2$	276.5	1.285		$C_{10}H_8O_2$	s
		110.11	110			160.16	190
		col. rhomb. pl. (w, Bz), v.e.s.w., acet., eth., Bz.				need. (w), s.w., acet., eth; chl., Bz

10	Hydroquinone structure	Hydroquinone 1,4-dioxybenzene			23	Anthranol 9-oxyanthracene
		$C_6H_6O_2$	286.2	1.538		$C_{14}H_{10}O$	170d
		110.11	170.5			194.22	76
		col. hex. pr., (w) s.w., e.s. acet., eth.				pa. yel. need. (pet.) s.h.w., acet., eth.

11	Pyrogallol structure	Pyrogallol 1,2,3-Trioxybenzene			24	Anthrahydroquinone 9,10-anthradiol
		$C_6H_6O_3$	309	1.453		$C_{14}H_{10}O_2$	d
		126.11	133…4			210.22	180
		need. or leaf; v.e.s.w., acet., eth; Bz., chl., CS₂				yelsh. need. fluroes; s. acet; al. (grn.)

12	1,2,4-Trioxybenzene structure	1,2,4-Trioxybenzene			25	Morphol 3,4-phenanthrenediol
		$C_6H_6O_3$				$C_{14}H_{10}O_2$
		126.11	140.5			210.22	143
		col. monocl. leaf. (acet.), e.s. acet., eth; Bz.				col. need., s. acet., eth.

* both OH—above the ring plane   ** 1 OH above, 1 OH below the ring plane

No.	Formula	Name	MF	Mol. wt.	MP	D	nD and other properties
1	$CH_3-O-CH_2-CH_2OH$	2-methoxyethanol / Methylcellosolve	$C_3H_8O_2$	76.09	124.3	0.9660	
2	$CH_3-CH_2-O-CH_2-CH_2OH$	2-ethoxyethanol / Cellosolve	$C_4H_{10}O_2$	90.12	135.1	0.9311	
3	$CH_3-CH_2-CH_2-O-CH_2-CH_2OH$	2-n-propoxyethanol	$C_5H_{12}O_2$	104.15	~150	0.9112	1.41328
4	$\begin{array}{c}H_3C\\H_3C\end{array}>CH-O-CH_2-CH_2OH$	2-isopropoxyethanol	$C_5H_{12}O_2$	104.15	~142	0.9030	1.40954
5	$CH_3-CH_2-CH_2-CH_2-O-CH_2-CH_2OH$	2n-butoxyethanol, Butylcellosolve	$C_6H_{14}O_2$	118.17	170.6	0.9027	
6	$HOCH_2-CH_2-O-CH_2-CH_2-OH$	Diethyleneglycol / 2,2-oxydiethanol	$C_4H_{10}O_3$	106.12	244.5	1.7177	MP ~11°
7	$CH_3-O-CH_2-CH_2-O-CH_2-CH_2OH$	Diethyleneglycolmonomethylether, 2-(2-methoxyethoxy) ethanol	$C_5H_{12}O_3$	120.15	193.2	1.0354	1.4264
8	$CH_3-CH_2-O-CH_2-CH_2-O-CH_2-CH_2OH$	Diethyleneglycol monoethylether, 2-(2-ethoxyethoxy) ethanol	$C_6H_{14}O_3$	134.17	201.9	0.9902	(Carbitol)
9	$HOCH_2-CH_2-O-CH_2-CH_2-O-CH_2-CH_2-OH$	Triethyleneglycol	$C_6H_{14}O_4$	150.17	280...90	1.1254	MP-5°
10	$CH_3-NH-CH_2-CH_2OH$	2-methylaminoethanol	C_3H_9ON	75.11	~150	0.937	14385
11	$CH_3-CH_2-NH-CH_2-CH_2OH$	2-ethylaminoethanol	$C_4H_{11}ON$	89.14	~170	0.914	1.444
12	$CH_3-CH_2-CH_2-NH-CH_2-CH_2OH$	2-butylaminoethanol	$C_6H_{15}ON$	117.19	200	0.8907	1.4437
13	$\begin{array}{c}H_3C\\H_3C\end{array}>N-CH_2-CH_2OH$	2-dimethylaminoethanol / β-dimethylaminoethylalcohol	$C_4H_{11}ON$	89.14	135	0.8866	1.43
14	$\begin{array}{c}H_3C-H_2C\\H_3C-H_2C\end{array}>N-CH_2-CH_2OH$	2-diethylaminoethanol / β-diethylaminoethylalcohol	$C_6H_{15}ON$	117.19	163	0.8601	1.4400
15	$HOCH_2-CH_2-NH-CH_2-CH_2OH$	Diethanolamino	$C_4H_{11}O_2N$	105.14	268	1.0966	1.4776, MP28°
16	$CH_3-N<\begin{array}{c}CH_2-CH_2OH\\CH_2-CH_2OH\end{array}$	2,2-methyliminodiethanol / β,β-dihydroxy-N-methyldiethylamine	$C_5H_{13}O_2N$	119.16	~250	1.0377	1.4678
17	$CH_3-CH_2-N<\begin{array}{c}CH_2-CH_2OH\\CH_2-CH_2OH\end{array}$	2,2,ethyliminodiethanol / β,β-dihydroxytriethylamine	$C_6H_{15}O_2N$	133.19	~252	1.0135	1.4663 yel. liq.
18	$HOCH_2-CH_2-N<\begin{array}{c}CH_2-CH_2OH\\CH_2-CH_2OH\end{array}$	Triethanol amine	$C_6H_{15}O_3N$	149.19	277...9 (150mm)	1.1242	1.4852 visc., col. liq.

19	2-pyridol, α-pyridone 2-pyridone

19 — 2-pyridol, α-pyridone 2-pyridone

C_5H_5ON	281
95.10	107

col. need (Bz), e.s.w., acet., eth., Bz. ligr.

20 — 3-pyridol, β-pyridone 3-hydroxypyridine

C_5H_5ON	
95.10	129

need., e.s.w., acet., eth.

21 — 4-pyridol, γ-pyridone 4-pyridone

C_5H_5ON	>350	col.
95.10	148.5ww	monocl.

(H_2O-MP92°), e.s.w., acet., eth., i. Bz

22 — 2,4-pyridinediol 2,4-dihydroxypyridine

$C_5H_5O_2N$		yelsh.
111.10	265	rhomb.

cr. (w. acet.), e.s.w., acet., v.e.s., eth.

23 — 2,6-pyridinediol 2,6-dihydroxypyridine

$C_5H_5O_2N$	
111.10	195

col. need, H_2O (w), e.s.w., acet., v.e.s. eth.

24 — 2,4,6-pyridinetriol 2,4,6-trihydroxypyridine

$C_5H_5O_3N$		need or
127.10	230d	powd.,

e.s.w., acet., s. eth., e.s. lgr.

25 — Uracil; 2,4-dihydroxypyrimidine 2,4-(1,3)-pyrimidinedione

a)   b)

$C_4H_4O_2N_2$		
112.09	338	need (w).

e.s.w., e. eth., i. acet., s. NH_4OH

26 — a) 2,5-dihydroxypiperazine b) 2,5-(1,4) piperazinedione diglycil diamide

a)   b)

$C_4H_6O_2N_2$	s
114.10	275d

pl. e.s. acet., s.h.w.

27 — 2,4,6-trioxytriazine 2,4,6-triazinetrione cyanic acid

a)   b)

$C_3H_3O_3N_3$	d	1.768
129.08	>360	col. monocl.,

($1H_2O$), sl.s.w., acet. s. conc. H_2SO_4

28 — Xanthine 2,6-dioxiparine (a) 2,6-(1,3) furindione (b)

a

b

$C_5H_4O_2N_4$	sd	
152.11	>150d	($-H_2O$)

yelsh. wh. powd. or thin pl., v. sl. s. acet., e.s. nialk.

29 — Uric acid 2,6,8-trioxypurine (a) 2,6,8-(1,3.9) purine trione (b)

a

b

$C_5H_4O_3N_4$	-	1.893
168.11	d	

acidic, v.sl. s.w., i. acet., eth., s. glycerol, conc. H_2SO_4

For keto-enol tantomerism see p. 115

Oxy compounds of carbon rings with side chains — OH

No.	Formula	Name / MF / Mol. wt. / Other properties	BP/CP / MP/FP	D / n_D
		Name	BP/CP	D
		MF	MP/FP	n_D
		Mol. wt.		
		Other properties		
1	CH₃ / OH (ring)	o-cresol, o-methylphenol; o-hydroxytoluene; C_7H_8O ; 108.13; col. cr., sl. s.w., e.s. acet., eth. chl. org. solv.	191.5 / 30	1.0465 / 1.5453
2	CH₃ / OH (ring)	m-cresol, m-methylphenol; m-hydroxytoluene; C_7H_8O ; 108.13; col. liq., sl. misc. w., misc. acet., eth., chl. org. solv.	202.8 / 18—12	1.034 / 1.5398
3	CH₃ / OH (ring)	p-cresol, p-methylphenol; p-hydroxytoluence; C_7H_8O ; 108.13; col. prism., sl. s.w., e.s. acet., eth., chl. org. solv.	202.5 / 36	1.0347 / 1.5395
4	CH₃ / OH / H₃C—CH—CH₃ (ring)	Carvacrol, Cymophenol; 2-p-cymenol; $C_{10}H_{14}O$; 150.21; col. oily liq. misc. eth., acet; alk.	237.9 / 0.5	0.976 / 1.52295
5	CH₃ / OH / H₃C—CH—CH₃ (ring)	Thymol; 3-p-cymenol; $C_{10}H_{14}O$; 150.21; pl. (ac.a), v.s. s.w., e.s. acet; eth. chl. ac.a. CS_2, alk.	233.5 / 51.5	0.9485 liq / 1.51893
6	CH=CH—CH₃ / OH (ring)	Anol; p-propenylphenol; $C_9H_{10}O$; 134.17; col. leaf. (h.w), s.h.w., s. acet., org. solv.; alk.	250d / 93	
7	CH₂—CH=CH₂ / OH (ring)	Chavicol; p-allylphenol; $C_9H_{10}O$; 134.17; liq., misc. acet.; eth. chl.	237 / <—25	1.033 / 1.5441
8	CH₃ / HO ... OH (ring)	Oroin, Oreinol, 5-methyl, resorcinol 5-methyl-1,3-benzene diol; $C_7H_8O_2$; 124.13; col. monocl. cr., (chl.) [MP (H₂O) 58*], s.w., e.s. acet., eth.	289...90w / 107...80w	1.290
9	CH₂OH (ring)	Benzylalcohol, Phenyl carbinol, α-hydroxytoluene; C_7H_8O ; 108.13; col. liq., misc. acet., eth., chl. ac. a. me. al	205.2 / —15.3	1.050 / 1.53955
10	CH₂—CH₂OH (ring)	2-phenylethanol, 2-phenyl-ethylalcohol or B-P.E.A.; $C_8H_{10}O$; 122.16; col. liq., misc. acet., eth., sl. misc. w	219...21 / —27	1.6235 / 1.5240
11	CH=CH—CH₂OH (ring)	Cinnanylalcohol; 3-phenyl-2-propen-1-ol; $C_9H_{10}O$; 134.17; Need., e.s.w. acet., eth.	257.5 / 33	1.0440 / 1.58140
12	CH₂OH / OH (ring)	Saligenin; o-hydroxybenzylalcohol; $C_7H_8O_2$; 124.13; rhomb. (w), sl. s.w., e.s. acet., eth	s / 86	1.161
13	menthol structure	1-menthol; $C_{10}H_{20}O$; 156.26	215 / 35.5	0.890 / 1.460 liq
14	terpineol structure	α-terpineol (d,l); dl-1-p-methene-8-ol; $C_{10}H_{18}O$; 154.25; col. liq., i.w., e.s. acet., eth., chl.	219.8 / 35	0.9357 / 1.4287
15	borneol structure	dl-borneol; dl-2-hydroxycamphane; dl-2-hydroxybornane; $C_{10}H_{18}O$; 154.24; col., pl. (pet.), v.e.s.w., e.s. acet., eth., chl., Bz., pet. d[α] +37.44 (acet.) 1[α] -37.74 (acet.)	~210s / 208.6	1.011
16	(C₆H₅)₃C—OH structure	Triphenylcarbinol; Triphenylhydroxymethane; $C_{19}H_{16}O$; 260.32; hex. pr. (Bz) (acet.), e.s. acet., eth., Bz., s in H_2SO_4 (yel).	380 / 162.5	1.188
17	CH₂-O-CH₂-CH₂OH (ring)	2-benzyloxyethane; Benzylcellosolve; $C_9H_{12}O_2$; 152.19; col. liq., v. sl. s.w.	256	1.068
18	CH₂-CH=CH₂ / O—CH₃ / HO (ring)	Eugenol, 1-hydroxy-2-methoxy-4-allylbenzene; $C_{10}H_{12}O_2$; 164.20; col. liq., misc. eth., acet., chl.	252...3 / 10.3	1.0664 / 1.5416
19	CH=CH—CH₃ / O—CH₃ / OH (ring)	Isoeugenol, 1-hydroxy-2-methoxy-4-propenylbenzol; $C_{10}H_{12}O_2$; 164.20; pa. yelsh. liq., misc. acet., eth.	267.5 / —10	1.0852 / 1.5680
20	CH=CH-CH₂OH / O—CH₃ / OH (ring)	Coniferylalcohol, 3-(4hydroxy-3-methoxyphenyl)-2-prope-1-ol; $C_{10}H_{12}O_3$; 180.20; pr, e.s. h.w, s. acet., eth., alk.	73...4	
21	H₂C—CH / O / CH₂OH	Glycidol; 2,3-epoxy-1-propanol; $C_3H_6O_2$; 74.08; col. liq., misc. w. acet., eth.	162d	1.165
22	tetrahydrofurfuryl structure —CH₂OH	Tetrahydrofurfurylalcohol; tetrahydro-2-furancarbinol; $C_5H_{10}O_2$; 102.13; col. misc. w., acet., eth.	~180	1.0495 / 1.4502
23	furan ring —CH₂OH	Furfuryl alcohol; 2-furan carbinol; $C_5H_6O_2$; 98.10; col. yelsh. liq., misc. w., acet., eth.	171	1.1296 / 1.4850
24	thiophene ring —CH₂OH	2-Thiophenecarbinol; α-thienylalcohol; C_5H_6OS ; 114.16; col. liq., i.w., misc. acet., eth.	207	

Oxo compounds (aldehydes-ketones)

No.	Formula	Name	MF	Mol. wt	MP/FP	BP/CP	D/sp. gr. nD and other properties
1	HCHO	Formaldehyde, Methanal	CH_2O	30.03	−118.3	−19.3	0.815 liq. col; gas; e.s.w:
2	CH_3–CHO	Acetaldehyde, Ethanal	C_2H_4O	44.05	−123.5	20.2	0.783 liq. n 1.3316, misc. w
3	CH_3–CH_2–CHO	Propionaldehyde, Propanal	C_3H_6O	58.08	−81	48.8	0.807 n 1.36356
4	CH_3–CH_2–CH_2–CHO	n-Butyraldehyde, Butanal	C_4H_8O	72.10	−99.0	75.7	0.817 n 1.38433
5	H_3C>CH–CHO / H_3C	Isobutyraldehyde 2-Methyl propanal	C_4H_8O	72.10	−65.9	61.5,3.5	0.7938 n 1.37302
6	CH_3–$(CH_2)_3$–CHO	n-Valeraldehyde, Pentanal	$C_5H_{10}O$	86.13	−91.5	103.4	0.8185 n 1.3952
7	H_3C>CH–CH_2·CHO / H_3C	iso-Valeraldehyde 3-Methyl butanal	$C_5H_{10}O$	86.13	−51	92.5	0.803 n 1.3902
8	CH_3–$(CH_2)_4$–CHO	Caproaldehyde, n-Hexanal	$C_6H_{12}O$	100.16	liq.	131	0.8335 immisc. w; misc. acet; eth
9	CH_3–$(CH_2)_5$–CHO	Onanthaldehyde, Heptanal	$C_7H_{14}O$	114.18	−45	155	0.850 n 1.4131
10	CH_3–$(CH_2)_6$–CHO	Caprylaldehyde, Octanal	$C_8H_{16}O$	128.21	liq.	163.4	0.821 n 1.4217 misc. w
11	CH_3–$(CH_2)_7$–CHO	Pelargonaldehyde, Nonanal	$C_9H_{18}O$				
12	CH_3–$(CH_2)_{16}$–CHO	Stearaldehyde, Octadecanal	$C_{18}H_{36}O$	268.47	63.5	261[100]	i.w; s. acet; eth
13	CH_2=CH–CHO	Actolin, propenal	C_3H_4O	56.06	−87.7	52.5	0.841 n 1.39975
14	CH_3–CH=CH–CHO	Crotonaldehyde-2-Butenal	C_4H_6O	70.09	−69	104...5	0.8575 n 1.43838
15	CH≡C–CHO	Propargylaldehyde, Propynal	C_3H_2O	54.05	oily liq.	61	misc. w:
16	OHC–CHO	Glyoxal, Ethandial	$C_2H_2O_2$	58.04	15	50.4	1.14 yelsh. cr.; n 1.3828

No.	Formula	Name					
		MF	BP/CP	D			
		Mol. wt	MP/FP	nD			
		other properties					
17	H–C=O (benzaldehyde structure)	Benzaldehyde Formylbenzene	C_7H_6O	178.9	1.0498		
		106.12	−26	1.54629			
		col. liq., misc. acet., eth., v. sl. misc. w					
18	CH_3 / C=O / H (o-tolylaldehyde structure)	o-tolylaldehyde o-methylbenzaldehyde	C_8H_8O	199	1.019		
		120.14	liq.	1.54068			
		col. misc. w., acet., eth.					
19	CH_3 / H–C=O (p-tolylbenzaldehyde structure)	p-tolybenzaldehyde p-methylbenzaldehyde	C_8H_8O	204...5	1.019		
		120.14	liq.	1.54693			
		col. misc. w., acet., eth.					
20	HC=CH–C=O / H (cinnamaldehyde structure)	Cinnamaldehyde 2-phenylpropenal	C_9H_8O	251	1.1119		
		132.15	−7.5	1.61949			
		col. liq., misc. w., acet., eth.					
21	O–CH_3 / H–C=O (anisaldehyde structure)	Anisaldehyde p-methoxybenzaldehyde	$C_8H_8O_2$	247	1.123		
		136.14	2.5	1.57641			
		col. liq. misc. acet., eth., v. sl. misc. a.					
22	H_2C<O₂ ring–C=O/H (piperonal structure)	Piperonal, Heliotropin 3,4-methylenedioxybenzaldehyde	$C_8H_6O_3$	263			
		150.13	37				
		wh. yelsh. cr., s. acet., eth., v. sl. misc.					
23	O ring–C=O/H (furfural structure)	Furfurol, 2-furfural 2-formylfuran	$C_5H_4O_2$	161.6	1.1598		
		96.08	−36.5	1.52608			
		col. yelsh. liq., i.w., misc. acet., eth					
24	H_2 ring H_2 H_2 –C=O/H (tetrahydrofurfural structure)	Tetrahydrofurfurol Tetrahydrofurfural	$C_5H_8O_2$	148	1.10947		
		100.11					
		col. liq., i.w., misc. acet.					
25	S ring–C=O/H (2-formylthiophene structure)	2-formylthiophene 2-thiophenecarbonal	C_5H_4OS	198	1.215		
		112.14					
		yelsh. oil., i.w., misc. acet., eth.					

No.	Formula	Name	MF	Mol. wt	MP/FP	BP/CP	D	nD and other
26	CH_3–CO–CH_3	Acetone, 2-Propanone	C_3H_6O	58.08	−94.8	56.2	0.7905	n 1.35886 misc. w; eth; acet
27	CH_3–CO–CH_2–CH_3	2-Butanone, Methyl ethyl ketone	C_4H_8O	72.10	−86.4	79.6	0.81010	n 1.38071
28	CH_3–CO–$(CH_2)_2$–CH_3	2-Pentanone; Diethyl ketone	$C_5H_{10}O$	86.13	−77.8	101.7	0.812	n1.38946 misc. w;
29	CH_3–CH_2–CO–CH_2–CH_3	3-Pentanone, Diethyl ketone	$C_5H_{10}O$	86.13	−42	102.7	0.81590	n1.3939 sl.misc. w
30	CH_3–CO–$(CH_2)_3$–CH_3	2-Hexanone, Methyl-n-proply ketone	$C_6H_{12}O$	100.16	−56.9	127.2	0.830	n1.39694 misc. w.
31	CH_3–CH_2–CO–CH_2–CH_2–CH_3	3-Hexanone ethyl-n-proply ketone	$C_6H_{12}O$	100.16	liq.	124	0.813	n 1.39899 misc. w;
32	H_3C>C<CH_3 / CO–CH_3	Pinacolin, Pincolone 3,3-Dimethyl-2-butanone	$C_6H_{12}O$	100.16	−52.5	106.5	0.7999	Col. liq; misc. acet eth; al; sl. misc. w
33	H_3C>C=C< H / CO–CH_3	Mesityloxide 4-Methyl-3-penten-2-one	$C_6H_{10}O$	98.14	−59.0	128.7	0.8539	w; misc. al acet; eth
34	H_3C>C=CH–C–HC=C<CH_3 / CH_3	Phorone, 2,6-Dimethyl-2,5-heptadiene-4-one	$C_9H_{14}O$	138.20	28	198.5 (79.8v)	0.885	n 1.49982 yel. grn. pr; s. acet., eth.
35	(Methylheptenone structure)	Methylheptenone 6-methyl-5-hepten-2-one	$C_8H_{14}O$	126.19	−67.3	173.2	0.860	col. liq., misc. acet., eth.
36	(Pseudoionone structure)	Pseudoionone, 6,10-dimethyl-3,5,9-undecatriene-2-one	$C_{13}H_{20}O$	192.29		165v	0.8979	1.53116 liq.

Oxo compounds—(ketones-quinones) =O

No.	Formula and Name	M.F.	Mol. wt.	MP	BP	D/sp.gr.	nD and other properties
1	CH$_3$-CO-CO-CH$_3$ Diacetyl; 2,3-Butanedione	C$_4$H$_6$O$_2$	88.09	liq.	88	0.9904	grn. yel; misc. w; acet; eth
2	CH$_3$-CO-CH$_2$-CO-CH$_3$ Acetylacetone; 2,4-Pentanedione	C$_5$H$_8$O$_2$	100.11	−23.2	140	0.976	nD = 1.45178, col. misc. w; acet; eth; Bz; a.
	CH$_3$-CO-CH=COH-CH$_3$ ↔ [CH$_3$-CO-C=COH-CH$_3$] ↔ [CH$_3$-CO-CH=CO-CH$_3$]						
3	CH$_3$-CO-CH$_2$-CH$_2$-CO-CH$_3$ Acetonylacetone	C$_6$H$_{10}$O$_2$	114.14	−9	192...4	0.970	col; nD = 1.449
	CH$_3$-C(OH)=CH-CH=C(OH)-CH$_3$ 2,5-Hexadione						misc. w; acet; eth
4	H$_2$C=C=O Ketene; Ethene	C$_2$H$_2$O	42.04	−151	−56		col. gas; d.w; s. acet; eth
5	H$_3$C >C=C=O Dimethyl ketene; 3 Methyl-1-oxo-1-propene	C$_4$H$_6$O	70.09	−97.5	35		yel. liq. polym.
6	O=C=C=C=O Carbon suboxide, Allenedione	C$_3$O$_2$	68.03	−107	+6.3	1.114 liq.	nD = 1.454

No.	Formula	Name	BP	D
		MF		
		Mol. wt.	MP	nD
		Other properties		

No.	Name / MF / Mol.wt.	BP	D	MP	nD	Other properties
7	Cyclobutanone; 1-oxocyclobutane; C$_4$H$_6$O	99...101				
8	Cyclopentanone; 1-oxocyclopentane; C$_5$H$_8$O; 84.11	.136.6	0.941	−58.2	1.4366	col. liq., oil. s.w., misc. acet., eth.
9	Cyclohexanone; 1-oxocyclohexane; C$_6$H$_{10}$O; 98.14	156.7	0.948	−45	1.4507	col. liq., oily, s.w., misc. acet., eth.
10	Cycloheptanone; suberone; C$_7$H$_{12}$O; 112.17	179.5	0.951	oily liq.	1.46027	misc. w., acet., eth.
11	Cyclooctanone; azelone; C$_8$H$_{14}$O; 136.20	195...97	0.959	110...1		
12	Cycloheptadecanone; C$_{17}$H$_{32}$O; 252.43	145 hv	0.910	63		
13	9-cycloheptadecene-1-one; civetone; C$_{17}$H$_{30}$O; 250.41	159 hv		32.5		
14	Quinone, p-quinone; 1,4-dioxocyclohexadiene; C$_6$H$_4$O$_2$; 108.09	s	1.318	115.7		yel. monocl. es.w., acet., eth., lgr. alkalies.
15	1,2-naphthaquinone; C$_{10}$H$_6$O$_2$; 158.15			115-120d		yelsh. red. need (eth.), s. acet., eth; Bz., H$_2$SO$_4$
16	1,4-naphthaquinone; C$_{10}$H$_6$O$_2$; 158.15	100s	1.422	125		yel. e.s.w. acet., Bz., chl. ac. a., CS$_2$
17	2,6-naphthaquinone; amphi-naphthaquinone; C$_{10}$H$_6$O$_2$; 158.15			135		or. pr., s. acet., eth., alkali.
18	Anthraquinone; 9,10-dioxoanthracene; C$_{14}$H$_8$O$_2$; 208.20	286s	1.438	377...81		yelsh. rhomb. cr., s. acet., Bz., toluene
19	Anthrone [→Anthronol]; C$_{14}$H$_{10}$O; 194.22			154...5		col; need., s. acet., eth., h. NaOH
20	Phenanthraquinone; 9,10-dioxophenanthrene; C$_{14}$H$_8$O$_2$; 208.20	>360s	1.405	207		yelsh. or need., e.s. acet., eth., h. ac. a. s. s.Bz.
21	Benzanthrone; C$_{17}$H$_{10}$O; 230.25			170...11		yelsh. need., e.s. acet., H$_2$SO$_4$ (orange)

Polycyclic quinones and their derivatives

No.	Name	MF	Mol. wt.	Colour	Absorption maxima λ$_{max}$ in xylene
22	Violanthrone*	C$_{34}$H$_{16}$O$_2$	456.50	copper redsh. vlt.	5865, 5300, 4970 Å
23	Isoviolanthrone*	C$_{34}$H$_{16}$O$_2$	456.50	Copper vlt.	5845, 5440, 5045 A°

*The numbering of C-atoms follows the IUPAC system. In literature, the numbering begins with the topmost ring C-atom and follow the order as in anthracene. For example, for violanthrone, 5 = 10, 10 = 10, 18 = 11, 17 = 12, 16 = 11', 15 = 12'.

Oxo compounds of hetero chains (esters of organic acids)

No.	Formula	Name	MF.	M.W	MP/FP	BP/CP	D/sp.gr.	nD and other properties
1	$H_3CO-C\stackrel{\nearrow OH_3C-O-C(O)-R^*}{\searrow HH_3C-COO-R}$	Methyl formate*	$C_2H_4O_2$	60.05	−99.0	31.5	0.98149	nD1.344
2	$H_3C-H_2C-O-C(O)-H$	Ethyl formate	$C_3H_6O_2$	74.08	−80.5	54.3	0.9236	nD1.35975
3	$H_3C-O-C(O)-CH_3$	Methyl acetate	$C_3H_6O_2$	74.08	−98.1	57.1	0.92740	nD1.35935
4	$H_3C-H_2C-H_2C-O-C(O)-H$	n-Propyl formate	$C_4H_8O_2$	88.10	−92.9	81.3	0.9006	nD1.3771
5	$H_3C-H_2C-O-C(O)-CH_3$	Ethyl acetate	$C_4H_8O_2$	88.10	−83.6	77.15	0.901	nD1.37216
6	$H_3C-O-C(O)-CH_2CH_3$	Methyl-n-propionate	$C_4H_8O_2$	88.10	−87.5	79.9	0.9108	nD1.37767
7	$H_3C-(H_2C)_3-O-C(O)H$	n-Butyl formate	$C_5H_{10}O_2$	102.13	−90.0	106.8	0.9108	nD1.3891
8	$H_3C-H_2C-H_2C-O-C(O)-CH_3$	n-Propyl acetate	$C_5H_{10}O_2$	102.13	−92.5	101.6	0.887	nD1.38438
9	$H_3C-H_2C-O-CH_2-CH_3$	Ethyl-n-propionate	$C_5H_{10}O_2$	102.13	−73.9	99.10	0.89574	nD1.38385
10	$H_3C-O-C(O)-CH_2-CH_2-CH_3$	Methyl-n-butyrate	$C_5H_{10}O_2$	102.13	<−95	102.3	0.898	nD1.3879
11	$H_3C-H_2C-O-C(O)-CH_2-CH_2-CH_3$	Ethyl-n-butyrate	$C_6H_{12}O_2$	116.16	−93.3	121.3	0.79	nD1.39302
12	$H_3C-(CH_2)_3-H_2C-OC(O)-CH_3$	n-Amyl acetate	$C_7H_{14}O_2$	130.18	liq.	~150	0.879	nD1.4012
13	$\stackrel{H_3C}{H_3C}>CH-CH_2-H_2C-O-C\stackrel{\nearrow O}{\searrow CH_3}$	Isoamyl acetate	$C_7H_{14}O_2$	130.18	−78.5	142.5	0.8699	nD1.40170
14	$H_3C-H_2C-O-C\stackrel{\nearrow O}{\searrow}CH_2-CH<\stackrel{CH_3}{CH_3}$	Ethyl-iso valerate	$C_7H_{14}O_2$	130.18	−99.3	135	0.8657	nD1.39671
15	$\stackrel{H_3C}{H_3C}>CH-CH_2-CH_2-O-C\stackrel{\nearrow O}{\searrow CH_2-CH_2-CH_3}$	Isoamyl butyrate	$C_9H_{18}O_2$	158.24	−73.2	159..79	0.882	col. liq
16	$H_3C-(CH_2)_4-H_2C-O-C(O)-(CH_2)_2-CH_3$	n-Hexyl-n-butyrate	$C_{10}H_{20}O_2$	172.26		205		
17	$\stackrel{H_3C}{H_3C}>CH-CH_2-CH_2-O-C\stackrel{\nearrow O}{\searrow}CH_2-CH<\stackrel{CH}{CH}$	isoanyl iso-valerate	$C_{10}H_{20}O_2$	172.26	liq.	194	0.8584	nD 1.41311
18	$H_3C-H_2C-O-C(O)-(CH_2)_7-CH_3$	Ethyl pelargonate	$C_{11}H_{22}O_2$	186.29	− 36.7	227.5	0.8657	nD 1.42200
19	$H_3C-(CH_2)_6-CH_2-O-C(O)-(CH_2)_2-CH_3$	Octyl-n-butyrate	$C_{12}H_{24}O_2$	200.32		244		
20	$H_3C-(CH_2)_6-CH_2-O-C(O)-(CH_2)_4-CH_3$	Octyl-caporonate	$C_{14}H_{28}O_2$	228.36		275		
21	$H_2C=HC-O-C(O)-CH_3$	Vinyl acetate	$C_4H_6O_2$	86.09	−100.2	72...3	0.9317	nD 1.3953
22	$\stackrel{H_3C}{O=}C-CH_2-O-C\stackrel{\nearrow O}{\searrow CH_2-CH_3}$	Acetoaceticester Ethylacetoacetate	$C_6H_{10}O_3$	130.14	<−80	180	1.025	nD 1.42092
23	$H_3C-C-CH-O-C-CH_2-CH_3$ (O CH_3 O)	Methylacetoaceticester	$C_7H_{12}O_3$	144.17	liq.	186.8	1.019	nD 1.42066

BP °C diagram with curves labelled: Acids, Alcohols, Ketones, C–H substances, Esters, Ethers.
Vertical axis BP °C: 30 40 50 60 70 80 90 100 110 120 130 140 150 160 170 180
Horizontal axis Molecular weight: 46 60 74 88 102 116 130

Molecular weight
Boiling point diagram for derivatives of ALKANES

24	$O=C-O-CH_2-CH_3$ / $O=C-O-CH_2-CH_3$	Diethyloxalate
	$C_6H_{10}O_4$	185.4	1.0785
	146.14	−40.6	1.41011
	oily liq., misc. acet., eth., Bz.		

25	$O=C-O-CH_2-CH_3$ / HCH / $O=C-O-CH_2-CH_3$	Diethylmalonate "Malonicester"
	$C_7H_{12}O_4$	198.9	1.0553
	160.16	−49.8	1.4143
	col. liq., misc. acet., eth., Bz., chl., v.s. s.w.		

26	$O=C-O-CH_2-CH_3$ / $C=O$ / CH_2 / $O=C-OCH_2-CH_3$	Oxalacetic ester Diethyl ester
	$C_8H_{12}O_5$	132v	1.159
	188.18		1.45614
	oily liq., misc. acet., eth., Bz., col. [keto-enol-tantomer]		

27	$H_2C-O-C\stackrel{\nearrow O}{\searrow CH_2-(CH_2)_{13}-CH_3}\rightarrow R_1$ (Rest1) / $HC-O-C\stackrel{\nearrow O}{\searrow CH_2-(CH_2)_{13}-CH_3}\rightarrow R_2$ (Rest2) / $H_2C-O-C\stackrel{\nearrow O}{\searrow CH_2-(CH_2)_{13}-CH_3}\rightarrow R_3$ (Rest3)	
	Glyceroltripalmitate, "Tripalmitin" (Palmin)		
	$C_{51}H_{98}O_6$	B.P.	310–320	D 0.866 liq.
	807.30	MP$_1$ 46**	MP$_2$ 65.1	n 1.4381
	col. need. (eth.), e.s. acet., eth., s. chl.		

28	Glyceroltrioleate R$_1$ = R$_2$ = R$_3$ = oleicacid	
	$C_{54}H_{104}O_6$	240v	DO. 915
	885.41	MP$_1$-17 MP$_2$-6	
	col. oil., misc. acet., eth., s. chl., i.w.		

29	Glyceroltristearate R$_1$ = R$_2$ = R$_3$ stearic acid	
	$C_{54}H_{110}O_6$		D 0.862 liq.
	891.46	MP$_1$ 54.5 MP$_2$ 70.8	n 1.4399 liq.
	col. Cr. (eth.), e.s. acet., eth.		

* The structural and the rational formula are once more arranged for illustration.

** On heating the molten crystals, the solid is regenerated and once again melts. The reason is not yet explained.

No.	Name	M.F	Mol. wt.	MP/FP	BP/CP	D/sp.gr.	nD and other properties
1	Aceticanhydride	$C_4H_6O_3$	102.09	73.1	140.0	1.0820	1.39038
				col. liq; misc. acet; eth, Bz, chl, d. in w			
2	Propionic acid-anhydride	$C_6H_{10}O_3$	130.14	−45	169.3	1.0336	1.4038
				col. liq; misc. acet; eth; d. in acet			
3	Butyric acid-anhydride	$C_8H_{14}O_3$	158.19	−75.0	198	0.9946	
				col. d. in W; acet; misc. eth			
4	n-Valeric acid-anhydride	$C_{10}H_{18}O_3$	186.25	−56.1	215	0.929	
				col. liq; misc. eth; d. in w; acet			
5	n-Caproic acid-anhydride	$C_{12}H_{22}O_3$	214.30	−40.6	241m3(d)	0.9279	
				col. oil; misc. acet; eth; d. in w			
6	Stearic acid-anhydride	$C_{36}H_{70}O_3$	550.93	71.5		0.8368 liq.	
				col. cr; s. eth; d. in hot w; acet			
7	Diacetyl peroxide, Acetyl peroxide	$C_4H_6O_4$	118.09	30	60 v		
				col. leaf s. eth; d. in w; exp			

No.	Formula	Name / MF / Mol. wt. / and other properties	BP/CP / MP/FP	D/sp.gr. / nD
8	$H_2C{-}C{=}O$; $H_2C{-}O$	Propiolactone, 2-oxo-oxetane, $C_3H_4O_2$	51v / −33.4	
9	Butyrolactone ring	Butyrolactone, 2-oxotetrahydrofuran, $C_4H_6O_2$	206	
10	1,4-pyrone ring	1,4-pyrone, 4-oxo-1,4-pyran, $C_5H_4O_2$, 96.08	217.7 / 32.5	1.190 liq. / 1.5238 liq.
		col. pri., s. acet., e.s. eth., s. KOH yelsh. sol.		
11	Succinic anhydride ring	Succinic anhydride, 2,5-dioxotetrahydrofuran, $C_4H_4O_3$, 100.07	261 / 119.6	1.104
		col. need., (acet.), s. acet., eth.		
12	Maleic anhydride ring	Maleic anhydride, 2,5-dioxo-2,5-dihydrofuran, $C_4H_2O_3$, 98.06	202s / 52.8	0.934
		col. rhomb. need., s. eth., chl.		
13	Glycolide ring	Glycolide, 2,5-dioxotetrahydro-1,4-dioxane, $C_4H_4O_4$, 116.07	86...7	col. leaf (acet.)
		s.h.w., e.s. acet., eth., ac. a., hot. chl.		
14	Cumarin ring	Cumarin, 1,2-benzopyran, $C_9H_6O_2$, 146.14	301.72 / 67...68	0.935
		col. need. or rhomb.; cr.; s.h.w, acet, eth, chl.		
15	Chromane ring	Chromane, 1,4-benzopyran, $C_9H_6O_2$, 146.14	s / 59	
		wh. need (pet.), s. acet., eth., Bz. chl.		
16	Phthalic anhydride ring	Phthalic anhydride, $C_8H_4O_3$, 148.11	284.5s / 130.8	1.527
		col. rhomb. need., (acet.), s. acet.		

No.	Formula	Name / MF / Mol. wt. / and other properties	BP/CP / MP/FP	D/sp.gr. / nD
17	Caprolactam ring	2-Caprolactam, $C_6H_{11}ON$, 113.16	139v / 70	1.02 liq.
		wh. scales., e.s.w., acet., eth.		
18	Glycine anhydride ring	Glycine anhydride, 2,5-diketopiperazine, $C_4H_6O_2N_2$, 114.10	s / 275d	
		pl., s.h.w., s. acet.		
19	Parabanic acid ring	Parabanic acid, $C_3H_2O_3N_2$, 114.06	243d	col. monocl., pl.
		(w.), sl. s.w., e.s. acet., eth.		
20	Barbituric acid ring	Barbituric acid, 2,4,6-trioxopyrimidine, $C_4H_4O_3N_2$, 128.09	260d / 245	
		wh. rhomb., (H_2O); s.w., eth.		
21	Alloxan ring	Alloxan, 2,4,5,6-tetroxopyrimidine, $C_4H_2O_4N_2$, 142.07	170d	dk. yel.
		or col. rhomb. pr., e.s.w., s. acet.		
22	Indoxyl ring	Indoxyl, 3-oxoindoline (3-oxyindole), C_8H_7ON, 133.14	110 / 85	
		leaf., s. alkalis.		
23	Oxindole ring	Oxindole, 2-oxoindoline (2-oxyndole), C_8H_7ON, 133.14	d / 120	
		col. need. (w)., e.s.w., s. acet., eth., alk.		
24	Isatin ring	Isatin, 1,3-dioxoindoline, $C_8H_5O_2N$, 147.13	s / 201	redsh. monocl.
		need., (acet.), s.w., acet., eth., alk.		
25	Phthalimide ring	Phthalimide, 1,3-dioxoisoindoline, $C_8H_5O_2N$, 147.13	s / 234	need (w) hex. pris.
		(eth.), v.sl. s.w., e.s. ac. a., KOH		
25a	Potassium phthalimide ring	Potassium phthalimide, $C_8H_4O_2NK$, 185.22		

Oxo compounds of carbon and hetero rings with side chains

No.	Formula	Name	MF	BP/CP	D/sp.gr.	Mol. wt.	MP/FP	nD	other properties
1	(structure)	l-menthone	$C_{10}H_{18}O$	207	0.896	154.24	−6.6		col. liq., misc. acet., eth., Bz., CS_2
2	(structure)	d-carvone	$C_{10}H_{14}O$	230	0.9608	150.21		1.49994	col. liq., misc. acet., eth., chl.
3	(structure)	β-ionone, 4-(2,6,6-trimethyl-1-cyclohexalyl)-3-butene-2-one	$C_{13}H_{20}O$	140v	0.944	129.29		1.51977	col. liq.
4	(structure)	Acetophenone	C_8H_8O	202.0	1.026	120.14	19.6	1.53418	col. liq., or col. leaf., sl. s.w., s. acet., eth., Bz., chl.
5	(structure)	Muscone	$C_{16}H_{30}O$	311…27	0.9268	238.4			col. oil., misc. eth.
6	(structure)	Camphor	$C_{10}H_{16}O$	204s	1.000	152.23	176.7	1.532	col. pl., e.s. acet., eth., al., Bz., chl., ac. a., v. sl. s.w.
7	(structure)	Benzophenone Diphenylketone	$C_{13}H_{10}O$	306.1		128.21	a) MP48.0 D_1 1.0846; b) MP26 D_2 1.108		e) rhomb, pri. e.s. acet., eth., chl. c) monocl. Cr.
8	(structure)	Benzil Diphenyldiketone	$C_{14}H_{10}O_2$	346-348d	1.521	210.22	95	yelsh.	rhomb. need., (acet.) e.s. acet., eth., chl.
9	(structure)	Fuchsone	$C_{19}H_{14}O$			258.30	168	or. vlt.	blk. need. (Bz) e.s. Bz., chl., ac. a
10	HN−C(=O)−CH₃	Acetanilide "Antifebrin"	C_8H_9ON	305	1.211	135.16	114		wh. leaf. (w. acet.), e.s.h. w., acet., eth.
11	$H_2C=C-CH_2$, $O-C=O$	Diketene	$C_4H_4O_2$	67-69 (92 tor)		84.08			unpleasant odour
12	(structure)	Dimer of dimethylketene 2,2,4,4-tetramethyl-1,3 dioxocyclobutane	$C_8H_{12}O_2$						
13	(structure)	5-methyl-2-oxo-4,5-dihydrofuran	$C_5H_6O_2$	55-56v	1.084	98.10	18-18.5	144.76	
14	(structure)	Lactide, 3,6-dimethyl 2,5-dioxo-1,4-dioxane	$C_6H_8O_4$	255	0.862	144.12	125		col. monocl. pl., e.s.w., acet.
15	(structure)	Triacetoneanine, 2,2,6,6-Tetramethyl-1,4-piperdone	$C_9H_{17}ON$ (H_2O)	173.25	OW 40 NW 58				tetra. need., (w.), s.w., acet., eth
16	(structure)	Thymine	$C_5H_6O_2N_2$	s		126.11	270d		need (acet.). s. acet., H_2SO_4., v. sl. s.w.
17	(structure)	Veronal 5,5-diethylbarbituric acid	$C_8H_{12}O_3N_2$	s		184.19	191		saporific, cr. (w), s.w., e.s. acet., alkalis.
18	(structure)	Tropinone Tropanone-3	$C_8H_{13}ON$	218.0…5	0.9872	139.20	41	1.4621	glistn. need. (pt. eth.) s. all org. sol.
19	(structure)	Pseudopelletirine	$C_9H_{15}ON$	246	1.001	153.22	48…9	1.47596	pl. (pet. eth.), e.s. acet., eth., s.w., chl. Bz
20	(structure)	Coramine Nicotinic acid diethylamide	$C_{10}H_{14}ON_2$	280		178.23			yelsh. oil., misc. w. and with all org. solv.
21	(structure)	Dibenzoylperoxide	$C_{14}H_{10}O_4$	ex	1.837	242.22	103.5	1.545	col. rhomb. cr., s. acet. eth. Bz. CS_2
22	(structure)	Antipyrine	$C_{11}H_{12}ON$	319(174torr)	1.19	188.22	113	1.5697	leaf. (eth.), s. Bz., tol., eth.
23	(structure)	Pyramidone, 1-Phenyl-4-dimethylamino-2,3-dimethylpyrazolone	$C_{13}H_{17}ON_2$		zolone	231.30	108		leaf. (liq.), e.s.w., acet., eth., Bz., dil. acids

No.	Formula	Name	MF	M W	MP	BD	D/sp. gr. n_D and other props
1	H–C≤O HCOOH / OH	Formic acid	CH_2O_2	46.03	8.40	100.5	1.220 n 1.37137 / kd $2.1410×10^{-4}$
2	$CH_3–COOH$	Acetic acid	$C_2H_4O_2$	60.05	16.6	118.1	1.0492 n 1.37182
3	$CH_3–CH_2–COOH$	Propionic acid	$C_3H_8O_2$	74.08	−22	141.3	0.9930 n 1.38736
4	$CH_3–CH_2–CH_2–COOH$	n-Butyric acid	$C_4H_8O_2$	88.10	−7.9	164.0	0.9640 n 1.39906
5	$CH_3–(CH_2)_2–CH_2–COOH$	n-Valeric acid	$C_5H_{10}O_2$	102.13	−34.5	186.4	0.942 n 1.4086
6	$CH_3–(CH_2)_3–CH_2–COOH$	n-Caproic acid	$C_6H_{12}O_2$	116.16	−15…20	205.3	0.9290 n 1.41635
7	$CH_3–(CH_2)_4–CH_2–COOH$	Heptoic acid-Heptanoic acid	$C_7H_{14}O_2$	130.18	−10.5	223	0.9127 n 1.42162
8	$CH_3–(CH_2)_5–CH_2–COOH$	Caprylic acid	$C_8H_{16}O_2$	144.20	16.3	237.5	0.910 n 1.4285
9	$CH_3–(CH_2)_6–CH_2–COOH$	n-Pelangonic acid	$C_9H_{18}O_2$	158.23	12.5	254	0.9055 n 1.4330
0	$CH_3–(CH_2)_7–CH_2–COOH$	Cupric acid	$C_{10}H_{20}O_2$	172.26	31.3	268.4	0.8858liq n 1.42855
1	$CH_3–(CH_2)_8–CH_2–COOH$	n-Undecylacid (Undecanoicacid)	$C_{11}H_{22}O_2$	186.29	28	212*	0.8905liq n 1.4294
2	$CH_3–(CH_2)_9–CH_2–COOH$	n-Lauric acid (Dodecanoicacid)	$C_{12}H_{24}O_2$	200.31	44	225*	0.8679liq n 1.4183
3	$CH_3–(CH_2)_{10}–CH_2–COOH$	n-Tridecyl acid (Tridecanoicacid)	$C_{13}H_{26}O_2$	214.34	40.5	236*	
4	$CH_3–(CH_2)_{11}–CH_2–COOH$	n-Myristicacid (Tetradecanoicacid)	$C_{14}H_{28}O_2$	228.37	58	250.5*	0.8622liq n 1.4308
5	$CH_3–(CH_2)_{12}–CH_2–COOH$	n-Pentadecyl acid (Pentadecanoic acid)	$C_{15}H_{30}O_2$	242.40	52.1	257*	
6	$CH_3–(CH_2)_{13}–CH_2–COOH$	n-Palmitic acid (Hexadecanoic acid)	$C_{16}H_{32}O_2$	256.42	64	271.5*	0.853liq. n 1.4273
7	$CH_3–(CH_2)_{14}–CH_2–COOH$	n-Margarinic acid (Heptadecanoic acid)	$C_{17}H_{34}O_2$	274.45	60.66	277*	0.8578
8	$CH_3–(CH_2)_{15}–CH_2–COOH$	n-Stearic acid (Octadecanoic acid)	$C_{18}H_{36}O_2$	284.46	70.1	291*	0.8386liq n 1.4299
9	$CH_3–(CH_2)_{16}–CH_2–COOH$	n-Nonadecyl acid (Nonadecanoic acid)	$C_{19}H_{38}O_2$	298.49	66.5	298*	
0	$CH_3–(CH_2)_{17}–CH_2–COOH$	n-Arachinic acid (Eicosanoic acid)	$C_{20}H_{40}O_2$	312.52	76.3	328d	0.842 liq
1	$CH_2=CH–COOH$	Acrylic acid (1-propenoicacid)	$C_3H_4O_2$	72.06	12.3	141.9	1.051 misc. w; acet; eth.
2	$H_2C=C<^{COOH}_{CH_3}$	Methacrylic acid / 2-Methyl-1-propenoic acid	$C_4H_6O_2$	86.09	16	163	1.015 n 1.43143 / col. liq. misc. w. acet.
3	$^H_{H_3C}>C=C<^{COOH}_H$	Crotonic acid / 2-Butenoic acid (transform)	$C_4H_6O_2$	86.09	72	189	1.018 n 1.4228 / col. need. s.w. li.
4	$^{H_3C}_H>C=C<^{COOH}_H$	Iso-crotonic acid (cis form) / 2-Butenoic acid	$C_4H_6O_2$	86.09	14…15	171.9d	1.0312 n 1.4457 / col. need. (pet.) s. w., acet., pet.
5	$H_3C–(CH_2)_7$ $>C=C<^{(CH_2)_7–COOH}_H$ / H	Oleic acid / cis-9-Octadecanoic acid	$C_{18}H_{34}O_2$	282.46	14	286*	0.8905 n 1.463 / col. need., s. acet., eth., Bz., chl.
6	$H_3C–(CH_2)_7$ $>C=C<^H_{(CH_2)_7–COOH}$	Elaidic acid / transform	$C_{18}H_{34}O_2$	282.45	51.5	288*	0.851liq / col. leaf. (acet.). s. acet., eth., Bz., chl.
7	$H_3C–(CH_2)_4$ $>C=C<^{CH_2}_{H H}>C=C<^{(CH_2)_7–COOH}_H$	Linoloid acid 9,12-Octa-decadienoic acid	$C_{18}H_{32}O_2$	280.44	−11	228v	0.9025
8	$H_3C–CH_2–CH=CH–CH_2–CH=CH–CH_2–CH=CH–(CH_2)_7–COOH$ / 3.6.9-Octadecatrienoic acid; α-Linolenic acid		$C_{18}H_{30}O_2$	278.42	col. liq.	231v	0.905 misc. acet; eth;
9	$H_3C–CH_2–CH_2–CH_2–CH=CH–CH=CH–CH=CH–(CH_2)_7–COOH$ / 5.7.9-Octadecatrienoic acid, α-Eleostearic acid		$C_{18}H_{30}O_2$	278.42	48…9	235v	leaf. or need; e.s. acet; eth;
0	$HC≡C–COOH$	Propiolic acid, Prorgylic acid	$C_3H_2O_2$	70.05	9	144d	1.139 col. liq. misc. w., acet., eth.
1	$HOOC–COOH$	Oxalic acid	$C_2H_2O_4$	90.04	189.5	s	1.653m wh. cr., s.w., acet., eth.
2	$HOOC–CH_2–COOH$	Malonic acid	$C_3H_4O_4$	104.06	135.6	s(d)	1.631 s.w., acet., eth
3	$HOOC–(CH_2)_2–COOH$	Succinic acid	$C_4H_6O_4$	118.09	185	235(d)	1.572 s.w., sl. s. eth.
4	$HOOC–(CH_2)_3–COOH$	Glutaric acid	$C_5H_8O_4$	132.11	97…8	302…4d	1.429 s.w., acet., eth., Bz., chl.
5	$HOOC–(CH_2)_4–COOH$	Adipic acid	$C_6H_{10}O_4$	146.14	151…3	265*	1.360 sl. s.w., eth., e.s. acet.
6	$HOOC–(CH_2)_5–COOH$	Pimelic acid	$C_7H_{12}O_4$	160.17	103	272(s)*	1.329 sl. s.w., e.s. acet., eth.,
7	$HOOC–(CH_2)_6–COOH$	Suberic acid	$C_8H_{14}O_4$	174.19	140	279*	col. need. sl. s.w. s. acet; eth.
8	$HOOC–(CH_2)_7–COOH$	Azelaic acid	$C_9H_{16}O_4$	188.22	106.5	226v	1.029 n 1.403. s. acet., eth.
9	$HOOC–(CH_2)_8–COOH$	Sebacic acid	$C_{10}H_{18}O_4$	202.25	133	295*	n 1.422 col. leaf., s. acet.; e.s. eth.
0	$HOOC$ $>C=C<^{COOH}_H$ / H	Maleic acid / cis-2-Butendicarboxyl acid	$C_4H_4O_4$	116.07	130.5	135d	1.590 / col. pr., e.s.w., al., s. eth., ac. a. acet.
1	$HOOC$ $>C=C<^H_{HOOC}$ / H	Fumaric acid / trans-form	$C_4H_4O_4$	116.07	286…7	290	1.635 / col. pr., sl. s.w., s. al., i. eth.
2	$HOOC–C≡C–COOH$	Acetylene dicarboxylic acid	$C_4H_2O_4$	114.06	179		long. pr., e.s.w., al., eth.

* Hydroxy groups on a **C-atom** only in exceptional cases see pp. 118/8; 10; 126/20; 55/22.

* Oxogroups on a **C-atom** only in Carbondioxide O = C = O pp. 40/2.

Hydroxy: 1-Oxogroup on a **C-atom** gives rise to Carboxyl group. (The C-atom of Carboxyl derived from oxidation of hydro carbon, so that the C-atom belongs not to the functional group but to the structural framework see p. 105.) The double bond of the Oxo-group activates the H-atom of the immediate neighbourhood and liberates so as to cause acidic character of the Carboxyl group and other groups (e.g., > NH see p. 108) as like in "keto-enol tautomerism" (see p. 108 and 118).

Carboxy compounds (organic acids)

No.	Formula	Name / MF / Mol. wt. / Other properties	BP/CP / MP/FP	D/sp. gr. / nD
1	$H_2C-COOH$ $HC-COOH$ $H_2C-COOH$	Tricarballylacid; 1,2,3-propanetricarboxylic acid; $C_6H_8O_6$; 176.12; col. rhomb. pri., w., e.s.w., acet., sl.s. eth	d; 162...3	
2	COOH (cyclohexane/benzene ring)	Benzoic acid; $C_7H_6O_2$; 122.12; col. need or leaf., sl.s.w., e.s. acet., eth., Bz and other org. solv.	249.2; 122.4	1.2659
3	$H_2C-COOH$ (ring)	Phenylacetic acid; α-tolylacid; $C_8H_8O_2$; 136.14; col. leaf. (w.) v.sl.sw., acet., eth., Bz., chl., ac.a	265.5; 76.7	1.228
4	$HC=CH-COOH$ (ring)	Cinnamic acid; trans-β-phenyl-acrylic acid; $C_9H_8O_2$; 148.15; col. leaf. (acet.) v.sl. s.w., e.s.acet., eth. Bz., ac. a.	300; 133	
5	COOH (naphthalene)	β-naphthoic acid; $C_{11}H_8O_2$; 172.17; col. monocl. need., (liq.) e.s.acet., eth., ligr.	>300; 185	1.077
6	H_3C $COOH$ ... CH_3 CH_3 ... $COOH$	d-camphoric acid; $C_{10}H_{16}O_4$; 200.23; col. leaf or pri. (acet.), sl. s.h.w., e.s.al., acet., eth.	187	1.186
7	$H_2C-CH-COOH$ ... $H_2C-CH-COOH$	Hexahydrophthalic acid (cis-form); 1,2-cyclohexanedicarboxylic acid; $C_8H_{12}O_4$; 172.18; (acet.), v.sl. s.w., s.acet., eth.	>192d(w); 192	col. pri. or pl.
8	$H_2C-C-COOH$ ... $H_2C-C-COOH$	Cyclohex-1-ene-1,2-dicarboxylic acid 3,4,5,6-tetrahydrophthalic acid; $C_8H_{10}O_4$; 170.16; col. leaf. (w.), e.s.w.	120	
9	COOH COOH (ring)	o-phthalic acid; 1,2-benzenedicarboxylic acid; $C_8H_6O_4$; 166.13; wh. amor. cr. need., v.sl. s.w., acet., eth.	350d; s300	1.510
10	COOH HOOC COOH HOOC COOH COOH	Mellitic acid; Benzenehexacarboxylic acid; $C_{12}H_6O_{12}$; 342.17; col. need. (acet.), s. acet., H_2SO_4., e.s.w	d; 286	
11	$HN-CH_2-COOH$ (ring)	N-phenylglycine; $C_8H_9O_2N$; 151.16; col. cr., e.s.w., acet.	d; 127	
12	COOH (furan)	Pyromucic acid; 2-furanoic acid; $C_5H_4O_3$; 112.08; wh. need or leaf. (w.), s. acet., e.s. eth.	230...2 (s100); 133...4	
13	CH_3 $HOOC\cdot CH_2\cdot C\cdot COOH$ $H_3C\cdot C\cdot CH_3$ $COOH$	l-campharonic acid; $C_9H_{14}O_6$; 218.20; hyg. need. (w.), e.s.w., acet., chl., s. eth.	164...5; 195...210(v)	
14	$HC-COOH$ $C-COOH$ $H_2C-COOH$	Aconitic acid; 1,2,3-propenetricarboxylic acid; $C_6H_6O_6$; 174.11; col. leaf or need. (w.), s.w., e.s. acet.	192...5d	
15	$HOOC$—furan—$COOH$	Dehydromucic acid; 2,5-furandicarboxylic acid; $C_6H_4O_5$; 156.09; need. (w.), s.w., acet., eth.	s, >320	
16	$HC=C\langle^H$ ... $C=C\langle^H$ ... $C\cdot COOH$ ... O ... H_2C-O	Piperinic acid; 5(3,4-methylene-dioxyphenyl)-2,4-pentadienicacid; $C_{12}H_9O_4$; 218.20; yelsh. need. (acet.), s. acet., eth.	s220(d); 217	
17	H_2N $COOH$ H_2 H_2 (pyrroline)	L-proline; Pyrroline-2-carboxylic acid; $C_5H_9O_2N$; 115.13; col. cr. (acet.), e.s.w., i. acet., s. eth.	220...2d	
18	N—COOH (pyridine)	Picolinic acid; Pyridine-2-carboxylic acid; $C_6H_5O_2$; 123.11; col. need. (w. acet), e.s.w., acet.	s(d); 137.8	
19	N COOH (pyridine)	Nicotinic acid; Pyridine-3-carboxylic acid; $C_6H_5O_2N$; 123.11; col. need. (w. acet.), e.s. h.w., sl.s. acet.	s; 235...6	
20	N COOH (pyridine)	Isonicotinic acid; Pyridine-4-carboxylic acid; $C_6H_5O_2N$; 123.11; col. need. (w.), e.s. h.w.	s(d); 317	
21	N COOH (quinoline)	Quinoline-4-carbolic acid; $C_{10}H_7O_2N$; 173.16; monocl. pri. or need. (w.)	253...4	
22	$HOOC$ N COOH (quinoline)	Quinoline-4,6-dicarboxylic acid; $C_{11}H_7O_4N$; 203.19; yelsh. leaf. (acet.), s. acet.	s; 280	
23	N COOH (quinoline)	Atophane; 2-phenylquinoline-4-carboxylic acid; $C_{16}H_{11}O_2N$; 249.26; wh. need., s. acet., eth., me-al., h.w.	209	
24	$O=C-O-O-H$ (ring)	Perbenzoic acid; $C_7H_6O_3$; 138.12; leaf. (higr.), s. acet., eth.	ex. 80...100; 41...3	

Compounds with several oxygen functions (keto acids, hydroxyacids, etc.)

No.	Formula	Name	MF	MW	MP	BP	D/sp.gr. nD and other props.
1	$H_3C-C-C\overset{H}{\underset{O}{\gtrless}}$ (2-oxo)	Methylglyoxal 2-oxopropanal	$C_3H_4O_2$	72.06	yel. liq.	72	yelsh. liq., misc. w., acet., eth., strong tendency to polymer.
2	$H_3C-C-CH_2-CH_2-C\overset{H}{\underset{O}{\gtrless}}$	Laevulinic aldehyde 4-oxopentanal	$C_5H_8O_2$	100.11	< −21	186...8d	1.0181 n 1.4263 strong odour, col. liq., s.w., al., eth.
3	H_2C OH–CHO	2-hydroxyethanal, glycolaldehyde	$C_2H_4O_2$	60.5	96...7		1.366 col. pl., e.s.w., acet.
4	$H_3C-HCOH-CH_2-CHO$	Aldol, 3-hydroxybutanal	$C_4H_8O_2$	88.10			1.103 col. oil., misc. w., eth.
5	$H_2COH-HCOH-CHO$	Glyceraldehyde, 2,3-dihydroxypropanal	$C_3H_6O_3$	90.08	138	140...50V	1.453 need., s. me. al., sl.s.w.

6		Salicylaldehyde o-hydroxybenzaldehyde				7		Vannilin 3-methoxy-4-hydroxybenzaldehyde

6	$H-C=O$	Salicylaldehyde o-hydroxybenzaldehyde			7	$H-C=O$	Vannilin 3-methoxy-4-hydroxybenzaldehyde
		$C_7H_6O_2$	196.5	1.4669		$O-CH_3$	$C_8H_8O_2$ 285(in CO_2) 1.056
		122.12	−7	1.57358		OH	152.14 81...2
		col. oily. liq., misc. acet. eth.					need., sl. s.w., s.acet., eth., chl., B.P., 140

8	$H_3C-HCOH-CO-CH_3$	Acetoin, 3-hydroxybutan-2-one	$C_4H_8O_2$	88.10	15	148	1.002 n1.4194, s.w., acet. eth.
9	$H_3C\underset{H_3C}{>}C\overset{OH}{\underset{CH_2}{<}}C\overset{=O}{<}CH_3$	Diacentone alcohol 4-hydroxy-4-methylpentan-2-one	$C_6H_{12}O_2$	116.16	−54...7	169...6	0.938 n1.43. col. liq. misc. w., eth., acet.
10	$H_2COH-CO-H_2COH$	Dihydroxy acetone, 1,3-Dihydroxy propan-2-one	$C_3H_6O_3$	90.08	80		Dimer., cr., e.s. hot w., acet. eth.

11		Benzoin Benzoylphenyl alcohol			13		Dihydroxyindole, 3-hydroxy-2-ketoindole
		$C_{14}H_{12}O_2$	343	1.079			$C_8H_7O_2N$
		212.24	137				238.23 169...70
		B.P. 194° d., acidic, s.h. al.					pa. yel. need. (acet.), s. acet.

12		Ninhydrin Hydrate of triketohydrindene			14		Dihydroxyindole 3-hydroxy-2-ketoindole
		$C_9H_6O_4$	178.14	139d			$C_{14}H_{10}O_3$ 2195
		pri., (w.), s.w.					149.14 180 Cr., (acet.), s. acet., eth., alkali.

5	$H_2COH-COOH$	Glycolic acid, 1-hydroxyacetic acid	$C_2H_4O_3$	76.05	63 79		Need. (w.), e.s.w., acet., eth.
6	$H_3C-HCOH-COOH$	Lacticacid, α-hydroxypropionicacid	$C_3H_6O_3$	90.08	16...8	119v	1.249 Syrap., misc. w.
7	$H_2COH-CH_2-COOH$	β-Hydroxy propionic acid; or 3-hydroxyprop. acid	$C_3H_6O_3$	90.08[10]		d	Syrap., misc. w., eth., acet.
8	$H_3C-HCOH-CH_2-COOH$	β-hydroxybutyric acid	$C_4H_8O_3$	104.10	49...50	130(d)	Syrap., misc. w., acet., eth.
9	$H_2COH-HCOH-COOH$	Glyceric acid, 2,2-Dihydroxy propionic acid	$C_3H_6O_4$	106.08			Syrap., misc. w., acet., eth.

No	Formula	Name			25	COOH COOH	Malic acid
		MF	BP	D		HCH HCH	$C_4H_6O_5$ 140d 1.595
		Mol. wt.	MP	nD		HCOH HOCH	134.09 100...1 $H_2C-COOH$
		and other properties				COOH COOH	col., cr., glit., e.s.w., acet., s. eth.
20	$H_2C-COOH$ $HO-C-COOH$ $H_2C-COOH$	Citric acid			26	COOH COOH HCOH HOCH HCOH HCOH COOH COOH	Tartaric acid
		$C_6H_8O_7$	d	1.542			$C_4H_6O_6$ 1.7782
		192.12	153	1.493			150.09 170 1.4955
		wh., rhomb. 1.H_2O at 70–75°–H_2O, e.s.w., acet. eth.					col. monocl., e.s.w., acet., v. sl. s. eth.
21	$HO-\overset{H}{\underset{}{C}}-COOH$	Mandelic acid			27	Gallic acid	3,4,5-trihydroxybenzoic acid
		$C_8H_8O_3$	d	1.300			$C_7H_6O_5$ d 1.694
		152.14	118.1				170.12 220d
		pl. (w.), s., acet., eth., col. rhomb cr. (Bz)					col., monocl. need (w) s.w., acet, eth., gly
22	$HOCH_2-\overset{H}{\underset{}{C}}-COOH$	Tropic acid			28	Quinic acid	
		$C_9H_9O_3$	160d				$C_7H_{12}O_6$ 200d 1.637
		166.17	117...8				192.16 161.6
		pl., s.w., acet., eth.					col. pri., (w.) v. sl. s. eth., s. ac. a., al.
23	COOH OH	Salicyclic acid o-hydroxybenzoic acid			29	Syringic acid	3,5-dimethoxy-4-hydroxybenzoic acid
		$C_7H_6O_3$	76s	1.443			$C_9H_{10}O_5$
		138.12	158.3	1.565			203
		col., monocl., need.(w.), s. acet., eth. chl					
24	COOH OH	Protocatechuic acid 3,4-dihydroxybenzoic acid			30	Benzilic acid Diphenylglycolic acid	
		$C_7H_6O_4$	d	1.542			$C_{14}H_{12}O_3$ 180d
		154.12	199d				228.24 150
		need. or pl. (1H_2O) (w.), sl.s.w., s.acet., eth.					need. (w.), s.h.w., acet., eth., s. H_2SO_4

Compounds with several oxygen functions

No.	Formula	Name	MF	MW	MP	BP	DnD and other properties
1	O=C-C(=O)OH ; H	Glyoxalic acid	$C_2H_2O_3$	74.05	d		
					Prism; (H_2O); e.s.w; s. acet.		
2	O=C-C(=O)OH ; H_3C	Pyruvic acid ; 2-Oxopropionic acid	$C_3H_4O_3$	88.06	13.6	165d	1.2646
					col. liq; misc; w; acet; eth.		
3	O=C-CH₂-C(=O)OH ; H_3C	Acetoacetic acid ; 3-Oxoputyric acid	$C_4H_6O_3$	102.09	< 100 (d) very unstable		
					liq; col; syrup; misc; w; acet; eth.		
4	O=C-CH₂-CH₂-C(=O)OH ; H_3C	Levulinic acid ; 4-Oxopentanoic acid	$C_5H_8O_3$	116.11	37.2	246,154 D1.1395	
					col. leaf; e.s.w; acet; eth.		
5	O=C-COOH ; H₂-C-COOH	Oxalacetic acid (keto form) ; 2-Oxobutane dicarboxylic acid	$C_4H_4O_5$	132.08	In free state not known → enol; aboniester 80% of enol.		
6	HO-C-COOH ; H-C-COOH	Hydroxy maleic acid (enol form) ; cis-2-Hydroxy-2-butenedicarboxylic acid	$C_4H_4O_5$	132.08	152		
					fine cr; (acet.) e.s. acet; ac. a.		
7	HOOC-C-OH ; H-C-COOH	Hydroxy fumaric acid trans-2- ; Hydroxy-2-butene dicarboxylic acid	$C_4H_4O_5$	132.08	184d		
					s.w., al., eth., acet + Bz.		
8	COOH COOH ; C=O C<OH OH ; COOH COOH ; a) b)	a) Keto malonic acid ; Mesoxalic acid	$C_3H_2O_5$	118.05	121d	glist. cr. (H_2O), e.s.w; acet; eth.	
		b) Diketomalonic acid ; Hydrate ketomalonic acid	$C_3H_6O_5$	136.06	stable hydrate with 2 (OH) on same C		
9	HOOC-CH₂-CO-CH₂COOH	Acetonedicarboxylic acid	$C_5H_6O_5$	146.10	135d	need (ac. a). e.s.w; acet; eth; ac.	

10	COOH ; HOCOH ; HOCOH ; COOH	Diketotartaric acid ; Stable hydrate ; $C_4H_6O_8$; 182.09 . 110 d ; wh. Cr. powd; e.s.w; d. in h.w.

11	COOH ... CH₃	Acetylsalicylic acid ; Asprin ; $C_9H_8O_4$; 180.15 143 (d) 1.645 ; col. need (w). e.s.acet; Eth; sl. s.w.

12	[O··H-O ... O-H··O] ⇌ O + OH ... OH	Quinhydrone = Quinone + Hydroquinone ; $C_{12}H_{10}O_4$ B.P. subl. D 1.401 ; 218.20 M.P. 171° ; dk. grn. rhomb. pri; e.s.w. (diss); s.acet; eth

Alcoholates

No.	Formula	Name	Mol. wt.	MP	D nD and other properties
13	[CH₃-CH₂-O]₃Al	Aluminium alcoholate	162.15	134	BP 205 (14) D1.142 wh; cr. d. in w; e.s. acet; eth
14	[CH₃-CH₂-CH₂O]₃Al	Aluminium-n-propoxide	204.23	106	BP 248 (14) D1.0578 wh; cr; d. in w; s. acet.
15	[CH₃-C(O)-CH₃]₃Al	Aluminium iso proxide	204.23	118.5	BP 145.5 D 1.0346 wh; cr; d. in w; s. Bz

Salts of organic acids

No.	Formula	Name	Mol. wt.	MP	D nD and other properties
16	HC(=O)ONa [HC OO]Na	Sodium formate	68.02	253	D 1.92. col. monocl; cr; e.s.w; s. acet; i. eth.
17	[HCOO]NH₄	Ammonium formate	63.06	116	BP 180 (d) D 1.266, wh; monocl; e.s.w; s. acet
18	[H₃C-COO]Na	Sodium acetate	82.04	324	wh; monocl; cr; (3H₂O) 120°–3H₂O, e.s.w; s. ac
19	[H₃C-COO]NH₄	Ammonium acetate	77.08	114	D 1.073. wh; hygr. cr; e.s.w; s. ac. a. acet
20	[H₃C-COO]₃Al	Aluminium triacetate	204.10	d	Solid. d. in h. w.
21	[H₃C-COO]₂Pb	Lead (II) acetate	325.30	280	D 3.25 wh; cr; e.s.w; acet; s. glyc.
22	[H₃C-COO]₄Pb	Lead(IV) acetate, lead tetra acetate	443.39	175	D 2.228 monocl; wh; cr; d in w; acet; s. chl; s.h. a
23	[H₃C-COO]₂Cu·H₂O	Cupric acetate	199.64	115	D 1.882 dirty grn; s.w; acet; eth
24	[H₃C-(CH₂)₂-COO]₂Zn2H₂O	Zinc butyrate	303.66	d	wh; glit; sc. or powd; sl. s.w; sh. w
25	[H₃C-(CH₂)₁₆-COO]Na	Sodium stearate (soap)	306.46	d	wh; featherly powd; s.h. w
26	[H₃C-(CH₂)₁₆-COO]₂Ca	Calcium stearate	607.00	179...80	cr; powd; v.sl. s.w. (0.004) i. acet; al; eth.
27	[H₃C-(CH₂)₇-CH=CH-(CH₂)₇-COO]Na	Sodium oleate	304.44	232...5	wh; cr; or yelsh. amorp; s.w. acet; e.s. eth. s. ace
28	[H₃C-(CH₂)₇-CH=CH-(CH₂)₇-COO]₂Ca	Calcium oleate	602.97	83...4	wh; cr; waxy. sl. s.w. (0.04) e.s. eth;
29	Na[OOC-COO]Na	Sodium oxalate	134.01	d	D 2.34 sl. s.w; col. cr.
30	H[OOC-COO]Na·H₂O	Sodium hydrogen oxalate	130.04	d	monocl. wh; cr; sl. s.w; s.h. w
31	K[OOC-COO]K·H₂O	Potassium oxalate	184.23	d	D 2.08, nD 1.440. wh; monocl. cr; e.s.w
32	H[OOC-COO]K	Potassium hydrogen oxalate	128.12	d	D 2.0 nD 1.415 col. monocl; s.w; e.s. acet
33	(NH₄)[OOC-COO](NH₄)·H₂O	Ammonium oxalate	142.12	d	D 1.50 nD 1.439 col. cr; i.s. NH₄OH; s.w
34	[OOC-COO]Ca	Calcium oxalate	128.10	d	D 2.2 col. cub. cr; i.w; s. acet; i. ac. a.
35	[OOC-COO]Sr·H₂O	Strontium oxalate	193.67	d	150–H₂O; col. cr; s. h. w; R → HCl HNO₃
36	[OOC-COO]Ba	Barium oxalate	225.38	d	D 2.658 cr; i.w. s. acet; NH₄Cl solv
37	K[OOC-HCOH-HCOH-COO]H *		188.18		D 1.954 col. monocl. cr.
38	Na[OOC-HCOH-HCOH-COO]K **		282.23	70...80	BP 215 (d) D 1.750 col. cr.
39	K[OOC-HCOH-HCOH-COO]K ***		333.94	d100	D 2.607 (–H₂O). col. cr.
40	[C₆H₅-COO]K·3H₂O	Potassium benzoate	214.25	d110	(–3H₂O). wh; cr; powd. e.s.w; s. acet
41	[C₆H₅-COO]₂Cu·2H₂O	Copper benzoate	341.79	d110	(–2H₂O). blsh. cr; powd. e.s.w

* Potassium hydrogen tartrate   ** Sodium potassium tartrate   *** Dipotassium tartra

Amino compounds (primary amines) · $-NH_2$

No.	Formula	Name	MF	Mol. wt.	MP	BP	D/sp.gr. nD and other properties
1	H_3C-NH_2	Methylamine	CH_5N	31.06	−93.46	−6.32	0.7691 (vap) nD 1.54, s.w., acet., eth.
2	$CH_3-CH_2-NH_2$	Ethylamine	C_2H_7N	45.08	−80.6	16.6	0.7059 (vap) col; misc. w, eth; acet.
3	$CH_3-CH_2-CH_2-NH_2$	n-Propylamine	C_3H_9N	59.11	−83	48.7	0.719 nD 1.39006, misc. w; acet; eth.
4	$CH_3-(CH_2)_2-CH_2-NH_2$	n-Butylamine	$C_4H_{11}N$	73.14	−50.5	77.8	0.7401 nD 1.401 misc. w., acet., eth.
5	$CH_3-(CH_2)_3-CH_2-NH_2$	n-Amylamine	$C_5H_{13}N$	87.16	−55	104	0.7614 col. misc. w., acet., eth.
6	$CH_3-(CH_2)_4-CH_2-NH_2$	n-Hexylamine	$C_6H_{15}N$	101.19	−19	132.7	col. liq. misc. w., acet., eth.
7	$CH_3-(CH_2)_{10}-CH_2-NH_2$	Laurylamine	$C_{12}H_{27}N$	185.35	28	135v	col. cr., e.s.w., s. acet., eth.
8	$CH_2=CH-CH_2-NH_2$	Allylamine	C_3H_7N	57.09	liq.	53.2	0.761 nD 1.41943 misc. w., acet., eth
9	$H_2N-CH_2-CH_2-NH_2$	Ethylenediamine	$C_2H_8N_2$	60.10	8.5	116.1	0.8994 nD 1.454, col. liq. misc. w, acet
10	$H_2N-(CH_2)_3-NH_2$	Trimethylendiamine	$C_3H_{10}N_2$	74.13	liq.	135.5	0.884 col. liq. misc. w., acet., eth.
11	$H_2N-CH_2-(CH_2)_2-CH_2-NH_2$	Tetramethylendia.	$C_4H_{12}N_2$	88.15	27…8	158…60	0.877 cr. e.s.w., acet., eth.
12	$H_2N-CH_2-(CH_2)_3-CH_2-NH_2$	Cadaverine*	$C_5H_{14}N_2$	102.18	9	178…80	0.9178 syrap. e.s.w., acet.
13	$H_2N-CH_2-(CH_2)_4-CH_2-NH_2$	Hexamethylendia.	$C_6H_{12}N_2$	116.21	39…40	s196	silky leaf., e.s.w., d. in air.

No.	Formula	Name			
14	H N NH₂ / H₂ ring (Cyclohexylamine structure)	Cyclohexylamine / 1-aminocyclohexane			
		$C_6H_{13}N$	134		0.8191
		99.17			1.43716
		col. liq., misc. w., acet. eth.			

No.	Formula	Name			
			MF	BP	D/sp. gr.
			Mol. wt.	MP	nD
			Other properties		

15	NH₂ (benzene ring)	Aniline / Aminobenzene			
		C_6H_7N	184.4		1.0217
		93.12	−6.0		1.5863
		col. oily liq. sl. misc. w, misc. acet. eth. Bz			

18	NH₂ / NH₂ (benzene ring, ortho)	o-phenelene diamine / 1,2-diaminobenzene			
		$C_6H_8N_2$	252		
		108.14	102		
		brn. yelsh. monocl. cr. or pl. (chl.) e.s. acet., sl. s.w.			

16	NH₂ (naphthalene)	α-naphthylamine / 1-aminonaphthalene			
		$C_{10}H_9N$	301		1.131
		143.18	50		1.6703
		col. rhomb. need. (acet), sl. s.w, e.s. acet. eth.			

19	NH₂ / NH₂ (benzene ring, meta)	m-phenelenediamine / 1,3-diaminobenzene			
		$C_6H_8N_2$	237		1.107
		108.14	62.8		1.63390
		col. rhomb. need., (acet), s.w., e.s. acet., eth.			

17	NH₂ (naphthalene)	β-naphthylamine / 2-aminonaphthalene			
		$C_{10}H_9N$	306.1		1.061
		143.18	110.2		1.64927
		leaf (w), s.h.w., s. Bz, acet., eth.			

20	NH₂ / NH₂ (benzene ring, para)	p-phenelenediamine / 1,4-diaminobenzene			
		$C_6H_8N_2$	267		
		108.14	139.7		
		col. monocl. (eth. w.) s. acet., eth., chl. sl. s.w			

21	$CH_3-CH_2-O-CH_2-CH_2-NH_2$	2-Ethoxyethylamine	$C_4H_{10}ON$	89.4	liq.	108	0.8512 nD 1.4101, i.w., acet., eth.
22	$\begin{matrix}CH_3-CH_2-O\\CH_3-CH_2-O\end{matrix}>CH-CH_2-NH_2$	2,2-Diethoxy ethylamine	$C_6H_{15}O_2N$	133.19	liq.	163	0.9161 nD 1.4120, i.w., w., acet., eth., chl.
23	$H_2N-CH_2-CH_2-NH-CH_2-CH_2-NH_2$	Diethylenetriamine	$C_4H_{13}N_3$	103.17	liq.	207.1	0.9586 col. yel., misc. w., acet., i. eth.
24	$H_2N-CH_2-CH_2-CH_2-NH-CH_2-CH_2-CH_2-NH_2$	Dipropylenetriamine	$C_6H_{15}N_3$				

25	N—NH₂ (pyridine)	2-aminopyridine / α-pyridylamine					
		$C_5H_6N_2$	204			
		94.11	56			
		leaf (liq), e.s. acet., s. eth., e.s. liq.				

29	N / NH₂ (quinoline)	4-aminoquinoline / γ-quinolylamine				
		$C_9H_8N_2$			
		144.17	154 without w. MP 69…70		
		need (1H₂O) (w) 508 - H₂O e.s.w; acet; chl; liq., CS₂			

26	N—NH₂ (pyridine)	3-aminopyridine / β-pyridylamine				
		$C_5H_6N_2$	252		
		94.11	64		
		leaf (Bz), e.s.w., acet., eth., liq., s. Bz			

30	H N / N / NH₂ (purine)	Adenine / 6-aminopurine				
		$C_5H_5N_5$	s		
		135.15	365		
		need (3H₂O) (w), sl. s.w., s.h. NH₄OH			

27	N / NH₂ (pyridine)	4-aminopyridine / γ-pyridylamine			
		$C_5H_6N_2$			
		94.11	158		
		col. need., (Bz), s.w. acet. Bz. liq			

31	S—NH₂ (thiophene)	Thiophenine / 2-aminothiophene			
		C_4H_5NS			
		99.15	61…2d		
		yel. oil., misc. w. acet., i. eth.			

28	N—NH₂ (quinoline)	2-amino-quinoline / α-quinolyl amine			
		$C_9H_8N_2$			
		144.17	129		
		leaf (w), s. acet., eth., Bz. chl., liq. e.s.w.			

32	S / N—NH₂ (thiazole)	2-aminothiazole / 2-thiazolylamine			
		$C_3H_4N_2S$			
		100.14	90		
		yel. leaf. (acet.) e.s. acet., eth.			

*Pentamethylendiamine

Amino compounds and hydroxyamino compounds

No.	Formula	Name / MF / Mol.wt. / Other properties	MF	BP / FP	D/sp.gr. / n_D
1	NH₂, –CH₃ (benzene ring)	o‑toludine; 2‑amino‑1‑methylbenzene	C_7H_9N	199.84 / β‑16.3	1.004 / 1.57276
		107.15; col. liq., β‑16.3, sl. misc.w; misc. acet; eth.		α‑24.4	
2	NH₂, –CH₃ (benzene ring)	m‑toludine; 3‑amino‑1‑methylbenzene	C_7H_9N	203.3 / −31.5	0.989 / 1.571
		107.15; liq., misc. w. acet., eth.			
3	NH₂, CH₃ (benzene ring)	p‑toludine; 4‑amino‑1‑methylbenzene	C_7H_9N	200.3 / 45	1.046 / 1.55324
		107.15; leaf (w.), e.s. acet., eth., sl. s.w			
4	NH₂, –CH₃, –CH₃	2,3‑xylidine; 3‑amino‑1,2‑dimethylbenzene	$C_8H_{11}N$	223.8 / <−15	0.991 / 1.570
		121.18; misc. w., acet., eth.			
5	NH₂, –CH₃, CH₃	2,4‑xylidine; m‑xylidine; 4‑amino‑1,3‑dimethylbenzene	$C_8H_{11}N$	liq.	1.559
		121.18; misc. w., acet., eth. Bz.			
6	NH₂, –CH₃, H₃C–	2,5‑xylidine; p‑xylidine; 3‑amino‑1,4‑dimethylbenzene	$C_8H_{11}N$	217 / 15.5	0.980
		121.18; oil., sl. misc. acet., misc. eth. w			
7	NH₂, H₃C–, –CH₃	3,6‑xylidine; 2‑amino‑1,3‑dimethylbenzene	$C_8H_{11}N$	216.9 / col. liq.	0.979 / 1.561
		121.18; immisc. w., misc. eth.			
8	NH₂, –CH₃, CH₃	3,4‑xylidine; 5‑amino‑1,2‑dimethylbenzene	$C_8H_{11}N$	226 / 49	1.076
		121.18; monocl. pl. (liq.), e.s.w., liq.			
9	NH₃, H₃C–, –CH₃	3,5‑xylidine; 5‑amino‑1,3‑dimethylbenzene	$C_8H_{11}N$	221 / liq.	0.972 / 1.558
		121.18; mis. w.			
10	NH₂, –NH₂, CH₃ (benzene ring)	m‑toluylenediamine; 2,4‑diamino‑1‑methylbenzene	$C_7H_{10}N_2$	105	
		122.17; pri (w), s.w. acet.			
11	H₂CNH₂ (benzene ring)	Benzylamine	C_7H_9N	185	0.9826 / 1.5401
		107.15; col. liq. misc. w., acet. eth.			
12	H₂C–CH₂–NH₂ (benzene ring)	β‑phenyl ethylamine; 2‑phenyl ethylamine	$C_8H_{11}N$	195…8 / liq.	0.958 / 1.575
		121.18; misc. w., acet., eth.			
13	H₂N– –NH₂ (biphenyl)	Benzidine; 4,4‑diaminobiphenyl	$C_{12}H_{14}N_2$	401.7 / 116…129	1.251
		184.23; wh. pa. redsh. Cr (w), s.w. acet. eth			
14	CH₃, CH₃, H₂N–, –NH₂ (biphenyl)	o‑toludine, 4,4‑diamino‑3,3‑dimethylbiphenyl	$C_{14}H_{16}N_2$	126.5…29	
		212.28; col. leaf., e.s.acet., eth., Bz., ac.a			
15	OCH₃, OCH₃, H₂N–, –NH₂ (biphenyl)	o‑Dianisidine, 4,4‑diamino‑3,3‑dimethoxy biphenyl	$C_{14}H_{16}O_2N_2$	13	
		244.28; col. leaf, e.s.acet., eth., Bz., ac.a			
16	H₂N––CH ‖ H₂N––CH	4,4'‑diamino stilbene	$C_{14}H_{14}N_2$	s / 227…8	210.17
		yel. leaf. (acet.), s.h.w., acet., me. al. eth.			
17	H₃C>N–, H₃C> –NH₂	N,N‑dimethyl‑p‑phenylene di amine	$C_8H_{12}N_2$	262 / 53	1.036
		136.19; col. need., s. chl., w.			
18	–CH₂–CH₂–NH₂ (imidazole, HC–HN)	Histamine; 5 (β‑aminoethyl)‑imidazole	$C_5H_9N_3$	209v / 83…4	111.15
		wh., pl. (acet.), s.w., acet.			
19	–N=N– –NH₂	p‑aminobenzene	$C_{12}H_{11}N_2$	>360 / 126…7	197.23
		yel. need. (acet.), s.h. acet., s. eth., Bz. chl.			

No.	Formula	Name	MF	Mol.wt.	FP	BP	D/spgr.	n_D and other properties
20	H₃C–NHOH	Methyl hydroxylamine	CH_5ON	47.06	42	62.5v	1.003	Hygr. powd; e.s.w; eth; acet.
21	H₃C–O–NH₂	Methoxyamine	CH_5ON	47.06		49…50		cr:
22	H₃C–H₂C–NHOH	Ethyl hydroxyamine	C_2H_7ON	61.08	59d	—	0.908	col. need, e.s.w; acet; eth.
23	H₃C–H₂C–O–NH₂	Ethoxyamine	C_2H_7ON	61.08	liq.	68	0.883	col. misc. w; acet; eth.

No.	Formula	Name	MF / Mol.wt. / FP / other properties
24	–NHOH (benzene ring)	Phenylhydroxylamine; Hydroxyaminobenzene	C_6H_7ON; 109.12; 82d; col. need (w)., e.s.w., acet., eth. chl.
25	H₃C– –NHOH (benzene ring)	p‑tolylphenyl hydroxylamine 1‑hydroxylamino‑4‑methylbenzene	C_7H_9ON; 123.15; 94; col. leaf (Bz), sl.s.w., e.s.acet. eth. Bz.

Hydrazino-, Diazo, Imino-, and Hydroxylimino-compounds

No.	Formula	Name			MF	Mol. wt.	MP	BP n_D and other properties	
1	$H_3C-NH-NH_2$	Methyl hydrazine			CH_6N_2	46.07	col. liq.	87.5 pa. gr; misc. w; acet; eth.	**-NH-NH₂**
2	$H_3C-H_2C-NH-NH_2$	Ethyl hydrazine			$C_2H_8N_2$	60.10	col. liq.	101.5 misc. w; acet; eth.	

3		Phenylhydrazine Hydrazinobenzene		No.	Formula		Name		
	⬡-NH-NH₂	$C_6H_8N_2$	243.5d	1.0978			MF	BP	D
		108.14	19.6				Mol. wt.	MP	n_D
		pl. or yelsh. oil., immisc acet., eth., Bz., chl.					Other properties		

4	$H_2^{(-)}C-N^{(+)}\equiv N \approx H_2C=N^{(+)}=N^{(-)}$ Diazomethane	CH_2N_2	42.04	−145	−23 yelsh. gas; s. acet; eth	**−N₂**
5	$[⬡-N^{(+)}=N]+OH^- \rightleftharpoons ⬡-N=N-OH$	Diazonium hydroxide			chl. at 200° **exp**	
		free, not stable only in solv. (eth.) stable				

6		p-quinonediimine	7			Indamine., p-phenyl-quinonedimine	**-NH**
	HN=⬡=NH	$C_6H_6N_2$	106.13		⬡-N=⬡=NH	$C_{12}H_{10}N_2$	182.22

8	C=NOH Fulminic acid (oxime of carbon monoxide)		CHON	43.03	In ethereal solv. stable; strong smell; **ex**	**-NOH**
9	$H_2C=NOH$ Formal dehydeoxime; Formal doxime		CH_3ON	45.04	liq.	84	col; misc. w; d. in h. w.	
10	$CH_3HC=NOH$ Acetal dehydeoxime, Ethanal oxime		C_2H_5ON	59.07	47.13	114...5 D 0.9656 col. need; or liq. mis. w., acet., eth.	
11	$\begin{array}{c}H_3C\\H_3C\end{array}>C=NOH$ Acetoxime; 2-propanoxime; Acetoneoxime		C_3H_7ON	73.09	61	136.3 D 0.97 n_D 1.4156 col. cr., e.s.w., acet., eth., s. liq.	
12	$H_3C-C=NOH$ $H_3C-C=NOH$ Dimethyl glyoxime Diacetyldioxime		$C_4H_8O_2N_2$	116.12	234.5 s	col. cr; e.s. acet; eth. forms complexes e.g. Ni Nickel complexes see page. 221	

13		Benzal dehyde hydrazone Benzol hydrazine	$C_7H_8N_2$	120.15	16	140v	**=N-NH₂**
	⬡-CH=N-NH₂				col. leaf. or liq; d. in w; al; eth; chl. Bz. (see also phyenyl hydrazone)		

Nitroso- and Nitro compounds

14		Nitrosobenzene	15		Nitrosodimethylaniline	**-NO**
	⬡-NO	C_6H_5ON	59v	$\begin{array}{c}H_3C\\H_3C\end{array}>N-⬡-NO$	N,N-dimethyl-p-nitrosoaniline	
		107.11	68		$C_8H_{10}ON_2$ —	
		col. rhomb. Cr., e.s. acet., eth., Chl.			150.18 85−87	
					grn. leaf. (acet). s. acet., eth.	

16	H_3C-NO_2	Nitromethane	CH_3O_2N	61.04 MP−28.6, BP 101.2, D 1.3818, oil. misc. w; acet; eth	**-NO₂**
17	$H_2C(NO_2)_2$	Dinitro methane	$CH_2O_4N_2$	106.04 MP ~ −15, BP 100 **ex.** yelsh. unst. oil; misc. w; acet; eth	
18	$HC(NO_2)_3$	Trinitro methane, Nitroform	CHO_6N_3	151.04 MP 15, BP ~ 45 **exp**, D1.6, col. oil. or Cr; misc. w.	
19	$C(NO_2)_4$	Tetra nitromethane	CO_8N_4	196.04 MP13, BP125.7, D1.650 and 1.44, col; misc. acet; eth	
20	$H_2C=CH-NO_2$	Nitro ethylene	$C_2H_3O_2N$		

21		Nitrobenzene	25		o-nitrotoluene
	⬡-NO₂				1-methyl-2-nitrobenzene
		$C_6H_5O_2N$ 210.8 1.2032		⬡(CH₃)(NO₂)	$C_7H_7O_2N$ 222.3
		123.11 5.7 1.55291			137.13 α−10.6
		yelsh. liq., sl. misc. w., misc. acet., eth., Bz.			yelsh. liq., β−4.1, misc. acet. eth., chl., Bz; pet eth

22		m-dinitrobenzene 1,3-dinitrobenzene	26		2,4-dinitrototulene 1-methyl-2,4-dinitrobenzene
	⬡(NO₂)(NO₂)	$C_6H_4O_4N_2$ 300...2 1.575		⬡(CH₃)(NO₂)(NO₂)	$C_7H_6O_4N_2$ 300d 1.521
		168.11 89.8 yelsh. pl. or need.			182.13 70.1 1.442
		(acet.), s.Bz., tol., acet., eth.			need. (CS₂, acet.), s. acet., Bz, chl., Bz, CS₂, eth

23		1,3,5-trinitrobenzene Sym-trinitrobenzene	27		2,4,6-trinitrotoluene (TNT)
	O₂N-⬡(NO₂)(NO₂)	$C_6H_3O_6N_3$ d 1.688		O₂N-⬡(CH₃)(NO₂)(NO₂)	$C_7H_5O_6N_3$ 240**ex** 1.654
		213.11 61.1 yelsh. rhomb. pl.			227.13 80.7
		(Bz), e.s. Bz., s. eth. acet.			col. monocl. cr., s. acet., eth.

24		1-nitronaphthalene d-nitronaphthalene	28		2-nitronaphthaline β-nitronaphthalene
	⬢⬢-NO₂	$C_{10}H_7O_2N$ 304 1.2226liq.		⬢⬢-NO₂	$C_{10}H_7O_2N$ 165v
		173.16 58.8			173.16 79
		yelsh. need. (acet.), s. acet., eth., chl. CS₂			col. rhomb. need. (acet.), e.s. acet., eth

Cyanato- and Isocyanato compounds-Azido compounds

No.	Formula	Name	MF	MW	MP	BP	D/sp. gr.	nD and other properties
1	H3C-O-C≡N	Methyl cyanate	C2H3ON	71.08	liq.	162 d	0.89	misc. w; misc. eth. acet.
2	H3C-N=C=O	Methyl isocyanate	C2H3ON	71.08	liq.	60	0.898	n1.3794 misc. acet., eth.

3. Phenylisocyanate — (benzene ring)–N=C=O

MF	BP	D	MW	MP	nD
C7H5ON	165.6	1.095	119.12	liq.	1.5368

Other properties: e.s. eth.; misc.

4. Phenylazide / Triazobenzene — (benzene ring)–N3

MF	BP	D	MW	MP	nD
C6H5N3	59v	1.078	119.12	liq.	1.56421

yelsh. oil. immisc. w., misc. eth., acet. **ex**

5. Benzylazide / α-triazotoluene — (benzene ring)–CH(H)–N3

MF	BP	D	MW	MP	nD
C7H7N3	108v	1.0655	133.75	liq.	1.53414

oil., immisc. w., misc. acet., eth., **ex**

No.	Formula	Name	MF	MW	MP	BP	D/sp. gr.	nD and other properties
6	H-C≡N; H-CN	Hydrocyanic acid	CHN	27.03	-13.3	25.7	0.6876	nD 1.2675
7	H4C-C≡N	Acetonitrile, methylcyanide	C2H3N	41.05	-41...4	82	0.7828	col. misc. w; acet; eth.
8	H3C-N≡C	Methyl isocyanide	C2H3N	41.05	-45	59.6	0.7464	col. misc. w., acet; eth.
9	H3C-H2C-C≡N	Propionitrite, ethylcyanide	C3H5N	55.08	-91.9	97.1	0.783	col. misc. w; acet; eth.
10	H3C-H2C-N=C	Ethyl isocyanide	C3H5N	55.08	<-66	79	0.7402	col. misc. w; acet; eth.
11	NC-CH2-CN	Malanoic acid dinitrite, methylene dicyanide	C3H2N2	66.06	32.1	220	1.049	col. misc. w; acet; eth.
12	NC-CH2-CH2-CN	Glutaric acid dinitrite, ethylene dicyanide	C4H4N2	94.11	-29	287.4	0.995	col. sl.s.w; misc; acet; eth
13	NC-(CH2)4-CN	Adiponitrile	C6H8N2	108.14	2.48	295	0.951	col. misc. w; acet; eth.
14	H2C=CH-C≡N	Acrylonitrile, vinylcyanide	C3H3N	58.06	-82	78...9	0.797	col. misc. w; acet; eth.
15	H2C=CH-CH2-CN	Allylcyanide, But.-3-ene-nitrite	C4H5N	67.09	liq.	116...9	0.8316	col. misc. w; acet; eth.
16	H2C=CH-CH2-N=C	Allyl isocyanide	C4H5N	67.09	liq.	106	0.794	col. misc. w; acet; eth.

17. Benzonitrite / Phenyl cyanide — (benzene ring)–C≡N

MF	BP	D	MW	MP	nD
C7H5N	190.7	1.0102	103.12	-13	1.52892

col., sl. s.w., misc. acet., eth.

18. Benzylcyanide — (benzene ring)–CH2–C≡N

MF	BP	D	MW	MP	nD
C8H7N	234(107v)	1.015	117.14	-23.8	1.52105

col., immisc. w., misc. acet., eth.

Mixed functions with oxygen and nitrogen

No.	Formula	Name	MF	MW	MP	BP	D/sp. gr.	nD and other properties
19	H2NCH2-CH2OH	Amino alcohol, 2-Amino ethanol	C2H7ON	61.08	10.5	172.2	1.0880	nD 1.454, misc. w, acet; Bz

20. p-aminophenol "Rodinal" — HO–(benzene ring)–NH2

MF	MP	MW	
C6H7ON	s		
109.12	184d		

wh. leaf., sl. s.w., s. eth.

21. HO·CH·CH2NH2 (with OH) — Arterenol, Aminomethyl [3,4-dihydroxyphenyl carbinol]

MF	
C8H11O3N	
169.18	191d

wh. cr. powd., e.s. alk.

22. Cytosine / 1-hydroxy-4-aminopyridine — (ring with NH2 and OH, N)

MF	
C5H6ON2	
111.10	320...5

pl. (1H2O). (w). s.w

23. Guanine / 2-amino-6-hydroxypurine

MF	
C5H5ON5	
151.13	>360d

col. need. (me). s. me. al. KOH

No.	Formula	Name	MF	MW	MP	BP	D/sp. gr.	nD and other properties
24	H-C(=O)-NH2, H-CONH2	Formamide	CH3ON	45.04	2.55	210.7	1.13339	nD 1.44530, col. liq; misc. w; acet.
25	CH3-CONH2	Acetamide	C2H5ON	59.07	81	222	1.159	e.s.w; acet; eth; chl.
26	H2NOC-CONH2	Oxamide	C2H4O2N2	88.07	418d	—	1.667	wh. monocl; sl. s.w
27	O=C(NH2)(NH2)	Urea, carbamide	CH4ON2	60.06	132.7	d	1.335	nD 1.484; 1.602, d (w; acet.), e.s.w; s. acet; s. conc. HCl
28	H2N-OC-NH-C(=O)NH2	Biuret	C2H5O2N3	103.08	193 d			col. need. (1H2O) (w; acet), e.s.w; acet.

29. Benzamide — (benzene ring)–C(=O)–NH2

MF	BP	D	MW	MP
C7H7ON	290	1.341	121.13	130

col. monocl., sl.s.w., s. acet., e.s. eth

30. Nicotinamide — (pyridine ring)–C(=O)–NH2

MF		
C6H6ON2		
122.12	129...31	

wh. cr. powd., col. need (Bz). e.s.w., Bz., glyc. acet.

31. o-aminobenzaldehyde / Anthranilaldehyde — O=C-H on (benzene ring) with NH2

MF	
C7H7ON	d
121.13	39...40

silvery leaf. e.s.w. acet. eth. s. Bz. chl.

32. p-aminobenzaldehyde — O=C=H on (benzene ring)–NH2

MF	
C7H7ON	
121.13	71

pl. or leaf (w.). e.s.w., acet., eth.

Mixed functions with oxygen and nitrogen

No.	Formula	Name	MF	MW	MP/FP	BP/CP	D/sp. gr.	nD and other properties
1	$HN=C{<}^{NH_2}_{NH_2}$	Guanidine	CH_5N_3	59.07	~50			col. glit. cr;
			strong base absorbs CO_2 from air, e.s.w; acet.					

No.	Formula	Name	2	Keratinine			
		MF	BP	D	$HN=C$ ring $N-CH_3$, CH_2, $C=O$, H	$C_4H_7ON_3$	d
		Mol. wt.	MP	nD		113.12	260d
		Other properties				col. rhomb. pri. or leaf., (+2H_2O) (w) s.w., sl., s. acet.	

3	$H_3C{-}N{-}C{<}^{O}$, $ON{-}{}_{O-CH_2\cdot CH_3}$	Nitrosomethylurethane	$C_4H_8O_3N_2$	132.12	<-20	59…61V	1.1224	n 1.43632
			yelsh. red. liq; misc; acet; eth; Bz; (skin, eyes, lung)					☠

4	Picric acid 2,4,6-trinitrophenol	5	Styphnic acid 2,4,6-trinitroresorcinol
$C_6H_3O_7N_3$ — >300ex — 1.767; 229.11 — 121.8; yelsh. leaf. (w). s.w., acet., eth., Bz		$C_6H_4O_8N_3$ — s — 1.829; 245.11 — 180; yelsh. hex. pri. (ac. a), sl. s.w., s. acet., e.s.w. eth.	

6	o-nitrobenzaldehyde	7	α-nitroanthraquinone
$C_7H_5O_3N$ — 156v; 151.12 — α40 β37.9; pa. yel. need (w)., sl. s.w., e.s. acet; eth; s. Bz.		$C_{14}H_7O_4N$ — 270v(s); 253.20 — 230; yelsh. need., (ac. a), s. ac. a.	

8	o-nitraniline / o-nitroaniline	9	p-nitraniline / p-nitroaniline
$C_6H_6O_2N_2$ — 284.11 — 1.442; 138.12 — 71.5; yel. or. leaf. or need. (w)., sl. s.w., e.s. acet., eth. chl.		$C_6H_6O_2N_2$ — 331.73 — 1.424; 138.12 — 147.5; yel. monoci. need. (acet.), sl. s.w., e.s. eth., acet.	

10	Cyanic acid	CHON	43.03	-86	gas(d)	1.140vap	col; s.ac.a; eth

11	Mandelonitrite values for DL form	12	Mandelonitrite values for D form
C_8H_7ON — 170d — 1.124; 133.14 — 10(22); yel. pr. or oil., e.s. acet., eth.		C_8H_7ON — 170d; 133.14 — 28…9	

13	$N≡C-NH_2 \leftrightarrow HN=C=NH$ Cyanamide CH_2N_2 42.04 43…44 140(d) 1.072v col. need; e.s.w; acet., eth., chl., Bz

Amino acids

No.	Formula	Name	MF	MW	F	[αD]°	Pka	and other properties
14	$H{-}C(COOH)(NH_2){-}H$	Glycine, Aminoacetic acid	$C_2H_5O_2N$	75.07	232…6		6.20	wh; monocl; Cr; s.w. 25.3 (25°), 57.5 (75°) acet. 0.043 (25°); py 0.61 (21°)
15	$H_3C\cdot C(COOH)(NH_2){-}H$	L-α-Alanine; L-2-Amino propionic acid	$C_3H_7O_2N$	89.09	295d	+33.0°	6.11	col. need. (w); s.w. 16.6 (25°), 32.2 (75°), acet. 0.57 (75°)
16	$H_2N{-}C(H)(COOH)(H)$	β-Alanine 3-Amino propionic acid	$C_3H_7O_2N$	89.09	196d	—	—	rhomb. pri. (acet.), e.s.w; optically inactive, does not occur in nature
17	$(H_3C)_2 H{-}C{-}C(COOH)(NH_2)$	L-Valine; α-Amino valeric acid	$C_5H_{11}O_2N$	117.15	292…8d	+62.0°	6.00	bl; s.w; 7.44(25°), 13.31 (75°) aq. acet. 0.014-0.71
18	$(H_3C)_2 H{-}C{-}CH_2{-}C(COOH)(NH_2)$	L-Leucine; 2-Amino-4-Methyl-n-valeric acid	$C_6H_{13}O_2N$	131.17	>294d	+22.5°	2.2	col. hex. leaf (w.acet); n1.535, 1.525, 1.535, 1.560, s.w. 2.43 (25°), 3.82 (75°) ac. a. 10.9
19	$H_3C{-}H_2C$, H_3C $H{-}C{-}C(COOH)(NH_2)$	L-Isoleucine; 2-Amino-3-Methyl-n-valeric acid	$C_6H_{13}O_2N$	131.17	292d	+49.0°	4.1	rhomb. monocl. leaf; s.w. 4.1 (25°), s.h. acet. h. ac. a

Mixed functions with oxygen and nitrogen (amino acids)

No.	Formula	Name				
		MF	MW	MP	[α]D	P k
		s. in 100g.w and other properties				
1	H₂N-CH₂-CH₂-CH₂-$\overset{COOH}{\underset{NH_2}{C}}$-H	L-Ornithine; L-2,5-Diamino-n-valeric acid				
		C₅H₁₂O₂N₂	132.16			
		syrup; e.s.w; acet.				
2	H₂N-CH₂-CH₂-CH₂-CH₂-$\overset{COOH}{\underset{NH_2}{C}}$-H	L-Lysine; L-2,6-Diamino hexanoic acid				
		C₆H₁₄O₂N₂	146.19	224d	+25.9*	9.74
		flat. need. (w); hex. pl. (acet.), e.s.w.* in 5 NHCl.				
3	HOOC=$\overset{COOH}{\underset{NH_2}{C}}$H₂-C-H	L-Aspartic acid; L-2-Amino butanedicarboxylic acid				
		C₄H₇O₄N	133.10	278…80d	+25.4*	2.98
		col. monocl. pri; s.w; 0.82 (25°), 4.79 (75°)* in 5 NHCl				
4	HOOC-CH₂-CH₂-$\overset{COOH}{\underset{NH_2}{C}}$-H	L-Glutanic acid; L-2-Aminopentanedicarboxylic acid				
		C₅H₉O₄N	~200 d		+31.8*	3.08
		tetra. pl; s.w 0.89 (25°); v.sl. s.acet; me. al;* in 5 NHCl.				
5	$\overset{O}{\underset{H_2N}{C}}$-CH₂-$\overset{COOH}{\underset{NH_2}{C}}$-H	L-Aspargine; L-2-Aminobutane dicarboxyl-4-amide				
		C₄H₈O₃N₂	132.12	226…35 d	+34.3*	4.38
		col. rhomb. cr. s.w. 2.46 (25°), 86.6 (100°) s. NH₄OH,* 3–4 NHCl.				
6	$\overset{O}{\underset{H_2N}{C}}$-CH₂-CH₂-$\overset{COOH}{\underset{NH_2}{C}}$-H	L-Glutamine				
		C₅H₁₀O₃N₂	146.15	186	+6.1*	4.4
		need. (w); s.w: 4.2 (25°) in w.				
7	HOCH₂-$\overset{COOH}{\underset{NH_2}{C}}$-H	L-Serine; 2-Amino-3-hydroxy propionic acid				
		C₃H₇O₃N	105.09	246 d	+15.1*	5.0*
		leaf (w); s.w: 5.02 (25°); 19.21 (75°)* 5 NHCl.				
8	H₃C-HOCH-$\overset{COOH}{\underset{NH_2}{C}}$-H	L-Threonine; 2-Amino-3-hydroxybutyric acid				
		C₄H₉O₃N	119.12	229…30 d	−30	5.59
		rhomb. leaf (½H₂O), s.w. 20.5 (25°)				

No.	Formula	Name			9	O=$\overset{}{C}$-OH, —NH₂	Anthranilic acid / o-Aminobenzoic acid
		MF	BP	D			C₇H₇O₂N	s	1.412
		MW	MP	nD			137.13	145	
		Other properties					col. leaf. (acet.), sl. s.w., s. acet., eth.
10	O=C-OH, NH₂	p-aminobenzoic acid			11	O=C-OH, OH, NH₂	p-aminosalicyclic acid / 4-amino-2-hydroxybenzoic acid
		C₇H₇O₂N					C₇H₇O₃N		
		137.13	187				153.13	220d	
		yelsh. red. need., sl. s.w., s. acet., eth.					redsh. brn. cr. powd., s.w., acet., eth.
12	H₂C-$\overset{COOH}{\underset{NH_2}{C}}$-H	L-phenylalanine			13	H₂C-$\overset{COOH}{\underset{NH_2}{C}}$-H, OH	L-thyrosine / Lp-hydroxyalanine
		C₉H₁₁O₂N	s	pl 5.91			C₉H₁₁O₃N		pl 5.63
		165.19	318…20d	[α]D -7.5°			181.19	316d	[α]D-10.0°
		monocl. leaf. (w. acet), s.w. 3.0 (25°)					leaf. or short need., sl. s.w. 0.05 in 5nHC
14	H₂C-$\overset{COOH}{\underset{NH_2}{C}}$-H, N, NH	L-histidine / L-α-amino-5-imidazolepropionic acid			15	H₂C-$\overset{COOH}{\underset{NH_2}{C}}$-H, NH	t-Tryptophane
		C₆H₉O₂N₂	209d	pl 7.64			C₁₁H₁₂O₂N₂		
		155.16	277d	[α]D +7.5			204.22	289d	
		leaf. (acet. w.). s.w. 4.19					col. hex. pl. (acet). sl. s.w., s.h.w., acet.

16	O=C$\overset{NH-NH_2}{\underset{NH_2}{}}$	Semi carbazide	CH₅ON₃	75.07	96
					pri. (acet.), e.s.w; acet.
17	H₂N$\overset{NO}{\underset{O}{C-N}}$CH₃	N-Nitroso-N-methylurea	C₂H₅O₅N₃	103.09	121 d
					col. cr. (eth.), s.h.w; e.s. acet; eth; s. Bz. chl.
18	HOOC-H₂C$\overset{NH_2}{\underset{H_3C}{N-C}}$NH	Keratine	C₄H₉O₂N₃	131.13	295 (ww) at 100° −H₂O
					monocl. pri. (1H₂O) (w), sl. s.w; acet.
19	HOOC$\overset{}{\underset{H_2N}{H-C-CH_2-CH_2-CH_2-NH-C}}$$\overset{NH_2}{\underset{NH}{}}$	L-Arginine	C₆H₁₄O₂N₄	174.20	238 d [α]D +29.4 pk 10.76
					Tf (acet); pri. (w), s.w; 15 (21°)
20	$\overset{}{\underset{N}{}}$O	Diphenylnitrogen oxide	C₁₂H₁₀ON	128.21	64d
					redsh. cr. from me, al.
21	N=N$\overset{O}{}$	Azoxybenzene	C₁₂H₁₀ON₂	198.22	d 1.246
					36 1.6644
					yel. need., i.w., e.s. acet., eth., liq.

Fluoro-, chloro- and bromo - compounds

F

No.	Formula	Name	MF	MW	MP/FP	BP/CP	D/sp.gr.	nD	Other properties
1	H_3CF	Methylfluoride, fluoromethane	CH_3F	34.04	−141.8	−78.1			+44.9; 62 s.w; acet.
2	H_2CF_2	Methylenefluoride, difluoromethane	CH_2F_2	52.03		−52			i.w; s.acet.
3	HCF_3	Fluoroform, methane trifluoride	CHF_3	70.02	−160	−84.2			+25;50; s.acet.
4	CF_4	Carbon tetrafluoride	CF_4	88.01	−183.69	−128.02			−8;37; s.w
5	H_3C-H_2CF	Ethylfluoride, fluoroethane	C_2H_5F	48.06	−143.2	−37.7	2.198		
6	H_3C-HCF_2	1,1 - Difluoroethane, "Freon152"	$C_2H_4F_2$	66.05	−117	−24.7			
7	H_3C-CF_3	1,1,1 - Trifluoroethane, "Freon143"	$C_2H_3F_3$	84.04	−117	−46.8	3.784		
8	$H_2C=HCF$	Vinylfluoride, fluoroethylene	C_2H_3F	46.04		−51			i.w; s.acet.
9	$F_2C=CF_2$	Tetrafluoroethylene	C_2F_4	100.02	−142.5	−78.4			i.w
10	$H_2C=CH-H_2CF$	Allylfluoride, 3-fluoropropene	C_3H_5F	60.07		−10			sl.s.w; acet; eth.

No.	Formula	Name			11		Phenylfluoride, fluorobenzene		
		MF	BP	DₒDD			C_6H_5F	84.85	1.024
		MW	MP	nD	−F		96.10	−49.9	1.4646
		Other properties					col., misc. acet., eth, sl. misc. w		

−Cl

No.	Formula	Name	MF	MW	MP/FP	BP/CP	D/sp.gr.	nD	Other properties
12	H_3CCl	Methylchloride; chloromethane	CH_3Cl	50.49	−97.72	−24.22			+143; 68 s.w
13	H_2CCl_2	Methylene chloride; dichloromethane	CH_2Cl_2	84.94	−96.7	40.1	1.3255	1.4237	
14	$HCCl_3$	Chloroform, trichloromethane	$CHCl_3$	119.39	−63.5	61.26	1.4890	1.44643	
15	CCl_4	Carbontetrachloride	CCl_4	153.84	−22.8	76.8	1.595	1.46305	
16	H_3C-H_2CCl	Ethylchloride	C_2H_5Cl	64.52	−138.7	12.2	0.9214		sl.s.w; s.acet; eth.
17	$Cl_2CH-HCCl_2$	1,1,2,2-Tetrachloroethane	$C_2H_2Cl_4$	167.86	−43.8	146.2	1.600	1.4942	misc.acet; eth.
18	$H_2C=HCCl$	Vinylchloride; chloroethylene	C_2H_3Cl	62.50	−153.6	−13.9	1.9692		e.s. acet; eth.
19	Cis-1,2-Dichloroethylene / trans-1,2-Dichloroethylene		$C_2H_2Cl_2$ / $C_2H_2Cl_2$	96.95 / 96.95	−80.5 / −50	60.25 / 48.4	1.2818 / 1.265	1.4519 / 1.4490	misc.acet; eth; misc. acet; eth.
20	$ClCH=CCl_2$	Trichloroethylene "Tri"	C_2HCl_3	131.40	−73	87.2	1.4660	1.4777	misc. acet. eth.
21	$H_2C=CH-H_2CCl$	Allylchloride	C_3H_5Cl	76.53	−136.4	44.6	0.938	1.41538	misc. acet.

22	Chlorobenzene, phenylchloride			25	Benzylchloride; α-chlorotoluene		
	C_6H_5Cl 132	1.1066	col. liq.,		C_7H_7Cl 179	1.1026	col. liq.,
	112.56 −45	1.52479	misc.acet., eth.		126.58 −43	1.5415	misc. acet., eth

23	o-dichlorobenzene, 1,2-dichlorobenzene			26	Benzalchloride, Benzylidenechloride		
	$C_6H_4Cl_2$ 180...3	1.3048	col. liq.		$C_7H_6Cl_2$ 207	1.2557	col. oil.,
	147.01 −17.5	1.5518	misc. acet. eth.		161.03 −16	1.5502	misc. acet., eth

24	p-dichlorobenzene, 1,4-dichlorobenzene			27	Benzotrichloride, n-trichlorobenzene		
	$C_6H_4Cl_2$ 173.4	1.4581	monocl		$C_6H_5Cl_3$ 214	1.38	col. oil.,
	147.01 53	1.52104	e.s. acet., eth., Bz.		195.48 −22		misc. acet., eth., Bz.

28	γ - Hexachlorocyclohexane "HCH", gammaxane		30	Triphenylchloromethane	
	$C_6H_6Cl_6$			$C_{19}H_{15}Cl$	310 (~230v)
	290.85	112...3		278.77	112
	need., (acet)			col. need. d. in w., e.s. eth., chl., CS_2 Bz.	

29	Bornylchloride		31	Dichlorodiphenyltrichloromethane "D.D.T"	
	$C_{10}H_{17}Cl$	BP 207.4		$C_{14}H_9Cl_5$	d
	172.69	MP 131...2		354.50	107...9
	col. cr. s. acet. (26.04), eth.			col. need. (acet.), s. acet., eth. Bz.	

32	Epichlorohydrin	
	C_3H_5OCl 117	1.1801 col. liq., misc
	92.53 −25.6	1.44195 acet., eth.

−Br

No.	Formula	Name	MF	MW	MP/FP	BP/CP	D/sp.gr.	nD	Other properties
33	H_3CBr	Methylbromide, bromethane	CH_3Br	94.95	−93.7	3.56	1.732	liq.	+194'; 52v.sl.s.w.
34	H_2CBr_2	Methylenebromide, dibromomethane	CH_2Br_2	173.86	−52.8	98.2	2.4953		col; misc.acet; eth;
35	$HCBr_3$	Bromoform, tribromomethane	$CHBr_3$	252.77	8.3	149.5	2.8904	1.5980	col. pl.; misc; acet; eth. Bz.
36	CBr_4	Carbontetrabromide, tetrobromomethane	CBr_4	331.67	94° 48.4°	1895(d)	2.9609	liq.1.6	col. monocl; tab; acet. eth. ch.
37	H_3C-H_2CBr	Ethylbromide, bromoethane	C_2H_5Br	108.98	−119	38.0	1.430	1.42386	col. misc. acet. eth.
38	$BrCH_2-H_2CBr$	1,2 - Dibromoethane	$C_2H_4Br_2$	187.88	9.97	131.6	2.1701	1.53789	col. misc. acet. eth.
39	$H_2C=HCBr$	Vinylbromide, bromoethene	C_2H_3Br	106.96	−137.8	15.8	1.5167	1.4462	col. misc. acet. eth.
40	$BrCH=HCBr$	1,2 - Dibromoethylene (Cis.)	$C_2H_2Br_2$	185.87	−53	110	2.271	1.5428	misc. acet; eth.
41	$HCBr=HCBr$	1,2 - Dibromoethylene (trans.)	$C_2H_2Br_2$	185.87	−6.5	108			
42	$H_2C=CH-H_2CBr$	Allylbromide	C_3H_5Br	120.99	−119.4	71.3	1.398	1.46545	misc. acet; eth; Bz.

43	Bromobenzene, phenylbromide			44	p - Dibromobenzene, 1,4 - dibromobenzene		
Br	C_6H_5Br 155...6	1.4991		Br—⬡—Br	$C_6H_4Br_2$ 218...19	2.261	
	157.02 −30.6	1.55977			235.92 86.9	1.57425	
	col. oily liq., s. acet., eth., Bz.				col. monocl. cr. (acet.) s. acet., eth, CS_2, ac. a. lig.		

Iodo-, iodoso- and iodyl- compounds

No.	Formula	Name	MF	MW	MP	BP	D/sp.gr.	nD	Other properties
1	H_3CI	Methyliodide, Iodomethane	CH_3I	141.95	−66.1	42.5	2.2790	1.5293	col. brn. liq. misc. acet., eth.
2	H_2CI_2	Methyleneiodide, Diiodomethane	CH_2I_2	267.87	6.1	180d	3.3212	1.7559	leaf. col. misc. acet., eth.
3	HCI_3	Iodoform, Triiodomethane	CHI_3	393.78	119	210ex	4.008	1.800	yel. hex. fl., s. acet., chl.
4	CI_4	Carbontetraiodide, Tetraiodomethane	CI_4	519.69	171d	s(~100	4.32		red. cub. cr., s. acet., eth.
5	H_3C-H_2CI	Ethyliodide, Iodoethane	C_2H_5I	155.98	−108.5	72.2	1.933	1.5222	col. liq. misc. acet., Bz., eth.
6	$H_2C=HCI$	Vinyliodide	C_2H_3I	153.96	liq.	56	2.08		misc. acet; eth; chl; Bz, CS_2
7	$H_2C=CH-H_2CI$	Allyliodide, 3-Iodopropene	C_3H_5I	167.99	−99.3	103.1	1.848		yel. liq; misc; acet; eth; chl.

No.	Formula	Name			8	Phenyliodide, iodobenzene		
		MF	BP	D		C_6H_5I	188.6	1.832
		Mol. wt.	MP	nD		204.02	−31.4	1.62145
		Other properties				col. liq. misc. acet., eth., Bz., Chl.		
9	Iodosobenzene				10	Iodoxy benzene		
	C_6H_5OI					$C_6H_5O_2I$		
	220.02	ex. 210				236.02	ex.	236...7
	yelsh. powd; s.w. acet., h. eth., h. chl.					need(w); e.s.w; i. acet; s. eth; Bz; h. ac. a.		

Different halogen functions

No.	Formula	Name	MF	MW	MP	BP	D/sp.gr.	nD	Other properties
11	$HFC Cl_2$	Fluorodichloromethane "Frigen 21"	$CHFCl_2$*	102.93	−135	8.9			+ 178.5; 5.1
12	F_2CCl_2	Difluorodichloromethane "Frigen 12"	CF_2Cl_2	120.93	−155	−30.5	1.486 liq.		+ 112, 40, s. acet; eth.
13	F_3CCl	Trifluorochloromethane "Frigen 13"	CF_3Cl	104.47	−181.6	−81.2			+ 28.8, 38.2
14	F_3CI	Trifluoroiodomethane	CF_3I	195.92	—	−22.5			
15	$F_2CCl-ClCF_2$	1,1,2,2-Tetrafluoro-1,2-dichloroethane	$C_2F_4Cl_2$	170.93	−94	3.8	1.5312		col; s. acet; eth; i.w
16	$F_3C-ClCF_2$	Pentafluoro-2-chloroethane	C_2F_5Cl	154.48	−106	−38			col; s. acet; eth; i.w

17	Benzoylfluoride			18	Fluoro-2,4-dinitrobenzene		
	C_7H_5OF	159	>1		$C_6H_3O_4N_2F$	296	1.4718
	124.11				186.10	25.8	
	col. penetrating odour; liq; d. in w; misc. acet; eth.				pa. yel., cr.		

Mixed functions with halogens

No.	Formula	Name	MF	MW	MP	BP	D/sp.gr.	Other properties
19	$ClCH_2-H_2COH$	Glycol chlorohydrin	C_2H_5OCl	80.52	−69	128.8	1.213	col. misc. acet., eth.
20	$Cl_3-C-C\underset{OH}{\overset{H\diagup OH}{}}$	Choral hydrate, 2,2,2-Trichloro-,1,-ethanediol	$C_2H_3O_2Cl_3$	165.42	51.6	96	1.600	col. monocl. pl. s.w. chl. acet. eth., pyr.
21	$Cl_3C-C\underset{H}{\overset{\diagup O}{}}$	Chloral, Trichloro acetaldehyde	C_2HOCl_3	147.40	−57.5	98	1.512	n 1.45572 col. liq.. misc. w., acet., eth. chl.
22	$H_3C-C\underset{Cl}{\overset{\diagup O}{}}$	Acetylchloride	C_2H_3OCl	78.50	−112.0	51...2	1.1052	n 1.38976 pentr. odour, liq. d. in w., misc. eth., acet. chl., Bz. ac. a.
23	$O=C<\overset{Cl}{Cl}$	Phosgene, carbon oxychloride	$COCl_2$**	98.92	−127.79	7.56	1.392gas	t_k 181.7, p_x 56atm, e.s. eth., ac. a., s $\sim$ 270 tol.
24	$O=C<\overset{O-CH_3}{Cl}$	Methylester of chloroformic acid	$C_2H_3O_2Cl$	94.50	liq.	71.6	1.236	misc. w., acet., Bz, chl. col. liq.

25	p-chlorophenol			26	Benzoylchloride		
	C_6H_5OCl	217	1.306		C_7H_5OCl	197	1.2188
	128.56	43	1.5579		140.57	−96	1.55369
	need., (acet.), e.s. acet., eth., Bz., alk				col. penetrating smell; d. in w; misc; eth; Bz; CS_2		

No.	Formula	Name	MF	MW	MP	BP	D/sp.gr.	Other properties
27	$ClCH_2-COOH$	Monochloroacetic acid	$C_2H_3O_2Cl$	94.50	61.2	189	1.58	col. cr. e.s.w. acet, eth; Bz.
28	$Cl_2CH-COOH$	Dichloroacetic acid	$C_2H_2O_2Cl_2$	128.95	5...6	194	1.5634	col. misc. w., acet., eth
29	$Cl_3C-COOH$	Trichloroacetic acid	$C_2HO_2Cl_3$	163.40	57.5	197.5	1.6298	pri. (h. pet), e.s. acet; eth; Bz.
30	Br_3C-H_2COH	Avertin, 2,2,2-Tribromoethanol	$C_2H_3OBr_3$	282.79	80	92		pri. (h. pet), e.s. acet; eth; Bz.
31	Br_3C-CHO	Bromal, 2,2,2-Tribromoacetaldehyde	C_2HOBr_3	208.78	yel. liq.	174d	2.665	misc. acet., eth; d. in w
32	$H_3C-CO-CH_2Br$	Bromoacetone; 1-bromopropanone-2	C_3H_5OBr	136.99	−54	~140	1.634	misc. acet., eth.
33	$H_3C-HCBr-COOH$	DL, 2-Bromopropionic acid	$C_3H_5O_2Br$	125.99	25.7	203.5	1.700	col. pri; e.s.w; acet., eth.
34	$H_2N-\underset{O}{C}-NH-\underset{O}{C}-\overset{HCBr}{\underset{CH}{}}-\overset{CH_3}{}$	Bromural	$C_6H_{11}O_2N_2Br$	BP = s; 223.08	MP = 160			leaf. (tol.); e.s. h.w; acet., eth.

35	α-acetobromoglucose 2,3,4,6-tetracetyl -α-d-glucose-1-bromide		36	α-acetodibromoglucose 2,3,4-triaceto-α-d-glucose-1,6-dibromide	
	$C_{14}H_{19}O_9Br$ MP = 89 need (eth)			$C_{12}H_{16}O_7Br_2$ MP = 176.5 need	
	411.21	e.s. me. ch. Bz. eth.		432.98 (chl + liq.)	e.s. chl, acet
	Ac = O−C−CH_3			Ac = −O−C−CH_3	

37	3,5-diiodotyrosine (D, L)		38	L-Thyroxine	
	$C_9H_9O_3NI_2$	MP = 204 (d)		$C_{15}H_{11}O_4NI_4$	MP 235...6(d)
	433.01	pri. (w).		776.93	
				wh. pa. yel. need; sl. sw; i. acet; eth.	

* see p. 51; ** see p. 61

Thio compounds (mercaptans)-sulphinic and sulphonic acids

Formula	Name	MF	MW	MP	BP	D	nD and other properties	
H₃C-SH	Methanethiol	CH_4S	48.10	-123.1	7.6	0.868	col; d. in w; e.s. acet; eth.	-SH
H₃-H₂CSH	Ethanethiol	C_2H_6S	62.13	-121	34.7	0.840	nD1.43055, sl. misc. w; misc. acet; eth.	
H₃-CH₂-H₂CSH	1-Propanethiol	C_3H_8S	76.15	-111.5	68	0.8357	col. misc.w; acet; eth.	
H₃-H₂CSH-CH₃	2-Propanethiol	C_3H_8S	76.15	-130.7	60	0.8055	col. misc.w; acet; eth.	
H₃-(CH₂)₂-H₂CSH	n-Butanethiol	$C_4H_{10}S$	90.18	-115.9	98	0.858	col. misc.w; acet; eth..	
H₃-(CH₂)₃-H₂CSH	n-Pentanethiol	$C_5H_{12}S$	104.21	-75.7	126	0.857	nD1.44366, col. misc.w; acet. eth.	
H₃-(CH₂)₄-H₂CSH	n-Hexanethiol	$C_6H_{14}S$	118.23	-81.03	148	0.849	col. immsc. w., misc. acet; eth.	

Formula	Name		8		Thiophenol, Benzolethiol		
	MF	BP	D		C_6H_6S	169.5	1.078
	Mol. wt.	MP	nD	⬡-SH	110.17	liq.	1.58613
	Other properties				col., immisc. w., misc. acet., eth.		

	Benzylthiol		10		p-thioeresol, 4-toluethiol		
⬡-CH₂SH	C_7H_8S	194...5	1.058	HS-⬡-CH₃	C_7H_8S	195	
	124.19	liq.			124.19	42...3	
	immisc. w., misc. acet., eth., CS₂				leaf. (eth.), i.w. s.acet., eth.		

=C=S	Carbondisulphide*	CS₂*	76.13	-108.6	46.2	1.2632	nD1.62950, col; misc. acet; eth; Bz ⚗	-SO₂H
C-H₂C-SO₂H	Ethyl sulphinic acid	$C_2H_6O_2S$	94.13	......	......		syrupy. liq; s.alk.	-SO₂H
C-H₂C-SO₃H	Ethyl sulphonic acid	$C_2H_6O_3S$	110.13	......	......		hygr. cr. d. in w; s.acet; alk.	

	Benzenesulphinic acid		15		Benzene sulphonic acid		
⬡-SO₂H	$C_6H_6O_2S$	100d		⬡-SO₃H	$C_6H_6O_3S$	d	col.
	142.17	84			158.17	43 (anhy.)	leaf or
	Pri. (w.), e.s.w., s. acet., eth.				need. (1.5H₂O), e.s.w., acet, eth, Bz		

Different mixed functions with sulphur

C-C=N-S	Methyl thiocyanate	C_2H_3NS	73.11	-51	133	1.068	col; immisc.w; misc. acet; eth.
C-N=C=S	Methyl isothiocyanate	C_2H_3NS	73.11	35	119	1.069	col. cr. e.s.w., acet. eth.
C-H₂C-CNS	Ethyl thiocyanate	C_3H_5NS	87.14	-85.5	144.4	0.996	col; immisc. w; misc. acet; eth.
C-H₂C-NCS	Ethyl isothiocyanate	C_3H_5NS	87.14	-5.9	132	1.004	col.; immisc.w; misc. acet; eth.
C=CH-H₂C-CNS	Allyl thiocyanate	C_4H_5NS	99.15	liq.	161	1.056	col. oil; misc. acet; eth.
C=CH-H₂C-NCS	Allyl isothiocyanate	C_4H_5NS	99.15	-100	150.7	1.0155	col. oil; v.sl. s.w; misc. acet; eth. Bz

	Phenylthiocyanate		23		Phenyl isothiocyanate		
⬡-C≡N-S	C_7H_5NS	232	1.1228	⬡-N=C=S	C_7H_5NS	218.5	1.135
	135.18	liq.			135.18	-21	1.64918
	immisc. w., misc. acet., eth.				col. liq., immisc. w., misc. acet., eth.		

C< NH₂ / NH₂	Thiourea, Thiocarbamide	CH_4N_2S	76.12	182	d	1.405	rhomb. pri. (acet.), s.acet.
C< NH-NH₂ / NH₂	Thiosemicarbazide	CH_5N_3S	91.13	181...3			need. (w), s.w; acet.
COOH / CH₂-C-H / NH₂	L-Cysteine	$C_3H_7O_2NS$	121.15				Cr. powd; e.s.w; s.ac.a; NH₄OH
N-CH₂-H₂C-SO₃H	Taurine-2, Amino-ethane sulphonic acid	$C_2H_7O_3NS$	125.14	328.9d			tetr; acidic, s.w; v.sl. s; acet.
COOH / H₃SCH₂-C-H / NH₂	L-Cysteinic acid, 2-Amino-3-sulpho-propionic acid	$C_3H_7O_5NS$	169.15	d ~250			with 1H₂O, prismat. Na. salt. without water

SO₃H ⬡ NH₂	Metanilic acid 3-amino-benzene sulphonic acid		32	OH ⬡⬡ OH (HO₃S...SO₃H)	R-acid, 2-naphthol-3 6 disulphonic acid		
	$C_6H_7O_3NS$				$C_{10}H_8O_7S_2$		
	173.18	280d			304.28	d	
	Pl. (1½H₂O) or wh. need; sl. s.w; s. acet.				deliq., col., e.s.w., Na. salt.		

SO₃H ⬡ NH₂	Sulphanilic acid 4-amino-benzenesulphonic acid		33	SO₃H ⬡⬡ OH (HO₃S)	G-acid, 2 naphthol-6 8-disulphonic acid		
	$C_6H_7O_3NS$				$C_{10}H_8O_7S_2$		
	173.18	288d			304.28		
	rhomb. pl. (1 H₂O) monocl. 2 H₂O sl. s.w.				s.w., Na. salt, v.e., s.w.		

SO₃H ⬡⬡ NH₂	Naphthionic acid 1-amino-4-naphthalene sulphonic acid		34	OH OH ⬡⬡ (HO₃S...SO₃H)	Chromotoropic acid 4, 5-Dihydroxy-2, 7-naphthalene disulphonic acid		
	$C_{10}H_9O_3NS$				$C_{10}H_8O_8S_2$ Na salt with 2 (H₂O)		
	223.24	d			320.28	need or leaf; e.s.w.	
	need., (1½H₂O), sl. s.w.						

Different mixed functions with sulphur

No.	Name (MF / BP / D ; Mol. wt. / MP / nD ; Other properties)
1	H-acid, 1-naphthol-8-amino-3,6-disulphonic acid — $C_{10}H_9O_7NS_2$; 319.30, d ; col. need., e.s.w., eth. (Formula: NH_2 OH / HO_3SH SO_3H)
2	Flavianic acid, 2,4-Dinitronaphthol-7-sulphonic acid — $C_{10}H_6O_8N_2S$, ~175d ; 314.24, 150.../ ; Cr. (w.), e.s.w., acet. (Formula: O_3S–naphthalene–OH, NO_2, NO_2)
3	Azobenzene-4-sulphonic acid — $C_{12}H_{10}O_3N_2S$, looses water, **ex** ; 184.23, 127 misc. w ; deep or. red. leaf ($3H_2O$), s.w. (Formula: ⬡–N=N–⬡–SO_3)
4	p-aminobenzene sulphonamide, Prontalbin, Prontosilalbum — $C_6H_8O_2N_2S$; 172.20, ~165 ; col. leaf. (acet.), e.s. h.w., s. acet., eth. (Formula: H_2N–⬡–S(=O)(=O)–NH_2)
5	Prontosil; 2,3-Diamino-azo-benzene-4-sulphonamide — $C_{12}H_{13}O_2N_5S$; 291.35 (Formula: H_2N–⬡–N=N–⬡–S(=O)(=O)–NH_2, NH_2)
6	o-sulphanylbenzoic acid, o-sulphamidobenzoic acid — $C_7H_7O_4NS$; 201.19, 165...7 ; rhomb. Cr., (acet.), e.s.w., acet., eth. (Formula: ⬡ with C=O–OH and S(=O)(=O)–NH_2)
7	Saccharin, o-sulphobenzamide — $C_7H_5O_3NS$, s, 0.828 ; 183.18, 224...8 ; col. monocl., sl. s.w., s. acet., Bz. (Formula: benzene fused ring, C=O, NH, SO_2)
8	Sulphathiazole, 2-sulphanilamidothiazole — $C_9H_9O_2N_3S_2$; 255.31, 200...2 ; col. Cr., sl. s.w., acet., s. dil. HCl., alk. (Formula: H_2N–⬡–S(=O)(=O)–NH–thiazole)
9	Pyrimal, sulphapyrimidine, Sulphadiazine — $C_{10}H_{10}O_2N_4S$; 250.27, 250...5 ; v. sl. s.w. (Formula: H_2N–⬡–S(=O)(=O)–NH–pyrimidine)

Functions with phosphorus

No.	Formula	Name	MF	MW	MP	BP	D and other properties
10	H_3C–PH_2	Methyl phosphine	CH_5P	48.03		−14	col. gas; e.s.w; acet; eth; ☠
11	H_3C–H_2C–PH_2	Ethyl phosphine	C_2H_7P	62.06	liq.	25	col. d in w.
12	⬡–PH_2	Phenyl phosphine, phosphaniline	C_6H_7P	110.10	liq.	160	D 1.001
13	H_3C–PO_3H	Methyl phosphonic acid	CH_4O_3P	96.03	105	—	
14	O=C(HO)–⬡–PO(OH)	p-phosphono-benzonic acid	$C_7H_6O_5P$	202.11	>300	—	Need; (w.), s.w; acet; dil. HCl.

Functions with arsenic

No.	Formula	Name	MF	MW	MP	BP	D and other properties
15	H_3C–AsH_2	Methyl arsine	CH_5As	91.96		2	col. gas; vl. sl. s.w; e.s. acet; eth.
16	H_3C–H_2C–AsH_2	Ethyl arsine	C_2H_7As	105.99	liq.	36	D 1.217, col. v.sl.s.w; s. acet; eth
17	H_3C–AsO_3H	Methyl arsanoic acid	CH_4O_3As	139.96	161		monocl. leaf. (acet.), s.w; acet.
18	H_2N–⬡–AsOH(O)	Arsanilic acid "Atoxyl", p-Amino phonyarsonic acid	$C_6H_8O_3As$	217.04	232		wh.,need; s. acet.

Esters of inorganic acids

No.	Formula	Name	MF	MW	MP	BP	D and other properties
19	H_3C–O–SO_3H	Methyl sulphate	CH_4O_4S	112.10	<−30	d	Hygr. oil; misc. w; acet. eth.
20	H_3C–H_2C–O–SO_3H	Ethyl sulphate	$C_2H_6O_4S$	126.13		d280	Hygr. syr. liq.; misc. w; acet; eth.
21	H_3C–O, H_3C–O SO_2	Dimethyl sulphate	$C_2H_6O_4S$	126.13	−31.8	188.5	1.3322 and 1.3874 col. liq; misc. w; eth.
22	H_3C–H_2C–O, H_3C–H_2C–O SO_2	Diethyl sulphate	$C_4H_{10}O_4S$	154.18	−24.5	208(d)	1.842 and 1.401, col. oily liq. liq; d. in w; acet; s. al; ∝ s. eth ☠
23	$(H_3CO)_2$ SO	Dimethyl sulphite	$C_2H_6O_3S$	110.13	liq.	126	1.242, col. d. in w; misc. acet; eth
24	$(H_3C$–$H_2CO)_2$SO	Diethyl sulphite	$C_4H_{10}O_3S$	138.18	liq.	158	1.077, col. d. in w; misc. acet; eth

No.	Formula	Name	MF	MW	MP	D and other properties
25	H_2C–O·NO_2 / H C–O·NO_2 / H_2C–O·NO_2	Glycerol trinitrite, Nitroglycerine	$C_3H_5O_9N_3$	160v ; 227.09	13.1	1.5939 ; 1.482 ; col. yelsh. oil., misc. w., acet., Chl., **ex**
26	H_3C–H_2C–O–NO	Ethylorthosilicate coi. yelsh liq.,	$C_2H_5O_2N$	75.07	BP17	misc. w., acet., et
27	H_3C–$(CH_2)_3$–H_2C–O–NO	Amylnitrite, pa. yel. liq.	$C_5H_{11}O_2N$	117.15	BP104	misc. acet., eth
28	$(H_3CO)_3$PO	Trimethyl phosphine oxide	$C_3H_9O_4P$	140.08	liq.	193 — D1.220, misc. w; acet; eth
29	$(H_3C$–$H_2CO)_3$PO	Triethyl phosphine oxide	$C_6H_{15}O_4P$	182.16	liq.	216 — D1.0686, d. in w; misc. acet; eth
30	H_2C–O·PO_3H_2 / H COH / H_2COH	Glycerylphosphate	$C_3H_9O_6P$	172.08	−20	1.59 ; col. oily. liq., misc. w., acet.
31	H_5C_2O, H_5C_2O –Si– OC_2H_5, OC_2H_5	Ethylorthosilicate col. liq.	$C_8H_{20}O_4Si$ 208.30 BP 165.5 misc. w., acet., eth
32	H_3C–O, H_3C–O –B–O–CH_3	Methylborate, boric acid trimethylester	$C_3H_9O_3B$ 103.92 BP 65 col. liq., d. in w., misc. acet., d. in eth

Mixtures: survey and introduction

Substance A in Substance B	Gases	Liquid	Solid	Distribution grade
Gas	Gaseous mixture Air	(Real) solution air in water	occlusion- Mixed crystal H in metals	Homogeneous
	- - -	Lather	Solid foam	Colloids
	- - -	Soap lather	pumice stone	Heterogeneous
Liquid		Liquid mixture (solution) water ethanol	Inclusion compound Zeolite water	Homogeneous
	Aerosol	Emulsion Milk	In solid substances interlocked liquid	Colloids
	Smoke	Emulsion formation through creaming water - Benzene	droplets	Heterogeneous
Solid	- - -	True solution sugar - water	Mixed crystals (k, Rb) Cl	Homogeneous
	Aerosol	Suspension water glass	Solid suspension	Colloid
	Dust (pollens (solar particles)	Emulsion formation through sedimentation	Mixture or conglomerate. Granite	Heterogeneous

Sedimentation	Homogeneous		Colloids		Heterogeneous		
equilibrium	Air	Bromine gas smoke = water droplets in air					
Particles Φ cm. Particle weight g Height from	3×10^{-8} 5×10^{-23} 6 km	5×10^{-8} 2.7×10^{-22} 1 km	2.2×10^{-6} $\sim 10^{-20}$ 30 m	10^{-5} $\sim 10^{-18}$ 30 cm	10^{-4} $\sim 10^{-15}$ 0.3 mm	2×10^{-2} $\sim 10^{-10}$ 3×10^{-6} mm	
	the concentration has been expressed as half the density or the partial pressure.						

A phase is a homogeneous constituent of the mixture which is separated from other constituents of the mixture either by a visible or making into a visibly (e.g. Tyndall effect) separable layer.

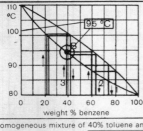

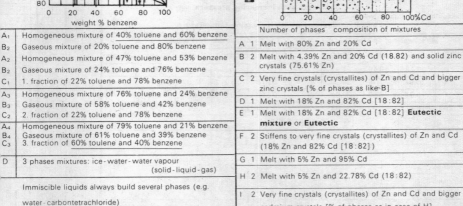

—Liquidus curve: Beginning of melting process with increasing temperatures
—Solidus curves: Beginning of solidification with decreasing temperature.

≠ Zn 419°
† Cd 321°
E = Eutectic temp. 264°

	Number of phases	composition of mixtures
A₁	Homogeneous mixture of 40% toluene and 60% benzene	
B₂	Gaseous mixture of 20% toluene and 80% benzene	
A₂	Homogeneous mixture of 47% toluene and 53% benzene	
B₂	Gaseous mixture of 24% toluene and 76% benzene	
C₁	1. fraction of 22% toluene and 78% benzene	
A₃	Homogeneous mixture of 76% toluene and 24% benzene	
B₃	Gaseous mixture of 58% toluene and 42% benzene	
C₂	2. fraction of 22% toluene and 78% benzene	
A₄	Homogeneous mixture of 79% toluene and 21% benzene	
B₄	Gaseous mixture of 61% toluene and 39% benzene	
C₃	3. fraction of 60% toulene and 40% benzene	
D	3 phases mixtures: ice - water - water vapour (solid - liquid - gas)	

Immiscible liquids always build several phases (e.g. water - carbontetrachloride)

	Number of phases	composition of mixtures
A	1	Melt with 80% Zn and 20% Cd
B	2	Melt with 4.39% Zn and 20% Cd (18.82) and solid zinc crystals (75.61% Zn)
C	2	Very fine crystals (crystallites) of Zn and Cd and bigger zinc crystals [% of phases as like-B]
D	1	Melt with 18% Zn and 82% Cd [18:82]
E	1	Melt with 18% Zn and 82% Cd [18:82] **Eutectic mixture or Eutectic**
F	2	Stiffens to very fine crystals (crystallites) of Zn and Cd (18% Zn and 82% Cd [18:82])
G	1	Melt with 5% Zn and 95% Cd
H	2	Melt with 5% Zn and 22.78% Cd (18:82)
I	2	Very fine crystals (crystallites) of Zn and Cd and bigger cadmium crystals [% of phases as in case of H]

Mixtures

Gas/Gas

The composition of air

Formula	Name	Volume%	Weight%	M.W.	BP °K	BP °C	T_x °K	T_x °C	P_x	D_x
CO_2	Carbon dioxide	0.03	0.04.	44.011	194	− 78.5s	303.3	31.1	73	0.460
Xe	Xenon	0.000009	0.00004	131.3	166	−107.1	288.8	16.6	58.2	1.155
Kr	Krypton	0.0001	0.0003	83.80	120	−152.9	209.2	− 63	54	0.78
O_2	Oxygen	20.93	23.01	32.000	90	−182.96	153.4	−118.8	49.7	0.430
Ar	Argon	0.9325	1.286	39.944	87	−185.7	150.2	−122	48	0.531
N_2	Nitrogen	78.10	75.51	28.016	77	−195.8	131.1	−147.1	33.5	0.3110
Ne	Neon	0.0018	0.0012	20.183	37	−245.9	43.5	−228.7	25.9	0.484
H_2	Hydrogen	0.01	0.001	2.016	20	−252.8	32.3	−239.9	12.8	0.0310
He	Helium	0.0005	0.00007	4.003	4	−268.9	5.3	−267.9	2.26	0.0693
—	Air	100%	100%	~30	—	—	131.5	−140.7	37.2	0.35

1 lit. of dry air at 0°C and 760 Torr weighs 1.2928 g

Maximum concentration of health hazardous gases in industries (MAK-Values)

According to existing experience associated with these substances or/and out of the basic experiments conducted on animals, in case of the undermentioned concentrations it can be taken that by exposing oneself daily to about 8 hours health is not affected generally.

Composition (vol.%) and calorific value (kcal/Nm³) of industrial gases

Type of gas	H_2	CH_4	C_mH_n	CO	CO_2	O_2	N_2	kcal/Nm³
Natural gas*	—	81.1	3.4	—	0.8	—	14.7	7500
Refinery gas**	55.0	16.0	7.0	2.9	4.8	2.2	12.1	4200
City gas	51.3	29.6	1.4	3.8	3.3	0.6	10.0	4200
Coke oven gas	57.5	24.0	2.1	6.0	1.7	0.9	7.8	6000
Water gas	50.0	0.6	0.1	40.0	5.0	—	4.3	2500
Producer gas	17.2	2.6	0.5	24.6	4.8	0.3	48.2	1200
Blast furnace gas	1.5	0.2	—	32.8	8.7	—	56.8	900

* Natural gas contains mostly H_2S. (Natural gas springs in lac 15%).
** The compositions are example; the calorific value is the mean value.

Gas/Liquids

The solubility of a gas in water increases with increasing pressure / decreases with increasing temperature

	Air in water at 760 torr				Carbondioxide in 1l water				
t°C	cm³ air	cm³O_2	Vol.%O_2	t°C	1atm	30 atm	40 atm	50 atm	100 atm
0	29.18	10.19	34.91	0	1.713	—	—	—	—
14	20.97	7.19	34.20	20	0.878	18.2	22.0	25.7	—
21	18.34	6.23	33.99	35	0.592	10.6	14.2	18.0	21.4
28	16.21	5.46	33.68	60	0.359	—	8.5	10.2	9.7

Gas/Solids

1 mole of metal takes up n moles of H_2 at t°C [1]

t°C	La	Ce	Ti	Zr	Th	V	Nb	Ta	Pd
300	1.19	1.15	0.85	1.14	1.10	0.35	0.401	0.285	0.142
600	1.01	1.00	0.68	0.75	0.91	0.022	0.041	0.053	0.085
900	0.83	0.86	0.09	0.56	0.79	0.007	0.009	0.002	0.075
1200	0.25	0.33	—	0.03	0.18	—	—	—	0.072

Water vapour in air (aqueous tension)

At t°C, the partial pressure of water vapour amounts to p Torr.
This corresponds to the quantity of g/Nm³

t	p	g
−10	1.95	2.14
− 9	2.13	2.33
− 8	2.32	2.54
− 7	2.53	2.76
− 6	2.76	2.99
− 5	3.01	3.24
− 4	3.28	3.51
− 3	3.57	3.81
− 2	3.88	4.13
− 1	4.22	4.47
0	4.58	4.84
+ 1	4.9	5.2
2	5.3	5.6
3	5.7	6.0
4	6.1	6.4
5	6.5	6.8
6	7.0	7.3
7	7.5	7.8
8	8.0	8.3
9	8.6	8.8
10	9.2	9.4
11	9.8	10.0
12	10.5	10.7
13	11.2	11.4
14	12.0	12.1
15	12.8	12.8
16	13.6	13.6
17	14.5	14.5
18	15.5	15.4
19	16.5	16.3
20	17.5	17.3
21	18.7	18.3
22	19.8	19.4
23	21.1	20.6
24	22.4	21.8
25	23.8	23.0
26	25.2	24.4
27	26.7	25.8
28	28.3	27.2
29	30.0	28.7
30	31.8	30.3

[1] see also hydrides—p. 5

MIXTURES: Gas/Gas and Gas/Solids

MAK-values of inorganic and organic substances (see p. 130)

Name	cm³/m³	mg/m³	see page	Name	cm³/m³	mg/m³	see page
Fluorine	0.1	0.2	10	Hydrogen chloride	5	7	39/35
Chlorine	1	3	10	Hydrogen cyanide (HCN)	10	11	68/11
Bromine	1	7	10	Hydrogen fluoride	3	2	39/34
Iodine	0.1	1	10	Hydrogen peroxide 95%	1	1.4	43/2
Ozone	0.1	0.2	43/7	Hydrogen selenide	0.05	0.2	39/32
Mercury	—	0.1	10	Hydrogen sulphide	20	30	30/31
				Hydrazine	1	1.3	39/24
Ammonia	100	70	39/22	Nickel carbonyl	0.1	0.7	68/55
Arsine	0.05	0.2	39/26	Nitric acid	10	25	55/42
Borontrifluoride	1	3	45/1	Nitrogen dioxide	5	9	40/7
Carbondioxide	5000	9000	40/2	Phosgene	0.1	0.4	61/1
Carbon monoxide	100	110	40/1	Phosphorus oxychloride	0.5	3	61/10
Carbon disulphide	20	60	51/58	Phosphorus trichloride	0.5	3	46/12
Chlorine dioxide	0.1	0.3	40/18	Phosphine	0.1	0.15	39/25
Decaborane	0.05	0.3	39/9	Sulphur dioxide	5	13	40/11
Diborane	0.1	0.1	39/5	Sulphur hexafluoride	1000	6000	45/12
Disulphur dichloride	1	6	46/18	Stibine	0.1	0.5	39/27
Hydrogen bromide	5	17	39/36				

Name	cm³/m³	mg/m³	see page	Name	cm³/m³	mg/m³	see page
Acetal dehyde	200	360	110/2	Ethylene diamine	10	30	119/9
Acetic acid	25	65	115/2	Ethylene imine	5	9	91/13
Acetic anhydride	5	20	113/2	Ethylene oxide	50	90	90/1
Acetone	1000	2400	110/26	Ethyl formate	100	300	112/4
Acrolein	0.5	1.2	110/13	Ethyl mercaptan	250	640	127/2
Acrylonitrile	20	45	122/14	Formaldehyde	5	6	110/1
Allyl alcohol	5	12	106/25	Furfurol	5	20	110/23
Allyl chloride	5	15	125/21	Furfuryl alcohol	50	200	109/23
Amyl acetate	200	1050	112/12	n-Heptane*	500	2000	80/14
Aniline	5	19	119/15	n-Hexane	500	2000	80/9
Benzene	25	80	83/12	Isoamyl alcohol	100	360	106/10
Benzyl chloride	1	5	125/25	Isopropyl alcohol	400	980	106/4
Butadiene	1000	2200	80/14	Methanol	200	260	106/1
n-Butanol	100	300	106/5	Methyl acetate	200	610	112/3
t-Butanol	100	300	106/8	Methyl ethyl ketone	250	740	110/27
Butyl amine	5	15	119/5	Methyl bromide	20	80	125/33
Butyl cellosolve	50	240	108/5	Methyl butyl ketone	100	410	110/30
Carbon tetrachloride	200	740	108/2	Methyl cellosolve	25	80	108/1
Cellosolve	0.1	0.4	111/14	Methyl chloride	50	105	125/12
Chlorobenzene	75	350	125/22	Methyl cyclohexane	500	2000	95/7
Chloroform	50	250	125/14	Methylene chloride	500	1750	125/13
Cresol (o, m, p)	5	22	109/1...3	Methyl formate	100	250	112/1
Cyclo hexane	400	1400	83/8	Methyl mercaptan	50	100	127/1
Cyclo hexanol	100	410	107/2	Methyl propyl ketone	200	700	110/28
Cyclo hexanone	100	400	111/9	N-monomethyl aniline	2	9	100/3
Cyclo hexene	400	1350	83/19	p-nitro aniline	1	6	123/9
Cyclo propane	400	690	83/2	Nitrobenzene	1	5	121/21
Diacetone alcohol	50	240	117/9	Nitroglycerine	0.5	5	128/25
Diethylamine	25	75	88/21	Nitromethane	100	250	121/16
1,2-Dibromoethane	25	190	125/38	Nitrotoluene	5	30	121/25
1,2-Dichloro ethylene	200	790	125/19	Octane*	500	2350	80/23
o-Dichloro benzene	50	300	125/23	Pentane	1000	2950	80/6
p-Dichloro benzene	75	450	125/24	Phenol	5	19	107/7
Dichlorodifluro methane	1000	4950	126/12	Phenyl hydrazine	5	22	121/3
Dichloro methane	500	1750	125/13	Propyne-1	1000	1650	82/28
Dichloromonofluoro methane	1000	4200	126/11	Propylacetate	200	840	112/8
Dichloro tetrafluoro ethane	1000	7000	126/15	Propylene oxide	100	240	101/1
Diisopropyl ether	500	2100	87/11	Pyridine	10	30	91/20
Dimethoxy methane	1000	3100	87/26	Quinone	0.1	0.4	111/14
N-dimethyl amiline	5	25	100/4	Styrene	100	420	96/22
Dimethyl sulphate	1	5	128/21	1.1,2,2-tetrachloro ethane	1	7	125/17
Dioxane	100	360	90/9	Tetra hydrofuran	200	590	90/3
Epichloro hydrin	5	18	125/32	Tetra nitromethane	1	8	121/3
Ethyl acetate	400	1400	112/5	Toluene	200	750	96/1
Ethyl ether	400	1200	87/4	o-Toludine	5	22	120/1
Ethyl alcohol	1000	1900	106/2	Triethyl amine	25	100	88/36
Ethyl amine	25	45	119/2	Trichloro ethylene	200	1050	125/20
Ethyl benzene	200	870	96/5	Trimethyl imine	25	60	91/14
Ethyl bromide	200	890	125/37	Vinyl chloride	500	1300	125/18
Ethyl chloride	1000	2600	125/16	Xylidene	5	25	120/4...9
Ethylene chlorohydrin	5	16	126/19	Xylenes (o, m, p)	200	870	96/2...4

Name (see page)	mg/m³	Name (see page)	mg/m³	Name (see page)	mg/m³
Antimony (10)	0.5	Manganese	6	Thallium compd. sol.	0.15
Arsenic trioxide (41/19)	0.5	Molybdenum compd. sol. 5; insol.	15	Titanium dioxide (42/32)	15
Barium compounds soluble	0.5	Pentaborane (possibly)	0	Uranium compd.** sol. 0.05 insol.	0.25
Cadmium oxide (41/31)	0.1	Phosphorus (cycl.) (10)	0.1	Vanadium pentoxide (dust) 0.5v	0.1
Chronic acid and chromate (60)	0.1	Phosphorus pentachloride (46/13)	1	Yttrium & compd.	15
Cyanide (as CN) (62)	5	Phosphorus pentoxide (40/11)	1	Zinc oxide-vapour (42/35, 36)	15
Fluoride (44 and 45)	2.5	Phosphorus pentasulphide (52/60)	1	Zirconium compd. (as Zr)	5
Iron oxide-vapours (41/54)	15	Selenium compd. (as Se)	1	D.D.T. (125/31)	1
Lead (10)	0.2	Sodium hydroxide (54/2)	2	Hydroquinone (107/10)	2
Lithium hydride (41/6)	0.025	Sulphuric acid (55/36)	1	Organic mercury compd. (as Hg)	0.01
Magnesium oxide-vapour (41/6)	15	Tellurium (10)	0.1	Picric acid (123/4)	0.1

Mixture	Ferrovanadium-alloys 1 mg/m³	Gasoline 500 cm³/m³ or 2000 mg/m³ (see above heptane* and octane**)

The MAK values of mixtures do not arise from the additive MAK-values of the constituents of the mixture

** see also radiation protections.

Density, volume and weight per cent of mixtures

Water/methanol ------ and water/ethanol — —

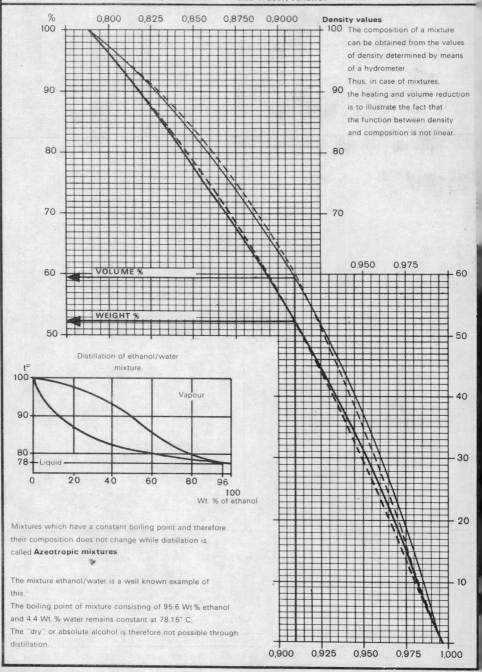

Density values

The composition of a mixture can be obtained from the values of density determined by means of a hydrometer.

Thus, in case of mixtures, the heating and volume reduction is to illustrate the fact that the function between density and composition is not linear.

VOLUME %

WEIGHT %

Distillation of ethanol/water mixture

Vapour

Liquid

Wt. % of ethanol

Mixtures which have a constant boiling point and therefore their composition does not change while distillation is called **Azeotropic mixtures**

The mixture ethanol/water is a well known example of this.

The boiling point of mixture consisting of 95.6 Wt % ethanol and 4.4 Wt. % water remains constant at 78.15° C.

The "dry" or absolute alcohol is therefore not possible through distillation.

Solution of salts in water (examples)

Sea water

Sea water contains about 30 cations and 10 anions in the following given abundance ratio.
The aqueous solution in vegetables, animals and human bodies (blood) are similarly complex.

Hardness of water

$1°$ H = 10 mg CaO/l or 1g CaO/100l

Soaploss

1g CaO or $1°$H/100l combines with about 16g of soap.

Boiler scale formation

Decomposition of hydrogen carbonates.
e.g. saturated solution

$$Ca(HCO_3)_2 \rightarrow CaCO_3$$
$$160g/l \qquad 15mg/l$$

The decrease in solubility of gypsom in water with increasing pressure is an exception (high pressure boilers)

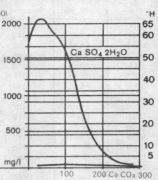

1000 l. of sea water contains	
g	of the substances
21694,000	Na Cl
2807,000	Mg Cl$_2$
1877,000	Mg SO$_4$
1202,000	Ca SO$_4$
538,300	Mg (HCO$_3$)$_2$
469,500	KCl
67,630	Na Br
64,400	Na BO$_2$
4,275	K NO$_3$
2,800	Ca (HCO$_3$)$_2$
1,400	F
1,300	H$_3$ SiO$_3$
0,944	(NH$_4$)Cl
0,919	LiCl
0,6	Al
0,445	Fe (HCO$_3$)$_2$
0,2	Rb
0,224	K NO$_2$
0,174	Ca HPO$_4$
0,05	Ba
0,045	Na I
0,028	Mn (HCO$_3$)
0,005	Zn
0,005	Cu
0,004	Pb
0,004	Se
0,003	Sn
0,002	Ca (HAsO$_3$)
0,002	Cs
0,002	U
0,001	Th
0,0005	Mo
0,0005	Ga
0,0004	Ce
0,0003	La
0,0003	Y
0,0003	V
0,0003	Ag
0,0001	Ni
0,00004	Au
0,00003	Hg

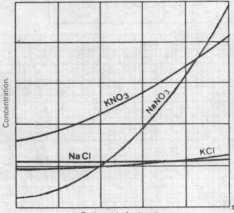

Reciprocal of salt pairs

In a solution of NaNO$_3$ (Chile salt peter) and KCl (Sylvine) a temperature dependent equilibrium sets in:

$$NaNO_3 + KCl \rightleftharpoons KNO_3 + NaCl$$

Higher temperatures $\rightleftharpoons$ lower temperatures

Formerly, potassium nitrate was manufactured in the above manner
Another example: Basic process of sodium bicarbonate preparation.

$$NaCl + (NH_4)HCO_3 \rightleftharpoons Na(HCO_3) + (NH_4)Cl$$

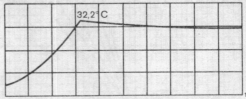

Solubility in case of Na$_2$SO$_4$·10H$_2$O

At 32.2°C the salt loses its inherent water of hydration. The second part of the curve is the solubility of water free substance.

133

Mixtures: dilute acids and bases

Densities and concentrations of mixtures in weight percentage

H₂SO₄		HNO₃	
Density ρ_{20}	Wt.%	Density ρ_{20}	Wt.%
1,400	50,50	1,330	53,41
1,425	53,01	1,340	55,13
1,450	55,45	1,350	56,95
1,475	57,84	1,360	58,78
1,500	60,17	1,370	60,67
1,525	62,45	1,380	62,70
1,550	64,71	1,390	64,74
1,575	66,91	1,400	66,97
1,600	69,09	1,410	69,23
1,625	71,25	1,420	71,63
1,650	73,37	1,430	74,09
1,675	75,49	1,440	76,71
1,700	77,63	1,450	79,43
1,725	79,81	1,460	82,39
1,750	82,09	1,470	85,50
1,775	84,61	1,480	89,07
1,800	87,69	1,490	93,49
1,825	92,25	1,500	96,73
1,833	94,72	1,510	99,26

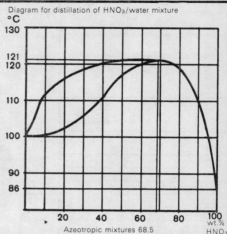

Diagram for distillation of HNO₃/water mixture

Azeotropic mixtures 68.5

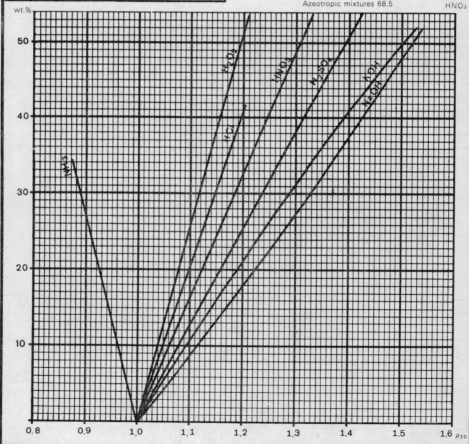

Densities and concentrations of mixtures in mole/l

H₂SO₄ Density ρ_{20}	Mol/l	HNO₃ Density ρ_{20}	Mol/l
1,400	7,208	1,330	11,27
1,425	7,702	1,340	11,72
1,450	8,198	1,350	12,20
1,475	8,699	1,360	12,68
1,500	9,202	1,370	13,19
1,525	9,711	1,380	13,73
1,550	10,23	1,390	14,29
1,575	10,74	1,400	14,88
1,600	11,27	1,410	15,49
1,625	11,80	1,420	16,14
1,650	12,34	1,430	16,81
1,675	12,89	1,440	17,53
1,700	13,46	1,450	18,28
1,725	14,04	1,460	19,09
1,750	14,65	1,470	19,95
1,775	15,31	1,480	20,92
1,800	16,09	1,490	22,11
1,825	17,17	1,500	23,02
1,833	17,70	1,510	23,79
		1,513	24,01

A very high concentration of H₂SO₄ cannot be more accurately determined with a hydrometer.

In solutions, depending on the concentrations, the following equilibriums result:

$$NaOH \rightleftharpoons Na^+ + OH^-$$
$$KOH \rightleftharpoons K^+ + OH^-$$
$$NH_3 + H_2O \rightleftharpoons (NH_4)OH \rightleftharpoons (NH_4)^+ + OH^-$$
$$HCl + H_2O \rightleftharpoons (H_3O)^+ + Cl^-$$
$$HNO_3 + H_2O \rightleftharpoons (H_3O)^+ + NO_3{}^-$$
$$H_2SO_4 + H_2O \rightleftharpoons (H_3O)^+ + (HSO_4)^-$$
$$(HSO_4)^- + H_2O \rightleftharpoons (H_3O)^+ + (SO_4)^{2-}$$

Vapour pressure diagram of mixture: HCl/water

Azeotropic mixture HCl/H₂O
BP of substances °C
HCl − 83.7 } 108.6   20.2%
H₂O + 100.0 }   79.8% ρ_{20} 1.102

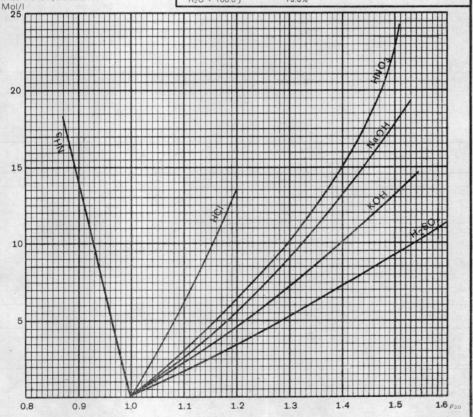

Mixtures: solids / solids - alloys

Lattice structure of elements in solid state

The three frequent and therefore important lattice types are:

Legend:

- X = Elements' symbol
- Ai = Lattice type (see below)
- ML = Molecular lattice

Length in Å (1Å = 10^{-8} cm)

- =a Cubical or hexagonalfaces in A1, A2 or A3
- =c Height (altitude) of hexagonal side of the prisms in A3
- =c/a Ratio of altitude to side in A3

Period	1a	2a	3a	4a	5a	6a	7a	8a	8a	8a	1b	2b	3b	4b	5b	6b	7b	8b
I	H2 A3 ML																	He A3
II	Li A2 3.502	Be A3 2.281 3.577 1.568											B A4	C A4 3.56 (A9)	N2 A3	O2 A1 ML	F2 ML	Ne A1 ML
III	Na A2 4.282	Mg ML A3 3.203 5.200 1.623											Al A1 4.041	Si A4 5.42	P4 A17 ML	S8 A16 ML	Cl2 A18 ML	Ar A1 ML
IV	K A2 5.33	Ca A1* 5.56 (A3)	Sc A1	Ti A3* 2.95 4.73 1.603 A2 3.32	V A2 3.034	Cr A2 2879 (A12) (A3)	Mn A6 (A12) (A13)	Fe A2* 2.860 3.564	Co A3* 2.51 A1* 4.07 1.622 A1* 3.54	Ni A1 3.517	Cu A1 3.608	Zn A3 2.659 4.93 1.85	Ga A11	Ge A4	As A7	Se A8	Br A14 ML	Kr A1 ML
V	Rb A2	Sr A1 6.07	Y A3 A3	Zr A2 A3	Nb A2	Mo A2 3.140	Tc	Ru A3	Ru A1	Pd A1	Ag A1 4.078	Cd A3 2.973 5.60 1.884 A4 6.46	In A6	Sn A5 5.82 3.17	Sb A7	Te A8	i A14 ML	Xe A1 ML
VI	Cs A2	Ba A2	La A1 A3	Hf A3	Ta A2 3.296	W A2 3.158 A15	Re A3	Os A3	Ir A1	Pt A1 3.916	Au A1 4.070	Hg A10	Tl A1	Pb A1 4.939	Bi A7	Po A19	At	Ru A1 ML
	1a	2a	3a	4a	5a	6a	7a		8a		1b	2b	3b	4b	5b	6b	7b	8b

Lanthanides / 2 Actinides →

Ce	Pr	Nd	Pm	Sm	Eu	Gd	Tb	Dy	Ho	Er	Tm	Yb	Lu	Th	U
A1 A3	A3	A3		A3	A2	A3	A3	A3	A3	A3	R3	A1	A3	A1	A20

No.	Type	Description	Frequency		No.	Type	Frequency
A1	Cu	Cubical face centred (see Fig. above)	26		A11	Ga	1
A2	W	Cubical body centred (see Fig. above)	16		A12	α-Mn	2
A3	Mg	Mexagonal close sphere packing (see Fig. above)	31		A13	β-Mn	1
A4		Diamond (Illustration see p.21)	5		A14	I2	2
A5		β-Sn	1		A15	β-W	1
A6		In	2		A16	S8	1
A7		As deformed simple cubical (transition towards network)	3		A17	Black	1
A8		Se deformed simple cubical (transition towards chain)	2		A18	Cl2	1
A9		Graphite network lattice (Fig. p.21)	1		A19	Po	1
A10		Hg	1		A20	U	1

*Note:	Ca lattice below 300°C	Ti	A3 = α Ti upto 880° A2 = β Ti above 880°	Fe	A2 = α Fe upto ~910° A1 = γ Fe upto 1400°	Co	A3 = α Co upto 400° A1 = β Co upto 400°C

Knowledge of lattice structures for all metallic elements are basic foundations towards understanding of structures and properties of alloys (see also p.24).

In the following pages only a few examples are illustrated. For further examples technical literature should be referred.

Abbreviations used on the following pages:

Pmisc = partially miscible...

misc = miscible with.....

liq. alloy = liquid alloy

E = Eutectic

MP/FP = Melting point/ (freezing point) always in °C

MC = Mixed crystals

HRR = Hume-Rothary's Rule (see p.24)

Mixtures: solids / solids alloys

Alloy type structures								Salt type structures
P misc. Li liq. Alloy		Be		B	C	N	O	NaF
Na		pmisc. Mg		pmisc. Al	Si	P	S	NaCl
misc. K Na₂K	Cu	Ca	NaZn₃ NaZn₄	Ga	NaGe	Na₃As	Na₂Se	NaBr
misc. Rb Na₂Rb	Ag	Sr	NaCd₂ NaCd₅	NaIn	Na₁₅Sn₄ Na₂Sn Na₄Sn₃ NaSn NaSn₂	Na₃Sb	Na₂Te	NaI
misc. Cr Na₂Cr	NaAu₂ NaAu	Ba	Na₃Hg Na₅Hg₂ Na₃Hg₄ NaHg Na₇Hg₈ NaHg₂ NaHg₄	Na₆Tl Na₂Tl NaTl	Na₄Pb Na₁₅Pb₄ Na₂Pb Na₅Pb₂ NaPb Na₂Pb₅ NaPb₃	Bi	Pr	At
1a	1b	2a	2b	3b	4b	5b	6b	7b

Melting point diagram of Na/Pb alloys

Note: The MP of alloys lies higher than that of lead.

326 MP lead

Na₄Pb is used for the preparation of lead tetraethyl Pb(CH₅)₄ (see p. 89/26) The formula Na₃₁Pb₈ corresponds to the H.R.R. (β-brass)

95 MP Na

For numbers see below

1	Na₄Pb and Na
	Intermetallic compounds (Phase) **Na₄Pb** or Na₁₅Pb₄ or Na₃₁Pb₈ (surface max.)
2	Mixed crystals Na₄Pb/Na₂Pb
3	Mixed crystals Na₄Pb/Na₂Pb and Eutectic mixture E₁
4	Mixed crystals Na₂Pb/Na₄Pb and Eutectic mixture E₁
E₁	Eutectic mixture of 75 Wt.% Pb/25 Wt.% Na.
5	Mixed crystals Na₂Pb/Na₄Pb
	Intermetallic compounds (Phase) **Na₂Pb** (First Na₅Pb₂, below 182° transformation)
6	Na₂Pb and Eutectic mixture E₂
E₂	Eutectic mixture of ~ 86 Wt% Pb/ ~ 14 Wt% Na
7	NaPb and Eutectic mixture E₂
	Intermetallic compounds (Phase) **NaPb**
8	NaPb and Eutectic mixture E₃
E₃	Eutectic mixture of ~94 Wt.% Pb/ ~ 6 Wt.% Na
9	Na₂Pb₅ and Eutectic mixture E₃
	Intermetallic compounds (Phase) **Na₂Pb₅** or NaPb₃ (surfaces max.)
10	Na₂Pb₅ and Eutectic mixture E₄
E₄	Eutectic mixture of ~95 Wt.% Pb/ ~ 3 Wt.% Na
11	Mixed crystals Pb/Na₂Pb₅ and Eutectic mixture E₄
12	Mixed crystals Pb/Na₂Pb₅

Mixtures: solids / solids - alloys

Alloy type structures							Salt type structures		
Li Mg₂		Be			B	C	N	O	F
Pmisc. Na		**Mg**			Mg₁₇Al₁₂ Mg Al Mg₂ Al₃	Mg₂Si	Mg₃P₂	Mg S	Mg Cl₂
Pmisc. k	Mg₂ Cu Mg Cu₂* (see fig.)	Mg₂ Ca*	Mg Zn Mg Zn₂* Mg₂ Zn₁₁	Sc	Mg₅ Ga₂ Mg₂ Ga Mg Ga Mg Ga₂	Mg₂ Ge	Mg₃As₂	Mg Se	MgBr₂
Rb	Mg₃ Ag Mg Ag	Mg₃ Sr Mg₄ Sr Mg₂ Sr*	Mg₃ Cd Mg Cd₃	Y	Mg₅ In₂ Mg₂ In Mg In Mg In₃	Mg₂ Su	Mg₃ Sb₂	Mg Te	Mg I₂
Cs	Mg₃ Au Mg₅ Au Mg₂ Au Mg Au	Mg₃ Ba Mg₄ Ba Mg₂ Ba*	Mg₃ Hg Mg₅ Hg₂ Mg₂ Hg Mg₅ Hg₃ Mg Hg Mg Hg₂	Mg₃ La Mg₂ La Mg La Mg La₄	Mg₅ Tl₂ Mg₂ Tl Mg Tl	Mg₂ Pb (see fig.)	Mg₃ Bi₂	Po	At
1a	1b	2a	2b	3a	3d	4b	5b	6b	7b

* Phases (complicated latice structures see p. 24)

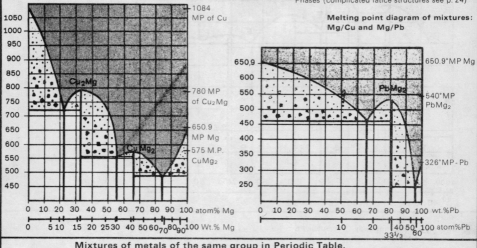

Melting point diagram of mixtures: Mg/Cu and Mg/Pb

Mixtures of metals of the same group in Periodic Table.

Mixture Pb/Sn (Lead/Tin)   Mixture Ag/Au (Silver/Gold)

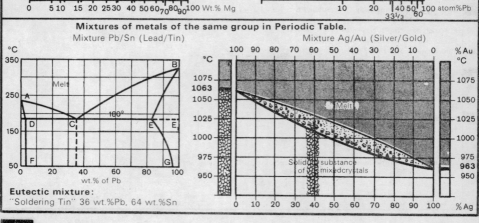

Eutectic mixture:
"Soldering Tin" 36 wt.%Pb, 64 wt.%Sn

138

Mixtures: solids / solids — alloys

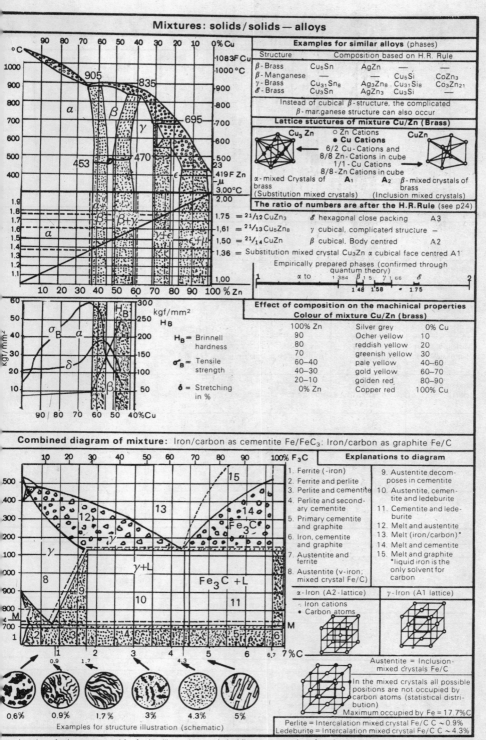

Examples for similar alloys (phases)

Structure	Composition based on H.R. Rule			
β - Brass	Cu₅Zn	AgZn		
β - Manganese	—		Cu₅Si	CoZn₃
γ - Brass	Cu₃₁Sn₈	Ag₃Zn₈	Cu₃₁Si₈	Co₃Zn₂₁
δ - Brass	Cu₃Sn₃	AgZn₃	Cu₃Si	

Instead of cubical β - structure, the complicated β - manganese structure can also occur

Lattice stuctures of mixture Cu/Zn (Brass)

Cu₃Zn
○ Zn Cations
● Cu Cations
CuZn

6/2 Cu - Cations and 8/8 Zn - Cations in cube
1/1 - Cu Cations
8/8 - Zn Cations in cube

α - mixed Crystals of brass (Substitution mixed crystals) **A₁**
β - mixed crystals of brass (Inclusion mixed crystals) **A₂**

The ratio of numbers are after the H.R. Rule (see p24)

1.75 = ²¹/₁₂ CuZn₃	δ hexagonal close packing	A3	
1.61 = ²¹/₁₃ Cu₅Zn₈	γ cubical, complicated structure	—	
1.50 = ²¹/₁₄ CuZn	β cubical, Body centred	A2	
1.36 = Substitution mixed crystal Cu₃Zn	α cubical face centred A1		

Empirically prepared phases (confirmed through quantum theory)

α to   1.384   β 1.5   γ 1.66   δ
1.46   1.58   1.75

Effect of composition on the machinical properties

H_B = Brinnell hardness
σ_B = Tensile strength
δ = Stretching in %

Colour of mixture Cu/Zn (brass)

%Zn	Colour	%Cu
100% Zn	Silver grey	0% Cu
90	Ocher yellow	10
80	reddish yellow	20
70	greenish yellow	30
60–40	pale yellow	40–60
40–30	gold yellow	60–70
20–10	golden red	80–90
0% Zn	Copper red	100% Cu

Combined diagram of mixture: Iron/carbon as cementite Fe/FeC₃; Iron/carbon as graphite Fe/C

Explanations to diagram

1. Ferrite (-iron)
2. Ferrite and perlite
3. Perlite and cementite
4. Perlite and secondary cementite
5. Primary cementite and graphite
6. Iron, cementite and graphite
7. Austentite and ferrite
8. Austentite (ν-iron; mixed crystal Fe/C)
9. Austentite decomposes in cementite
10. Austentite, cementite and ledeburite
11. Cementite and ledeburite
12. Melt and austentite
13. Melt (iron/carbon)*
14. Melt and cementite
15. Melt and graphite
*liquid iron is the only solvent for carbon

α-Iron (A2-lattice)
γ-Iron (A1 lattice)
○ Iron cations
● Carbon atoms

Austentite = Inclusion-mixed crystals Fe/C

In the mixed crystals all possible positions are not occupied by carbon atoms (statistical distribution)
Maximum occupied by Fe = 17.7%C

Perlite = Intercalation mixed crystal Fe/C ~0.9%
Ledeburite = Intercalation mixed crystal Fe/C ~ 4.3%

Examples for structure illustration (schematic)
0.6% 0.9% 1.7% 3% 4.3% 5%

explanation of other names and for further illustrations technical literature must be referred.

NUMBERS

Mathematical symbols

$+$	plus and	∞	infinity	$\leqq$	less than or equal
$-$	minus, few	$\parallel$	parallel	$\geqq$	greater than or equal
$\cdot$	times	⧣	parallel and equal	$\rightarrow$	tends to
$\times$	times	⇈	parallel and equally directed	lim	limit, e.g. lim. x = a
$:$	divided by	⇅	parallel and oppositely coupled	$a_{\log}$	a is limiting value of x
$/$	divided by	$\perp$	perpendicular to		logarithm to base a
$-$	divided by	$\sphericalangle$	angle	lg	Brigg's logarithms
$=$	equal to	$\triangle$	triangle	ln	natural logarithms
$\equiv$	congruent	$\overline{AB}$	Bar AB	$/$	per (m/s = meter per second)
⧧	not equal to	$\overParenB$	arc AB	$^{o}/_{o}$	percentage
$\approx$	nearly equal, about	$\sqrt{\ }$	square root of	$^{o}/_{oo}$	per mile (by the 1000)
$\sim$	similar, proportional	$\cdots$	upto	1°	1° degree
$<$	smaller than	Σ	sum of	$1'$	1' minitue
$>$	greater than	Δ	infinitesimally small change	$1''$	1" second
$\triangleq$	corresponds to		e.g. t = temp. change		

Mode of calculations

Mode	Example	Arithmetical parts			Symbols	Calculation Grade or Gradation	Combination by symbol
Addition (count on)	8 + 2 = 10 8 plus 2 equals 10	8 summand	2 summand	10 or 8 + 2 sum	$+$	first	weak
Subtraction (take away)	10 − 2 = 8 10 minus 2 equals 8	10 minuend	2 subtrahend	8 or 10 − 2 difference	$-$		
Multiplication (times taken)	4·3 = 12 4 times 3 equals 12	4 multiplicand	3 multiplicator	12 or 4·3 product	$\times$	second	strong
Division (part)	12:3 = 4 12 divided by 3 equals 4	12 dividend	3 divisor	4 or 12:3 quotient	$\div$ or $-$ or/ $1/n$		
Raise to a high power	2^3 = 8 2 power 3 equals 8	base or basic number	3 exponent or power number	8 or 2^3 power	a^n	third	very strong
Extract the root of (root extraction)	$^3\sqrt{8}$ = 2 cube root of 8 equals 2	8 radicand	3 root exponent	2 or $^3\sqrt{8}$ root	$\sqrt{\ }$		
Logarithmic	2 log 8 = 3 logarithm of 8 to the base 2 equals 3	8 numerus	2 base	3 or 2 log 8 logarithms	^{a} log lg ln		

Definite numbers

Calculations with fractions

$$12 : 3 = \frac{12}{3} = 4 \qquad \frac{12}{3}$$

Another way of writing for division problems 12÷3
Number above the fraction line is numerator
Number below the fraction line is denominator

Conversions

Common fractions in decimal fractions	Decimal fractions in common fractions
$\frac{3}{4} = 3 : 4 = 0,75$	$0,4 = \frac{4}{10} = \frac{2}{5}$
$2\frac{1}{4} = 2 + (1 \div 4) = 2 + 0,25 = 2,25$	$1,2 = \frac{12}{10} = \frac{6}{5} = 1\frac{1}{5}$

Expanding (expansion)	Shorten(ing)
$\frac{3}{5} = \frac{3 \cdot 2}{5 \cdot 2} = \frac{6}{10}$ Fractions are expanded in their numerators and denominators when they are multiplied by the same number. The values of fractions do not change by such expansions.	$\frac{6}{8} = \frac{6 : 2}{8 : 2} = \frac{3}{4}$ Fractions are shortened in their numerators and denominators by dividing with the same number. The values of the fractions do not change by such shortening.

Addition				Subtraction	
Similar fractions	$\dfrac{1}{5} + \dfrac{2}{5} = \dfrac{3}{5}$			$\dfrac{4}{7} - \dfrac{3}{7} = \dfrac{1}{7}$	
	Similar fractions can be added or subtracted when their number is added or subtracted and the denominator is retained.				
Dissimilar fractions	$\dfrac{1}{2} + \dfrac{1}{3} = \dfrac{3}{6} + \dfrac{2}{6} = \dfrac{5}{6}$			$\dfrac{2}{3} - \dfrac{1}{4} = \dfrac{8}{12} - \dfrac{3}{12} = \dfrac{5}{12}$	
	Dissimilar fractions, before addition or subraction are converted into similar fractions by expansion. The equal denominator is called "Common denominator" and it is the smallest common multiple of the individual denominators.				

Multiplication

With whole numbers or vice versa	$\dfrac{2}{5} \cdot 2 = \dfrac{2 \cdot 2}{5} = \dfrac{4}{5}$	When a fraction is multiplied with a whole number, the numerator is multiplied with that number
With a fraction	$\dfrac{1}{3} \cdot \dfrac{3}{5} = \dfrac{1 \cdot 3}{3 \cdot 5} = \dfrac{3}{15} = \dfrac{1}{5}$	When a fraction is multiplied with another fraction, then the numerator is multiplied with the numerator and the denominator is multiplied with denominator.

Division

Fraction by a whole number	$\dfrac{4}{7} \div 2 = \dfrac{4 \div 2}{7} = \dfrac{2}{7}$	When a fraction is divided by a whole number, then the numerator is divided by that number or the denominator is multiplied by that number.
Whole number by a fraction	$\dfrac{3}{4} \div 4 = \dfrac{3}{4 \cdot 4} = \dfrac{3}{16}$ $3 \div \dfrac{2}{5} = 3 \cdot \dfrac{5}{2} = \dfrac{15}{2}$	A whole number is divided by a fraction, then it is multiplied by the reciprocal value of the fraction.
Fraction by another fraction	$\dfrac{3}{8} \div \dfrac{2}{5} = \dfrac{3}{8} \cdot \dfrac{5}{2} = \dfrac{15}{16}$	When a fraction is divided by another fraction, then it is multiplied by the reciprocal value.

Transformation of double fractions

$$\frac{\dfrac{2}{3} + \dfrac{1}{5}}{\dfrac{2}{5} + \dfrac{1}{6}} = \frac{\dfrac{10 + 3}{15}}{\dfrac{12 + 5}{30}} = \frac{\dfrac{13}{15}}{\dfrac{17}{30}} = \frac{13}{15} \times \frac{30}{17} = \frac{26}{17}$$

Common numbers: Alphabets:

$3 + 3 + 3 + 3 = 4 \cdot 3 = 12$ $a + a + a + a = 4 \cdot a = 4a$	The numbers 3. 5. 2. ½ … are called definite numbers The letters a, b, c. … are called common numbers. With letters, one can calculate as precisely as with definite numbers.

Calculations with letters

Addition	Subtraction
$a + b + c = b + c + a$	$a - b = a - b$
In a sum, the summands can exchange positions.	In a difference, the minued and subtrahend retain their characters.
By enclosing in brackets, becomes collective.	A bracket always signifies the whole.
$a + (x + y) = a + (x + y)$	$a - (x + y) = a - x - y$
$a + (x - y) = a + x - y$	$a - (x - y) = a - x + y$
If (+) sign is placed before a bracket, then the bracket can be removed.	If (−) sign is placed before a bracket, then the bracket can only be removed when the previous sign in the bracket is inverted.

Multiplication

$a \cdot b = b \cdot a$ $ab = ba$ $a \cdot (x + y) = ax + ay$ $a \cdot (x - y) = ax - ay$	In a product, the factors can be changed. When a sum or difference is multiplied by a number then each term in the bracket is multiplied by that number.

Division

$(ax + ay + az) \div a = x + y + z$ $(ax + ay + az) \div b = \dfrac{ax}{b} + \dfrac{ay}{b} + \dfrac{az}{b}$	If a sum is to be divided by a number, then each summand is divided by that number.

Enclosing in brackets

$ax + ay + az = a (x + y + z)$ $a + ax = a (1 + x)$	A sum is transformed into a product in which the common factor is placed before a bracket and enclosing the rest.

Product with equal products: Powers

$a \cdot a \cdot a = a^3$ A product with equal factors can be written as power

a^3 Exponent or power number
  Base or functional number
Power

The exponent indicates how many times the base is multiplied with itself.

Multiplication of sums and differences

$(a + b)(x + y) = ax + bx + ay + by$

$(a + b)(x - y) = ax + by - ay - by$

$(a - b)(x - y) = ax - bx + ay - by$

$(a - b)(x - y) = ax - bx - ay + by$

Each term in the first bracket is multiplied with each term of the second bracket

Special cases:

Binomial formula

1. $(a + b)^2 = a^2 + 2ab + b^2$
2. $(a - b)^2 = a^2 - 2ab + b^2$
3. $(a + b)(a - b) = a^2 - b^2$

Calculation with relative numbers

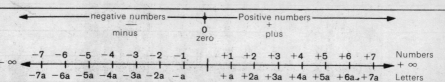

Relative numbers have a sign (plus or minus). The sign belongs to that number, and they are not the arithmetic signs.

Addition	Subtraction
$(+3) + (+2) = (+5)$	$(+6) - (+4) = (+2)$
$(+4) + (-2) = (+2)$	$(+4) - (-2) = (+6)$
$(-5) + (+3) = (-2)$	$(-7) - (+3) = (-10)$
$(-3) + (-2) = (-5)$	$(-5) - (-3) = (-2)$
	$(-4) - (-6) = (+2)$
If a relative number is added, then it carries its sign.	If a relative number is subtracted it changes its sign.

Multiplication	Division
$(+x) \cdot (+y) = + xy$	$(+x) \div (+y) = + \frac{x}{y}$
$(-x) \cdot (+y) = - xy$	$(-x) \div (+y) = - \frac{x}{y}$
$(+x) \cdot (-y) = - xy$	$(+x) \div (-y) = - \frac{x}{y}$
$(-x) \cdot (-y) = + xy$	$(-x) \div (-y) = + \frac{x}{y}$
The product of two numbers is positive when their signs are similar and it is negative when the signs are different.	The quotient of two numbers is positive when their signs are similar; it is negative when they have different signs.

Zero and infinity

$a - a = 0$	$0 \cdot a = 0$	$a \div 0 = \infty$	$\infty \div a = \infty$
$a + 0 = a$	$0 \div a = 0$	$a \div \infty = 0$	$0 \div 0 = $ Indefinite

Calculation with powers

Relative bases

$(+a)^n = + a^n$ A positive base always gives rise to a positive power.

$(-a)^{2n} = + a^{2n}$ A negative base with an even exponent gives rise to a positive power.

$(-a)^{2n \pm 1} = - a^{2n \pm 1}$ A negative base with an odd exponent gives rise to a negative power.

Relative exponents

$a^n = a^n$ A positive exponent embodies to the normal power values

$a^{-n} = \frac{1}{a^n}$ A negative exponent embodies to the reciprocal power values (reciprocal value or power).

$a^0 = 1$ A power with the exponents as zero has a value one.

Additions Subtractions

$$2a^m + 3a^m = 5a^m$$
$$3a^m - 2a^m = a^m$$

Powers can be added or subtracted only when they have same bases and exponents.

Multiplications Divisions

Powers with same base
$$a^m \cdot a^n = a^{m+n}$$
$$a^m \div a^n = a^{m-n}$$

Powers with equal exponents
$$a^m \cdot b^m = (ab)^m$$
$$a^m \div b^m = \left(\tfrac{a}{b}\right)^m$$

Power with equal bases are multiplied or divided, while one can add or subtract their exponents and retain the common base.
If powers with different bases are multiplied or divided, then one can multiply or divide their bases and retain the common exponents.

Raising to higher powers

$$(a^m)^n = a^{mn}$$

A power is raised to a higher power in which one multiplies the exponents and raises to the power the product of their bases.

Transformation of numbers to powers with base 10

Setting the number of zeros as tenth power:

$$100000 = 10 \cdot 10 \cdot 10 \cdot 10 \cdot 10 = 10^5$$

$$\frac{1}{100000} = \frac{1}{10 \cdot 10 \cdot 10 \cdot 10 \cdot 10} = \frac{1}{10^5}$$

5 zeros

$$360000 = 36 \cdot 10000 = 36 \cdot 10^4; \quad \frac{1}{360000} = \frac{1}{36 \cdot 10000} = \frac{1}{36 \cdot 10^4}; \quad \frac{48}{12000} = \frac{48}{12 \cdot 10^3} = \frac{4}{10^3}$$

By transforming above or below to the fraction stroke the sign changes!

$$\frac{1}{10^4} = 10^{-4}; \qquad 10^4 = \frac{1}{10^{-4}}; \qquad 5 \cdot 10^{-3} \quad \frac{5}{10^3}; \qquad \frac{5,4}{10^{-6}} = 5,4 \cdot 10^6 = 54 \cdot 10^5;$$

Calculation with roots

Powers with fraction exponent-root
$$a^{\frac{1}{n}} = \sqrt[n]{a^1} = \sqrt[n]{a}$$
$$a^{\frac{m}{n}} = \sqrt[n]{a^m} = \left(\sqrt[n]{a}\right)^m$$
power root

Powers with fraction exponents — roots.
The number of the fraction exponent is the power exponent of the radicand and the denominator is the root exponent.

Expansions—Shortenings

expand ⟶
$$\sqrt[n]{a^m} = \sqrt[n \cdot r]{a^{m \cdot r}}$$
⟵ shorter

Roots are expanded or shortened when the root exponent and power exponent of the radicand are multiplied or divided respectively with equal numbers.

Additions—Subtractions

$$\sqrt[m]{a} + 2\sqrt[m]{a} = 3\sqrt[m]{a}$$
$$4\sqrt[m]{a} - 2\sqrt[m]{a} = 2\sqrt[m]{a}$$

Roots can be added or subtracted when they have the same radicand and root exponents.

Multiplications—Divisions

$$\sqrt[m]{a} \cdot \sqrt[m]{b} = \sqrt[m]{a \cdot b}$$
$$\sqrt[m]{a} \div \sqrt[m]{b} = \sqrt[m]{a : b}$$
$$\sqrt[m]{a} \cdot \sqrt[n]{b} = \sqrt[mn]{a^n} \cdot \sqrt[mn]{b^m} = \sqrt[mn]{a^n \cdot b^m}$$
$$\sqrt[m]{a} \div \sqrt[n]{b} = \sqrt[mn]{a^n} \div \sqrt[mn]{b^m} = \sqrt[mn]{a^n : b^m}$$

Roots with same root exponents are multiplied or divided while out of the product or quotient their radicand is extracted into the roots.
Roots with different exponents are made of the same kind by expanding, thereafter they can be multiplied or divided.

Raising to powers—Extracting into roots

$$\left(\sqrt[m]{a}\right)^n = \sqrt[m]{a^n}$$

$$\sqrt[m]{\sqrt[n]{a}} = \sqrt[mn]{a}$$

A root is raised to the higher power, while the radicand is raised in higher power and from this power extracted into root.

A root is extracted into the root, while one extracts into root with the product of root exponents.

Calculation with logarithms

$x = a^y$ — x or a^y are powers. The value of power (x) and the base (a) are known. The exponent y is unknown. The problem is only solved with the help of a new calculating method. (Logarithms)

$$x = a^y \longrightarrow y = {}^a\log x$$

written in power → written in logarithm

y is the logarithm of x to the base a

Example: $y = {}^4\log x$

x	16	4	2	1	0,5
y	2	1	0,5	0	-0,5

The logarithm of x to the base a is that number with which one must raise it to higher power in order to obtain x.

Logarithms are exponents of powers with base a

Limiting cases	${}^a\log 1 = 0$	${}^a\log a = 1$	${}^a\log (a^n) = n$	${}^a\log o = -\infty$
Since	$a^o = 1$	$a^1 = a$	$a^n = a^n$	$a^{-\infty} = 0$

Logarithmic systems

Natural logarithm (Napierean logarithm)

Base $a = e = 2.7182818\ldots$

$${}^e\log x = \ln x$$

Application in higher mathematics

Decadic logarithm (Brigg's logarithm) (Common logarithm)

Base $a = 10$

$${}^{10}\log x = \log x$$

is preferred in case of logarithmic number calculations.

Transformation

$$\ln x = 2.3026 \cdot \log x$$

Calculations with decadic logarithm (common logarithm)

Multiplications	Divisions
$\log (a \cdot b) = \log a + \log b$	$\log\left(\dfrac{a}{b}\right) = \log a - \log b$
The logarithm of a product is equal to the sum of logarithms of individual factors.	The logarithm of a fraction is equal to the difference of the logarithm of numerator and denominator.

Raise to higher power	Extract into root
$\log a^n = n \cdot \log a$	$\log \sqrt[n]{a} = \dfrac{\log a}{n} = \dfrac{1}{n} \cdot \log a$
The logarithm of a power is equal to the product of exponent and the logarithm of the base	The logarithm of a root is equal to the product of the reciprocal value of root exponent and the logarithm of the radicand.

Calculation method of a problem	Calculation in logarithmic way is reduced to	The logarithmic calculation brings a simplification with which it signifies a reduction out of the following secondary steps.
Multiplication Division Raising to higher power Extracting to root	Addition Subtraction Multiplication Division } to which stand for the numbers logarithm	

Numbers	Numbers written in power	Logarithms of numbers (log)	Calculating process
10000	10^4	4,00000	Number of Calculating part
1000	10^3	3,00000	Search for logarithm
100	10^2	2,00000	
10	10^1	1,00000	Logarithm of the number
1	10^0	0,00000	Calculation with the log.
0,1	10^{-1}	0,00000 −1	Logarithm of numbers for that result
0,01	10^{-2}	0,00000 −2	
0,001	10^{-3}	0,00000 −3	Search for Antilog.
0,0001	10^{-4}	0,00000 −4	(Antilog) Result

Numerus	Logarithms

Transformation towards logarithms

Transformation towards antilogarithms

log 400 $=$ 2, 60206

Number | Characteristic | Mantissa

Logarithm

By transformation towards logarithms
yields digits of numbers = Characteristic
Digit series of numbers = Mantissa
The negative characteristic can be
written behind Mantissa.

Characteristic = The positive characteristic is smaller by 1 than the position before the point.
The negative characteristic is bigger than the position of 1 digit after the point.

Examples for calculation

For logarithm, a five figure (or 4 figure) table can be used.

a) $346 \cdot 14{,}24 = 4927$

log 346 = 2,53908
+ log 14,24 = 1,15351
log of Number = 3,69259
Antilog = 4927

b) $\dfrac{2500 \cdot 0{,}955}{36{,}5 \cdot 0{,}056} = 11680$

Numerator	Denominator
log 2500 = 4,39794	log 36,5 = 1,56229
+log 0,955 = 0,98000 −1	+log 0,056 = 0,74819 −2
log numerator = 5,37794 −1	log denominator = 2,31048 −2
= 4,87794	= 0,31048

log Denominator = 0,31048
log Number = 4,06746
Antilog = 11680

c) $15{,}3^4 = 54797$

log($15{,}3^4$) = 4·log15,3
4 log 15,3 = 1,18469 · 4
log Number = 4,73876
Antilog = 54797

but c') $0{,}153^4 = 0{,}00054797$

log($0{,}153^4$) = 4 ·log0,153
4 log 0,153 = (0,18469 − 1) · 4
log Number = 0,73876 − 4
Antilog = 0,00054797

d) $\sqrt[3]{55{,}2} = 3{,}807$

log $\sqrt[3]{55{,}2}$ = 1/3·log 55,2
$\frac{1}{3}$ · log 55,2 = 1,74194 ·1/3
log Number = 0,58064
Antilog = 3,807

but d') $\sqrt[3]{0{,}552} = 0{,}8203$

log$\sqrt[3]{0{,}552}$ = 1/3·log 0,552
$\frac{1}{3}$ · log 0,552 = (0,74194 − 1) · 1/3
= (2,74194 − 3) · 1/3 *
log Number = 0,91398 − 1
Antilog = 0,8203

*Negative characteristic must be divisible without remainder

Calculation with the decadic supplement (DE log)

The problem $\dfrac{35 \cdot 20{,}5}{12{,}3 \cdot 25{,}2}$ can be split up as $35 \cdot 20{,}5 \cdot \dfrac{1}{12{,}3} \cdot \dfrac{1}{25{,}2}$

$\log \frac{1}{12{,}3}$ = log1 − log12,3
log 1 = 0,00000
− log 12,3 = 1,08991
= 0,91009 −2

$\log \frac{1}{25{,}2}$ = log 1 − log25,2
log 1 = 0,00000
− log 25,2 = 1,40140   can be calculated
= 0,59260 −2

$\log\frac{1}{12{,}3}$ = DE log 12,3 = 0,91009 − 2 $\log \frac{1}{25{,}2}$ = DE log 25,2 = 0,59260 − 2

Problem : $\dfrac{35 \cdot 20{,}5}{12{,}3 \cdot 25{,}5} = 35 \cdot 20{,}5 \cdot \dfrac{1}{12{,}3} \cdot \dfrac{1}{25{,}2} = 2{,}283$

log 35 = 1,54407
log 20,5 = 1,31175
+ DE log 12,3 = 0,91009 − 2
+ DE log 25,2 = 0,59260 − 2
log Number = 4,35851 − 4
= 0,35851
Antilog = 2,283

Among the application of decadic supplement, the calculating method is simplified while in respect to the position of subtraction of logarithm, it passes into addition.

Equations

Type	Example	Explanations
Identical equations	$a + b = b + a$ $a^m \cdot b^m = (ab)^m$	Expression of a general regularity, all quantities stand fixed. Both sides of the equations are equally complete.
Definite equations	$x + 4 = 6$ $x^2 - 2 = 14$	First, one or several quantities are unknown. The unknowns are to be separated and their values determined.
Functional equations	$y = 3x + 7$ $y = x^2 - 3$	The quantities on either side are dependent on one another; they can take up arbitrary values while retaining interdependent relationship.

Definite equations

Degree	Number of unknowns			Usual designation
	One	Two	Three	
1	$x + 4 = 6$	$x - y = 4$	$x + y - z = 0$	Linear equation
2	$x^2 + x - 4 = 0$	$x^2 - y^2 = 21$	$x = y + z^2 + 6$	Quadratic equation
3	$x^3 - x^2 = x - 1$	$x^2 = 2x - y^3 - 7$	$x^3 + y^3 + z^3 = 0$	Cubic equation

Solution in case of equations with one unknown

Equal operation on either side	Isolation of unknowns by transposing the equation elements from the other equation side.	
	Procedure	**Note**
$x + a = b$ $x + a - a = b - a$ $x = b - a$	$x + a = b$ $x = b - a$	Summand becomes subtrahend
$x - a = b$ $x - a + a = b + a$ $x = a + b$	$x - a = b$ $x = b + a$	Subtrahend becomes summand
$ax = b$ $\frac{ax}{a} = \frac{b}{a}$ $x = \frac{b}{a}$	$ax = b$ $x = \frac{b}{a}$	Factor becomes division
$\frac{x}{a} = b$ $a \frac{x}{a} = b \cdot a$ $x = ab$	$\frac{x}{a} = b$ $x = b \cdot a$	Division becomes factor
$x^a = b$ $\sqrt[a]{x^a} = \sqrt[a]{b}$ $x = \sqrt[a]{b}$	$x^a = b$ $x = \sqrt[a]{b}$	Power exponent becomes root exponent
$\sqrt[a]{x} = b$ $(\sqrt[a]{x})^a = b^a$ $x = b^a$	$\sqrt[a]{x} = b$ $x = b^a$	Root exponent becomes power-exponent
$a^x = b$ $\log(a^x) = \lg b$ $x \cdot \log a = \lg b$ $x = \frac{\lg b}{\lg a}$		See calculations with logarithms

Equations: 1. Grades with two unknowns

Solution of an equation system with 2 unknowns under three conditions	1. It is necessary to have 2 equations 2. Both equations must be independent of one another 3. Both equations may not be specified again	**Example** I $ax + by = c$ II $dx + ey = f$

Procedure of solution

Substitution method	Equating method	Addition method
I $\; y = \dfrac{c - ax}{b}$ I in II $dx + e \cdot \dfrac{c - ax}{b} = f$	I $\; y = \dfrac{c - ax}{b}$ II $\; y = \dfrac{f - dx}{e}$ I = II $\dfrac{c - ax}{b} = \dfrac{f - dx}{e}$	First I×e; II × b I $aex + bey = ce$ II $bdx + bey = bf$ I–II $aex - bdx = ce - bf$

It continues to remain an equation with one unknown, even after the unknown is solved. The probe must be carried through with both equations.

Examples

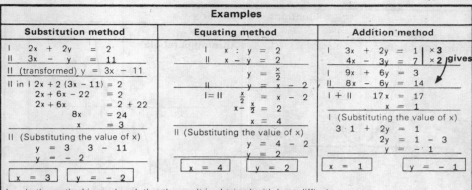

Substitution method	Equating method	Addition method

Substitution method

I $2x + 2y = 2$
II $3x - y = 11$

II (transformed) $y = 3x - 11$

II in I $2x + 2(3x - 11) = 2$
$2x + 6x - 22 = 2$
$2x + 6x = 2 + 22$
$8x = 24$
$x = 3$

II (Substituting the value of x)
$y = 3 \cdot 3 - 11$
$y = -2$

$\boxed{x = 3}$ $\boxed{y = -2}$

Equating method

I $x : y = 2$
II $x - y = 2$

$y = \frac{x}{2}$

II $y = x - 2$

I = II $\frac{x}{2} = x - 2$
$x - \frac{x}{2} = 2$
$x = 4$

II (Substituting the value of x)
$y = 4 - 2$
$y = 2$

$\boxed{x = 4}$ $\boxed{y = 2}$

Addition method

I $3x + 2y = 1$ $\times 3$
II $4x - 3y = 7$ $\times 2$ gives

I $9x + 6y = 3$
II $8x - 6y = 14$

I + II $17x = 17$
$x = 1$

I (Substituting the value of x)
$3 \cdot 1 + 2y = 1$
$2y = 1 - 3$
$y = -1$

$\boxed{x = 1}$ $\boxed{y = -1}$

In rule the method is used such that the result is obtained with least difficulty.
Examples from chemical calculation: Calculation of indirect analysis.

Functions

$y = f(x)$ y is a function of x.
y is dependent on x. The lawful
dependency is not known,
Functional equations in a co-ordinate
system can be graphically represented
after one has drawn up a table of values for
variable quantities through substitution of numbers
For better understanding the empirical values of
numbers are often represented graphically.

Functional equations

In a functional equation the lawful
dependency is given $y = 2x$.
y = Dependent variable
2 = Constant (proportionality factor)
x = Independent variable
A definite equation can be transformed
through insertion of $y = 0$
Example: $x^2 = x + 4$ $x^2 - x - 4 = 0$ $y = x^2 - x - 4$

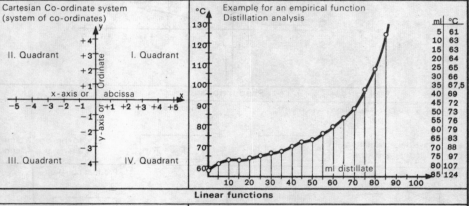

Cartesian Co-ordinate system
(system of co-ordinates)

II. Quadrant
I. Quadrant
x-axis or abcissa
III. Quadrant
IV. Quadrant

Example for an empirical function
Distillation analysis

ml distillate

ml	°C
5	61
10	63
15	63
20	64
25	65
30	66
35	67,5
40	69
45	72
50	73
55	76
60	79
65	83
70	88
75	97
80	107
85	124

Linear functions

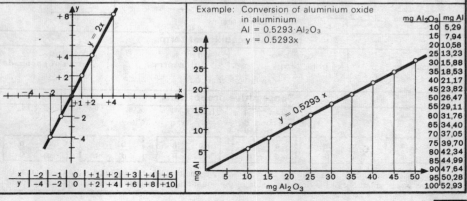

Example: Conversion of aluminium oxide
in aluminium
$Al = 0.5293 \cdot Al_2O_3$
$y = 0.5293x$

$y = 2x$

$y = 0.5293 x$

mg Al

mg Al_2O_3

x	-2	-1	0	+1	+2	+3	+4	+5
y	-4	-2	0	+2	+4	+6	+8	+10

mg Al₂O₃	mg Al
10	5,29
15	7,94
20	10,58
25	13,23
30	15,88
35	18,53
40	21,17
45	23,82
50	26,47
55	29,11
60	31,76
65	34,40
70	37,05
75	39,70
80	42,34
85	44,99
90	47,64
95	50,28
100	52,93

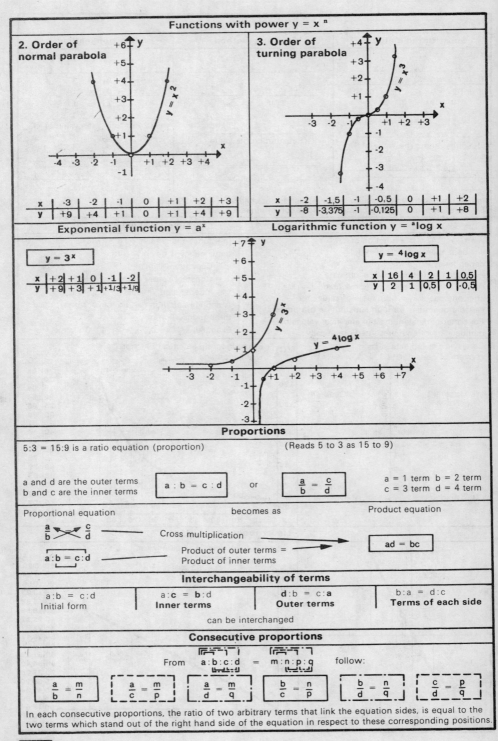

Functions with power y = x^n

2. Order of normal parabola

$y = x^2$

x	-3	-2	-1	0	+1	+2	+3
y	+9	+4	+1	0	+1	+4	+9

3. Order of turning parabola

$y = x^3$

x	-2	-1,5	-1	-0,5	0	+1	+2
y	-8	-3,375	-1	-0,125	0	+1	+8

Exponential function y = a^x

$y = 3^x$

x	+2	+1	0	-1	-2
y	+9	+3	+1	+1/3	+1/9

Logarithmic function y = ^{a}log x

$y = {}^4\log x$

x	16	4	2	1	0,5
y	2	1	0,5	0	-0,5

$y = 3^x$

$y = {}^4\log x$

Proportions

5:3 = 15:9 is a ratio equation (proportion)   (Reads 5 to 3 as 15 to 9)

a and d are the outer terms
b and c are the inner terms

$$a : b = c : d \quad \text{or} \quad \frac{a}{b} = \frac{c}{d}$$

a = 1 term   b = 2 term
c = 3 term   d = 4 term

Proportional equation   becomes as   Product equation

$$\frac{a}{b} \times \frac{c}{d}$$

Cross multiplication

$$a : b = c : d$$

Product of outer terms =
Product of inner terms

$$ad = bc$$

Interchangeability of terms

a:b = c:d	a:c = b:d	d:b = c:a	b:a = d:c
Initial form	**Inner terms**	**Outer terms**	**Terms of each side**

can be interchanged

Consecutive proportions

From $a:b:c:d$ = $m:n:p:q$ follow:

$$\frac{a}{b} = \frac{m}{n} \quad \frac{a}{c} = \frac{m}{p} \quad \frac{a}{d} = \frac{m}{q} \quad \frac{b}{c} = \frac{n}{p} \quad \frac{b}{d} = \frac{n}{q} \quad \frac{c}{d} = \frac{p}{q}$$

In each consecutive proportions, the ratio of two arbitrary terms that link the equation sides, is equal to the two terms which stand out of the right hand side of the equation in respect to these corresponding positions.

Rounding off in case of numbers

The rules for rounding off are employed when a decimal form of a given number is to be specified upto few digits.

Example:
$3.14159 \approx 3.142$

If before rounding off the last place, which is yet to be given, contains a 0, 1, 2, 3, 4, then the last place remains unchanged.

Example:
$3.12 \approx 3.1$
$5.234 \approx 5.23 \approx 5.2$
$8.2734 \approx 8.273 \approx 8.27$

If before rounding off the last place which is yet to be given, contains a 9, 8, 7, 6, the last digit increases by 1.
Same applies for 5 when followed by an irregular number of 0.

Example:
$2.17 \approx 2.2$
$6.369 \approx 6.37 \approx 6.4$
$8.2758 \approx 8.276 \approx 8.28$
$3.14159 \approx 3.142$

If before rounding off the last place which is yet to be given, contains an exact 5 then it is rounded off such that the last place becomes an even number [even number rule]. This also applies when the rounded number is further calculated.

Example:
$1/16 = 0.0625 \approx 0.062$
$3\frac{3}{4} = 3.75 \approx 3.8$

If before rounding off at the last place, which is yet to be given, a 5 of which we know how it has arisen, then it is rounded off as 5 was taken to be an approximate one and therefore it is an approximate figure.

Example:
6.3149 becomes 6.31$\underline{5}$
which further becomes 6.3$\dot{1}$
4.1852 becomes 4.18$\dot{5}$
which further becomes 4.1$\underline{9}$

Marking of the rounded last digit:

A rounded figure is marked by underlining of the last digit an approximate figure with a dot over it.
Example: $3.14159\ldots \approx 3.141\underline{6} \approx 3.14\underline{2} \approx 3.1\dot{4}$

Marking of an uncertain last digit:

If the uncertainty is not known at the same time numerically then at the most so many places are to be given so that the uncertainty lies at the last given place. If the uncertainty is maximum ± 0.5 of the last place, then it is written in the usual size of the script.

1st the uncertainty in last digit bigger than ± 0.5, then the last digit written as index number

When the marking through index number is not sufficient, so the uncertainty is given in addition to the value.

Example: Density of an oil at 15°C
$d_{15} = 0.884_6$ g/ml.

$d_{15} = 0.8840$ g/ml. ± 0.0006 g/ml.

Formation of mean values: Error calculation

Every measurement is affected with errors whether the result of measurements appear to be small or big. Besides these random errors, the results of measurement can be falsified by methodical errors, i.e. the errors in methods of measurement and personal or subjective errors.

Methodical and personal errors operate unilaterally and therefore they can be removed by suitable corrections. Towards the reduction of random errors, one carries out a series of single measurements (m_1, m_2, m_3 ...) and forms therefrom a mean value (M). The number of single measurements (n) is required to be atleast 3, better 5 or bigger numbers.

No/·· of single measurements	Single measurements ml. Measuring solution	Single error $f = m - M$	Square of single errors f^2
n = 5	$m_1 = 20.15$	$f_1 = -0.02$	$f_1^2 = +0.0004$
	$m_2 = 20.12$	$f_2 = -0.05$	$f_2^2 = +0.0025$
	$m_3 = 20.24$	$f_3 = +0.07$	$f_3^2 = +0.0049$
	$m_4 = 20.18$	$f_4 = +0.01$	$f_4^2 = +0.0001$
	$m_5 = 20.16$	$f_5 = -0.01$	$f_5^2 = +0.0001$
	$\Sigma m = 100.85$		$\Sigma(f^2) = 0.008$

Mean value $M = \frac{\Sigma m}{n} = \frac{100.85}{5} = 20.17$ Mean error or single measurement $f_m = \sqrt{\frac{\Sigma(f^2)}{n-1}} = \sqrt{\frac{0.008}{4}} = \pm 0.045$

Mean error of mean value $F_m = \sqrt{\frac{\Sigma(f^2)}{n(n-1)}} = \sqrt{\frac{0.008}{20}} = \pm 0.02$

Result = Mean value $\pm$ Mean error of mean values   R = 20.17 ml. $\pm$ 0.02 ml.
For obtaining mean values, all performed single measurements must be referred to. The so-called "wild shots" or stray measurements may be only separated after rigorous examination.

Areas

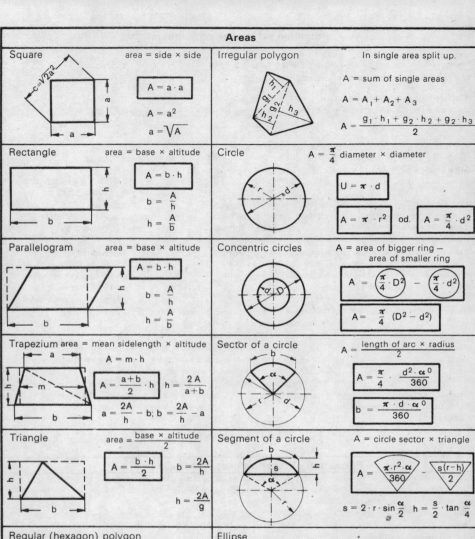

Square
area = side × side

$$A = a \cdot a$$
$$A = a^2$$
$$a = \sqrt{A}$$

$c = \sqrt{2a^2}$

Irregular polygon
In single area split up.

A = sum of single areas

$$A = A_1 + A_2 + A_3$$

$$A = \frac{g_1 \cdot h_1 + g_2 \cdot h_2 + g_2 \cdot h_3}{2}$$

Rectangle
area = base × altitude

$$A = b \cdot h$$
$$b = \frac{A}{h}$$
$$h = \frac{A}{b}$$

Circle
$$A = \frac{\pi}{4} \text{ diameter} \times \text{diameter}$$

$$U = \pi \cdot d$$

$$A = \pi \cdot r^2 \quad \text{od.} \quad A = \frac{\pi}{4} \cdot d^2$$

Parallelogram
area = base × altitude

$$A = b \cdot h$$
$$b = \frac{A}{h}$$
$$h = \frac{A}{b}$$

Concentric circles
A = area of bigger ring − area of smaller ring

$$A = \left(\frac{\pi}{4} \cdot D^2\right) - \left(\frac{\pi}{4} \cdot d^2\right)$$

$$A = \frac{\pi}{4}(D^2 - d^2)$$

Trapezium
area = mean sidelength × altitude

$$A = m \cdot h$$

$$A = \frac{a+b}{2} \cdot h \qquad h = \frac{2A}{a+b}$$

$$a = \frac{2A}{h} - b; \quad b = \frac{2A}{h} - a$$

Sector of a circle
$$A = \frac{\text{length of arc} \times \text{radius}}{2}$$

$$A = \frac{\pi}{4} \cdot \frac{d^2 \cdot \alpha^0}{360}$$

$$b = \frac{\pi \cdot d \cdot \alpha^0}{360}$$

Triangle
$$\text{area} = \frac{\text{base} \times \text{altitude}}{2}$$

$$A = \frac{b \cdot h}{2} \qquad b = \frac{2A}{h}$$

$$h = \frac{2A}{g}$$

Segment of a circle
A = circle sector × triangle

$$A = \frac{\pi \cdot r^2 \cdot \alpha}{360} - \frac{s(r-h)}{2}$$

$$s = 2 \cdot r \cdot \sin\frac{\alpha}{2} \qquad h = \frac{s}{2} \cdot \tan\frac{\alpha}{4}$$

Regular (hexagon) polygon
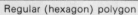

Area = area of triangle × no. of sides (n)

$$A = \frac{b \cdot h}{2} \cdot n$$

$$s = b$$

Ellipse

$$A = \frac{\pi}{4} \cdot D \cdot d$$

$$A = \pi \cdot R \cdot r$$

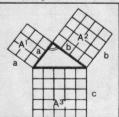

In a right angled triangle the square of the hypotenuse is equal to the sum of squares on the other two sides.

Example: a = 3, b = 4

$$c = \sqrt{3^2 + 4^2} = \sqrt{25}$$
$$c = 5$$

$$c^2 = a^2 + b^2$$
$$c = \sqrt{a^2 + b^2}$$
$$a = \sqrt{c^2 - b^2}$$
$$b = \sqrt{c^2 - a^2}$$

Pythagoras Theorem

	Bodies	
	Volumes	Surface areas

Prism

Volume = base area × altitude

$$V = A \cdot h$$

$$V = A_\square \cdot h$$
$$\boxed{V = A \cdot h} \quad V = A_\square \cdot h$$
$$V = a \cdot b \cdot h$$

$$\boxed{A_0 = 2 A_\square + 4 A_\square} \quad A = \text{square}$$

$$\boxed{A_0 = 2 A_\square + 2 A_{\square 1} + 2 A_{\square 2}}$$

$$A = \text{rectangle}$$

Prism

$$\boxed{V = A \cdot h} \quad V = A_\triangle \cdot h$$

$$V = \frac{g \cdot h'}{2} \cdot h$$

$$\boxed{A_0 = 2 A_\triangle + 3 A_{\square 1}}$$
$$A = \text{equilateral triangle}$$

$$\boxed{A_0 = 2 A_\triangle + A_{\square 1} + A_{\square 2} + A_{\square 3}}$$
$$A = \text{unequal sided triangle}$$

$$\boxed{A_0 = 2 A_\triangle + n \cdot A_\square}$$
$$A = \text{equal sided n-corners}$$

Cylinder

$$\boxed{V = A \cdot h} \quad V = A_O \cdot h$$

$$V = \frac{\pi}{4} \cdot d^2 \cdot h$$

$$\boxed{A_M = \pi \cdot d \cdot h} \quad \text{generated surface area}$$

$$\boxed{A_0 = (\pi \cdot d \cdot h) + \left(2 \cdot \frac{\pi}{4} \cdot d^2\right)}$$

Pyramid

$$\text{Volume} = \frac{\text{base area} \times \text{height}}{3}$$

$$\boxed{V = \frac{A \cdot h}{3}} \quad V = \frac{A_\square \cdot h}{3}$$

$$V = \frac{a \cdot b \cdot h}{3}$$

$$\boxed{A_0 = A_\square + 4 A_{\triangle 1}}$$
$$A = \text{square}$$

$$\boxed{A_0 = A_\square + 2 A_{\triangle 1} + 2 A_{\triangle 2}}$$

$$A = \text{rectangle}$$

Pyramid

$$\boxed{V = \frac{A \cdot h}{3}} \quad V = \frac{A_\triangle \cdot h}{3}$$

$$V = \frac{b \cdot h' \cdot h}{3}$$

$$\boxed{A_0 = A_\triangle + 3 A_{\triangle 1}}$$
$$A = \text{equilateral triangle}$$

$$\boxed{A_0 = A_\triangle + A_{\triangle 1} + A_{\triangle 2} + A_{\triangle 3}}$$
$$A = \text{triangle with unequal sides}$$

$$\boxed{A_0 = A_\triangle + n \cdot A_{\triangle 1}}$$
$$A = \text{equal sided n-corners}$$

Cone

$$\boxed{V = \frac{A \cdot h}{3}} \quad V = \frac{A_O \cdot h}{3}$$

$$V = \frac{\pi}{4} \cdot d^2 \cdot \frac{h}{3}$$

$$\boxed{A_0 = \frac{\pi}{4} \cdot d(d + 2s)}$$

$$\boxed{A_M = \frac{\pi \cdot d \cdot s}{2}} \quad \text{generated surface area}$$

$$\boxed{A_M = \pi \cdot r \sqrt{r^2 + h^2}}$$

Surface area	Volume	
A_0 Sum of all areas	$V = \frac{h}{3} \cdot (a^2 + a \cdot b + b^2)$ A = square $V \approx A_M \cdot h \quad V \approx (\frac{a+b}{2})^2 \cdot h$ A = square $V = \frac{h}{3} (A + \sqrt{A + A_1} + A_1)$ A = n-corners	**Truncated pyramid**
$A_0 = A + A_1 + A_M$ generated surface area $A_M = \pi \cdot \frac{d+D}{2} \cdot s$ $s = \sqrt{h^2 + (R-r)^2}$	$V = \frac{\pi}{12} h \cdot (D^2 + D \cdot d + d^2)$ $\frac{\pi}{12} = 0{,}261$ $V = A_M \cdot h$ (approximate)	**Truncated cone**
$A_0 = \pi \cdot d^2$	$V = \frac{2}{3} \cdot \frac{\pi}{4} \cdot d^2 \cdot d$ $V = \frac{\pi}{6} \cdot d^3$ $V = 0{,}5236 \cdot d^3$	**Sphere**
generated area $A_M = 2 \cdot \pi \cdot r \cdot h$ $A_M = \frac{\pi}{4} (s^2 + 4h^2)$	$V = \pi \cdot h^2 \cdot (r - \frac{h}{3})$ or $V = \pi \cdot h (\frac{s^2}{8} + \frac{h^2}{6})$	**Spherical segment (cup or cap)**
Surface area = generated surface area $A_0 = A_M = U \cdot \pi \cdot D$ $A_0 = \pi \cdot d \cdot \pi \cdot D$	$V = \frac{\pi}{4} \cdot d^2 \cdot \pi \cdot D$	**Concentric circles (tubing)**
Volume of a barrel by assumption of a circular stave is approximate: By hypothesis, for parabolic stave, the calculation is accurate.	$V = \frac{\pi h}{12} (2 D^2 + d^2)$ $V = \frac{\pi h}{15} (2 D^2 + Dd + 0{,}75 d^2)$	

Calculations with mixtures

A solution is a mixture consisting of a solvent and a solute. The concentration of solution of the dissolved can be expressed in different ways.

Concentrations data in case of solutions

Concentration specification	Definition	Mass unit	Application
Weight percentage	Gram of substance in 100 gram solution	$\dfrac{g}{100g}$	Reagent solution
Volume percentage	Litre of substance (liq.) in 100 litres of solution	$\dfrac{l}{100l}$	Mixture of liquid substance (alcohol)
Gram in litre	Gram of substance to a litre of solution	$\dfrac{g}{l}$	Reagent solution
Milligram in litre	Milligram of substance to a litre of solution	$\dfrac{mg}{l}$	In case of very small concentrations
Molar solution	Moles in a litre of solution	$\dfrac{mol.}{l}$	Reagent solution
	Millimole in a litre of solution	$\dfrac{m.mole}{l}$	Reagent solutions in case of very dilute solutions
Normal solution	Equivalents in a litre of solution	$\dfrac{Equ.}{l}$	Volumetric analysis
	Milliequivalents in millilitre of solution	$\dfrac{m\ equ}{ml}$	Volumetric analysis
	Milliequivalents in litre of solution	$\dfrac{m\ equ}{l}$	Volumetric analysis in case of very dilute solution

Quantity	Weight percentage	Volume percentage
Strength S	Dissolved substance in gram (kg, mg) of pure substance)	Dissolved substance in litres (ml.) of pure substance
% Contents P	$g/100g$; ($kg/100kg$; $mg/100\ mg$)	$l/100l$ ($ml/100ml$)
Solution L	Solution in gram (kg, mg.)	Solution in litre (ml.)
Strength	$S = \dfrac{P \cdot L}{100}$	$S = \dfrac{P \cdot L}{100}$
Percentage	$P = \dfrac{S \cdot 100}{L}$	$P = \dfrac{S \cdot 100}{L}$
Solution	$L = \dfrac{S \cdot 100}{P}$	$L = \dfrac{S \cdot 100}{P}$

For the preparation of a solution, if the substance employed is impure (moist) then the necessary weight is

$$wt. = \frac{g \cdot 100}{\text{percentage number}}$$

where g is the weight of pure substance percentage number = % purity of the substance

Cross mixing

a ↘ ↗ (c-b) parts of solution a	**Example**
c	**60** ↘ ↗ **13** Parts of 60% solution
b ↗ ↘ + (a-c) parts of solution b	**38**
(c-b) + (a-c) parts of solution C	**25** ↗ ↘ 22 parts of 25% solution
	35 parts of 38% solution

a = Strength of initial concentrated solution — = 60%

b = Concentration of initial diluted solution — = 25%

c = Concentration of desired solution — = 38%

With the help of three sets of composition the quantity of the desired solution can be calculated

---

If a solution is to be brought to a desired concentration by dilution, then for the concentration of the diluted solution, the pure solvent is set zero.

In a cross mixing, when the concentration in weight percentage is given in partial weight parts it signifies the concentrations of volume parts in volume percentage.

If the weight (w) of a solution is to be calculated from the volume, then the weight is to be divided by density (γ)

$$V = \frac{W}{\gamma}$$

Molar solutions and normal solutions

For the proportion of a normal solution the necessary weight (w) of a substance is

1 mol = Molecular weight in gram

v = Valency or no. of electrons

$$w = \frac{1 \text{ mol}}{v} \cdot l \cdot n \quad \text{or} \quad W = E \cdot l \cdot n$$

n = normality of solution

l = litre of solution

when $E = 1$ equivalent $= \dfrac{1 \text{ mol}}{v}$

If a normal solution is to be prepared by weighing out of a per cent solution then

$$w = \frac{100}{P} \cdot E \cdot l \cdot n$$

$$V = \frac{E \cdot 100}{P \cdot \gamma} \cdot l \cdot n$$

If a normal solution is to be prepared by measuring the volume (V) of initial solution then it applies to formula

P = Per cent number of solution

γ = Density of solution

If a normal solution in case of bigger normalities is brought to smaller normalities by dilution, then the normalities as like volumes retain through inverting.

n_1 = normality of concentrated solution

n_2 = normality of diluted solution

v_1 = Volume of concentrated solution

v_2 = Volume of diluted solution

The same applies for molar solutions

m = Molarity

$$n_1 : n_2 = v_2 : v_1$$

$$v_2 = \frac{n_1 \cdot v_1}{n_2}$$

$$m_1 : m_2 = v_2 : v_1$$

$$v_2 = \frac{m_1 \cdot v_1}{m_2}$$

Formula symbols

Signs	Nomenclature	Signs	Nomenclature
Geometrical quantities		**Force and pressure**	
L	Length	F, P, K	Force
h, H	Height, altitude	G, W	Weight
b, B	Breadth	γ	Density
r, R	Radius	M, T	Moment of a force
d, D	Diameter		(force × length of a lever)
λ	Wave length	p	Pressure (force per unit area)
	Path length	σ	Tensile stress (compressible
	Wall thickness		normal)
A, S, F	Area (surface, cross	τ	Shear stress
	section etc.)	E	Elasticity Modulii
α, β, γ	Angle	G	Shear modulus
ω, Ω	Solid angle	α	Coefficient of expansion ($1/E$)
V, τ	Volume	β	Shear coefficient
J, S	Area-moment of Inertia ($fr^2 dA$)	ε	Expansion ($\Delta l/l$)
Masses		μ	Poisson number
m	Mass	γ	Surface tension
ρ, d	Density	μ, f	Coefficient of friction
	Specific volume (V/m)	η	Dynamic viscosity
Θ	Mass, moment of inertia ($fr^2 dm$)	ν	Kinematic viscosity
	Atomic weight	**Heat, Work, Energy**	
M	Molecular weight	Q, W	Heat quantity
	Concentration	A, W	Work
	Diffusion coefficient	W, E	Energy
	Valency	c	Specific heat
	Loschmidt's number.	c_p	Specific heat at constant
	Avagadro's no/.		pressure
Time		c_v	Specific heat at constant
τ, z	Time		volume
	Periodicity	$\varkappa$	Ratio of sp. heats.
T	Time constant		(c_p/c_v)
	Nullphase angle, Angle of lag	λ	Thermal conductivity
	and lead	S	Entropy
	Frequency of revolution	P, N	Power
ω, Ω	Angular velocity	R	Gas constant
r, v	Frequency	η	Efficiency
ω	Circular frequency ($2\pi f$)	σ, γ	Surface tension
u, w	Velocity	**Light**	
b	Acceleration	c	Velocity of light
	Gravity	n, μ	Refractive index
Temperature		f	Focal length
ϑ, y	Temperature from ice point of	Φ, F	Luminus flux
Θ, Y	Absolute temperature (thermodynamic	E	Illumination intensity of a
	temperature)		light area.
	Coefficient of linear expansion	B, L	Light density of lighted area (brightness)
	$[(dl/dt) \div l_0]$	I	Intensity of light (Φ/ω)
β	Coefficient of volumes expansion	R	Specific light radiations
	$[(dV/dr) \div V_0]$	Φ_i	Incident light
		Φ_e	Emergent light
		ϑ	Absorbance (Φ_e/Φ_i)
		E	Extinction ($\log l/v$)
		m	Extinction modules (E/d;
			d = density of medium)
		ε	Extinction co-efficient (m/c;
			c = concentration)

Electric and magnetic quantities

Symbol	Nomenclature	Symbol	Nomenclature
Q,q	Electrostatic quantity, charge	$\varkappa$	Electrical conductivity ($1/e$)
e	Electronic charge	A	Equivalent conductivity
o	Charge density area	D	Electrical lag
ϱ,η	Charge density volume		
E	Strength of electrical field	ε	Dielectrical constant
U	Electrical tension	C	Capacitance
E,U	Electromotive force	H	Magnetic field strength
φ,V	Electrical potential (voltage)	V	Magnetic stress
I	Electrical current strength	R,P	Magnetic resistance
Θ	Electrical flow or frequency	A	Magnetic conductance
S,G	Electric current density	N,w	Winding number
R	Electrical resistance	B	Magnetic induction
		A	Vector potential
X	Electrical reactance	μ	Permeability
Z	Impedance	Φ,Ψ	Magnetic induction flux
ϱ	Specific resistance	$\varkappa$	Magnetic susceptibility
G	Electrical conductance	L	Inductance
B	Susceptance value	M,L	Opposite inductance
Y	Admittance value	S,N	Electromagnetic radiation density
α	Degree of dissociation	Z,W,Γ	Wavelength resistance (drag)

Units

Prefixes to notation in case of multiples and parts of units

Parts ◄─────── Units **1** ───────► Multiples

b = Deci = $\frac{1}{10}$	$= 10^{-1}$	D = Deca = $10 = 10^{1}$
c = Centi = $\frac{1}{100}$	$= 10^{-2}$	h = Hecto = $100 = 10^{2}$
m = Milli = $\frac{1}{1000}$	$= 10^{-3}$	k = Kilo = $1000 = 10^{3}$
μ = Micro = $\frac{1}{1000\,000}$	$= 10^{-6}$	M = Mega = $1\,000\,000 = 10^{6}$
n = Nano = $\frac{1}{1\,000\,000\,000}$	$= 10^{-9}$	G = Giga = $1\,000\,000\,000 = 10^{9}$
p = Pico = $\frac{1}{1\,000\,000\,000\,000}$	$= 10^{-12}$	T = Tera = $1\,000\,000\,000\,000 = 10^{12}$

	Symbol	notation	Relation to other units and explanations	sym. used for prefi
Length	m	Metre		km, mm, μm, nm
	u	Micron	1 u = 1 um	
	Å	Angstrom	1 Å = 10^{-10} m	
	X·U	X-Units	not metric, 1 × U. ~ 10^{-13} m	
Area	m²	Square metre		cm², mm²
	a	Ar	1 a = 100m²	ha
Volume	m³	Cubic metre		cm³ (=10^{-6}m³)
	l	Litre	1 l ~ 1.000028 dm³	
Angle	L	Right angle	⎫	
	°	Degrees	⎪	
	´	Minutes	⎬ Symbols are placed on top	
	˝	Seconds	⎪	
	g	New degree Gon	⎭ $100^{g} = 90°$	
	rad	Radiant	1 rad = 2 L/π = 63.6...g = 57.29°	
	Sr.	Steradian		
Time	s	Seconds	A second is of 86400 st part of mean	ms, μs (not δ), ns
	min	Minutes	Sunday	
	h	Hours	1 h = 60 min = 3600 s	
	d	Day	3h signifies a time interval (span)	
	a	Year	3ʰ signifies a time point	
Frequency	Hz	Hertz	A Hertz and its derived units are not employed for circular frequencies	kHz, MHz

	Symbol	Nomenclature	Relation to other units. Explanations	Symbols with prefix
Mass	g t	Gram Tonne	1t = 1000 kg	kg, mg, μg (not γ) dt (= 100kg)
Force	dyn N p	Dyne Newton Pond	1 N = 10^5 dyne 1 kp = 9.80665 N For p, kp... g, kg can be employed if this does not result in confusion For Mp, under certain assumptions t is admissible.	kp, Mp, mp, μg
Energy	erg J	Erg Joule	1 erg = 1 dyne × cm 1 J = 1 N × m = 10^7 erg	1 kJ, mJ
Power	W	Watt	1 W = 1 J/s 1 kW = 1.359... PS	kW, MW, mW, μW
Pressure and Tension	bar atm Torr at	Bar Atmosphere Torr (Tech.) Atmosphere	1 bar = 10^5 N/m^2 1 atm = 1.01325 bar = 1.0333 kp/cm^2 1 Torr = 1 atm/760 1 at = 1 kp/cm^2	mbar, μbar mTorr, μTorr
Viscosity	P St	Poise Stokes	Dynamic viscosity 1 P = 1g/cm s Kinetic viscosity 1 St = 1 cm^2/S	cP, μP cSt, mSt
Temperature	°K °C °F	Degrees Kelvin Degrees Centigrade Degrees Farenheit	absolute temperature	
Heat-quality	J cal	Joule Calorie	1 J = 1 Nm 1 cal = 4.1868 J	kcal, Mcal
Electrical and magnetic units	V A W Ω S Wb C J H F G M Oe	Volt Ampere Watt Ohm Siemens Weber Coulomb Joule Henry Faraday Gauss Maxwell Oersted	 1W = 1VA 1Ω = 1V/A 1S = 1A/V 1Wb = 1Vs 1C = 1As 1J = 1Ws 1H = 1Ωs 1F = 1 s/Ω 1G = 10^{-4} Vs/m^2 1M = 10^{-4} Gm^2 1 Oe = $\frac{10^3}{4\pi}\frac{A}{m}$	MV, kV, mV, μV k A, mA, μA MW, kW, mW, μW kΩ, mΩ kS, mS kH, mH, μH μF, nF, pF
Photo metric units	cd sb asb lm lx ph	Candela Stilb Apostilb Lumen Lux Phot	Unit of light strength 1sb = 1cd/cm^2 1asb = $\frac{10^{-4}}{\pi}$ sb 1lm = 1cd sr 1lx = 1lm/m^2 1ph = 10^4 lx	

Conversion of Units — Constants

Notation	Unit
Litre atmosphere (physical)	l atm = 10.333 kpm = 1.0133 × 10^9erg = 24.205 cal = 101.25 J
Litre atmosphere (technical)	l at = 10 kpm = 0.980665 × 10^9erg = 23.42 cal = 98.017 J
Metre kilo pond	kmp = 9.8065 × 10^7erg = 2.342 cal = 9.8017 J
Horse power	HP = 75 kpm/s = 7.355 × 10^9erg/s = 735.26 W
Mechanical equivalent heat	cal = 4.184 J = 4.186 × 10^7erg = 0.4269 kpm = 0.04131 latm = 0.4269 lat
Universal gas constant	R = 8.313 × 10^7erg/°C = 0.8477kpm/°C = 0.8204 latm/°C = 1.986
Loschmidt's (Avogadro's) number	N = 6.02 × 10^{23} molecules in 1 mole cal/°C = 8.309 J/°C
Molar volume of an ideal gas in standard conditions (0°C and 760 Torr)	V_{mol} = 2241.4 ml
Boltzmann's constant	k = R/N = 1.380 × 10^{-16}erg/degree

Notation	Units
Charge of an electron	$e = 1.601 \times 10^{-19}$ C (Coulomb)
Velocity of an electron	$v = 5.932 \times 10^7$ cm/s
Energy units in electron volts*	eVolt $= 1.602 \times 10^{-12}$ erg $= 3.827 \times 10^{-20}$ca
Energy units in "rest mass" 1g	1g $= 8.9868 \times 10^{20}$ erg $= 2.147 \times 10^{13}$ cal
Mass equivalent of 1 erg	1 erg $= 1.1127 \times 10^{-21}$ g
Mass equivalent of 1 cal	1 cal $= 4.658 \times 10^{-14}$ g
Planck's constant	$h = 6.626 \times 10^{-27}$ erg $\times$ s.

Conversion Table

		erg	kWh	J (Joule)	L. Atm	L. at	kpm	k cal
1 erg	=	1	2.79×10^{-14}	1.00×10^{-7}	9.87×10^{-10}	1.02×10^{-9}	1.02×10^{-8}	2.39×10^{-1}
1 kWh	=	3.60×10^{13}	1	3.60×10^6	3.55×10^4	3.67×10^4	3.67×10^5	8.60×10^2
1 J	=	1.00×10^7	2.78×10^{-7}	1	9.87×10^{-3}	1.02×10^{-2}	1.02×10^{-1}	2.39×10^{-4}
1 L. atm	=	1.01×10^9	2.81×10^{-5}	1.01×10^2	1	1.0332	1.03×10^1	2.42×10^{-2}
L. at	=	9.81×10^8	2.72×10^{-5}	9.80×10^1	0.9679	1	1.00×10^1	2.34×10^{-2}
1 kpm	=	9.81×10^7	2.72×10^{-6}	9.80	9.68×10^{-2}	1.00×10^{-1}	1	2.34×10^{-3}
1 k cal	=	4.19×10^{10}	1.16×10^{-3}	4.18×10^3	4.13×10^1	4.27×10^1	4.27×10^2	1
R (Gas const)	=	8.31×10^7	2.31×10^{-6}	8.31	8.2×10^{-2}	8.47×10^{-2}	8.48×10^{-1}	1.99×10^{-3}

Physical Principles

Mechanics

Physical force-unit	1 Dyne (It is that force which when imparted to a normal body** ($= 1$ l water) produces an acceleration 1 m/s² $= 1$ ms^{-2}.)
Technical force unit	1 Force kilogram (new notation: kilopond) (It is that force which when imparted to a normal body** produces an acceleration $g = 9.81$ m/s² $= 9.81$ ms^{-2})
Weight gravity	The force with which the earth pulls a body is called gravity. 1 kg (kilogram) is therefore the weight of normal body** (1 l of water)
Mass	The ratio of force F to the produced acceleration a which is a definite quantity, and for same bodies which is always equal, is called mass m. For a normal body** m is $= \dfrac{F}{a} = \dfrac{1\,kg}{9.81\,m/s^2} = 0.102$ kg m^{-1} s²
	For any given body this also implies $\boxed{m = \dfrac{G}{g}}$ G = Weight in kg, g = Gravity due to acceleration = 9.81 m/s² $= 9.81$ ms^{-2} m results in kg m^{-1} s²
Dynamics basic principles	Force = Mass × acceleration $\boxed{F = m \times a}$ m = mass in kg s²/m = kg m^{-1} s² a = acceleration in m/s² = ms^{-2} F results in kg.

Equilibrium

Type of equilibrium	Stable	Neutral	Labile
Sphere on different supports			
With displacement the centre of gravity moves	upwards	remains unaffected	downwards
In case of displacement	work is to be done	friction is to be overcome	work is done

* MeV = Million electron volt (See pp. 16–29)

** The normal body is a cylinder cut platinum-iridium and corresponds to 1 dm³ water (4°C) = 1 litre water

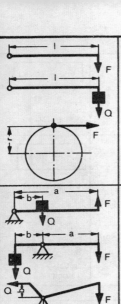

Moment

Moment M = Product of Force and lever arm.
Force moment M = Product of force and lever arm.
Load moment M = Product of load and lever arm.
Torsional moment M = Product of circumferential force and lever arm

Units: Force F and load Q in kilogram (kg)
Lever arm a, b, or r in metres (m)
Moment M in kilogram metres (kg)

Example: The circumference of a belt pulley of 200 Φ works on a belt of track force of 12 kg
$M = 12 \text{ kg} \times 0.1 \text{ m} = 1.2 \text{ kgm}$

$$M = F \times l \text{ (kgm)}$$

$$M = Q \times l$$

$$M = F \times r$$

$$F = \frac{M}{r} \text{ (kg)}$$

$$r = \frac{M}{F}$$

Lever

A lever remains in equilibrium when force moment = load moment
Lever arm is always perpendicular to force or load direction standing at a distance from the point of application of force towards fulcrum.

Example: At an angular lever are: a = 30 cm (a'=25cm), b=12cm (b'=10cm.) F=8kg

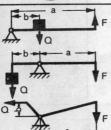

$$Q = \frac{8kg \times 25cm}{10cm} = 20 \text{ kg.}$$

$$F \times a = Q \times b$$

$$F = \frac{Q \times b}{a}$$

$$Q = \frac{F \times a}{b}$$

$$a = \frac{Q \times b}{F}$$

$$b = \frac{F \times a}{Q}$$

Wheel and axle

The radius R and r corresponds to the lever arms a and b at the two sides of the lever.

$$F \times R = Q \times r$$

$$F = \frac{Q \times r}{R}$$

Fixed pulleys

Equilibrium prevails when load Q is intercepted through an equal quantity of opposite force F. Through a fixed pulley, the direction of the opposite force is changed.

$$F = Q$$

Loose pulley

The load Q is collected on two cords.
The tension (opposing force) in each cord corresponds to half load.
The cord length s is twice the delivery head or pressure head h.

$$F = \frac{Q}{2}$$

$$Q = 2F$$

$$s = 2h$$

Set of pulleys; block and tackle

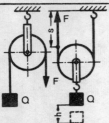

Through a fixed pulley results the directional change of force.
Through loose pulleys, the load Q is divided out of two tracked cords.
The track force F = load Q divided by number of tracks of cords (or no. of pulleys)

Example: A set of pulleys with 6 pulleys
$$Q = 72kg \qquad F = \frac{72kg}{6} = 12 \text{ kg}$$

$$F = \frac{Q}{n}$$

$$Q = F \times n$$

$$s = n \times h$$

Differential set of pulleys; block and tackle

Through a loose pulley the load Q is divided into two cords.
According to the valid rules of lever results the transmission of cord pull through a wheel and axle.

Example: Load Q = 60kg, R = 160mm Φ, r = 80mm Φ
$$F = \frac{60kg}{2} \times \frac{160mm - 80mm}{160mm} = 15kg$$

$$F = \frac{Q}{2} \; \frac{R - r}{R} \text{ (kg)}$$

$$Q = \frac{2 \times F \times R}{R - r} \text{ (kg)}$$

Motion

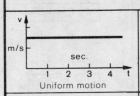

Uniform motion

Velocity

Velocity v = Path covered s divided by the time t taken to travel it.

Units: Distance s in metres (m) or kilometres (km)
Time t in seconds (s) or minutes (min.) or hours (h)
Velocity v in m/sec, m/min. or km/h.

$$v = \frac{s}{t} \ (m/s)$$

$$s = vt$$

$$t = \frac{s}{v}$$

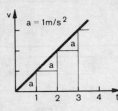

Uniform acceleration motion

Acceleration:

Acceleration a = velocity increase in a second
(Acceleration due to gravity see above)
Decelaration = negative acceleration = velocity decrease in a second

Units: Velocity v in m/sec.
time t in seconds (sec.)
acceleration a in m/sec²

Example: Automobile has 20 s after a drive at 72 km/h.

$$a = \frac{\frac{72 \times 1000}{3600} \ m/s}{20s} = 1 m/s^2$$

$$a = \frac{v}{t} \ (m/s^2)$$

$$v = at \ (m/s)$$

$$t = \frac{v}{a}$$

$$s = \frac{v}{2} \times t \ (m)$$

Initial velocity = 0

Energy

Energy is the capacity of body to carryout work.

Energy is a state of matter.

Potential energy = Mass times acceleration due to gravity times the height difference between two motionless bodies.

Kinetic energy = Half the mass times the square of instantaneous velocity of moving bodies.

Mechanical energy E = By virtue of inducing work of a body consequently their position or their motion.

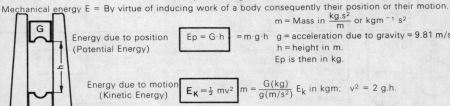

m = Mass in $\frac{kg.s^2}{m}$ or $kgm^{-1} s^2$

Energy due to position (Potential Energy) $\boxed{E_p = G \cdot h}$ = m·g·h   g = acceleration due to gravity = 9.81 m/s²
h = height in m.
Ep is then in kg.

Energy due to motion (Kinetic Energy) $\boxed{E_K = \frac{1}{2} mv^2}$ $m = \frac{G(kg)}{g(m/s^2)}$ E_k in kgm; $v^2 = 2 g.h.$

Example: The weight of a hammer G = 250kg of a pile driver falls from a height of h = 2m, Ep is = G × h 250kg × 2m = 500kg, m. This energy must also produce by the free fall (without any friction) of the forge bloom as kinetic energy.

$$E_k = \frac{m \times v^2}{2} = \frac{250kg \times 2m \times 9.81 m/s^2 \times 2}{9.81 m/s^2 \times 2} = 500 kgm$$

Work

Force = F kg

Work is a complete process with a change of position
Work w = Force F × Distance s (s measured in direction of force)

$\boxed{A = F \times S}$
A = kgm

A = G·h

Example: If a load of force F = 25kg s = 6m lifted and moved over with F = 25kg in an arbitrary direction s = 6m
then W = F × s
= 25kg × 6m = 150kgm

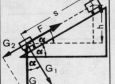

Along an inclined plane, an equal work is to be performed due to a perpendicular lifting.

Example: What work is necessary for a wagon of 5000 kg to move over an inclination 1:100, 1km wide?   G = 5000kg   h = 10m
Solution W = G.h. = 5000kg × 10m = 50000 kgm
or W = F × s = 50kg × 1000m = 50000 kgm
From the law of parallelogram of forces, we find F.
G can be split up into G₁ and G₂
F = G₂ and oppositely directed. Mathematically F = G₂ = G sin. α

Electrical work

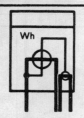

Electrical work is the product of electrical power and time while the power is active.

    a) Power in watt., time in seconds
    b) Power in kilowatt., time in hours.

Units
Electrical work in watt seconds (Ws)
or   in watt hours (Wh.)
or   in kilowatt hours (kWh)

Example: Heating oven of 1.5 kW. is switched on for 4 hrs.
    $W = 1.5 \text{ kW.} \times 4 \text{ h} = 6 \text{ kWh.}$

$$A = P \times t\,(Ws)$$

$$P = \frac{A}{t}, \quad t = \frac{A}{P}$$

$$A = P \times t\,(Wh)$$
$$A = P \times t\,(kWh)$$

$$A = U \times J \times t$$

$$t = \frac{A}{P} = \frac{A}{U \times J}$$

Power

Mechanical power

t sek

75 kg m/s = 1 PS
      = 0,736 kW
102 kg m/s = 1 kW
      = 1,36 PS

Power P = work W in time units

W in kgm; t in sec;
P becomes in kgm/s.

F in kg; s in m/s; t in sec.
P becomes kgm/s

F in kg; V in m/s
F becomes in kgm/s

Power P = Force F × velocity v

Example: If a load of
$F = 25$ kg; $s = 6$m in
5 sec (t) is to be raised high
then the power is
$$P = \frac{W}{t} = \frac{F \times S}{t} = F \times V$$
$$\frac{25 \text{kg} \times 6\text{m}}{5 \text{ sec.}} = 30 \text{kgm/sec.}$$

Example: what kW power a motor must have for lifting 1000 kg in 1 min a height of 12m.

$W = F \times s = 1000 \text{ kg.} \times 12\text{m} = 12000 \text{ kgm.}$

$P = \dfrac{W}{t} = \dfrac{12000 \text{ kgm}}{60s} = 200 \text{ kgm/s} = \dfrac{200}{102}\text{kW} \approx 2 \text{ kW}$

$$P = \frac{A}{t}$$

$$P = \frac{F \times s}{t}$$

$$P = F \times v$$

$$P_{ps} = \frac{F \times v}{75}$$

$$P_{kW} = \frac{F \times v}{102}$$

Electrical power

1. The electrical power is the product of current and voltage.

2. The electrical power is the product of resistance and the square of current strength.

3. P $\dfrac{\text{Proportional to the square of } U}{\text{Inversely proportional to } R}$

Units:
Power P in watt (W)
Current J in ampere (A)
Voltage U in volt (V)
Resistance R in ohm (Ω)

Example 1: A heating oven receives 220V, 10A
    $P = 220V \times 10A = 2200W$
Example 2: Through a resistance 500 Ω flows 0.2A
    $P = (0.2A)^2 \times 500 \ \Omega = 20W$
Example 3: In a resistance of 600 Ω, a current of 30V comes forth
$$P = \frac{(30V)^2}{600\,\Omega} = 1.5W$$

$$P = U \times J\,(W)$$

$$J = \frac{P}{U}, \quad U = \frac{P}{J}$$

$$P = J^2 \times R\,(W)$$

$$R = \frac{P}{J_2}, \quad J = \sqrt{\frac{P}{R}}$$

$$P = \frac{U^2}{R}\,(W)$$

$$R = \frac{U^2}{P}, \quad U = \sqrt{P \times R}$$

Efficiency

In case of every work and power, there is a loss through friction, radiation and so on. The useful work or the useful power is always smaller than the spent, and supplied work or power.

Efficiency $\eta = \dfrac{\text{useful work, Wu}}{\text{supplied work, Ws}}$ $\boxed{\eta = \dfrac{Wu}{Ws}}$ or $\eta = \dfrac{\text{Useful power, Pu}}{\text{supplied power, Ps}}$ $\boxed{\eta = \dfrac{Pu}{Ps}}$ $\eta = \dfrac{Pu}{Ps} \times 100$ in %

η is always smaller than 1, A machine is so good according to how close η approaches 1

Efficiency	Steam Turbine	Water Turbine	Steam Engine	IC Engine	Electric Motor
	0.18...0.22	0.85...0.9	0.15...0.2	0.25...0.30	0.8...0.9

Electrical Engineering

Spectrum of electromagnetic vibrations

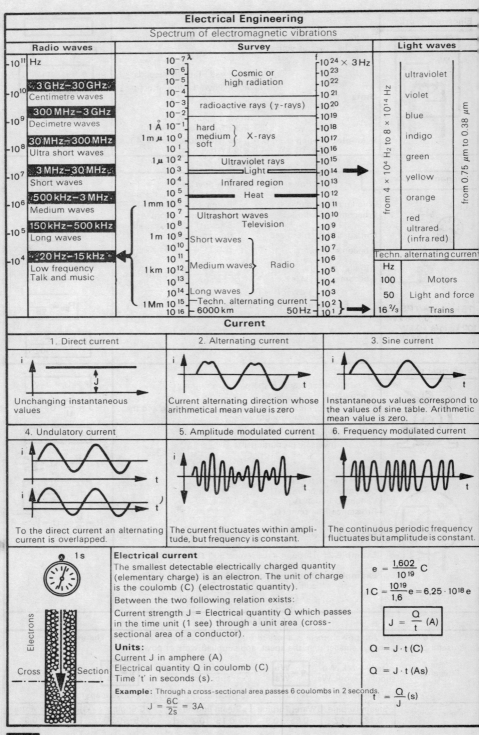

Radio waves	Survey	Light waves
10^{11} Hz	10^{-7} λ	

Radio waves column:
- 10^{10} — 3 GHz–30 GHz — Centimetre waves
- 10^{9} — 300 MHz–3 GHz — Decimetre waves
- 10^{8} — 30 MHz–300 MHz — Ultra short waves
- 10^{7} — 3 MHz–30 MHz — Short waves
- 10^{6} — 500 kHz–3 MHz — Medium waves
- 10^{5} — 150 kHz–500 kHz — Long waves
- 10^{4} — 20 Hz–15 kHz — Low frequency Talk and music

Survey column (λ and f):
- 10^{-7} ... 10^{-6}, 10^{-5}, 10^{-4} — Cosmic or high radiation — $10^{24} \times 3$ Hz, 10^{23}, 10^{22}, 10^{21}
- 10^{-3}, 10^{-2} — radioactive rays (γ-rays) — 10^{20}, 10^{19}
- 1 Å 10^{-1} — hard — 10^{18}
- 1 mμ 10^{0} — medium } X-rays — 10^{17}
- 10^{1} — soft — 10^{16}
- 1μ 10^{2} — Ultraviolet rays — 10^{15}
- 10^{3} — Light — 10^{14}
- 10^{4} — Infrared region — 10^{13}
- 10^{5} — Heat — 10^{12}
- 1 mm 10^{6} — 10^{11}
- 10^{7} — Ultrashort waves — 10^{10}
- 10^{8} — Television — 10^{9}
- 1 m 10^{9} — Short waves — 10^{8}
- 10^{10} — 10^{7}
- 10^{11} — 10^{6}
- 1 km 10^{12} — Medium waves } Radio — 10^{5}
- 10^{13} — 10^{4}
- 10^{14} — Long waves — 10^{3}
- 1 Mm 10^{15} — Techn. alternating current — 10^{2}
- 6000 km 10^{16} — 50 Hz — 10^{1}

Light waves column:
from 4×10^{14} Hz to 8×10^{14} Hz — from 0.75 μm to 0.38 μm
- ultraviolet
- violet
- blue
- indigo
- green
- yellow
- orange
- red
- ultrared (infra red)

Techn. alternating current

Hz	
100	Motors
50	Light and force
16 ⅔	Trains

Current

1. Direct current	2. Alternating current	3. Sine current
Unchanging instantaneous values	Current alternating direction whose arithmetical mean value is zero	Instantaneous values correspond to the values of sine table. Arithmetic mean value is zero.

4. Undulatory current	5. Amplitude modulated current	6. Frequency modulated current
To the direct current an alternating current is overlapped.	The current fluctuates within amplitude, but frequency is constant.	The continuous periodic frequency fluctuates but amplitude is constant.

Electrical current

The smallest detectable electrically charged quantity (elementary charge) is an electron. The unit of charge is the coulomb (C) (electrostatic quantity).

Between the two following relation exists:

Current strength J = Electrical quantity Q which passes in the time unit (1 see) through a unit area (cross-sectional area of a conductor).

Units:
Current J in amphere (A)
Electrical quantity Q in coulomb (C)
Time 't' in seconds (s).

Example: Through a cross-sectional area passes 6 coulombs in 2 seconds.

$$J = \frac{6C}{2s} = 3A$$

$$e = \frac{1{,}602}{10^{19}}\ C$$

$$1\,C = \frac{10^{19}}{1{,}6}\,e = 6{,}25 \cdot 10^{18}\,e$$

$$\boxed{J = \frac{Q}{t}\ (A)}$$

$$Q = J \cdot t\ (C)$$

$$Q = J \cdot t\ (As)$$

$$t = \frac{Q}{J}\ (s)$$

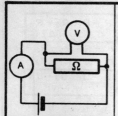

Ohm's Law:

$J \longleftarrow$ $\dfrac{\text{Proportional } U}{\text{Inversely proportional } R}$

Units: Current J in ampere (A), voltage U in volt (V)
Resistance R in ohm $\rightarrow$ (Ω)

Example: A resistance of 16 Ω registers a voltage of 4V.

$$J = \frac{4\,V}{16\,\Omega} = 0{,}25\,A$$

$$\boxed{J = \frac{U}{R}\ (A)}$$

$$R = \frac{U}{J}$$

$$U = J \cdot R$$

1m 2m 1m 2m
1 1 2 2
m² m² m² m²

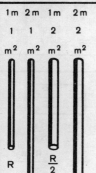

R $\dfrac{R}{2}$

2:R $\dfrac{2 \cdot R}{2}$

Resistance R of a conductor

$R \longleftarrow$ $\dfrac{\text{Proportional to } l}{\text{Inversely proportional to } A}$

Units:
Length l in metres (m)
Resistance R in ohm (Ω)

Conductor $\varkappa$ in metre per ohm and mm² ($\dfrac{m}{\Omega \times mm^2}$)
or
Metre times Siemens per mm² ($\dfrac{S \times m}{mm^2}$)

Unit resistance ϱ in ohm $\times$ mm² per metre ($\dfrac{\Omega \times mm^2}{m}$)

Cross-sectional area A in mm².

Example: A 114m long conductor of copper has cross-sectional area of 10mm².

$$R = \frac{114m}{10mm^2 \times 57 \dfrac{m}{\Omega \times mm^2}} = 0.2\ \Omega$$

or

$$R = \frac{114m \times 0.0175 \dfrac{\Omega \times mm^2}{m}}{10mm^2} = 0.2\ \Omega$$

$$\boxed{R = \frac{l}{A \cdot \varkappa}\ (\Omega)}$$

$$l = R \cdot A \cdot \varkappa$$

$$A = \frac{l}{R \cdot \varkappa}$$

$$\boxed{R = \frac{l \cdot \varrho}{A}\ (\Omega)}$$

$$l = \frac{R \cdot A}{\varrho}$$

$$A = \frac{l \cdot \varrho}{R}$$

Parallel connections of resistances

Several parallely connected resistances (R_1, R_2, R_3) can be thought of as one resistance of definite quantity.
This is termed as equivalent resistance R_E

1. All parallel resistances are of same size.
   Equivalent resistance = single resistance divided through the number of parallel resistances.

2. Two parallel resistances are of different sizes
   Equivalent resistance = product of resistances divided by sum of the resistances.

3. The conductance of an arbitrary parallel connection is equal to the sum of individual single conductivity value.

4. In a parallel connection, the passing current distributes into individual (single) branches corresponding to their conductances.

Units: Conductance G in Siemens (S) $G = \dfrac{1}{R}$

Example 1: $R_1 = 4\ \Omega$, $R_2 = 6\ \Omega$, $R_3 = 12\ \Omega$
They lie in parallel at 24V.

$$G = \frac{1}{4}S + \frac{1}{6}S + \frac{1}{12}S = \frac{1}{2}S \cdot R_E = 2\ \Omega \quad J = \frac{24\,V}{2\,\Omega} = 12\,A$$

Example 2: $R_1 = 4\ \Omega$, $R_2 = 12\ \Omega$, $J = 16\,A$.

$$J_1 = 16\,A \frac{12\ \Omega}{4\ \Omega + 12\ \Omega} = 12\,A$$

$$J_2 = 16\,A \frac{4\ \Omega}{4\ \Omega + 12\ \Omega} = 4\,A$$

$$\boxed{R_E = \frac{R}{n}\ (\Omega)}$$

$$R_E = \frac{R_1 \cdot R_2}{R_1 + R_2}\ (\Omega)$$

$$R_1 = \frac{R_2 \cdot R_E}{R_2 - R_E},$$

$$R_2 = \frac{R_1 \cdot R_E}{R_1 - R_E}$$

$$\boxed{\begin{array}{l} G = G_1 + G_2 + G_3 \\ \qquad\quad + \cdots (S) \end{array}}$$

$$G_1 = G - G_2 \\ \qquad\quad - G_3 - \cdots$$

$$G_2 = G - G_1 \\ \qquad\quad - G_3 - \cdots$$

$$\boxed{J_1 : J_2 = G_1 : G_2}$$

$$J_1 : J_2 = R_2 : R_1$$

$$J_2 = J_1 \cdot \frac{R_1}{R_2};$$

$$J_1 = J_2 \cdot \frac{R_2}{R_1}$$

$$J_1 = J \frac{R_2}{R_1 + R_2}$$

$$J_2 = J \frac{R_1}{R_1 + R_2}$$

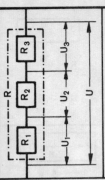

Resistances connected in series

1. Total resistance of the connection = sum of all the resistances
   At these resistances partial voltage appear.
2. The total voltage = sum of all the partial voltages.
3. The voltages are the product of current and resistance.
4. The voltages act according to resistances.

Example: Resistances of 5Ω, 10Ω, 15Ω connected in series at 60V

$$R = 5\,\Omega + 10\,\Omega + 15\,\Omega = 30\,\Omega$$
$$J = \frac{60\,V}{30\,\Omega} = 2A, \quad U_1 = 2A \times 5\,\Omega = 10V$$
$$U_2 = 2A \times 10\,\Omega = 20V, \quad U_3 = 2A \times 15\,\Omega = 30V$$

$$R = R_1 + R_2 + R_3 + \cdots \,(\Omega)$$
$$R_1 = R - R_2 - R_3 - \cdots$$
$$R_2 = R - R_1 - R_3 - \cdots$$

$$U = U_1 + U_2 + U_3 + \cdots \,(V)$$

$$U_1 = U - U_2 - U_3 - \cdots$$
$$U_1 = JR_1, \quad U_2 = R_2,$$
$$U_3 = JR_3$$
$$J = \frac{U}{R} = \frac{U_1}{R_1} = \frac{U_2}{R_2} = \frac{U_3}{R_3}$$

Bridge connections

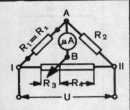

In an equilibrium condition between A and B there is no potential difference. Then the resistance in arm I–A–II is like that of arm I–B–II. $J_{A-B} = 0$.

Example: $R_2 = 50\,\Omega$, $R_3 = 30\,\Omega$, $R_4 = 60\,\Omega$

$$R_x = R_1 = \frac{50\,\Omega \times 30\,\Omega}{60\,\Omega} = 25\,\Omega$$

$$\frac{R_1}{R_2} = \frac{R_3}{R_4}$$
$$\frac{R_1}{R_3} = \frac{R_2}{R_4}$$
$$R_1 = R_x = \frac{R_2 R_3}{R_4}$$

Multiphase current or Alternating current

No phase lag between voltage and current

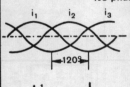

Sum of all cord currents = 0

when: a) the phase lag is 120° opposite to each other and
b) the current values are equal.

The power P of multiphase current consumers = sum of cord power P_{cor}

Cord power P_{cor} = cord voltage

times the cord current = $U_{cor} \times J_{cor}$

Power P of the multiphase current consumers = conducting voltage times the conducting current multiplied by 1.73

Star connection

Triangular connection

$$J_1 + J_2 + J_3 = O$$

$$P = 3 \cdot P_{cor}\,(W)$$
$$P = P_{u-x} + P_{v-y} + P_{w-z}$$
$$P_{cor} = U_{cor}\,J_{cor}$$
$$P = U \cdot J \cdot 1.73 \,(W)$$
$$J = \frac{P}{U \cdot 1.73}; \quad U = \frac{P}{J \cdot 1.73}$$

Star connections

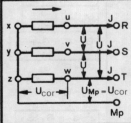

Conducting current J = Cord current J_{cor}

,, ,, voltage U = Cord voltage × 1.73

$U_{cor} = U_{u-x} = U_{v-y} = U_{w-z}$
= Cord voltage

$U = U_{RS} = U_{RT} = U_{ST}$
= Conducting voltage (supply voltage)

Example: $J = J_{cor} = 10A$, $U_{cor} = 200V$
$U = 200V \times 1.73 = 346V$
$P = 346V \times 10A \times 1.73 = 6000W$
$P = 3 \times 200V \times 10A = 6000W$

$$J = J_{u-x} = J_{v-y} = J_{w-z}$$
$$J = J_{cor}$$
$$U = 173 \cdot U_{cor}$$
$$U_{cor} = \frac{U}{1.73}$$

Triangular connections

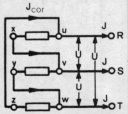

Conducting voltage U = Cord voltage U_{cor}

,, ,, current J = Cord current U_{cor} × 1.73

$J_{cor} = J_{u-x} = J_{v-y} = J_{w-z}$ = Cord current
$J = J_R = J_S = J_T$ = Conducting current
(current in outer conductor)

Example: J = 15A, U = 380V
$P = 380V \times 5A \times 1.73 = 9883W$
$J_{cor} = \frac{15A}{1.73} = 8.67A$
$P = 3 \times 380V \times 8.67A = 9883W$

$$U = U_{u-x} = U_{v-y} = U_{w-z}$$
$$U = U_{cor}$$
$$J = 1.73 \cdot J_{cor}$$
$$J_{cor} = \frac{J}{1.73}$$

Transformer

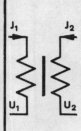

Transmission T = Ratio of voltage of a coil with larger winding number to the voltage of a coil with a smaller winding number in case of idle motion.

The **Voltage ratio** agrees with winding number sufficiently accurately in case of idle motion.

N_1 = greater winding number, N_2 = smaller winding number.

U_1 = **Rated primary voltage** = voltage for which the primary coil is measured.

U_2 = **Rated secondary voltage** = when at the primary coil arises a nominal voltage, at the terminals secondary coil an idle motion voltage.

The **Current**, primary and secondary behave inversely as the winding numbers of the coils.

J_2 = **Rated secondary current** = secondary side is built for the transformer coil.

J_1 = **Rated primary current** = Current which results when a rated secondary current with the ratio rated secondary voltage to the rated primary voltage is multiplied (does not correspond quite to the primary current in case of rated primary voltage and rated secondary voltage).

P_S = **Rated power** = "Type Power" = Calculation quantity for a secondary voltage, rated secondary current and phase factor. It does not correspond to the effective apparent output in case of rated performance, when secondary terminal voltage is very small than the rated voltage.

P_1 = **Input power** = Effective power at the terminals of primary coil.

P_2 = **Output power** = Effective power at the terminals of secondary coil.

n = Efficiency = $\dfrac{\text{output power}}{\text{output power + loss}}$

$$T = \frac{U_1}{U_2}, \quad T = \frac{N_1}{N_2}$$

$$\frac{U_1}{U_2} = \frac{N_1}{N_2}$$

$$\frac{J_1}{J_2} = \frac{N_2}{N_1}$$

$$J_1 = J_2 \times \frac{N_2}{N_1}$$

$$J_2 = J_1 \times \frac{N_1}{N_2}$$

Single phase current
$$P_S = U_2 \cdot J_2 \cdot \text{(VA)}$$

Multipurpose current
$$P_S = U_2 \cdot J_2 \cdot 1{,}73 \text{(VA)}$$

$$P_1 = U_1 \cdot J_1 \cdot \cos \varphi_1$$

$$P_2 = U_2 \cdot J_2 \cdot \cos \varphi_2$$

$$\eta = \frac{P_2}{P_2 + V}$$

Heat

Temperature scales

Symbols	Nomenclature	Zero point
°K	Degree kelvin, Absolute temperature	Absolute zero
°C	Degree Celsius	Ice point
°F	Degree Fahrenheit	−17.8° C

Conversion of degrees of temperature

Given temperature in	Desired temperature in °K	°C	°F
°K	°K	°K-273.1	(1.8°K)-459.4
°C	°C + 273.1	°C	(1.8°C) + 32
°F	(0.556°F)+255.3	(0.556°F)-17.8	°F

Example:
20°C = 293.1°K = 68°F
320°K = 46.9°C = 116.6°F
210°F = 98.9°C = 372°K

Thermometer fixed points in °C

B.P of oxygen	−183.0
B.P of carbondioxide	− 78.5
F.P of mercury	− 38.87
F.P of water	0.00
B.P of water	+100.0
B.P of naphthalene	+217.9
Softening point tin	+231.8
B.P. benzophenone	+305.9
S.P. cadmium	+320.9
S.P zinc	+419.4
B.P sulphur	+444.6
S.P antimony	+630.5
S.P silver	+960.5
M.P gold	+1063
S.P copper	+1083
M.P palladium	+1557
M.P platinum	+1770
M.P tungston	+3400

BP = Boiling point MP = melting point
FP = Freezing point
SP = Softening point.

Unit of heat quantity: 1 cal is the heat quantity which is necessary to heat 1 g of water to 1° C
[from 14.5 to 15.5°]
Heat quantities are measured in Calorimeters.

The **specific heat** c is the heat quantity (cal), taken up by heating 1 g of substance through 1° C.
The mean specific heat at a temperature interval between t_1 and t_2 is the heat quantity taken up
divided by the temperature difference $t_2 - t_1$ (Δt).

The specific heat can be given either at constant pressure (c_p) or at constant volume (c_v);
c_v is smaller than c_p when work done with container in case of expansion.
In case of solid and liquid substances the specific heat is given at constant pressure as a rule; in case
of gases either at constant pressure or at constant volume.

Atomic heat: is the product of specific heat and atomic weight.

Molar heat is the product of specific heat and molecular weight.

According to the law of Dulong and Petit, the atomic heat of elements in case of mean and high
temperature amounts to 6.2 cals/mole/degree. The elements with low atomic weights strongly
deviate from this rule.

Heat of fusion is the heat quantity (cal) that must be supplied at the melting temperature of 1 g of a substance changing over from solid to liquid state.

Heat of vaporization is the heat quantity (cal) that must be supplied at evaporation temperature of 1 g of
a substance changing over from liquid to vapour state.

Heat of cumbustion the heat quantity (kcal) which is liberated in case of burning of a mole of a
substance.

Heat value is the heat quantity which is liberated by burning of a kg. of a fuel.

High heat value: when the temperature of the fuel and of the combustion products amounts
                   to 20°C and in the available fuel the water formed due to combustion.
                   exists in liquid state.
Low heat value: when the temperature of the fuel and of combustion products
                   amounts to 20° C and is available in the water formed due to combustion exists
                   in the vapour state.

Mean specific heats of gases and vapours c_p in kcal/m³ from 0° to t° C							
t°	H₂	N₂	CO	O₂	H₂O	CO₂	air
0	0.310	0.310	0.310	0.312	0.354	0.382	0.311
100	0.310	0.311	0.311	0.314	0.358	0.406	0.312
200	0.310	0.311	0.313	0.319	0.362	0.429	0.313
300	0.310	0.313	0.315	0.324	0.367	0.448	0.315
400	0.310	0.315	0.318	0.329	0.372	0.464	0.318
500	0.311	0.318	0.321	0.333	0.378	0.478	0.321

Specific heats of metals c_p in kcal/kg and per 0° C at 20° C						Heat of vaporization at BP in kcal/kg	
Metal	c_p	Metal	c_p	Metal	c_p	Substance	
Na	0.288	W	0.032	An	0.031	Ammonia	300
Be	0.418	Mn	0.116	Zn	0.092	Sulphurdioxide	95
Mg	0.243	Fe	0.108	Cd	0.055	Freon 12	40
Al	0.214	Ni	0.106	Hg	0.033	Alcohol	210
V	0.12	Pt	0.032	Si	0.168	Ether	90
Cr	0.105	Cu	0.092	Sn	0.054	Benzene	94
Mo	0.06	Ag	0.056	Pb	0.031	Toluene	85

Mean specific heats c_p of solids and liquids in kcal/kg and per degree						Carbondisulphide	85
						Pentane	86
Substance	Temp. in °C	c_p	Substance	Temp. in °C	c_p	Hexane 30° C	80
Asbestos	20..98	0.195	Bakelite	20	0.383	Heptane 60° C	76
Dolomite	20..98	0.222	Cotton	18	0.304	Turpentine oil	70
Feldspar	20..100	0.191	Gascoke	0..420	0.278	Water	539
Mica	20..98	0.208	Graphite	20..480	0.297		
Copper pyrites	15..99	0.129	Charcoal	18	0.277		
Magneticpyrites	0..100	0.153	Machine oil	25..35	0.474		
Talcum	20..98	0.209	Petrolium	0..100	0.55		

Calorific value (H_u) of fuels in kcal/kg

Solid fuels		Liquid fuels	
Carbon	8100	Alcohol, absolute	6400
γ-Graphite	7860	Alcohol, 90% by weight	5700
Sugar charcoal	8060	Benzine	10400
Wood (air dried, 16% water and 1.5% ash)	3500	Gasoline	10300
Peat (with about 25% water and 50% ash)	3200–3800	Petroleum	10500
Crude lignite (with about 48% water and 7% ash)	2600	Diesel oil	10000–10200
Lignite briquettes (15% water and 8% ash)	4800	Fuel oil	
Mineral coal		extralight	10000
fresh charcoal with 4% water and 8% ash	6900	light	9800
sicilious with 3% water and 5% ash	7000	medium	9600
old coal with 2.5% water and 5% ash	7500	heavy or hard	9400
Mineral coal briquettes	7800	Coal tar oil	8800–9100
Anthracite, 2% water and 4% ash	7800	Benzene	9600
Gas coke, 3% water and 8% ash	7000	Paraffin oil	9800

Caloric values of industrial gases (kcal/m³) under normal condition

Gas	Formula	H_u	H_o	Gas	Formula	H_u	H_o
Carbon monoxide	CO	3020		Propane	C_3H_8	22350	24320
Hydrogen	H_2	2570	3050	Propylene	C_3H_6	21070	22540
Methane	CH_4	8550	9520	Acetylene	C_2H_2	13600	14090
Ethane	C_2H_6	15370	16820	Benzene vapour	C_6H_6	33500	35000
Ethylene	C_2H_4	14320	15290	Hydrogen sulphide	H_2S	5660	6140

Heating and cooling

In order to heat a substance, a quantity of heat Q is to be supplied. The heat is given through the combustion of a fuel, through electrical current or through conversion of other energies into heat.

Q is proportional to the weight of substance G, the temperature difference and the specific heat c.

Q = heat quantity in cal or kcal
G = Weight in gram or kilogram
Δt = Final temperature—Initial temperature
c = Sp. heat in cal/g/°C or kcal/kg/°C

If a substance is to be converted into vapour a large amount of heat is to be fed first till the boiling point is reached.
Thereafter no additional heat of evaporation is conveyed.
By cooling a substance, the heat quantity set free is equal to that supplied.

$$Q = G \times c \times \Delta t \text{ (cal or kcal)}$$

$$c = \frac{Q}{G \times \Delta t}$$

$$G = \frac{Q}{c \times \Delta t}$$

$$Q = G \times c \times \Delta t + G \times \text{Heat of vaporization}$$

1 Ws produces 0.24 cal
1 kWh produces 860 kcal

The heat Q generated through the electrical current is directly proportional to the electrical power P and the input time t.
Units: The heat quantity Q in calories (cal)
Electrical power P in watt (W)
Input time t in seconds (s) or:
Heat quantity in kilo cal (kcal)
Power P in kilowatts (kW)
Input time t in hours (h)
In case of electrical heating instruments, the loss of heat through conduction or radiation can occur. This aspect leads to the consideration of efficiency

Example: A 2 litre container filled with water at 10° has to be heated to 100° in 15 minutes.
Here $\eta = 0.90$
$Q = 2l \times 90° = 180$ kcal
$$P = \frac{180 \text{ kcal}}{860 \text{ kcal/kW} \times 0.25h \times 0.90}$$
$$= 0.930 \text{kW}$$

$$Q = 0.24 \times P \times t \text{ (cal)}$$

$$Q = 0.24 \times U \times J \times t$$

$$P = \frac{Q}{0.24 \times t}$$

$$t = \frac{Q}{0.24 \, P}$$

$$Q = 860 \times P \times t \text{ (kcal)}$$

$$P = \frac{Q}{860} \, t$$

$$t = \frac{Q}{860} \, P$$

$$Q = 0.24 \times P \times t \times \eta \text{ (cal)}$$
$$Q = 860 \times P \times t \times \eta \text{ (kcal)}$$

Temperature of mixtures

The heat content of a substance is given by the weight of the substance w, specific heat c and its temperature t.

If different substances are mixed, it does not alter thereby the heat content. The temperature t_m of the mixture changes when the different substances have different temperatures.

$$Q = W \times c \times t \text{ (cal or kcal)}$$

Units

t_m = Temperature of mixture

w_1 = Weight
c_1 = Sp. heat $\Big\}$ of first substance
t_1 = Temperature

w_2 = Weight
c_2 = Sp. heat $\Big\}$ of second substance
t_2 = Temperature

$$t_m = \frac{w_1 \cdot c_1 \cdot t_1 + w_2 \cdot c_2 \cdot t_2 + \dots}{w_1 \cdot c_1 + w_2 \cdot c_2 + \dots}$$

Example : Find t_m

given: First substance is iron
$w_1 = 1.5$ kg
$c_1 = 0.108$
$t_1 = 208°C$

$$t_1 = \frac{[t_m(w_1 \cdot c_1 + w_2 \cdot c_2)] \, w_2 \cdot c_2 \cdot t_2}{w_1 \cdot c_1}$$

Second substance is lubricating oil
$w_2 = 8$ kg
$c_2 = 0.474$
$t_2 = 20°C$

$$t_2 = \frac{[t_m(w_1 \cdot c_1 + w_2 \cdot c_2)] \, w_1 \cdot c_1 \cdot t_1}{w_2 \cdot c_2}$$

$$t_m = \frac{w_1 \times c_1 \times t_1 + w_2 \times c_2 \times t_2}{w_1 \times c_1 + w_2 \times c_2} = \frac{1.5 \times 0.108 \times 208 + 8 \times 0.474 \times 20}{1.5 \times 0.108 + 8 \times 0.474} = 30.7°C$$

Light

Light rays Φ

Luminous flux Φ = Power of light radiation (light quantity radiated from a light source in a unit time) Unit = lumen (lm)

$$\Phi = \frac{Q}{t} \text{ (lm)}$$

Intensity of light J = Light rays density inside a solid angle (portion of the total light radiations Φ escaping out of a solid angle). Unit = candela (cd)

$$J = \frac{\Phi}{\omega} \text{ (cd)}$$

1 cd

$1 m^2$ 1 lm

Solid angle ω = The angle in space at the tip of a cone. It is measured through the area in which the mantle of cone cuts out a unit cone (radius lm) from the cone surface area. The solid angle is described as I when the cross-section from the surface area of unit cone is lm².

1 Lumen = Luminous flux of a pointed light source of strength. 1 cd emitted inside of a unit solid angle.

1 candela (cd) = 1 New candle = 1.09 Hefner candle

Degree of temperature v = Ratio of emergent light rays Φ_e (emergent luminous flux) to the incident rays Φ_i.

Extinction E = negative logarithm of transparency degree v.

The extinction E is a calorimetric and photometric important quantity which in case of graphical record gives a linear curve.

$1cd = 1NK$

$\approx 1,16$ HK

$$\vartheta = \frac{\Phi_e}{\Phi_i}$$

$$E = -\log \vartheta$$

Φ_i Φ_e

Illuminating intensity E = that of luminous flux falling out of unit area; Unit of illuminating intensity E in lux (lx). **1 Lux** is available when a luminous flux of 1 Lumen is radiated on an area of 1 m².

The illumination intensity decreases in proportion to (a) the square of distance, (b) the angle of inclination of irradiated area opposite the normal surface.

$$E = \frac{\Phi}{A} \text{ lx}$$

$$\Phi = E \cdot A$$

$$A = \frac{\Phi}{E}$$

$$E = \frac{J}{r^2} \text{ (lx)}$$

Chemical principles	
Fundamental laws	
Law of conservation of matter and mass	Matter can be neither created nor destroyed. In case of chemical reactions the total weight of reactants is equal to the total weight of products.
Conservation of elements	An element cannot be converted into another element. [Exceptions: Nuclear reactions] In case of chemical reactions, the number of atoms remain unchanged, i.e. the number of atoms of elements before and after the reaction is equal.
Constant Proportion	The composition of a compound is constant, i.e. in a compound, the constituent elements stand to one another in a constant ratio of weight.
Definitions	
Atomic weight or atomic mass	The atomic weight of an element indicates how many times the atom of that element is heavier than $1/12$ atom of carbon isotope 12.
Molecular weight or molecular mass	The molecular weight indicates how many times the molecule of the substance is heavier than $1/12$ atom of carbon isotope 12. The molecular weight is equal to sum of the atomic weights of the atoms contained in that molecule.
Valency	The valency indicates the number of hydrogen atoms that can bind or substitute an atom. In case of oxidations and reductions medium the valency indicates the number of electrons an atom or a molecule or an atom group can receive or give up.
Gram atom (g-atom)	1 gram atom of an element is its atomic weight in gram.
Gram molecule (mole)	1 mole of a chemical compound is its molecular weight in gram. In case of substances which do not consist of molecules, e.g. salts, ions, radicals, in such cases the formula weight is used instead of molecular weight. 1 mole is then equal to the formula weight in gram.
Gram equivalent (val)	1 val or 1 gram of a substance is its equivalent weight in gram. 1 val is equal to the gram atom or mole of a substance divided by its valency. 1 val is equal to the mole of a substance divided by the number of electrons received or given for a redox reaction.
Molar volume	The volume of 1 mole of a gaseous substance under standard conditions (760 torr, $0°$ C) is always equal to 22.4 litres.
Ions	Ions are electrically charged atoms or group of atoms, its charge can be either positive or negative. Ions with positive charge have electron deficiency and ions with negative charge have excess electrons. [see pp. 9, 23, 32 and 33]

The symbol of an element or the formula of a compound signifies
      1. 1 atom or 1 molecule or 1 formula unit
      2. 1 gram atom or 1 mole of the substance.
  Example: H   1 atom of hydrogen 1 gram atom = 1.008 g hydrogen.
        $2H_2O$ 2 molecules of water 2 moles = $2 \times 18.016 = 36.032$ g water
        $SO_4{}^{2-}$ 1 sulphate ion 1 mole = 96 g sulphate ions.
        NaCl 1 formula unit of sodium chloride 1 mole = 58.5 g sodium chloride
        CO_2 1 molecule of carbon dioxide 1 mole 44 g = 22.4 l (NTP) CO_2
If in a molecule several atoms of the same element are contained, then it is expressed by an index number behind these corresponding element symbol. The same happens with an atom group in a molecule or formula unit. The group of atoms is bracketed and the index number is written behind the brackets.
A coeffecient indicates the number of times an atom or a molecule can combine.

Reactive equations

A chemical reaction equation is a small description of a chemical process. In case of reactions, the element symbols or formula of the concerned substances are inserted.
The starting materials of reaction are written on the left hand side and new products of reaction on the right hand side of the equation.
To formulate a reaction equation, the following must be known:
      1. Initial starting materials
      2. The reaction products
      3. The formula of the concerned reaction substances.
Amidst observing the basic laws and the definitions, the problem of writing of the reaction equation symbols or formulae.
The problem of writing a chemical reaction is to insert element symbols or coefficients of compounds according to basic laws and definitions.
Example: $4\,Fe + 3\,O_2 = 2\,Fe_2O_3$ Coefficient inserted to balance.

A reaction which is not so easy to examine can be split up into partial reactions. Through addition of partial reaction equations, one can obtain a complete reaction equation. It is immaterial whether the actual reaction proceeds by way of these intermediates. It is to be observed that no contradiction exists towards the chemical behaviour of reacting substances; besides through the choice of coefficients, it is possible to provide that by addition of the partial equations the intermediate reactions disappear.

Example: Dissolving copper in dilute nitric acid.

$2HNO_3$	$\rightarrow H_2O + 2NO + 3O$	Nitric acid acts as an oxidizing agent, since it gives oxygen.
$3O + 3Cu$	$\rightarrow 3CuO$	Copper oxidized by the oxygen.
$3CuO + 6HNO_3$	$\rightarrow 3Cu\,(NO_3)_2 + 3H_2O$	Copperoxide is dissolved in nitric acid.

$$3Cu + 8HNO_3 \rightarrow 3Cu\,(NO_3)_2 + 2NO + 4H_2O$$

Redox processes are thus characterized that in case of reaction, concerned substances (ions) electrons are given up or taken in.

The substance which is oxidized gives up electrons.

The substance which is reduced receives electrons.

Oxidation and reduction proceed always simultaneously, i.e. a substance can only give up electrons when another receives them.

Redox processes are suitably represented through ionic equations whereby the total process can also split into partial transformations. Ions whose charges do not change in case of reactions, can be left out. By addition of partial reaction equations, the electrons (e^-) must disappear.

Example: Fe^{2+} is oxidized in sulphuric acid solution by potassium permanganate to Fe^{3+} whereby the MnO_4^- is simultaneously reduced to Mn^{2+}.

$$Fe^{2+} \rightarrow Fe^{3+} + e^-$$
$$MnO_4^- + 8H^+ + 5e^- \rightarrow Mn^2 + 4H_2O$$

$$5Fe^{2+} \rightarrow 5Fe^{3+} + 5e^-$$
$$MnO_4^- + 8H^+ + 5e^- \rightarrow 5Fe^{3+} + Mn^{2+} + 4H_2O$$

Coefficients inserted, that by addition the electrons (e^-) are dropped.

$$5Fe^{2+} + MnO_4^- + 8H^+ \rightarrow 5Fe^{3+} + Mn^{2+} + 4H_2O$$

Through insertion of coefficients the ionic equation becomes the whole molecule of the molecular equation. In the above case, the ionic equation is to be multiplied with 2, so that in the molecular reaction equation the coefficients are integral.

$$10FeSO_4 + 2KMnO_4 + 8H_2SO_4 \longrightarrow 5Fe_2\,(SO_4)_3 + 2MnSO_4 + K_2SO_4 + 8H_2O$$

Calculation of weight ratios

A chemical reaction equation gives information not only about the substances that react with one another, but also, about the weight ratios in which they react with each other.

Condition: The quantity of concerned substances in that reaction must be known.

$$Zn + H_2SO_4 \rightarrow ZnSO_4 + H_2$$
$$65.38g + 98.08g \rightarrow 161.45g + 2.016g$$

Insertion of atomic or molecular weight with the weight notation gram.

Instead of mass unit g, other mass units mg, kg, can also be provided.

First the weight quantity of one substance is ascertained (e.g. Zn = 50g), so that with the help of triple set, the quantities of other substances can be calculated.

For 50g Zn,

$$H_2SO_4 = \frac{98.08 \times 50}{65.38} = 75g\ H_2SO_4$$

or,

It results,

$$ZnSO_4 = \frac{161.45 \times 50}{65.38} = 123.47g\ ZnSO_4$$

$$H_2 = \frac{2.016 \times 50}{65.38} = 1.54g\ H_2$$

$$Zn + H_2SO_4 \rightarrow ZnSO_4 + H_2$$
1 mole 1 mole → 1 mole 1 mole

50g Zn are $\frac{50}{65.38} = 0.7678$ moles of Zn

It is necessary:

0.7678 moles of H_2SO_4 = 75g H_2SO_4

This gives:

0.7678 moles of $ZnSO_4$ = 123.47g

0.7678 moles of H_2 = 1.54g H_2

Under normal conditions, the mole for a gaseous substance can be taken as molar volume (22.4 l).

1 mole of H_2 = 2.016g H_2 = 22.4l H_2 0.7678 moles of H_2 = 0.7678 × 22.4 = 17.13 l H_2

In case of substances which are not 100% pure the necessary weight W is

$$W = \frac{g \times 100}{\text{Percent number}}$$

g = Weight of pure substance

Percent number = purity of the substance in percent.

Composition of compounds

To find	given			
		A, B, C, D,...	Atom weight or atom symbol	A = 39.10; B = 26.98 C = 28.09; D = 16.00
Formula	Atom percent, weight percent			
Atom percent	Formula, weight percent	$i_1, i_2, i_3, i_4,...$	Indices	$i_1 = 1; i_2 = 1$ $i_3 = 3; i_4 = 8$
weight percent	Formula, atom percent	$A_{i_1}, B_{i_2}, C_{i_3}, D_{i_4},...$	Formula	$K[AlSi_3O_8]$

Atom percent = Atoms of that elements in a total of 100 atoms.

Mole percent = Molecules of that compound in a total of 100 molecules.

The calculation of mole percent follows analogously the calculation of atom percent. The indices in case of atom percent turns into coefficients in case of mole percent.

$a, b, c, d,...$	Atom percent	$a = 7.69; b = 7.69$ $c = 23.08; d = 61.54$
$u, v, w, x,...$	Weight percent	$u = 14.05; v = 9.69$ $w = 30.27; x = 45.98$
$A_{i_1}+B_{i_2}+C_{i_3}+D_{i_4}+...=M$	Molecular weight or formula weight	
$i_1 + i_2 + i_3 + i_4 + ... = \Sigma_i$	Sum of indices	$\Sigma_i = 13$

Given	Formula—$A_{i_1} B_{i_2} C_{i_3} D_{i_4}$	$K[AlSi_3O_8]$

Sought:	Atom percent $a, b, c, d,...$	or	Weight percent $u, v, w, x,...$

$$a = i_1 \times \frac{100}{\Sigma_i} = 1 \times \frac{100}{13} = 7.69 \qquad u = A_{i_1} \times \frac{100}{M} = 39.10 \times \frac{100}{278.35} = 14.05$$

$$b = i_2 \times \frac{100}{\Sigma_i} = 1 \times \frac{100}{13} = 7.69 \qquad v = B_{i_2} \times \frac{100}{M} = 26.98 \times \frac{100}{278.35} = 9.69$$

$$c = i_3 \times \frac{100}{\Sigma_i} = 3 \times \frac{100}{13} = 23.08 \qquad w = C_{i_3} \times \frac{100}{M} = 84.27 \times \frac{100}{278.35} = 30.27$$

$$d = i_4 \times \frac{100}{\Sigma_i} = 8 \times \frac{100}{13} = 61.54 \qquad x = D_{i_4} \times \frac{100}{M} = 128.0 \times \frac{100}{278.35} = 45.98$$

Given	Atom percent $a, b, c, d,...$	or	Weight percent $u, v, w, x,...$
To find	Formula—$A_{i_1} B_{i_2} C_{i_3} D_{i_4}$		$K_{i_1}Al_{i_2}Si_{i_3}O_{i_4}$

(There are in every case indices $i_1, i_2, i_3, i_4,...$ to seek for the atom symbols A, B, C, D,...)

$$a+b+c+d+...=100 \qquad i_1+i_2+i_3+i_4+...=\Sigma_i = 13$$
$$a:b:c:d = i_1:i_2:i_3:i_4$$

$$i_1 = a \times \frac{\Sigma_i}{100} = 7.96 \times \frac{13}{100} = 1 \qquad i_1 = \frac{u}{A} = \frac{14.04}{39.10} = 0.359 = 1$$

$$i_2 = b \times \frac{\Sigma_i}{100} = 7.69 \times \frac{13}{100} = 1 \qquad i_2 = \frac{v}{B} = \frac{9.69}{26.98} = 0.359 = 1$$

$$i_3 = c \times \frac{\Sigma_i}{100} = 23.08 \times \frac{13}{100} = 3 \qquad i_3 = \frac{w}{C} = \frac{30.28}{28.09} = 1.078 = 3$$

$$i_4 = d \times \frac{\Sigma_i}{100} = 61.54 \times \frac{13}{100} = 8 \qquad i_4 = \frac{x}{D} = \frac{45.98}{16.00} = 2.874 = 8$$

indices by division or by multiplication with the total factor, must be made integral.

$$A_1 B_1 C_3 D_8 = K[AlSi_3O_8]$$

Given:	Atom percent $a, b, c, d,...$	Weight percent $u, v, w, x,...$
To find:	Weight percent $u, v, w, x,...$	Atom percent $a, b, c, d,...$

$$u = Aa \times \frac{100}{Aa+Bb+Cc+Dd} = 300.68 \times 0.0467 = 14.05 \qquad a = \frac{100u}{u+\frac{Av}{B}+\frac{Aw}{C}+\frac{Ax}{D}} = \frac{1405}{14.05+14.05+42.13+112.36} = 7.69$$

$$v = Bb \times \frac{100}{Aa+Bb+Cc+Dd} = 207.48 \times 0.0467 = 9.69 \qquad b = \frac{100v}{\frac{Bu}{A}+v+\frac{Bw}{C}+\frac{Bx}{D}} = \frac{969}{9.69+9.69+29.07+77.53} = 7.69$$

$$w = Cc \times \frac{100}{Aa+Bb+Cc+Dd} = 648.64 \times 0.0469 = 30.28 \qquad c = \frac{100w}{\frac{Cu}{A}+\frac{Cv}{B}+w+\frac{Cx}{D}} = \frac{3027}{10.09+10.09+30.27+80.72} = 23.08$$

$$x = Dd \times \frac{100}{Aa+Bb+Cc+Dd} = 984.64 \times 0.0469 = 45.98 \qquad d = \frac{100x}{\frac{Du}{A}+\frac{Dv}{B}+\frac{Dw}{C}+x} = \frac{4598}{5.75+5.75+17.24+45.98} = 61.54$$

Gas laws

Boyle-Mariotte law	At a constant temperature, the product of pressure (p) and volume (v) of a gas is constant.	$p \times v = \text{constant}$ $p_1 \times v_1 = \text{constant}$
	Pressure in atmospheres, volume in litres; Dimension. = litre atmosphere.	$\boxed{p \times v = p_1 \times v_1}$
	The volume of a gas, at constant temperature is inversely proportional to pressure.	$\dfrac{v}{v_1} = \dfrac{p_1}{p}$
Gay-Lussac law	The Gay-Lussac law gives information about the variation of volume and pressure on heating.	a) $\quad v = v_0 \left(1 + \dfrac{t}{273} \right)$
	a) From 0°C onwards, at constant pressure per degree rise in temperature, the initial volume (v_0) increases by 1/273.	
	b) From 0°C onwards, at constant volume, per degree rise in temperature, the initial pressure (p_0) increases by 1/273.	b) $\quad p = p_0 \left(1 + \dfrac{t}{273} \right)$
	By inserting the absolute temperatures (T_1, T_2) both equations simplify:	a) $\boxed{\dfrac{v_1}{v_2} = \dfrac{T_1}{T_2}}$
	a) At constant pressure, the volume changes with the change in absolute temperature.	
	b) At constant volume, the pressure changes with the change in absolute temperature.	b) $\boxed{\dfrac{p_1}{p_2} = \dfrac{T_1}{T_2}}$
Equation of state	The combination of laws of Boyle-Mariotte and Gay-Lussac, results in the equation of state for gases. It states: when pressure and temperature in a gas change simultaneously then the volume is directly proportional to absolute temperature and the pressure is inversely proportional to absolute temperature.	$\boxed{\dfrac{p_1 \times v_1}{T_1} = \dfrac{p_2 \times v_2}{T_2}}$
	The equation of the state in this form serves, in particular, in the reduction of a gas-volume into normal conditions (°C, 760 Torr).	$\boxed{\dfrac{p_0 \times v_0}{T_0} = \dfrac{p \times v}{T}}$
	The volume in case of normal condition is designated as v_0; accordingly the dependent absolute temperature becomes T_0 (273°k) and the pressure p_0 (760 Torr).	$v_0 = \dfrac{p \times v \times T_0}{p_0 \times T}$
Universal equation of state	The equation of state for a gas is further simplified when one substitutes, for volume 22.4 l and for pressure in atmospheres (1 atm) in case of a mole of gas. It is $\dfrac{p_0 \times v_0}{T_0} = R = \dfrac{1 \times 22.4}{273} = 0.082$ litre atm/degree/mole R is the universal gas constant. R can take up other values when p, v mol, T are expressed in other units. Example: $R = 8.313 \times 10^7$ erg/degree/mole = 1.987 cal/degree/mole	$\boxed{p \times v = R \times T}$
	When v is the volume of n moles of gas then the universal equation of state becomes: When for $n = \dfrac{a}{M}$ is substituted then the equation of state takes up the form	$\boxed{p \times v = n \times R \times T}$
	a = weight of gas in gram; M = molecular weight These forms of universal equation of state of gases are used for the calculation of molecular weights of gaseous substances.	$\boxed{p \times v = \dfrac{a}{M} = R \times T}$

Standard cubic metre weight and molar volumes of industrial gases

Gas Name	Formula	Molecular weight M	Standard cubic-metre weight kg/Nm³	Specific density 0°C and 760° Torr	Molar volume Nm³/k mole
Helium	He	4.0026	0.1785	0.1381	22.42
Hydrogen	H_2	2.0156	0.08987	0.06952	22.43
Nitrogen	N_2	28.013	1.2505	0.9673	22.40
Atmospheric nitrogen			1.2567	0.9721	
Oxygen	O_2	32.00	1.42895	1.1053	22.40
Chlorine	Cl_2	70.90	3.22	2.49	22.02
Carbon monoxide	CO	28.01	1.2500	0.9669	22.41
Nitric oxide	NO	30.01	1.3402	1.0367	22.39
Carbon dioxide	CO_2	44.01	1.9768	1.5291	22.26
Sulphur dioxide	SO_2	64.06	2.9263	2.2635	21.90
Methane	CH_4	16.04	0.7168	0.5545	22.38
Acetylene	C_2H_2	26.04	1.1709	0.9057	22.23
Ethylene	C_2H_4	28.05	1.2605	0.9750	22.25
Ethane	C_2H_6	30.07	1.356	1.049	22.18
Propylene	C_3O_6	42.08	1.915	1.481	22.02
Propane	C_3H_8	44.10	2.019	1.562	21.84
η - Butane	C_4H_{10}	58.12	2.703	2.091	21.50
Iso - Butane	C_4H_{10}	58.12	2.668	2.064	21.79
Ammonia	NH_3	17.03	0.7714	0.5967	22.08
Hydrogen chloride	HCl	36.46	1.6391	1.2679	22.26
Hydrogen sulphide	H_2S	34.08	1.5392	1.1906	22.15
Air (dried and CO_2-free)		28.96	1.2928	1.00000	22.41

Note: The density is = 1.00000 (with reference to air).

$$\text{Molar volume} = \frac{\text{Molar weight}}{\text{Standard cubic metre weight}}$$

Osmotic pressure

The osmotic pressure of a dilute solution follows the laws similar to gas laws.
The osmotic pressure is

    inversely proportional to volume
    directly proportional to concentration (mole/l)
    directly proportional to absolute temperature

The laws applicable to gases are also valid for osmotic pressure.

$$p \times v = n \times R \times T$$

n = Number of mole of dissolved substance $n = \dfrac{a}{M}$ (mole)

v = Volume of solution in litres

R = 0.082 litre atmosphere/degree/mole $c = \dfrac{n}{v}$ (mole/l)

p = Osmotic pressure in atmosphere

a = dissolved substance in gram

$$p \times v = \frac{a}{M} \times R \times T$$

$$p = \frac{n}{v} \times R \times T$$

$$p = c \times R \times T$$

The laws of osmotic pressure applies only to dilute solutions and under the assumption that by heating the number of molecules do not change (association, dissociation)

Vapour pressure of solutions

Raoult's law	The vapour pressure of a solvent falls in case of a dissolved substance, (applies only when the vapour pressure of dissolved substance is very small in contrast to the solvent). The vapour pressure of a solution is equal to the vapour pressure of the solvent multiplied by the mole fraction of the solvent.		$p = p_0 \dfrac{N}{n + N}$
Lowering of vapour pressure	The relative lowering of vapour pressure caused by addition of a non-volatile solute is equal to the mole fraction of the dissolved surface	$\dfrac{N}{n + N}$ = Mole fraction of solvent	$\dfrac{p_0 - p}{p_0} = \dfrac{n}{n + N}$
	p_0 = vapour pressure of solvent p = vapour pressure of solution	$\dfrac{n}{n + N}$ = Mole fraction of dissolved substance	$\dfrac{\Delta \times p}{p_0} = \dfrac{n}{n + N}$
	N = number of moles of solvent n = number of moles of solute	$p_0 - p = \Delta \times p$	

	Mole fraction: Number of moles of a substance in ratio to the total number of moles in the solution (inclusive of moles of solvent) Mole fraction × 100 = Mole percent **Molal concentration:** Number of moles of dissolved substance in 1000g of solvent. in contrast to this: **Molar concentration:** Number of moles of dissolved substance in a litre of solution.	
Elevation in boiling point	The boiling point of a liquid is the temperature at which the vapour pressure of the liquid attains the pressure present in the gas phase above that liquid. When the vapour pressure of a solvent is lowered by the dissolution of a substance, then the boiling point of the solution must lie higher than the boiling point of a solvent.	$\Delta T = k \times m$
Depression of freezing point	The elevation boiling point is proportional to the osmotic pressure which in turn is proportional to the number of moles of dissolved substance. ΔT = Elevation in B.P when m = 1 then k = Constant for solvent $\Delta T = k$ = molal m = molal concentration elevation constant For the depression in freezing point these equations are equally applicable. k here is the molal depression constant. (see also molecular weight determination)	
	Law of mass action	
Guldberg and Waage's law	In case of reversible chemical reactions a state of equilibrium sets in. In the equilibrium condition, the concentrations of concerned substances in respect of the reaction stand in definite ratio to one another. Example: A+B $\rightleftharpoons$ AB The square brackets signify molar concentrations. If in the reaction equation the molecules have coefficients, then these coefficients in the equation of mass action law become the power exponents. K is the equilibrium constant, which is dependent on temperature.	$\dfrac{[AB]}{[A] \times [B]} = K$
Electrolytic dissociation	The electrolytic dissociation, i.e. the dissociation of a compound into ions in likewise an equilibrium reaction which sets in instantaneously in contrast to other equilibrium reactions. The equation of law of mass action takes up the form $AB \rightleftharpoons A^+ + B^-$ where K is the dissociation constant.	$\dfrac{[A^+] \times [B^-]}{[AB]} = K$ $\dfrac{C \times \alpha^2}{1 - \alpha} = K$
Degree of dissociation Oswald's dilution law	The degree of dissociation α gives the ratio of dissociated molecule to the undissociated molecules available at hand. In an equilibrium condition $[A^+] = [B^-] = C\alpha$ and $[AB] = C(1 - \alpha)$ C is total concentration of electrolyte in mole/l. $C = \frac{1}{V}$ (1 mole in V litres)	$\dfrac{\alpha^2}{V(1 - \alpha)} = K$
Ionic product of water	Water dissociates to a small extent to H^+ and OH^- $H_2O \rightleftharpoons H^+ + OH^-$ Since the concentration of H^+ and OH^- ions in ratio to the total number of water molecules is very small, the concentration of undissociated water can be considered as a constant. Therefore the equation for the dissociation of water is simplified. K_w = ionic product of water. The value of K_w changes with temperature.	$\dfrac{[H^+] \times [OH^-]}{[H_2O]} = K$ $[H^+] \times [OH^-] = K_w$ $K_w = 10^{-14}$

pH	The ionic product of water has particularly a greater significance, when a substance is dissolved in water which by dissociation produces H^+ ions (acids) or OH^- ions (bases). Knowledge of concentration in respect of H^+ − ions in gram ions/ litre is important and with it the concentration in respect of OH^- ions also as a measure of the H^+ − ion concentration is equal to: P_H = negative logarithm of H^+ ion concentration.	$pH = -\log [H^+]$
Solubility product	In case of sparingly soluble electrolytes which exist in the form of solid precipitates or substances at the bottom of a liquid, the concentration of the compound can be regarded as a constant and, therefore, can be incorporated in the dissociation constant. Thereby, the dissociation constant K changes to another new constant K_{sp}, which is known as solubility product.	$[A^+] \cdot [B^-] = [AB] \cdot K$ $K_{sp} = [A^+][B^-]$

Electrochemistry

Electrolysis	If two electrodes are dipped into a solution of an electrolyte and an EMF is applied an electric current flows. The ions migrate towards oppositely charged electrodes and are discharged there. The process is known as electrolysis. Example: Process at the cathode (negative pole) $Cu^{2+} + 2e^- \rightarrow Cu$ (Reduction) Process at the anode (positive pole) $2Cl^- \rightarrow 2\,Cl + e^-$ (Oxidation) Cl_2 (escapes as gas)	Redox − process
Faraday's Laws	During electrolysis, the quantities of substances discharged at the electrodes are proportional to the current quantity passed through the solution. By the same quantity of current the discharged substances in different electrotypes can be collected as their equivalent weights. To deposit a gram equivalent, a current quantity of 96500 coulombs (ampere seconds) is necessary.	96500 Coulomb = 1 Faraday (IF) 1 F = 26.8 ampere hours (Ah)

Quantities of deposited substances

Deposited substances	Process	A current of 1 amp. discharged in			
		1 sec = mg. subs.	1 min = mg subs	1 h = g subs	
Silver	$Ag^+ \rightarrow Ag$	1.1180 mg	67.080	4.0248 g	
Copper	$Cu^{2+} \rightarrow Cu$	0.3294 mg	19.76 mg	1.186 g	Standard
Hydrogen	$H_2O \rightarrow H_2$	0.1162 ml	6.974 ml	418.4 ml	condition of
Oxygen	$H_2O \rightarrow \frac{1}{2}O_2$	0.05801 ml	3.481 ml	208.8 ml	0° C and
Oxyhydrogen gas	$H_2O \rightarrow H_2 + \frac{1}{2}O_2$	0.1742 ml	10.455 ml	627.2 ml	760 Torr

Redox processes

In a redox system, the concerned reaction partners have the tendency either to give or to take such electrons. This tendency is measurable as an electric potential against a standard reference electrode. As a reference electrode, the normal hydrogen electrode is used.

The standard hydrogen electrode is defined as a platinum electrode which is dipped in a 1 molar hydrogen ion solution at 18°C with gaseous hydrogen flowing around it at 760 mm pr.

In case of redox system, in which no metals are concerned, the dipped platinum electrodes in solution acquire a potential.

Potential series of metals				EMF series of Redox system				The measured single electrode potential (potential difference against the standard hydrogen electrode) is dependent on the concentration of concerned reaction partners
Process Red. $\rightleftharpoons$ Ox.		+ e⁻	e_0 (Volt)	Process Red. $\rightleftharpoons$ Ox.		+ e⁻	e°(Volt)	at a constant temperature. This lawful dependency is expressed by the Nernst's equation
K	$\rightleftharpoons K^+$	+ e⁻	−2.92	Cr^{2+}	$\rightleftharpoons Cr^{3+}$	+ e⁻	− 0.41	$e = e_o + \dfrac{0.058}{n} \log \dfrac{{}^cOx}{{}^cRed}$ (Volt)
Na	$\rightleftharpoons Na^+$	+ e⁻	−2.71	Cu^+	$\rightleftharpoons Cu^{2+}$	+ e⁻	+ 0.17	
Mg	$\rightleftharpoons Mg^{2+}$	+ 2 e⁻	−2.40	Sn^{2+}	$\rightleftharpoons Sn^{4+}$	+ 2 e⁻	+ 0.20	e = measured potential e_o = standard potential
Al	$\rightleftharpoons Al^{3+}$	+ 3 e⁻	−1.69	Fe^{2+}	$\rightleftharpoons Fe^{3+}$	+ e⁻	+ 0.75	n = the number of given or taken electrons.
Mn	$\rightleftharpoons Mn^{2+}$	+ 2 e⁻	−1.10	Hg_2^{2+}	$\rightleftharpoons 2Hg^{2+}$	+ 2 e⁻	+ 0.91	c = concentration of reaction partners.
Zn	$\rightleftharpoons Zn^{2+}$	+ 2 e⁻	−0.76	Pb^{2+}	$\rightleftharpoons Pb^{4+}$	+ 2 e⁻	+ 1.69	The coefficients of reaction partners are
Cr	$\rightleftharpoons Cr^{3+}$	+ 3 e⁻	−0.51	Co^{2+}	$\rightleftharpoons Co^{3+}$	+ e⁻	+ 1.82	introduced as exponents corresponding to the
Fe	$\rightleftharpoons Fe^{2+}$	+ 2 e⁻	−0.44	S^{2-}	$\rightleftharpoons S$	+ 2 e⁻	− 0.51	law of mass action in case of these concentrations.
Cd	$\rightleftharpoons Cd^{2+}$	+ 2 e⁻	−0.40	$2I^-$	$\rightleftharpoons I_2$	+ 2 e⁻	+ 0.58	In case of metal electrodes when dipped into its
Co	$\rightleftharpoons Co^{2+}$	+ 2 e⁻	−0.29	$2Br^-$	$\rightleftharpoons Br_2$	+ 2 e⁻	+ 1.07	salt solution the Nernst's equation becomes:
Ni	$\rightleftharpoons Ni^{2+}$	+ 2 e⁻	−0.25	$2Cl^-$	$\rightleftharpoons Cl_2$	+ 2 e⁻	+ 1.36	n = number of
Sn	$\rightleftharpoons Sn^{2+}$	+ 2 e⁻	−0.16	$2F^-$	$\rightleftharpoons F_2$	+ 2 e⁻	+ 2.85	$e = e_o + \dfrac{0.58}{n} \log c\,Me^{n+}$ electrons
Pb	$\rightleftharpoons Pb^{2+}$	+ 2 e⁻	−0.13					Me^{n+} = Metal ions
H_2	$\rightleftharpoons 2H^+$	+ 2 e⁻	0.00	$H_2SO_3 + H_2O \rightleftharpoons SO_4^{2-} + 4H^+ + 2e^-$			+0.14	with charge n
Cu	$\rightleftharpoons Cu^{2+}$	+ 2 e⁻	+0.35	$Cr^{3+} + 4H_2O \rightleftharpoons HCrO_4^- + 7H^+ + 3e^-$			+1.36	In case of hydrogen electrodes (Pt. electrodes
Ag	$\rightleftharpoons Ag^+$	+ e¹	+0.81	$Mn^{2+} + 2H_2O \rightleftharpoons MnO_2 + 4H^+ + 2e^-$			+1.35	dipped into a solution with H^+ ions) the
Hg	$\rightleftharpoons Hg^{2+}$	+ 2 e⁻	+0.86	$Pb^{2+} + 2H_2O \rightleftharpoons PbO_2 + 4H^+ + 2e^-$			+1.44	Nernst's equation is further simplified:
Au	$\rightleftharpoons Au^{3+}$	+ 3 e⁻	+1.38	$Cl^- + 3H_2O \rightleftharpoons ClO_3 + 6H^+ + 6e^-$			+1.44	$e = 0.058 \log {}^cH^+$ or $e = -0.058\ pH$
Pt	$\rightleftharpoons Pt^{2+}$	+ 2 e⁻	+1.60	$Mn^{2+} + 4H_2O \rightleftharpoons MnO_4 + 8H^+ + 5e^-$			+1.52	

e_o = Standard potential. The standard potential can be considered with the assumption that the metal dipped into a 1 molar solution of the metal ions at 18°C measured against the standard hydrogen electrode with the standard potential = 0.

With the help of these equations, concentrations can be measured. Important field of application is the measurement of H.

Common table for determination of nitrogen

The table contains the densities of nitrogen at a given temperature t°C and pressure p Torr with reference to the standard density of nitrogen 1.2505 (0°C and 760 Torr).

In case of nitrogen determination according to Dumas method, the calculation of nitrogen, can be obtained:

$$\%N = \frac{VF\ 100}{\text{weighed portion in mg.}}$$

V = Volume of nitrogen in ml.

F = Density of nitrogen at t°C and p Torr (Table values)

In case of logarithmic calculation it results in further simplification

$$\%N = \log V + \log F + \log 100 - \log \text{weighed portion}$$

The calculation is further simplified by application of decadic supplement to logarithmic weighed portion.

$$\%N = \log V + \log F + 2,00,000 + DE \log \text{weighed portion}$$

The table for nitrogen determination according to Duma's method has been provided, since it represents the frequently used volumetric method for determination of gases. As the nitrogen is collected over 50% potassium hydroxide in an azometer, the vapour pressure can be disregarded. Therefore limited error signifies further below the remaining errors of these methods.

This table can also be used for the calculation of other gases that are determined volumetrically.

In this case, the factor F is to be divided by the normal density of nitrogen and is to be multiplied by the standard density of the gas whose volume is to be determined. Of course, under certain conditions, a correction factor is to be applied for the vapour pressure, when the gas is collected over a liquid, of which the vapour pressure is to be considered.

For calculation of other gases one can use the suitable equation of state of gas.

p\t°	670 F	670 log F	671 F	671 log F	672 F	672 log F	673 F	673 log F	674 F	674 log F	675 F	675 log F	676 F	676 log F	p\t°
10	1.0634	02869	1.0650	02734	1.0666	02799	1.0682	02864	1.0697	02928	1.0713	02992	1.0729	03057	10
11	1.0596	02516	1.0612	02581	1.0628	02646	1.0644	02711	1.0660	02775	1.0676	02839	1.0692	02904	11
12	1.0559	02363	1.0575	02428	1.0591	02493	1.0606	02558	1.0623	02622	1.0638	02686	1.0654	02751	12
13	1.0522	02211	1.0538	02276	1.0554	02341	1.0570	02406	1.0585	02470	1.0601	02534	1.0617	02599	13
14	1.0486	02059	1.0501	02124	1.0517	02189	1.0533	02254	1.549	02318	1.0564	02382	1.0580	02447	14
15	1.0449	01907	1.0465	01972	1.0480	02037	1.0496	02102	1.0511	02166	1.0527	02230	1.0543	02295	15
16	1.0413	01756	1.0428	01821	1.0444	01886	1.0460	01951	1.0475	02015	1.0490	02079	1.0506	02144	16
17	1.0377	01606	1.0392	01671	1.0408	01736	1.0423	01801	1.0439	01865	1.0454	01929	1.0470	01994	17
18	1.0341	01456	1.0356	01521	1.0372	01586	1.0387	01651	1.0403	01715	1.0418	01779	1.0434	01844	18
19	1.0306	01307	1.0321	01372	1.0336	01437	1.0351	01502	1.0367	01566	1.0382	01630	1.0398	01695	19
20	1.0270	01158	1.0286	01223	1.0301	01288	1.0316	01353	1.0332	01417	1.0347	01481	1.0362	01546	20
21	1.0235	01010	1.0251	01075	1.0266	01140	1.0281	01205	1.0296	01269	1.0312	01333	1.0327	01398	21
22	1.0200	00862	1.0216	00927	1.0231	00992	1.0246	01057	1.0261	01121	1.0277	01185	1.0292	01250	22
23	1.0166	00715	1.0181	00780	1.0197	00845	1.0212	00764	1.0227	00974	1.0242	01038	1.0257	01103	23
24	1.0132	00568	1.0147	00634	1.0162	00699	1.0177	00910	1.0193	00828	1.0207	00892	1.0223	00957	24
25	1.0098	00423	1.0113	00488	1.0128	00553	1.0143	00618	1.0158	00682	1.0173	00746	1.0189	00811	25
26	1.0064	00277	1.0079	00342	1.0094	00407	1.0110	00472	1.0124	00536	1.0139	00600	1.0154	00665	26
27	1.0030	00132	1.0045	00197	1.0060	00262	0.9976	00327	1.0090	00391	1.0105	00455	1.0120	00520	27
28	0.9997	99987	1.0012	00052	1.0027	00117	1.0042	00182	1.0057	00246	1.0072	00310	1.0087	00375	28
29	0.9964	99843	0.9979	99908	0.9994	99973	1.0009	00038	1.0024	00102	1.0038	00166	1.0053	00231	29
30	0.9931	99699	0.9946	99764	0.9961	99829	1.0076	99894	0.9990	99958	1.0005	00022	1.0020	00087	30

	677		678		679		680		681		682		683		
	F	log F	F	log F	F	log F	F	log F	F	log F	F	log F	F	log F	
10	1.0745	03121	1.0761	03185	1.0777	03249	1.0793	03313	1.0809	03377	1.0824	03440	1.0840	03504	10
11	1.0707	02968	1.0723	03032	1.0739	03096	1.0755	03160	1.0771	03224	1.0786	03287	1.0802	03351	11
12	1.0670	02815	1.0685	02879	1.0701	02943	1.0717	03007	1.0733	03071	1.0748	03134	1.0764	03198	12
13	1.0632	02663	1.0648	02727	1.0664	02791	1.0679	02855	1.0695	02919	1.0711	02982	1.0727	03046	13
14	1.0595	02511	1.0611	02575	1.0626	02639	1.0642	02703	1.0658	02767	1.0673	02830	1.0689	02894	14
15	1.0558	02359	1.0574	02423	1.0589	02487	1.0605	02551	1.0621	02615	1.0636	02678	1.0652	02742	15
16	1.0521	02208	1.0537	02272	1.0553	02336	1.0568	02400	1.0584	02464	1.0599	02527	1.0615	02591	16
17	1.0485	02058	1.0501	02122	1.0516	02186	1.0532	02250	1.0547	02314	1.0563	02377	1.0578	02441	17
18	1.0449	01908	1.0465	01972	1.0480	02036	1.0495	02100	1.0511	02164	1.0526	02227	1.0542	02291	18
19	1.0413	01759	1.0429	01823	1.0444	01887	1.0459	01951	1.0475	02015	1.0490	02078	1.0506	02142	19
20	1.0377	01610	1.0393	01674	1.0408	01738	1.0424	01802	1.0439	01866	1.0454	01929	1.0470	01993	20
21	1.0342	01462	1.0358	01526	1.0373	01590	1.0388	01654	1.0404	01718	1.0419	01781	1.0434	01845	21
22	1.0307	01314	1.0322	01378	1.0338	01442	1.0353	01506	1.0368	01570	1.0383	01633	1.0398	01697	22
23	1.0272	01167	1.0288	01231	1.0303	01295	1.0318	01359	1.0333	01423	1.0348	01486	1.0363	01550	23
24	1.0238	01021	1.0253	01085	1.0268	01149	1.0283	01213	1.0298	01277	1.0313	01340	1.0329	01404	24
25	1.0204	00875	1.0219	00939	1.0234	01003	1.0249	01067	1.0264	01131	1.0279	01194	1.0294	01258	25
26	1.0169	00729	1.0184	00793	1.0199	00857	1.0214	00921	1.0229	00985	1.0244	01048	1.0259	01112	26
27	1.0135	00584	1.0150	00648	1.0165	00712	1.0180	00776	1.0195	00840	1.0210	00903	1.0225	00967	27
28	1.0102	00439	1.0116	00503	1.0131	00567	1.0146	00631	1.0161	00695	1.0176	00758	1.0191	00822	28
29	1.0068	00295	1.0083	00359	1.0098	00423	1.0113	00487	1.0128	00551	1.0142	00614	1.0157	00678	29
30	1.0035	00151	1.0050	00215	1.0064	00279	1.0079	00343	1.0094	00407	1.0109	00470	1.0124	00534	30

	684		685		686		687		688		689		690		
10	1.0856	03568	1.0872	03631	1.0888	03694	1.0904	03758	1.0920	03821	1.0936	03884	1.0951	03946	10
11	1.0818	03415	1.0834	03478	1.0850	03541	1.0866	03605	1.0881	03668	1.0897	03731	1.0913	03793	11
12	1.0780	03262	1.0796	03325	1.0812	03389	1.0827	03452	1.0843	03515	1.0859	03578	1.0874	03640	12
13	1.0742	03110	1.0758	03173	1.0774	03236	1.0789	03300	1.0805	03363	1.0821	03426	1.0836	03488	13
14	1.0705	02958	1.0720	03021	1.0736	03084	1.0752	03148	1.0767	03211	1.0783	03274	1.0798	03336	14
15	1.0667	02806	1.0683	02869	1.0698	02932	1.0714	02996	1.0730	03059	1.0745	03122	1.0761	03184	15
16	1.0630	02655	1.0646	02718	1.0661	02781	1.0677	02845	1.0693	02908	1.0708	02971	1.0723	03033	16
17	1.0594	02505	1.0609	02568	1.0625	02631	1.0640	02695	1.0656	02758	1.0671	02821	1.0686	02883	17
18	1.0557	02355	1.0573	02418	1.0588	02481	1.0604	02545	1.0619	02608	1.0634	02671	1.0650	02733	18
19	1.0521	02206	1.0536	02269	1.0552	02332	1.0567	02396	1.0583	02459	1.0598	02522	1.0613	02584	19
20	1.0485	02057	1.0500	02120	1.0515	02183	1.0531	02247	1.0546	02310	1.0562	02373	1.0577	02435	20
21	1.0449	01909	1.0465	01972	1.0480	02035	1.0495	02099	1.0510	02162	1.0526	02225	1.0541	02287	21
22	1.0414	01761	1.0429	01824	1.0444	01887	1.0459	01951	1.0475	02014	1.0490	02077	1.0505	02139	22
23	1.0379	01614	1.0394	01677	1.0409	01740	1.0424	01804	1.0439	01867	1.0454	01930	1.0469	01992	23
24	1.0344	01468	1.0359	01531	1.0374	01594	1.0389	01658	1.0404	01721	1.0419	01784	1.0434	01846	24
25	1.0309	01322	1.0324	01385	1.0339	01448	1.0354	01512	1.0369	01575	1.0384	01638	1.0399	01700	25
26	1.0274	01176	1.0289	01239	1.0304	01302	1.0320	01366	1.0335	01429	1.0350	01492	1.0364	01554	26
27	1.0240	01031	1.0255	01094	1.0270	01157	1.0285	01221	1.0300	01284	1.0315	01347	1.0330	01409	27
28	1.0206	00886	1.0221	00949	1.0236	01012	1.0251	01076	1.0266	01139	1.0281	01202	1.0295	01264	28
29	1.0172	00742	1.0187	00805	1.0202	00868	1.0217	00932	1.0232	00995	1.0247	01058	1.0261	01120	29
30	1.0139	00598	1.0153	00661	1.0168	00724	1.0181	00788	1.0198	00851	1.0213	00914	1.0227	00976	30

t°	691 F	691 log F	692 F	692 log F	693 F	693 log F	694 F	694 log F	695 F	695 log F	696 F	696 log F	697 F	697 log F
10	1.0967	04009	1.0983	04072	1.0999	04135	1.1015	04197	1.1031	04260	1.1046	04322	1.1062	04385
11	1.0928	03856	1.0944	03919	1.0960	03982	1.0976	04044	1.0992	04107	1.1008	04169	1.1024	04232
12	1.0890	03703	1.0906	03766	1.0922	03829	1.0937	03891	1.0953	03954	1.0969	04016	1.0985	04079
13	1.0852	03551	1.0868	03614	1.0884	03677	1.0899	03739	1.0915	03802	1.0930	03864	1.0946	03927
14	1.0814	03399	1.0830	03462	1.0846	03525	1.0861	03587	1.0877	03650	1.0892	03712	1.0908	03775
15	1.0776	03247	1.0792	03310	1.0808	03373	1.0823	03435	1.0839	03498	1.0854	03560	1.0870	03623
16	1.0739	03096	1.0754	03159	1.0770	03222	1.0785	03284	1.0801	03347	1.0817	03409	1.0832	03472
17	1.0702	02946	1.0717	03009	1.0733	03072	1.0748	03134	1.0764	03197	1.0779	03259	1.0795	03322
18	1.0665	02796	1.0680	02859	1.0696	02922	1.0711	02984	1.0727	03047	1.0742	03109	1.0758	03172
19	1.0628	02647	1.0644	02710	1.0659	02773	1.0675	02835	1.0690	02898	1.0705	02960	1.0723	03023
20	1.0592	02498	1.0607	02561	1.0623	02624	1.0638	02686	1.0653	02749	1.0669	02811	1.0684	02874
21	1.0556	02350	1.0571	02413	1.0587	02476	1.0602	02538	1.0617	02601	1.0632	02663	1.0648	02726
22	1.0520	02202	1.0535	02265	1.0551	02328	1.0566	02390	1.0581	02453	1.0596	02515	1.0612	02578
23	1.0485	02055	1.0500	02118	1.0515	02181	1.0530	02243	1.0545	02306	1.0560	02368	1.0576	02431
24	1.0449	01909	1.0465	01972	1.0480	02035	1.0495	02097	1.0510	02160	1.0525	02222	1.0540	02285
25	1.0414	01763	1.0429	01826	1.0445	01889	1.0459	01951	1.0475	02014	1.0490	02076	1.0505	02139
26	1.0379	01617	1.0394	01680	1.0410	01743	1.0424	01805	1.0440	01868	1.0455	01930	1.0470	01993
27	1.0345	01472	1.0360	01535	1.0375	01598	1.0390	01660	1.0405	01723	1.0420	01785	1.0435	01848
28	1.0310	01327	1.0325	01390	1.0340	01453	1.0355	01515	1.0370	01578	1.0385	01640	1.0400	01703
29	1.0276	01183	1.0291	01246	1.0306	01309	1.0321	01371	1.0336	01434	1.0350	01496	1.0366	01559
30	1.0242	01039	1.0257	01102	1.0272	01165	1.0287	01227	1.0301	01290	1.0316	01352	1.0331	01415

t°	698 F	698 log F	699 F	699 log F	700 F	700 log F	701 F	701 log F	702 F	702 log F	703 F	703 log F	704 F	704 log F
10	1.1078	04447	1.1094	04509	1.1110	04571	1.1126	04633	1.1142	04695	1.1158	04757	1.1174	04819
11	1.1039	04294	1.1055	04356	1.1071	04418	1.1087	04480	1.1103	04542	1.1112	04604	1.1134	04666
12	1.1000	04141	1.1016	04203	1.1032	04265	1.1048	04327	1.1064	04389	1.1080	04451	1.1095	04513
13	1.0962	03989	1.0978	04051	1.0993	04113	1.1009	04175	1.1025	04237	1.1041	04299	1.1056	04361
14	1.0924	03837	1.0939	03899	1.0955	03961	1.0971	04023	1.0986	04085	1.1002	04147	1.1018	04209
15	1.0886	03685	1.0901	03747	1.0917	03809	1.0932	03871	1.0948	03933	1.0964	03995	1.0979	04057
16	1.0848	03534	1.0863	03596	1.0879	03658	1.0894	03720	1.0910	03782	1.0925	03844	1.0941	03906
17	1.0810	03384	1.0826	33446	1.0841	03508	1.0857	03570	1.0872	03632	1.0888	03694	1.0866	03756
18	1.0773	03234	1.0789	03296	1.0804	03358	1.0819	03420	1.0835	03482	1.0850	03544	1.0903	03606
19	1.0736	03085	1.0752	03147	1.0767	03209	1.0782	03271	1.0798	03333	1.0813	03395	1.0829	03457
20	1.0699	02936	1.0715	02998	1.0730	03060	1.0745	03121	1.0761	03184	1.0776	03246	1.0792	03308
21	1.0663	02788	1.0678	02850	1.0693	02912	1.0709	02974	1.0724	03036	1.0739	03098	1.0755	03160
22	1.0627	02640	1.0642	02702	1.0657	02764	1.0672	02826	1.0688	02888	1.0703	02950	1.0718	03012
23	1.0591	02493	1.0606	02555	1.0621	02617	1.0636	02679	1.0652	02741	1.0667	02803	1.0682	02865
24	1.0555	02347	1.0570	02409	1.0585	02471	1.0601	02533	1.0616	02595	1.0631	02657	1.0646	02719
25	1.0520	02201	1.0535	02263	1.0550	02325	1.0565	02387	1.0580	02449	1.0595	02511	1.0610	02573
26	1.0485	02055	1.0500	02117	1.0515	02179	1.0530	02241	1.0545	02303	1.0560	02365	1.0575	02427
27	1.0450	01910	1.0465	01972	1.0480	02034	1.0495	02096	1.0510	02158	1.0525	02220	1.0540	02282
28	1.0415	01765	1.0430	01827	1.0445	01889	1.0460	01951	1.0475	02013	1.0490	02075	1.0505	02137
29	1.0380	01621	1.0395	01683	1.0410	01745	1.0425	01807	1.0440	01869	1.0455	01931	1.0470	01993
30	1.0346	01477	1.0361	01539	1.0376	01601	1.0390	01663	1.0405	01725	1.0420	01787	1.0435	01849

p/t	705 F	705 log F	706 F	706 log F	707 F	707 log F	708 F	708 log F	709 F	709 log F	710 F	710 log F	711 F	711 log F	p/t
10	1.1189	04880	1.1205	04942	1.1221	05003	1.1237	05065	1.1253	05126	1.1269	05189	1.1285	05250	10
11	1.1150	04727	1.1166	04789	1.1182	04850	1.1197	04912	1.1213	04973	1.1229	05035	1.1245	05096	11
12	1.1111	04574	1.1127	04636	1.1142	04697	1.1158	04759	1.1174	04820	1.1190	04882	1.1205	04943	12
13	1.1097	04422	1.1088	04484	1.1103	04545	1.1119	04607	1.1135	04668	1.1151	04730	1.1166	04791	13
14	1.1033	04270	1.1049	04332	1.1064	04393	1.1080	04455	1.1096	04516	1.1112	04578	1.1127	04639	14
15	1.0995	04118	1.1010	04180	1.1026	04241	1.1041	04303	1.1057	04364	1.1073	04427	1.1089	04488	15
16	1.0957	03967	1.0972	04029	1.0988	04090	1.1003	04152	1.1019	04213	1.1035	04276	1.1050	04337	16
17	1.0919	03817	1.0934	03879	1.0950	03940	1.0965	04002	1.0981	04063	1.0997	04126	1.1012	04187	17
18	1.0881	03667	1.0897	03729	1.0912	03790	1.0927	03852	1.0943	03913	1.0959	03976	1.0974	04037	18
19	1.0844	03518	1.0859	03580	1.0875	03641	1.0890	03703	1.0905	03764	1.0921	03827	1.0937	03888	19
20	1.0807	03369	1.0822	03431	1.0837	03492	1.0853	03554	1.0868	03615	1.0884	03678	1.0899	03739	20
21	1.0770	03221	1.0785	03283	1.0800	03344	1.0816	03406	1.0831	03467	1.0847	63530	1.0862	03591	21
22	1.0733	03073	1.0749	03135	1.0764	03196	1.0779	03258	1.0794	03319	1.0810	03382	1.0825	03443	22
23	1.0697	02926	1.0712	02988	1.0727	03049	1.0743	03111	1.0758	03172	1.0773	03235	1.0788	03296	23
24	1.0661	02780	1.0676	02842	1.0691	02903	1.0707	02965	1.0722	03026	1.0737	03088	1.0752	03149	24
25	1.0625	02634	1.0640	02696	1.0655	02757	1.0671	02819	1.0686	02880	1.0701	02942	1.0716	03003	25
26	1.0590	02488	1.0605	02550	1.0620	02611	1.0635	02673	1.0650	02734	1.0665	02796	1.0680	02857	26
27	1.0554	02343	1.0569	02405	1.0584	02466	1.0599	02528	1.0614	02589	1.0629	02651	1.0644	02712	27
28	1.0519	02198	1.0534	02260	1.0549	02321	1.0564	02383	1.0579	02444	1.0594	02506	1.0609	02567	28
29	1.0484	02054	1.0499	02116	1.0514	02177	1.0529	02239	1.0544	02300	1.0559	02362	1.0574	02423	29
30	1.0450	01910	1.0465	01972	1.0479	02033	1.0494	02095	1.0509	02156	1.0524	02218	1.0539	02279	30

p/t	712 F	712 log F	713 F	713 log F	714 F	714 log F	715 F	715 log F	716 F	716 log F	717 F	717 log F	718 F	718 log F	p/t
10	1.1301	05311	1.1317	05372	1.1333	05433	1.1349	05494	1.1364	05554	1.1380	05615	1.1396	05675	10
11	1.1261	05157	1.1277	05218	1.1293	05279	1.1309	05340	1.1324	05400	1.1340	05461	1.1356	05521	11
12	1.1221	05004	1.1237	05065	1.1253	05126	1.1269	05187	1.1284	05247	1.1300	05308	1.1316	05368	12
13	1.1182	04852	1.1198	04913	1.1213	04974	1.1229	05035	1.1245	05095	1.1261	05156	1.1276	05216	13
14	1.1143	04700	1.1159	04761	1.1174	04822	1.1190	04883	1.1206	04943	1.1221	05004	1.1237	05064	14
15	1.1104	04549	1.1120	04610	1.1136	04671	1.1151	04732	1.1167	04792	1.1182	04853	1.1198	04913	15
16	1.1066	04398	1.1081	04459	1.1097	04520	1.1112	04581	1.1128	04641	1.1143	04702	1.1159	04762	16
17	1.1028	04248	1.1043	04309	1.1058	04370	1.1074	04431	1.1089	04491	1.1105	04552	1.1120	04612	17
18	1.0990	04098	1.1005	04159	1.1020	04220	1.1036	04281	1.1051	04341	1.1067	04402	1.1082	04462	18
19	1.0952	03949	1.0967	04010	1.0983	04071	1.0998	04132	1.1013	04192	1.1029	04253	1.1044	04313	19
20	1.0914	03800	1.0930	03861	1.0945	03922	1.0960	03983	1.0976	04043	1.0991	04104	1.1006	04164	20
21	1.0877	03652	1.0893	03713	1.0908	03774	1.0923	03835	1.0938	03895	1.0954	03956	1.0969	04016	21
22	1.0840	03504	1.0856	03565	1.0871	03626	1.0886	03687	1.0901	03747	1.0917	03808	1.0932	03868	22
23	1.0804	03357	1.0819	03418	1.0834	03479	1.0849	03540	1.0864	03600	1.0880	03661	1.0895	03721	23
24	1.0767	03210	1.0782	03271	1.0797	03332	1.0813	03393	1.0828	03453	1.0843	03514	1.0858	03574	24
25	1.0731	03064	1.0746	03125	1.0761	03186	1.0776	03247	1.0791	03307	1.0806	03368	1.0821	03428	25
26	1.0695	02918	1.0710	02979	1.0725	03040	1.0740	03101	1.0755	03161	1.0770	03222	1.0785	03282	26
27	1.0659	02773	1.0674	02834	1.0689	02895	1.0704	02956	1.0719	03016	1.0734	03077	1.0749	03137	27
28	1.0624	02628	1.0639	02689	1.0654	02750	1.0669	02811	1.0683	02871	1.0698	02932	1.0713	02993	28
29	1.0589	02484	1.0604	02545	1.0618	02606	1.0633	02667	1.0648	02727	1.0663	02788	1.0678	02848	29
30	1.0554	02340	1.0568	02401	1.0583	02462	1.0598	02523	1.0613	02583	1.0628	02644	1.0642	02704	30

p/t°	719 F	719 log F	720 F	720 log F	721 F	721 log F	722 F	722 log F	723 F	723 log F	724 F	724 log F	725 F	725 log F
10	1.1412	05736	1.1428	05796	1.1444	05857	1.1460	05917	1.1475	05977	1.1491	06037	1.1507	06097
11	1.1372	05582	1.1387	05642	1.1403	05703	1.1419	05763	1.1435	05823	1.1451	05883	1.1466	05943
12	1.1332	05429	1.1347	05489	1.1363	05550	1.1379	05610	1.1395	05670	1.1411	05730	1.1426	05790
13	1.1292	05277	1.1308	05337	1.1324	05398	1.1339	05458	1.1355	05518	1.1371	05578	1.1386	05638
14	1.1253	05125	1.1268	05185	1.1284	05246	1.1300	05306	1.1315	05366	1.1331	05426	1.1346	05486
15	1.1214	04974	1.1229	05034	1.1245	05095	1.1260	05155	1.1276	05215	1.1291	05275	1.1307	05335
16	1.1175	04823	1.1190	04883	1.1206	04944	1.1221	05004	1.1237	05064	1.1252	05124	1.1268	05184
17	1.1136	04673	1.1151	04733	1.1167	04794	1.1183	04854	1.1198	04914	1.1213	04974	1.1229	05034
18	1.1098	04523	1.1113	04583	1.1129	04644	1.1144	04704	1.1159	04764	1.1175	04824	1.1190	04884
19	1.1060	04374	1.1075	04434	1.1091	04495	1.1106	04555	1.1121	04615	1.1137	04675	1.1152	04735
20	1.1022	04225	1.1037	04285	1.1053	04346	1.1068	04406	1.1083	04466	1.1099	04526	1.1114	04586
21	1.0984	04077	1.0999	04137	1.1015	04198	1.1030	04258	1.1045	04318	1.1061	04378	1.1076	04438
22	1.0947	03928	1.0962	03989	1.0977	04050	1.0993	04110	1.1008	04170	1.1023	04230	1.1038	04290
23	1.0910	03782	1.0925	03842	1.0940	03903	1.0955	03963	1.0971	04023	1.0986	04083	1.1001	04143
24	1.0873	03635	1.0888	03695	1.0903	03756	1.0918	03816	1.0934	03876	1.0949	03936	1.0964	03996
25	1.0837	03489	1.0852	03549	1.0867	03610	1.0882	03670	1.0897	03730	1.0912	03790	1.0927	03850
26	1.0800	03343	1.0815	03403	1.0830	03464	1.0845	03524	1.0860	03584	1.0875	03644	1.0890	03704
27	1.0764	03198	1.0779	03258	1.0794	03319	1.0809	03379	1.0824	03439	1.0839	03499	1.0854	03559
28	1.0728	03053	1.0743	03113	1.0758	03174	1.0773	03234	1.0788	03294	1.0803	03354	1.0818	03414
29	1.0693	02909	1.0708	02969	1.0723	03030	1.0737	03090	1.0752	03150	1.0767	03210	1.0782	03270
30	1.0657	02765	1.0672	02825	1.0687	02885	1.0702	02946	1.0717	03006	1.0731	03066	1.0746	03126

p/t°	726 F	726 log F	727 F	727 log F	728 F	728 log F	729 F	729 log F	730 F	730 log F	731 F	731 log F	732 F	732 log F
10	1.1523	06157	1.1539	06216	1.1555	06276	1.1571	06336	1.1586	06395	1.1602	06455	1.1618	06514
11	1.1482	06003	1.1498	06062	1.1514	06122	1.1530	06182	1.1545	06241	1.1561	06301	1.1577	06360
12	1.1442	05850	1.1458	05909	1.1473	05969	1.1489	06029	1.1505	06088	1.1521	06148	1.1536	06207
13	1.1402	05698	1.1418	05757	1.1433	05817	1.1449	05877	1.1465	05936	1.1480	05996	1.1496	06055
14	1.1362	05546	1.1378	05605	1.1393	05665	1.1409	05725	1.1425	05784	1.1440	05844	1.1456	05903
15	1.1322	05395	1.1338	05454	1.1354	05514	1.1369	05574	1.1385	05633	1.1401	05693	1.1416	05752
16	1.1283	05244	1.1299	05303	1.1314	05363	1.1330	05423	1.1345	05482	1.1361	05542	1.1377	05601
17	1.1245	05094	1.1260	05153	1.1275	05213	1.1291	05273	1.1306	05332	1.1322	05392	1.1337	05451
18	1.1206	04944	1.1221	05003	1.1236	05063	1.1252	05123	1.1267	05182	1.1283	05242	1.1298	05301
19	1.1167	04795	1.1182	04854	1.1198	04914	1.1213	04974	1.1229	05033	1.1244	05093	1.1260	05152
20	1.1129	04646	1.1144	04705	1.1160	04765	1.1175	04825	1.1190	04884	1.1206	04944	1.1221	05003
21	1.1091	04498	1.1106	04557	1.1122	04617	1.1137	04677	1.1152	04736	1.1168	04796	1.1183	04855
22	1.1053	04350	1.1069	04409	1.1084	04469	1.1099	04529	1.1114	04588	1.1130	04648	1.1145	04707
23	1.1016	04203	1.1031	04262	1.1046	04322	1.1062	04382	1.1076	04440	1.1092	04501	1.1107	04560
24	1.0979	04056	1.0994	04115	1.1009	04175	1.1024	04235	1.1039	04294	1.1055	04354	1.1070	04413
25	1.0942	03910	1.0957	03969	1.0972	04029	1.0987	04089	1.1002	04148	1.1017	04208	1.1032	04267
26	1.0905	03764	1.0920	03823	1.0935	03883	1.0950	03943	1.0965	04002	1.0980	04062	1.0995	04121
27	1.0869	03619	1.0884	03678	1.0899	03738	1.0914	03798	1.0929	03857	1.0944	03917	1.0959	03976
28	1.0833	03474	1.0848	03533	1.0863	03593	1.0878	03653	1.0892	03712	1.0907	03772	1.0922	03831
29	1.0797	03330	1.0812	03389	1.0827	03449	1.0842	03509	1.0856	03568	1.0871	03628	1.0886	03687
30	1.0761	03186	1.0776	03245	1.0791	03305	1.0806	03365	1.0820	03424	1.0835	03484	1.0850	03543

	733		734		735		736		737		738		739		
	F	log F	F	log F	F	log F	F	log F	F	log F	F	log F	F	log F	
0	1.1634	06573	1.1650	06633	1.1666	06692	1.1682	06751	1.1698	06810	1.1714	06869	1.1729	06927	10
1	1.1593	06419	1.1609	06479	1.1625	06538	1.1641	06597	1.1656	06656	1.1672	06715	1.1688	06773	11
2	1.1552	06266	1.1568	06326	1.1584	06385	1.1600	06444	1.1615	06503	1.1631	06562	1.1647	06620	12
3	1.1512	06114	1.1528	06174	1.1543	06233	1.1559	06292	1.1575	06351	1.1590	06410	1.1606	06468	13
4	1.1472	05962	1.1488	06022	1.1503	06081	1.1519	06140	1.1534	06199	1.1550	06258	1.1565	06316	14
5	1.1432	05811	1.1448	05871	1.1463	05930	1.1479	05989	1.1494	06048	1.1510	06107	1.1525	06165	15
6	1.1392	05660	1.1408	05720	1.1423	05779	1.1439	05838	1.1454	05897	1.1470	05956	1.1485	06014	16
7	1.1353	05510	1.1368	05570	1.1384	05629	1.1399	05688	1.1415	05747	1.1430	05806	1.1446	05864	17
8	1.1314	05360	1.1329	05420	1.1345	05479	1.1360	05538	1.1376	05597	1.1391	05656	1.1406	05714	18
9	1.1275	05211	1.1290	05271	1.1306	05330	1.1321	05389	1.1337	05448	1.1352	05507	1.1367	05565	19
0	1.1236	05062	1.1252	05122	1.1267	05181	1.1282	05240	1.1298	05299	1.1313	05358	1.1328	05416	20
1	1.1198	04914	1.1213	04974	1.1229	05033	1.1244	05092	1.1259	05151	1.1275	05210	1.1290	05268	21
2	1.1160	04766	1.1175	04826	1.1191	04885	1.1206	04944	1.1221	05003	1.1236	05062	1.1251	05120	22
3	1.1122	04619	1.1138	04679	1.1153	04738	1.1168	04797	1.1183	04856	1.1198	04915	1.1213	04973	23
4	1.1085	04472	1.1100	04532	1.1115	04591	1.1130	04650	1.1145	04709	1.1160	04768	1.1175	04826	24
5	1.1048	04326	1.1063	04386	1.1078	04445	1.1093	04504	1.1108	04563	1.1123	04622	1.1138	04680	25
6	1.1011	04180	1.1026	04240	1.1041	04299	1.1056	04358	1.1071	04417	1.1086	04476	1.1101	04534	26
7	1.0974	04035	1.0989	04095	1.1004	04154	1.1019	04213	1.1034	04272	1.1049	04331	1.1064	04389	27
8	1.0937	03890	1.0952	03950	1.0967	04009	1.0982	04068	1.0997	04127	1.1012	04186	1.1027	04244	28
9	1.0901	03746	1.0916	03806	1.0931	03865	1.0946	03924	1.0960	03983	1.0975	04042	1.0990	04100	29
0	1.0865	03602	1.0880	03662	1.0895	03721	1.0909	03780	1.0924	03839	1.0939	03898	1.0954	03956	30

	740		741		742		743		744		745		746		
	F	log F	F	log F	F	log F	F	log F	F	log F	F	log F	F	log F	
10	1.1745	06986	1.1761	07045	1.1777	07103	1.1793	07162	1.1809	07220	1.1825	07279	1.1841	07337	10
11	1.1704	06832	1.1720	06891	1.1735	06949	1.1751	07008	1.1767	07066	1.1783	07125	1.1799	07183	11
12	1.1663	06679	1.1678	06738	1.1694	06796	1.1710	C6855	1.1725	06913	1.1741	06972	1.1757	07030	12
13	1.1622	06527	1.1637	06586	1.1653	06644	1.1669	06703	1.1685	06761	1.1700	06820	1.1716	06878	13
14	1.1581	06375	1.1597	06434	1.1612	06492	1.1628	06551	1.1644	06609	1.1659	06668	1.1675	06726	14
15	1.1541	06224	1.1557	06283	1.1572	06341	1.1588	06400	1.1603	06458	1.1619	06517	1.1635	06575	15
16	1.1501	06073	1.1517	06132	1.1532	06190	1.1548	06249	1.1563	06307	1.1579	06366	1.1594	06424	16
17	1.1461	05923	1.1477	05982	1.1492	06040	1.1508	06099	1.1523	06157	1.1539	06216	1.1554	06274	17
18	1.1422	05773	1.1437	05832	1.1453	05890	1.1468	05949	1.1483	06007	1.1499	06066	1.1514	06124	18
19	1.1383	05624	1.1398	05683	1.1413	05741	1.1429	05800	1.1444	05858	1.1460	05917	1.1475	05975	19
20	1.1344	05475	1.1359	05534	1.1374	05592	1.1390	05651	1.1405	05709	1.1421	05768	1.1436	05826	20
21	1.1305	05327	1.1320	05386	1.1335	05444	1.1351	05503	1.1366	05561	1.1382	05620	1.1397	05678	21
22	1.1267	05179	1.1282	05238	1.1297	05296	1.1312	05355	1.1327	05413	1.1343	05472	1.1358	05530	22
23	1.1228	05032	1.1244	05091	1.1259	05149	1.1274	05208	1.1289	05266	1.1304	05325	1.1320	05383	23
24	1.1191	04885	1.1206	04944	1.1221	05002	1.1236	05061	1.1251	05119	1.1266	05178	1.1282	05236	24
25	1.1153	04739	1.1168	04798	1.1183	04856	1.1198	04915	1.1213	04973	1.1228	05032	1.1244	05090	25
26	1.1116	04593	1.1131	04652	1.1146	04710	1.1161	04769	1.1176	04827	1.1191	04886	1.1206	04944	26
27	1.1079	04448	1.1094	04507	1.1108	04565	1.1123	04624	1.1138	04682	1.1154	04741	1.1168	04799	27
28	1.1042	04303	1.1057	04362	1.1072	04420	1.1086	04479	1.1101	04537	1.1116	04596	1.1131	04654	28
29	1.1005	04159	1.1020	04218	1.1035	04276	1.1050	04335	1.1064	04393	1.1080	04452	1.1094	04510	29
30	1.0969	04015	1.0983	04074	1.0998	04132	1.1013	04191	1.1028	04249	1.1043	04308	1.1058	04366	30

t°	747 F	747 log F	748 F	748 log F	749 F	749 log F	750 F	750 log F	751 F	751 log F	752 F	752 log F	753 F	753 log F
10	1.1856	07395	1.1872	07453	1.1888	07511	1.1904	07569	1.1920	07627	1.1936	07685	1.1951	07742
11	1.1814	07241	1.1830	07299	1.1846	07357	1.1862	07415	1.1878	07473	1.1894	07531	1.1909	07588
12	1.1773	07088	1.1788	07146	1.1804	07204	1.1820	07262	1.1836	07320	1.1852	07378	1.1867	07435
13	1.1732	06936	1.1747	06994	1.1763	07052	1.1779	07110	1.1794	07168	1.1810	07226	1.1826	07283
14	1.1691	06784	1.1706	06842	1.1722	06900	1.1738	06958	1.1753	07016	1.1769	07074	1.1785	07131
15	1.1650	06633	1.1666	06691	1.1681	06749	1.1697	06807	1.1712	06865	1.1728	06923	1.1744	06981
16	1.1610	06482	1.1625	06540	1.1641	06598	1.1656	06656	1.1672	06714	1.1687	06772	1.1703	06829
17	1.1570	06332	1.1585	06390	1.1601	06448	1.1616	06506	1.1632	06564	1.1647	06622	1.1663	06679
18	1.1530	06182	1.1545	06240	1.1561	06298	1.1576	06356	1.1592	06414	1.1607	06472	1.1623	06529
19	1.1490	06033	1.1506	06091	1.1521	06149	1.1536	06207	1.1552	06265	1.1567	06323	1.1583	06380
20	1.1451	05884	1.1466	05942	1.1482	06000	1.1497	06058	1.1512	06116	1.1528	06174	1.1543	06231
21	1.1412	05736	1.1427	05794	1.1443	05852	1.1458	05910	1.1473	05968	1.1488	06026	1.1503	06083
22	1.1373	05588	1.1388	05646	1.1404	05704	1.1419	05762	1.1434	05820	1.1449	05878	1.1464	05935
23	1.1335	05441	1.1350	05499	1.1365	05557	1.1380	05615	1.1395	05673	1.1410	05730	1.1426	05788
24	1.1297	05294	1.1312	05352	1.1327	05410	1.1342	05468	1.1357	05526	1.1372	05584	1.1387	05641
25	1.1259	05148	1.1274	05206	1.1289	05264	1.1304	05322	1.1319	05380	1.1334	05438	1.1349	05495
26	1.1221	05002	1.1236	05060	1.1251	05118	1.1266	05176	1.1281	05234	1.1296	05292	1.1311	05349
27	1.1183	04857	1.1198	04915	1.1213	04973	1.1228	05031	1.1243	05089	1.1258	05147	1.1273	05204
28	1.1146	04712	1.1161	04770	1.1176	04828	1.1191	04886	1.1206	04944	1.1221	05002	1.1235	05059
29	1.1109	04568	1.1124	04626	1.1139	04684	1.1154	04742	1.1169	04800	1.1184	04858	1.1198	04915
30	1.1072	04424	1.1087	04482	1.1102	04540	1.1117	04598	1.1132	04656	1.1147	04714	1.1161	04771

t°	754 F	754 log F	755 F	755 log F	756 F	756 log F	757 F	757 log F	758 F	758 log F	759 F	759 log F	760 F	760 log F
10	1.1967	07800	1.1983	07858	1.1999	07915	1.2015	07973	1.2030	08030	1.2045	08087	1.2062	08144
11	1.1925	07646	1.1941	07704	1.1957	07761	1.1973	07819	1.1988	07876	1.2003	07933	1.2020	07990
12	1.1883	07493	1.1899	07551	1.1915	07608	1.1931	07666	1.1946	07723	1.1962	07780	1.1978	07837
13	1.1842	07341	1.1857	07399	1.1873	07456	1.1889	07514	1.1904	07571	1.1920	07628	1.1936	07685
14	1.1800	07189	1.1816	07247	1.1831	07304	1.1847	07362	1.1863	07419	1.1878	07476	1.1894	07533
15	1.1759	07038	1.1775	07096	1.1790	07153	1.1806	07211	1.1822	07268	1.1837	07325	1.1853	07382
16	1.1718	06887	1.1734	06945	1.1750	07002	1.1765	07060	1.1781	07117	1.1796	07174	1.1812	07231
17	1.1678	06737	1.1694	06795	1.1709	06852	1.1725	06910	1.1740	06967	1.1755	07024	1.1771	07081
18	1.1638	06587	1.1653	06645	1.1669	06702	1.1684	06760	1.1700	06817	1.1715	06874	1.1730	06931
19	1.1598	06438	1.1613	06496	1.1629	06553	1.1644	06611	1.1660	06668	1.1675	06725	1.1690	06782
20	1.1558	06289	1.1574	06347	1.1589	06404	1.1604	06462	1.1620	06519	1.1635	06576	1.1650	06633
21	1.1519	06141	1.1534	06199	1.1549	06256	1.1565	06314	1.1580	06371	1.1595	06428	1.1610	06485
22	1.1480	05993	1.1495	06051	1.1510	06108	1.1526	06166	1.1541	06223	1.1556	06280	1.1571	06337
23	1.1441	05846	1.1456	05904	1.1471	05961	1.1487	06019	1.1502	06076	1.1517	06133	1.1532	06190
24	1.1402	05699	1.1417	05757	1.1432	05814	1.1448	05872	1.1463	05929	1.1478	05986	1.1493	06043
25	1.1364	05553	1.1379	05611	1.1394	05668	1.1409	05726	1.1424	05783	1.1439	05840	1.1454	05897
26	1.1326	05407	1.1341	05465	1.1356	05522	1.1371	05580	1.1386	05637	1.1401	05694	1.1416	05751
27	1.1288	05262	1.1303	05320	1.1318	05377	1.1333	05435	1.1348	05492	1.1363	05549	1.1378	05606
28	1.1250	05117	1.1266	05175	1.1280	05232	1.1295	05290	1.1310	05347	1.1325	05404	1.1340	05461
29	1.1213	04973	1.1228	05031	1.1243	05088	1.1258	05146	1.1273	05203	1.1288	05260	1.1302	05317
30	1.1176	04829	1.1181	04887	1.1206	04944	1.1221	05002	1.1235	05059	1.1250	05116	1.1265	05173

	761		762		763		764		765		766		767		
	F	log F	F	log F	F	log F	F	log F	F	log F	F	log F	F	log F	
0	1.2078	08201	1.2094	08258	1.2110	08315	1.2126	08372	1.2142	08429	1.2158	08486	1.2174	08543	10
1	1.2035	08047	1.2051	08104	1.2067	08161	1.2083	08218	1.2099	08275	1.2115	08332	1.2131	08389	11
2	1.1993	07894	1.2009	07951	1.2025	08008	1.2041	08065	1.2056	08122	1.2072	08179	1.2088	08236	12
3	1.1951	07742	1.1967	07799	1.1983	07856	1.1999	07913	1.2014	07970	1.2030	08027	1.2046	08084	13
4	1.1910	07590	1.1925	07647	1.1941	07704	1.1957	07761	1.1972	07818	1.1988	07875	1.2004	07932	14
5	1.1868	07439	1.1884	07496	1.1900	07553	1.1915	07610	1.1931	07667	1.1946	07724	1.1962	07781	15
6	1.1827	07288	1.1843	07345	1.1858	07402	1.1874	07459	1.1889	07516	1.1905	07573	1.1921	07630	16
7	1.1786	07138	1.1802	07195	1.1817	07252	1.1833	07309	1.1848	07366	1.1864	07423	1.1880	07480	17
8	1.1746	06988	1.1761	07045	1.1777	07102	1.1792	07159	1.1808	07216	1.1823	07273	1.1839	07330	18
9	1.1705	06839	1.1721	06896	1.1736	06953	1.1752	07010	1.1767	07067	1.1783	07124	1.1798	07181	19
0	1.1665	06690	1.1681	06747	1.1696	06804	1.1712	06861	1.1727	06918	1.1742	06975	1.1758	07032	20
1	1.1626	06542	1.1641	06599	1.1656	06656	1.1672	06713	1.1687	06770	1.1702	06827	1.1718	06884	21
2	1.1586	06394	1.1601	06451	1.1617	06508	1.1632	06565	1.1647	06622	1.1662	06679	1.1678	06736	22
3	1.1547	06247	1.1562	06304	1.1577	06361	1.1593	06418	1.1608	06475	1.1623	06532	1.1638	06589	23
4	1.1508	06100	1.1523	06157	1.1538	06214	1.1554	06271	1.1569	06328	1.1584	06385	1.1599	06442	24
5	1.1469	05954	1.1484	06011	1.1500	06068	1.1515	06125	1.1530	06182	1.1545	06239	1.1560	06296	25
6	1.1431	05808	1.1446	05865	1.1461	05922	1.1476	05979	1.1491	06036	1.1506	06093	1.1521	06150	26
7	1.1393	05663	1.1408	05720	1.1423	05777	1.1438	05834	1.1453	05891	1.1468	05948	1.1483	06005	27
8	1.1355	05518	1.1370	05575	1.1385	05632	1.1400	05689	1.1415	05746	1.1430	05803	1.1445	05860	28
9	1.1317	05374	1.1332	05431	1.1347	05488	1.1362	05545	1.1377	05602	1.1392	05659	1.1407	05716	29
0	1.1280	05230	1.1295	05287	1.1309	05344	1.1324	05401	1.1339	05458	1.1354	05515	1.1369	05572	30

	768		769		770		771		772		773		774		
0	1.2190	08599	1.2206	08656	1.2221	08712	1.2237	08768	1.2253	08825	1.2269	08881	1.2285	08937	10
1	1.2147	08445	1.2163	08502	1.2178	08558	1.2194	08614	1.2210	08671	1.2226	08727	1.2242	08783	11
2	1.2104	08292	1.2120	08349	1.2136	08405	1.2151	08461	1.2167	08518	1.2183	08574	1.2199	08630	12
3	1.2062	08140	1.2077	08197	1.2093	08253	1.2109	08309	1.2125	08366	1.2140	08422	1.2156	08478	13
4	1.2020	07988	1.2035	08045	1.2051	08101	1.2066	08157	1.2082	08214	1.2098	08270	1.2114	08326	14
5	1.1978	07837	1.1993	07893	1.2009	07949	1.2024	08005	1.2040	08062	1.2056	08118	1.2072	08174	15
6	1.1936	07686	1.1952	07743	1.1967	07798	1.1982	07854	1.1998	07911	1.2014	07967	1.2030	08023	16
7	1.1895	07536	1.1910	07593	1.1926	07648	1.1941	07704	1.1957	07761	1.1972	07817	1.1988	07873	17
8	1.1854	07386	1.1869	07443	1.1885	07498	1.1900	07554	1.1916	07611	1.1931	07667	1.1946	07723	18
9	1.1813	07237	1.1829	07294	1.1844	07349	1.1859	07405	1.1875	07462	1.1890	07518	1.1905	07574	19
0	1.1773	07088	1.1788	07145	1.1803	07200	1.1818	07256	1.1834	07313	1.1849	07369	1.1865	07425	20
1	1.1733	06940	1.1748	06997	1.1763	07052	1.1778	07108	1.1794	07165	1.1809	07221	1.1824	07277	21
2	1.1693	06792	1.1708	06849	1.1723	06904	1.1738	06960	1.1754	07017	1.1769	07073	1.1784	07129	22
3	1.1653	06645	1.1669	06702	1.1683	06757	1.1699	06813	1.1714	06870	1.1729	06926	1.1744	06982	23
4	1.1614	06498	1.1629	06555	1.1644	06611	1.1659	06667	1.1675	06724	1.1690	06780	1.1705	06836	24
5	1.1575	06352	1.1590	06409	1.1605	06465	1.1620	06521	1.1635	06578	1.1650	06634	1.1665	06690	25
6	1.1536	06206	1.1551	06263	1.1566	06319	1.1581	06375	1.1596	06432	1.1611	06488	1.1626	06544	26
7	1.1498	06061	1.1513	06118	1.1528	06174	1.1542	06230	1.1558	06287	1.1573	06343	1.1588	06399	27
8	1.1459	05916	1.1474	05973	1.1489	06029	1.1504	06085	1.1519	06142	1.1534	06198	1.1549	06254	28
9	1.1421	05772	1.1436	05829	1.1451	05885	1.1466	05941	1.1481	05998	1.1496	06054	1.1511	06110	29
0	1.1384	05628	1.1399	05685	1.1413	05741	1.1428	05797	1.1443	05854	1.1458	05910	1.1473	05966	30

p/t	775 F	775 log F	776 F	776 log F	777 F	777 log F	778 F	778 log F	779 F	779 log F	780 F	780 log F	781 F	781 log F	
10	1.2301	08993	1.2317	09049	1.2333	09105	1.2348	09161	1.2364	09217	1.2380	09272	1.2396	09328	10
11	1.2257	08839	1.2273	08895	1.2289	08951	1.2305	09007	1.2321	09063	1.2336	09118	1.2352	09174	11
12	1.2214	08686	1.2230	08742	1.2246	08798	1.2262	08854	1.2278	08910	1.2293	08965	1.2309	09021	12
13	1.2172	08534	1.2187	08590	1.2203	08646	1.2219	08702	1.2235	08758	1.2250	08813	1.2266	08869	13
14	1.2129	08382	1.2144	08438	1.2160	08494	1.2176	08550	1.2192	08606	1.2207	08661	1.2223	08717	14
15	1.2087	08230	1.2102	08286	1.2118	08342	1.2133	08398	1.2149	08454	1.2164	08509	1.2180	08565	15
16	1.2045	08079	1.2060	08135	1.2076	08191	1.2091	08247	1.2107	08303	1.2122	08358	1.2138	08414	16
17	1.2003	07929	1.2019	07985	1.2034	08041	1.2049	08097	1.2065	08153	1.2080	08208	1.2096	08264	17
18	1.1962	07779	1.1977	07835	1.1993	07891	1.2008	07947	1.2023	08003	1.2038	08058	1.2054	08114	18
19	1.1921	07630	1.1936	07686	1.1952	07742	1.1967	07798	1.1982	07854	1.1997	07909	1.2013	07965	19
20	1.1880	07481	1.1895	07537	1.1911	07593	1.1926	07649	1.1941	07705	1.1956	07760	1.1972	07816	20
21	1.1839	07333	1.1855	07389	1.1870	07445	1.1885	07501	1.1901	07557	1.1916	07612	1.1931	07668	21
22	1.1799	07185	1.1814	07241	1.1830	07297	1.1845	07353	1.1860	07409	1.1875	07464	1.1890	07520	22
23	1.1759	07038	1.1774	07094	1.1790	07150	1.1805	07206	1.1820	07262	1.1835	07317	1.1850	07373	23
24	1.1720	06892	1.1735	06948	1.1750	07004	1.1765	07060	1.1780	07116	1.1795	07171	1.1811	07227	24
25	1.1680	06746	1.1696	06802	1.1711	06858	1.1726	06914	1.1741	06970	1.1756	07025	1.1771	07081	25
26	1.1641	06600	1.1656	06656	1.1671	06712	1.1686	06768	1.1701	06824	1.1716	06879	1.1731	06935	26
27	1.1602	06455	1.1617	06511	1.1632	06567	1.1647	06623	1.1662	06679	1.1677	06734	1.1692	06790	27
28	1.1564	06310	1.1579	06366	1.1594	06422	1.1609	06478	1.1624	06534	1.1638	06589	1.1653	06645	28
29	1.1526	06166	1.1540	06222	1.1555	06278	1.1570	06334	1.1585	06390	1.1600	06445	1.1615	06501	29
30	1.1487	06022	1.1502	06078	1.1517	06134	1.1532	06190	1.1547	06246	1.1561	06301	1.1576	06357	30

p/t	782 F	782 log F	783 F	783 log F	784 F	784 log F	785 F	785 log F	786 F	786 log F	787 F	787 log F	788 F	788 log F	
10	1.2412	09384	1.2428	09439	1.2444	09495	1.2459	09550	1.2475	09605	1.2491	09660	1.2507	09716	10
11	1.2368	09230	1.2384	09285	1.2400	09341	1.2415	09396	1.2431	09451	1.2447	09506	1.2463	09562	11
12	1.2325	09077	1.2340	09132	1.2356	09188	1.2372	09243	1.2387	09298	1.2403	09353	1.2419	09409	12
13	1.2282	08925	1.2297	08980	1.2313	09036	1.2328	09091	1.2344	09146	1.2360	09201	1.2376	09257	13
14	1.2239	08773	1.2254	08828	1.2270	08884	1.2285	08939	1.2301	08994	1.2316	09049	1.2333	09105	14
15	1.2196	08621	1.2211	08676	1.2227	08732	1.2242	08787	1.2258	08842	1.2273	08897	1.2290	08953	15
16	1.2154	08470	1.2169	08525	1.2185	08581	1.2200	08636	1.2215	08691	1.2231	08746	1.2247	08802	16
17	1.2112	08320	1.2127	08375	1.2143	08431	1.2158	08486	1.2273	08541	1.2189	08596	1.2205	08652	17
18	1.2070	08170	1.2085	08225	1.2101	08281	1.2116	08336	1.2131	08391	1.2147	08446	1.2163	08502	18
19	1.2028	08021	1.2044	08076	1.2060	08132	1.2075	08187	1.2090	08242	1.2105	08297	1.2121	08353	19
20	1.1987	07872	1.2002	07927	1.2018	07983	1.2033	08038	1.2048	08093	1.2064	08148	1.2080	08204	20
21	1.1947	07724	1.1962	07779	1.1977	07835	1.1992	07890	1.2007	07945	1.2023	08000	1.2038	08056	21
22	1.1906	07576	1.1921	07631	1.1936	07687	1.1951	07742	1.1967	07797	1.1982	07852	1.1997	07908	22
23	1.1866	07429	1.1881	07484	1.1896	07540	1.1911	07595	1.1926	07650	1.1941	07705	1.1957	07761	23
24	1.1826	07283	1.1841	07338	1.1856	07394	1.1871	07449	1.1886	07504	1.1901	07559	1.1917	07615	24
25	1.1786	07137	1.1801	07192	1.1816	07248	1.1831	07303	1.1846	07358	1.1861	07413	1.1877	07469	25
26	1.1747	06991	1.1761	07046	1.1777	07102	1.1792	07157	1.1806	07212	1.1821	07267	1.1837	07323	26
27	1.1707	06846	1.1722	06901	1.1737	06957	1.1752	07012	1.1767	07067	1.1782	07122	1.1797	07178	27
28	1.1668	06701	1.1683	06756	1.1698	06812	1.1713	06867	1.1728	06922	1.1743	06977	1.1758	07033	28
29	1.1630	06557	1.1644	06612	1.1660	06608	1.1674	06723	1.1689	06778	1.1704	06833	1.1719	06889	29
30	1.1591	06413	1.1606	06468	1.1621	06524	1.1636	06579	1.1650	06634	1.1665	06689	1.1680	06745	30

General methods

Examples for application of standard joints

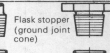
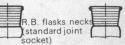

Tube connection (ground joint cone)

Flask stopper (ground joint cone)

Weighing glass stopper (ground joint cone)

Weighing glass cap (standard joint cone)

Tube-joint (ground glass socket)

R.B. flasks necks (standard joint socket)

Flask necks (standard joint socket)

Vessel or container tube (standard joint socket)

Standard joint cones
d_1 = bigger cone diameter
d_2 = smaller cone diameter
h = height or length

Interchangeable standard joints with ground glass joints

standard joint socket

standard joint cone

Standard ground glass joints "Quick fit" specifications

size designation	d_1 (mm)	d_2 (mm)	h (mm)	size designation	d_1 (mm)	d_2 (mm)	h (mm)
5/13	5.0	3.7	13.0	55/29	55.0	52.1	23.0
7/11	7.5	6.4	11.0	55/44	55.0	50.6	44.0
7/16	7.5	5.9	16.0	60/46	60.0	55.4	46.0
10/13	10.0	8.7	13.0	Spherical joints	(Ball & cup)		
10/19	10.0	8.1	19.0				
12/21	12.5	10.4	21.0	size designation	nominal dia.	Minimum dia.	Maximum dia.
14/15	14.5	13.0	15.0		(mm)	wide end (mm)	narrow end (mm)
14/23	14.5	12.2	23.0				
19/17	18.8	17.1	17.0	S 13	12.700	12.5	7.0
19/26	18.8	16.2	26.0	S 19	19.050	18.7	12.5
24/10	24.0	23.0	10.0	S 29	28.575	28.0	19.0
24/20	24.0	22.0	20.0	S 35	34.925	34.3	27.5
24/29	24.0	21.1	29.0	S 41	41.275	40.5	30.0
29/32	29.2	26.0	32.0	S 51	50.800	50.0	36.0
34/35	34.5	31.0	35.0				
40/13	40.0	38.7	13.0				
40/38	40.0	36.2	38.0				
45/40	45.0	41.0	40.0				
50/14	50.0	48.6	14.0				
50/42	50.0	45.8	42.0				

Quick fit is made from heat and chemical resistant borosilicate glass. Some properties and care of ground joints.

Composition: Low alkali content with no elements of alkaline earth, or Sb, As, Zn and heavy metals.

Stability : High resistant to attack of acids (except in HF and hot glacial H_3PO_4). Attacked by caustic alkali (a silicate) to a measurable extent.

$$\log_{10} W - 15.66 = \log_{10} (t \sqrt{C}) - \frac{5600}{T}$$

w is the weight loss in mg/dm² of glass surface, t is time in hrs, c is concentration of caustic soda % by wt. T is absolute temperature.

Thermal endurance: due to low coeff. of thermal expansion can withstand a high degree of thermal shock (depends on shape and dimension of article).

Temperature limits in service: As a general rule no article should be heated to temperature above 450°C particularly for prolonged time. Slow cooling is necessary to avoid possibility of leaving permanent strain in cases when heated above 300°C.

All ground joints must be free from dust which may cause leakage. Before use must be wiped free from foreign matter. After use must be separated preferable while it is still warm.

Lubrication fulfils one or more of three purposes: to prevent leakage, to protect surface of the joint, to facilitate separation of joint. Suitable lubricant must be used depending on the needs.

Petroleum jelly — not suited for high temperatures or for vacuum since it dries out, alternative to this is paraffin and rubber lubricant.

"Apiezon" Range — for severe conditions unless a hydrocarbon lubricant is unacceptable.

"Silicone" greases — useful for high temp. s; but not easily removable, spread readily; should not be left assembled for long time.

Polytetra fluoro ethylene (PTFE) —excellent chemical properties as a lubricant for conical ground joints. Care must be taken to avoid scratches, corrosion with alkalis, and while clamping, heating and cleaning.

Filtration

Funnel with an inserted round filter paper		Suction flask with Büchner funnel

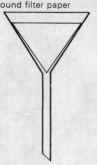

Funnel with an inserted round filter paper. The filter is twice folded and opened to form a cone. The funnel has an angle of 60°. The edge of the filter paper inserted is kept about 0.5 cm below the rim of funnel.

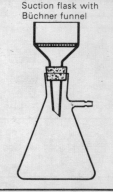

Filteration is a process of separation of a solid substance from a liquid (or gas) through a filter medium which must allow the liquid to pass through but holding back the solid substance.

In chemical laboratories, as a rule, filter papers in form of circles and sheets are used for filteration.

Besides one also employs a permeable disc of porcelain and sintered glass (or celain and glass sintered crucible). Filter paper in a funnel is inserted since it is not mechanically stable. Porcelain and sintered glass crucibles set up on a suction flask on which a vacuum can be applied.

Filter paper for qualitative work can contain also numeral substances whereas the papers for qualitative work must be extensively ash free since they are incinerated and is weighed together with the precipitate.

Porcelain filter crucibles

A 1

Sintered glass crucibles

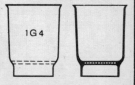

1 G 4

The number on the porcelain crucibles gives information about the pore (size) diameter.

Note: Smaller the number, smaller the pore width.

Porcelain filter crucibles can only be heated along with a "Crucible shoe" or with a protective crucible.

The mark on a glass sintered crucible gives information about the glass type, the form and the pore width.

Example: 1 G 4   1 = form
G = Jena apparatus glass
4 = pore width

Note: Bigger the number at the end of the notation, smaller the pore width. Sintered glass crucibles are not allowed to be heated.

Some typical applications

Application	Qualitative	wet strength	low ash	ashless	Hardened	Hardened ashless	Remarks
General purposes, medium precipitates	1 and 2	111	30	40	52	540	2 is more retentive
$BaSO_4$ (well precipitated)	2	111	30	40 & 44	52	540,544	44 & 544 are more retentive
$BaSO_4$ fine suspension	6	115	32	42	50	542	All are highly retentive
Silicia (coarse)	4	90	31	41	54	541	
Silicia (fine)	2	111	30	40	52	540	Use as for $BaSO_4$ (fine) it more retentive paper is required
Iron and aluminium - hydroxide	4	90	31	41	54	541	
Calcium oxalate	—	111	30	40	52	540	
Metastannic acid	6	115	32	42	50	542	
Ammonium magnesium orthophosphate	1 and 2	111	30	40	52	540	
Phosphomolybolic acid (yellow ppts)	—	111	30	40	52	540	all these grades are specifically designed for this work.
Büchner funnels	6	111,115	32	42	all hardened grades		
Assay of gold, etc.	—	—	—	44	—	544	
Caustic solutions	—	—	—	—	all hardened grades		
Agar Agar	15	—	—	—	—	—	
Gelatinous biological ppts.	—	90	—	—	—	—	

Filter papers (Whatman)

Relative filtration rates, retentivity and pore size

Classification	Fast			Fast medium	Medium	Slow	
Retaining properties	Coarse particles			Medium particles		Fine particles	
Mean pore size μ		5 to 3.4		2.8 to 2.1	2.9 to 1.4	1.1 to 0.4	
Relative filtration times	4–5	6–8	9–12	20–30	35–50	100–130	140–160
Qualitative grades		4		1	2;3		6
Qualitative wet strength grades	90			111			115
Low ash grades			31		30	32	
Ashless grades (qualitative)			41		40	44	42
Hardened grades		54			52		50
Hardened ashless grades		541			540	544	542

Note: For Whatman filter paper specifications refer to manufacturer's catalogues

Extraction

Extraction is dissolving of solid or liquid substances of a mixture with the help of suitable solvent medium.
In a boiling flask the solvent is evaporated and in a reflux condenser is condensed back to solvent.
The solvent falls into an extraction tube where the mixture is contained.
The extraction case must be permeable for the liquid. The extract solution runs back continuously or periodically into the distillation flask. The extraction serves in the dissolution of difficult to soluble substances of the mixture.

Application:
Fat determination, isolation of natural substances out of plant and animal materials.
Solvent:
Ether, chloroform, petroleum ether, benzene, etc.

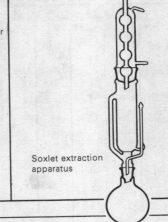

Soxlet extraction apparatus

Distillation

Distillation is a process of changing the substance into vapour state through heating and then condensing the vapours in an attached cooler.

One applies distillation for separation of a mixture solid and liquid substances (frequently in case of preparative work) or for the separation of a mixture of liquid substances of basically different distinguishable boiling points. Substances which easily change into vapours and do not decompose at their distillation temperatures, can be distilled at atmospheric pressure.

The fractional distillation is used for the separation of easily vapourable liquid substances.
The separation is improved by a distillation column attached, (fractionary column). Substances which form an azeotropic mixture cannot be separated by a simple fractional distillation.

One applies vacuum distillation for the separation of very high boiling liquids or in case of liquids which decompose at their boiling points.
By application of vacuum the distillation temperature can be considerably lowered (by vacuum of water pump to an average of 100 upto 120°C).

Steam distillation is a laboratory and often an industrial application of purification process (in laboratory for the recovery of preparation).
Steam distillation is accomplished only when the substance is vapourable with steam and is insoluble or only very lightly soluble in water.
Example: Nitrobenzene, bromobenzene, o-nitrophenol, organic bases.

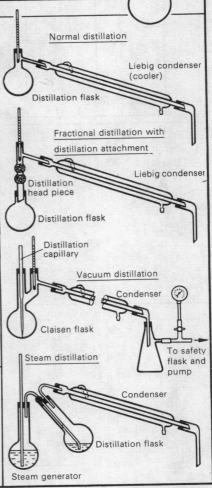

Normal distillation

Liebig condenser (cooler)

Distillation flask

Fractional distillation with distillation attachment

Liebig condenser

Distillation head piece

Distillation flask

Distillation capillary

Vacuum distillation

Condenser

Claisen flask

To safety flask and pump

Steam distillation

Condenser

Distillation flask

Steam generator

Measurement of pressure

Different pressure data

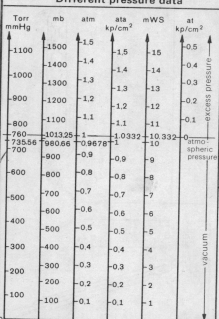

Torr mmHg	mb	atm	ata kp/cm²	mWS	at kp/cm²	
	1500	1,5	1,5	15	0,5	excess pressure
1100	1400	1,4	1,4	14	0,4	
1000	1300	1,3	1,3	13	0,3	
900	1200	1,2	1,2	12	0,2	
800	1100	1,1	1,1	11	0,1	
760	1013,25	1	1,0332	10,332	0	atmospheric pressure
735.56	980,66	0,9678	1	10		
700	900	0,9	0,9	9	0,1	
600	800	0,8	0,8	8	0,2	
500	700	0,7	0,7	7		
	600	0,6	0,6	6		
400	500	0,5	0,5	5		vacuum
300	400	0,4	0,4	4		
	300	0,3	0,3	3		
200	200	0,2	0,2	2		
100	100	0,1	0,1	1		

For the production of high pressures, compressors are employed in laboratories and in industries. The compressors are developed as blast apparatus, rotatory slide valve pumps or piston pumps. A seal at high pressure compression results mostly in stages.

For the production of vacuum in laboratories water jet pumps or rotatory slide valve pumps are employed; however, for producing high vacuum mercury or oil diffusion pumps are used. Normally rotatory slide valve pumps are employed first before vacuum is produced with diffusion pumps. High pressures are measured mostly with tube spring manometers. For the measurement of vacuum mercury manometers are used and for the measurement of high vacuums, the so called compression manometers are used.

Attainable vacuum:

Waterjet pumps	= 12 − 14	mm Hg
Rotatory slide valve pumps	= 0.01	mm Hg
Diffusion pumps	= 10^{-6}	mm Og

Tube spring manometers

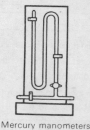

Mercury manometers

Measurement of gas flow

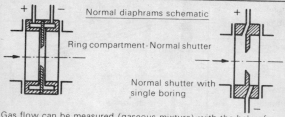

Normal diaphrams schematic

Ring compartment - Normal shutter

Normal shutter with single boring

Gas flow can be measured (gaseous mixture) with the help of normal shutters. Due to flow in the direction of arrow a high pressure develops at +Ve and a low pressure at −Ve. The pressure difference in this method frequently is measured with the help of ring balances.

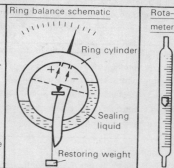

Ring balance schematic

Ring cylinder

Sealing liquid

Restoring weight

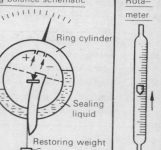

Rota- meter

Determination of density and specific gravity

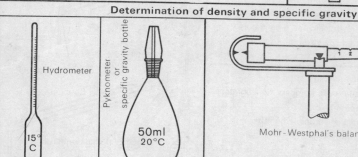

Hydrometer

15° C

Pyknometer or specific gravity bottle

50ml 20°C

Mohr - Westphal's balance

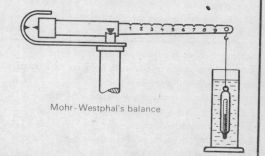

Density

The density ρ of a substance is the quotient of m/V m = mass and V = volume

$$\rho = \frac{m}{V}$$

In case of density, when necessary, suitable conditions about the substance to be given (temperature, pressure, etc.).

Specific Gravity

The specific gravity γ of a substance is the quotient of M/V where M is the weight of the substance, V the volume.

$$\gamma = \frac{M}{V}$$

In case of specific gravity, when necessary, suitable conditions of the substance to be given (pressure, temperature, etc.).

The mass (m) is independent of all outside conditions, e.g. temperature, buoyancy in air, and different accelerations due to gravity. In view of these independent characteristics mass stands in advantage to weight.

The numerical value of mass m is equal to the numerical value of weight M in vacuum (exact only when the weight of the standard value of acceleration due to gravity is stated 980.665 cm/sec^2.).

The commonly ascertained weight in air M' can be converted into the weight M in vacuum.

$M' = MA$ A = air buoyancy
or λ = density of air
M $M'(1 + \frac{\lambda}{d})$ mean value = 0.0012 g/ml at 20°C and 760 Torr

The determination of density or specific gravity is always a determination of weights (Mass) and of volumes.

For the density determinations for liquids we use a hydrometer or pyknometer (specific gravity bottle).

In case of Mohr-Westphal's balance for the density determination the Archimedes principle (buoyancy) is employed whereby a standardized immersion body is suspended into the liquid to be measured and weighed.

Measurement of electrical quantities

Symbols for measuring instruments

	Hotwire-instrument		Thermo-converters or Thermo-transformer (general)		Rectifier
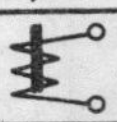	Rotating iron-apparatus		Moving coil apparatus with permanent magnet		Moving coil apparatus with thermo-transformer
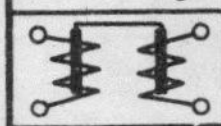	Rotating iron-quotient-meter		Moving coil quotient meter		Moving coil apparatus with measuring rectifier

Classification of measuring accuracy

	Precision-instruments			Industrial measuring instruments			
Class (in brackets = old notation)	0.1	0.2(E)	0.5(F)	1	1.5(G)	2.5(H)	5
Permissible error in reading in %	0.1	0.2	0.5	1	1.5	2.5	5

Test voltage

Star	Test volt in V	For industrial voltage in V
	500	upto 40
	2000	above 40 upto 650
	5000	above 650 upto 1500
	10000	above 1500 upto 3000

Apparatus

Voltmeter

P(R)
N(Mp)
Normal connection

Ammeter

P(R)
N(Mp)
Normal connection

Symbols for working conditions

	Perpendicular working position.
	Horizontal working condition.
30	Slant working position with given angle.

Example

Symbol	Meaning
	Moving coil measuring instrument of class 0.5 for perpendicular working condition
0.5 2	Direct current 2000 V of test voltage.

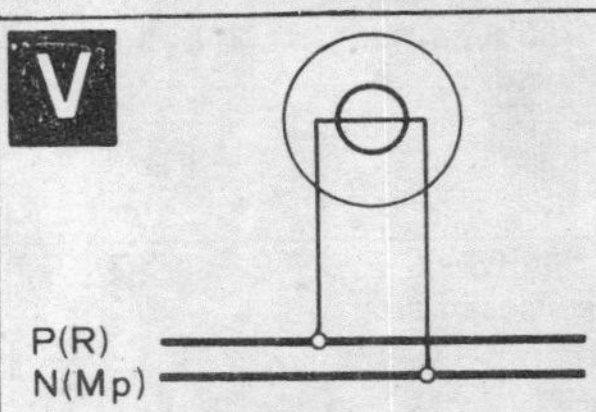

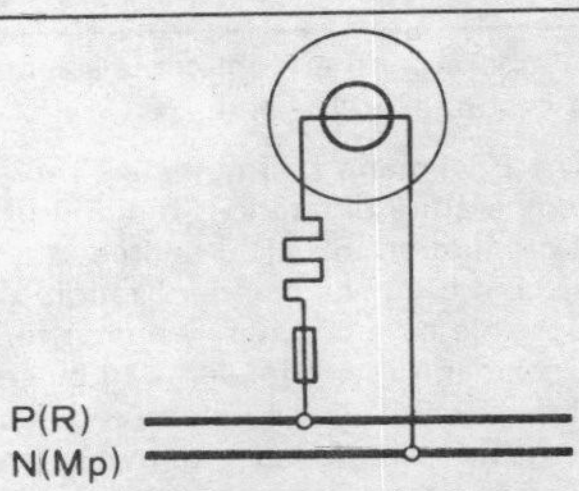

P(R)
N(Mp)
With series resistance

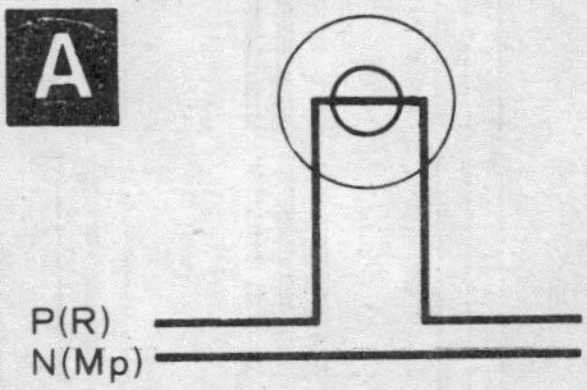

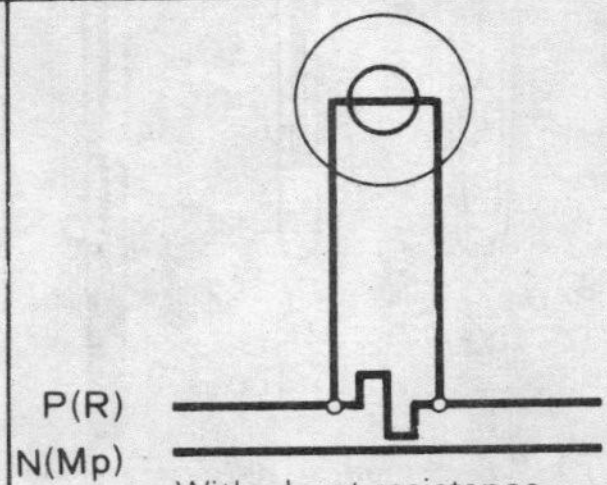

P(R)
N(Mp)
With shunt resistance

Temperature measurements

Temperature range for different thermometers

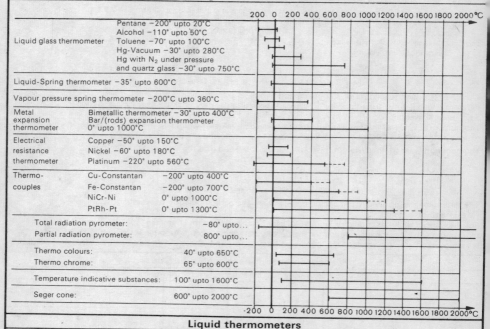

Liquid glass thermometer	Pentane −200° upto 20°C	
	Alcohol −110° upto 50°C	
	Toluene −70° upto 100°C	
	Hg-Vacuum −30° upto 280°C	
	Hg with N_2 under pressure and quartz glass −30° upto 750°C	
Liquid-Spring thermometer −35° upto 600°C		
Vapour pressure spring thermometer −200°C upto 360°C		
Metal expansion thermometer	Bimetallic thermometer −30° upto 400°C	
	Bar/(rods) expansion thermometer 0° upto 1000°C	
Electrical resistance thermometer	Copper −50° upto 150°C	
	Nickel −60° upto 180°C	
	Platinum −220° upto 560°C	
Thermo-couples	Cu-Constantan −200° upto 400°C	
	Fe-Constantan −200° upto 700°C	
	NiCr-Ni 0° upto 1000°C	
	PtRh-Pt 0° upto 1300°C	
Total radiation pyrometer:	−80° upto...	
Partial radiation pyrometer:	800° upto...	
Thermo colours:	40° upto 650°C	
Thermo chrome:	65° upto 600°C	
Temperature indicative substances:	100° upto 1600°C	
Seger cone:	600° upto 2000°C	

Liquid thermometers

Laboratory Thermometers

Type	Scale Range °C	Division °C	Standard deviation range °C
Rod type	0 to 100	1	0 to 50 ± 0.7
Enclosed type	0 to 100	1	50 to 100 ± 1
Rod type	0 to 250	1	0 to 50 ± 0.7
Enclosed type			50 to 100 ± 1
			100 to 200 ± 1.5
			200 to 250 ± 2
Rod type	0 to 360	1	0 to 50 ± 0.7
Enclosed type			50 to 100 ± 1
			100 to 200 ± 1.5
			200 to 300 ± 2
			300 to 360 ± 2.5
Rod type	0 to 360	2	0 to 50 ± 1
Enclosed type			50 to 100 ± 1.5
			100 to 200 ± 2
			200 to 300 ± 3
			300 to 360 ± 4

Laboratory Thermometers

100° 100°

0° 0°

100°

0°

Rod type
0 - 100°

Enclosed type
0 - 100°

Beckmann-Thermometer

A laboratory thermometer is selected according to the requirements it is supposed to serve.

The Beckmann thermometers serve for the measurement of very small temperature differences ranging from 0 to 100°C. They have a graduation in 1/100 degree at a temperature range of scale of around 5°C. The standardization for different temperature work is possible by a counter-reservoir for mercury.
Beckmann thermometer is to be set when small temperature differences are to be measured accurately and the temperature rise remains disregarded, e.g. measurement in calorimeters or molecular weight determinations.

Thermo-elements

Thermopotential of most common thermoelements at the reference temperature 0°C

Positive side* Negative side	Cu- Constantan		Fe Constantan		NiCr/ Ni		PtRh/ Pt	
Measuring Temp. °C	Theor. potential mV	Allowed deviation °C	Theor. potential mV	Allowed deviation °C	Theor. potential mV	Allowed deviation °C	Theor. potential mV	Allowed deviation °C
-200	-5.7		-8.15					
-100	-3.4		-4.75					
0	0	—	0	—	0	—	0	—
100	4.25	±3	5.37	±3	4.10	±3	0.643	±3
200	9.20	±3	10.95	±3	8.13	±3	1.436	±3
300	14.89	±3	16.55	±3	12.21	±3	2.316	±3
400	20.99	±3	22.15	±3	16.40	±3	3.251	±3
500	27.40	±3.75	27.84	±3.75	20.65	±3.75	4.221	±3
600	34.30	±4.5	33.66	±4.5	24.91	±4.5	5.224	±3
700			39.72	±5.25	29.14	±5.25	6.260	±3.5
800			46.23	±6	33.30	±6	7.329	±4
900			53.15	±6.75	37.36	±6.75	8.432	±4.5
1000					41.31	±7.5	9.570	±5
1100					45.16	±8.24	10.741	±5.5
1200					48.89	±9	11.935	±6
1300							13.138	±6.5
1400							14.337	±7
1500							15.530	±7.5
1600							16.716	±8

1st the reference temperature is 20°C. The thermopotential in the following values decrease

     Cu - Constantan   0.80 mV
     Fe - Constantan   1.05 mV
     Ni Cr - Ni   0.82 mV
     PtRh - Pt   0.11 mV

The gradation line limits the temperature range for permanent application of thermoelements in pure air. At high temperatures, bigger errors can occur. An accurate temperature limit cannot be given, since it depends on several factors, e.g. wire cross sectional area, type of protecting tubes, action of gases and other substances.

In principle, at the ends of the thermoelements wires, the temperature for comparative places is to be high, but is prolonged due to the so-called compensating circuit. These compensating circuits consist of either the same industrial material as the thermoelement wire or from cheap alloy which have the thermoelectric properties.

* Positive side is marked red or with a [+] sign.

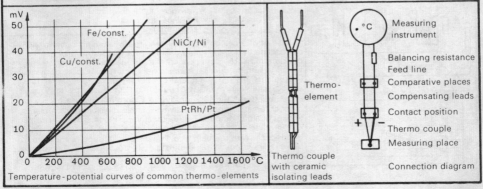

Temperature - potential curves of common thermo - elements

Thermo couple with ceramic isolating leads

Connection diagram

Resistance thermometers

Variation of resistance of Pt and Ni resistance thermometers with temperature

The accuracy of Pt resistance thermometers in case of application of sensitive measuring instruments is very high. One uses these, among others, for calibration of other thermometers.

Nickel		Platinum	
Measuring temp °C	Resistance Ω	Measuring temp. °C	Resistance Ω
– 60	69.5 ± 1.0	– 220	10.41 ± 0.7
0	100.0 ± 0.1	– 200	18.53 ± 0.5
100	161.7 ± 0.8	– 100	60.20 ± 0.3
180	223.1 ± 1.3	0	100.0 ± 0.1
		100	138.50 ± 0.2
		200	175.86 ± 0.4
		300	212.08 ± 0.6
		400	247.07 ± 0.8
		500	280.94 ± 1.0
		550	297.30 ± 1.1

Variation of resistance with increasing temperature.

— — Pt – resistance thermometer
– – – Ni – resistance thermometer

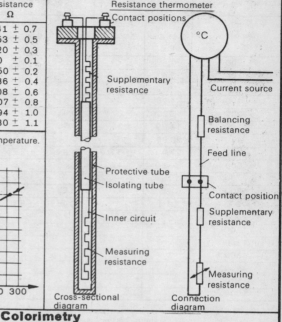

Resistance thermometer

Contact positions

Supplementary resistance

Protective tube
Isolating tube

Inner circuit

Measuring resistance

Cross-sectional diagram

°C

Current source

Balancing resistance

Feed line

Contact position

Supplementary resistance

Measuring resistance

Connection diagram

Colorimetry

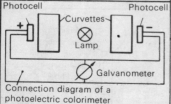

Immersion colorimeter
Turned prism
Immersion cylinder
Curvettes
s_2 c_2
c_1 s_1
Light source

Photocell
Curvettes
Photocell
Lamp
Galvanometer

Connection diagram of a photoelectric colorimeter

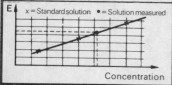

E
x = Standard solution   • = Solution measured
Concentration

With colorimetry, our understanding is of the method of measurement of concentration of a coloured solution by colour comparison with an equally coloured standard solution. For equally coloured solutions the Beer's Law is obeyed.
c_1 = concentration of the solution to be measured
c_2 = concentration of standard solution
s_1 = length (thickness) of solution to be measured
s_2 = length (thickness) of standard solution.
In case of visual methods, by changing the thickness or by setting up a diluted series of solution to be measured and the comparative solution, colour equality is obtained. The concentration stands inversely proportional to the ratio of thickness.
E = Extinction v = transparency degree
Φ_i = incident light rays
Φ_e = emergent light rays

$$c_1 s_1 = c_2 s_2$$

$$c_1 = \frac{c_2 \; s_2}{s_1}$$

$$\frac{c_1}{c_2} = \frac{s_2}{s_1}$$

$$E = -\log v$$

$$v = \frac{\phi_e}{\phi_i}$$

In case of photoelectric colorimeters, the weakening of light rays by passage through a solution is used for measuring the concentration. The light rays enter a photoelement or a photocell which in turn produces a corresponding current. In case of measuring instruments, often it is a matter of not only of colour comparator but also the weak intensity of light that is produced by its passage through the solution due to absorption. By use of monochromatic light the sensivity of these measuring instruments under certain conditions can be greatly increased. These instruments are not merely called as colorimeters but as spectrophotometers.
In principle, with the help of a comparative solution, a standard curve is drawn up. When the instrument possesses an absorbance scale, the measurement is simple especially, since the curve follows a linearity. In contrast, the absorption curves are curved strongly.

Electrolysis

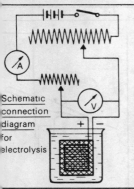

Schematic connection diagram for electrolysis

For demonstration, a 6 volt accumulator with an adequate capacity serves as a current source. The main resistance that is connected as potentiometer, serves for adjusting necessary potential; the smaller the adjustable resistance towards the reference, the smaller the requisite current strength. The current strength can be measured by an ammeter connected in series; while to control the voltage flowing between the electrodes, a voltmeter is attached.

If as a current source, an adjustable direct current source (direct current generator, regulating transformer with a rectifier) is used, then the connected sliding resistance as potentiometer is dropped.

As electrodes, one can use an appropriate platinum net electrode having a big surface area. Frequently one uses a platinum spiral as an electrode which can be simultaneously used as a stirrer.

The voltage normally required is 2–4 volts, with current strength 1–4 ampere. When electrolysis is started, one can increase the current strength. The same applies for an electrolysis at higher temperatures.

For the deposition, Faraday's law is applicable. [96500 coulomb = 1 Equivalent] In case of metal analysis, this principle remains disregarded. Through choice of voltage, current strength, temperature duration of electrolysis, and proper composition of electrolysis, the quantitative deposition becomes important.

Summary of quantitative electrolysis of metals

Metal	Electrolyte	Notes for the process
Cu	Nitric acid	Copper is deposited suitably from nitric acid solution. The solution is made free of NO_2 by boiling and addition of urea prevents the new formation of NO_2. The deposited Cu-quantity can amount upto 1 g.
Pb	Nitric-acid	1st, in addition to Cu, Pb is also present, Pb is quantitatively deposited as PbO_2 at the anode, when an excess in respect to Cu is present and the solution is strong HNO_3. The electrode with Cu is raised with alcohol and dried for a short time in an oven. The electrode with PbO_2 must be dried at 180°C for 2 hours in a drying oven.
Zn	NaOH-alkali	Zn is deposited out of a strong alkaline NaOH solution on a copper coated platinum net. The electrolyte should not contain any heavy metals as well as ammonium salts. Chlorides and Nitrates can be removed with H_2SO_4 by evaporation.
Ni	Ammoniacal solution	Ni can be deposited directly on a Pt-net from a strong ammoniacal solution. The electrolyte should not contain any heavy metals (under certain conditions separation of Ni with dimethyl glyoxime). Prolonged electrolysis is not done since the anode is attached.
Cd	NaOH + KCN	The solution if it contains Cd as sulphate, is first neutralized with NaOH, then an excess of KCN is added. Upto about 1 g weighable quantity is possible.

P_H measurement

P_H -measurement scheme

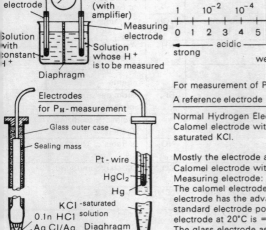

Reference electrode — Measuring instrument (with amplifier)

Measuring electrode

Solution with constant H⁺

Solution whose H⁺ is to be measured

Diaphragm

Electrodes for P_H -measurement

Glass outer case

Sealing mass

Pt - wire

$HgCl_2$

Hg

KCl -saturated solution

0.1n HCl

Ag Cl/Ag   Diaphragm

Glass membrane

Glass electrode with AgCl/Ag grounding and 0.1 NHCl

Calomel standard electrode with saturated KCl

P_H scale

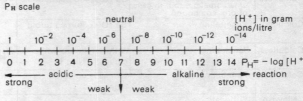

For measurement of P_H, the necessary electrodes are:

A reference electrode   and   a measuring electrode or (standard)

Normal Hydrogen Electrode Calomel electrode with saturated KCl.	Pt-Quinhydrone-Electrodes Antimony electrode Glass electrode-low or high resistance.

Mostly the electrode arrangements in usage are: Reference electrode: Calomel electrode with saturated KCl.
Measuring electrode: glass electrode, high resistance.
The calomel electrode in contrast to the standard hydrogen electrode has the advantage that it is easier to handle and its standard electrode potential e_o against the normal hydrogen electrode at 20°C is = + 0.249 volt.
The glass electrode as a measuring electrode has a wider range of application ($P_HO - 14$) than other electrodes and is largely insensitive towards oxidizing and reducing substances. It has the drawback that it (high resistance glass electrode) requires a measuring amplifier.

Qualitative inorganic analysis

Cations identification - Preliminary tests

Charcoal blowpipe test

The blowpipe test serves as a preliminary test for (heavy) metals. A small amount of substance is mixed with double the amount of soda (Na_2CO_3) and is placed in a cavity of a piece of charcoal and moistened. If the mixture is blown with a reducing flame, the metal is partly reduced. Thereby a metal grain (bead) results in case of easily melting metals and spangles in case of hard melting metals. If the metals are volatile then an incrustration of characteristic colour is formed around the melting zone. The metals can be identified in respect of the following properties.

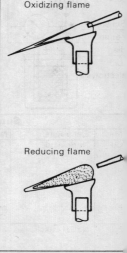

Oxidizing flame

Reducing flame

Melting zone	State and condition	Colour	Incrustration colour	Metal
Metals without incrustration	Ductile bead	White	—	Sn, Ag
	Metal spangles	yellow-red	—	Cu
	Metal spangles magnetic	grey	—	Fe, Ni, Co
Metals with incrustration	Ductile bead	—	yellow	Pb
	Brittle bead	—	yellow	Bi
	Brittle bead	—	white	Sb
Incrustration without metal	—	—	white (garlic odour)	As
	—	—	white, yellow when hot	Zn
	—	—	Brown	Cd

Ignition tube analysis

Small amount of substance taken in a glass test tube of 5–6 mm inner diameter and about 5 cm long, is heated. Observe whether a gas is evolved, a sublimate formed on the cooler parts of the test tube or whether the substance does not change at all.

By heating a gas is evolved with following properties:

Colour	Odour	Reactions	Gas	Evolved from
Colourless	Odourless (have an emyreumatic odour in case of org. compound	$Ba(OH)_2$ turned turbid	CO_2	Carbonates or org. compounds
Colourless	Penetrating	$Ba(OH)_2$ turned turbid	SO_2	Sulphites or sulphides in case of excess air blown
Colourless	Penetrating	With NH_3 forms white clouds of NH_4Cl	HCl	Chlorides
Colourless	Penetrating	Litmus is coloured blue	NH_3	Ammonium salts or org. N-containing compounds
Yellowish green	Suffocating		Cl_2	Chlorides + oxidizing agents
Brown	Suffocating		Br_2	Bromides + oxidation medium
Violet	Suffocating		I_2	Iodides + oxidizing agent
Brown	Suffocating		NO_2	Nitrates or Nitrites
Colourless	Odourless	burns with a pale blue flame	CO	Organic compounds
Colourless	bitter almonds	burns with blue flame	$(CN)_2$	Cynides ☠

By heating if a sublimate results:

White	:	Ammonium salts. $HgCl_2$. Hg_2Cl_2. As_2O_3, As_2O_5
Grey	:	Hg out of Hg compounds
Yellow	:	S, As_2S_3. As_2S_5 (reddish) HgI_2. (scarlet red)
Grey black	:	Ag, HgS, I_2, (also violet vapours)

Cation identification: Preliminary tests

Solubility of inorganic compounds in different solvents

The informations regarding the solubility are applied only with reservation since they are dependent upon different factors like concentration of concerned ions, the temperature, and other factors.

Solvent: H_2O

Owing to the reaction with solvent the precipitate are		In solution exclude the adverse	
	Anions	Cations	
BiOCl, SbOCl	OH^-	All cations except Na^+, K^+, Li^+, NH_4^+, Ba^{2+}, Sr^{2+}, Ca^{2+}	
Sn $(OH)_2$),	CO_3^{2-}	All cations except Na^+, K^+, (Li^+) NH_4^+	
Pb $(OH)_2$),	Cl^-, Br^-, I^-	Ag^+, Hg^+	
All white fluffy precipitates	SO_4^{2-}	Ba^{2+}, Sr^{2+}, Pb^{2+}, (Hg^+)	
	S^{2-}	All cations except Na^+, K^+, Li^+, NH_4^+, Ca^{2+}, Ba^{2+}, Sr^{2+}, Mg^{2+}	
	PO_4^{3-}, BO_3^{3-}	All cations except Na^+, K^+, NH_4^+	

Solvent: 2n HCl

Owing to the reaction with solvent, the precipitate are	Remain un-dissolved	In solution exclude the adverse	
		Anions	Cations
AgCl, Hg_2Cl_2, $(PbCl_2)$	Al_2O_3 Cr_2O_3	SO_4^{2-}	Ba^{2+}, Sr^{2+}, Pb^{2+}, (Hg^+)
		Br^-, I^-	Ag^+, Hg^+
	SnO_2 (Fe_2O_3)	S^{2-}	Hg^{2+}, Cu^{2+}, Bi^{3+}, Pb^{2+}, Cd^{2+}, $As^{3(5)+}$, $Sb^{3(5)+}$, $Sn^{2(4)+}$

Solvent: conc. $HCl + H_2O_2$

Owing to the reaction with solvent, precipitate are	
AgCl, $(PbCl_2)$	Remain undissolved
	(Al_2O_3), (Cr_2O_3), (Fe_2O_3), SnO_2, SiO_2 $BaSO_4$, $SrSO_4$, $(PbSO_4)$

Decomposing insoluble compounds

Insoluble compound	Decomposing medium	Method
$BaSO_4$ $SrSO_4$ $[CaSO_4]$	$Na_2CO_3 + K_2CO_3$	Decomposition takes place in a porcelain crucible on a blast flame, upto a clear molten liquid. The melt is extracted in soda solution and the residue filtered. Rinse with warm water till no more SO_4^{2-} can be detected. Residue: Alkaline earth carbonate
Fe_2O_3, Al_2O_3 Cr_2O_3, NiO CoO	$KHSO_4$ or $NaHSO_4$	Decomposition takes place in a Pt-crucible (Porcelain crucible) over a Bunsen flame. First heat only gently, later strongly heat till the evolution of SO_3. Do not ignite. After cooling the melt, extract with diluted H_2SO_4. The solution contains the corresponding metal ions.
Al_2O_3	$Na_2CO_3 +$ K_2CO_3	Decompose as in case of alkaline earth sulphates, extract the melt and thereafter acidify with HCl. Solution contains Al^{3+}
Cr_2O_3 $FeCr_2O_4$	$Na_2CO_3 + KNO_3$	Decompose in an ion or nickel crucible over a Bunsen flame. The melt is extracted with water and thereafter filtered Residue: Fe_2O_3 (By decomposition of chrom iron steel) Solution - CrO_4^{2-}
SiO_2 and silicates	$Na_2CO_3 + K_2CO_3$	Decompose as in case of alkaline earth sulphates in a Pt-crucible. The melt is dissolved and extracted in HCl from Pt-crucible. The solution is evaporated and boiled with HCl, thereafter filtered. Residue: SiO_2 (Indentification by water droplets test)
SnO_2	$Na_2CO_3 + S$	Soda and sulphur are mixed in the ratio 3:2. Decompose in a guarded Pt-crucible. First only to suiter, later heat strongly till the mass no more sparkles or foams. After cooling, extract with water, filter and the filtrate is weakly acidified with HCl. SnS_2 precipitates and in filtered.
$PbSO_4$	Hot saturated CH_3COONH_4 solution	$PbSO_4$ passes with hot saturated ammonium acetate solution into the filtered solution. The same solution is (poured) passed several times through the filter.
AgCl AgBr AgI	conc. NH_3 (warm)	Silver chloride (AgBr and AgI only partly) passes with conc. NH_3 into filter solution. From the filtrate, AgCl reprecipitated by acidification with HNO_3.

Qualitative inorganic analysis

Cation separation and identification by complete analysis

To analyse, one needs a solution of the substance in a suitable solvent. Such solvents are:

- Water
- Diluted HCl
- Concentrated HCl
- Concentrated HCl together with H_2O_2.

H_2SO_4 is not used as a solvent since it forms insoluble alkaline earth sulphates HNO_3 or aqua regia is as much avoided because HNO_3 is to be removed by evaporation and HCl group is to be separated first.

In case of full analysis, the examination of cations results not in an arbitrary succession but always in a sequential orders: HCl acid group, H_2S group, ammonium sulphide group, Alkaline earth group, Alkali group.

The insoluble residue is filtered and dissolved out in a known solvent.

Elimination of Interfering anions

Before analysis, F^-, CN^-, SCN^- and $(COO^-)_2$ can be eliminated through boiling with conc. H_2SO_4. If MnO_4^-, ClO_3^-, ClO_4^- exist before hand, by boiling with conc. H_2SO_4 explosion can occur.

Beware!

$(COO^-)_2$ can be decomposed by boiling repeatedly with HCl and H_2O_2.

BO_3^{3-} is to be removed from the filtrate of H_2S group by boiling twice with slightly conc. H_2SO_4 together with 1–2 ml. of methanol. Frequently it is appropriate to remove BO_3^- before the analysis process by evaporation with conc. H_2SO_4 and methanol. If small amount of BO_3 is present, it rarely interferes in cation separations..

Since AsO_4^{3-} give the same reaction, first from the filtrate of H_2S group PO_4^{3-} is examined. The examination follows with NH_4 molybdate in HNO_3 solution. If PO_4^{3-} is present it is precipitated from the filtrate and acidified with HCl of H_2S group with zirconium oxychloride.

Before the cation separation process is begun, the anions which interfere in the cation separations are to be examined. Such anions are:

F^-	interferes through complex salt formation or by formation of insoluble fluorides	BO_3^{3-}	forms insoluble alkaline earth borates which reach into ammonium sulphide group.
CN^-	interferes through complex salt formation	PO_4^{3-}	forms insoluble alkaline earth phosphates which reach into ammonium sulphide group.
SCN^-	interferes through complex salt formation	$(COO^-)_2$	interferes through complex salt formation or through formation of insoluble oxalates.

Cation separation and identification of HCl group

HNO_3 solution with the ions
Ag^+ Hg^{2+} Pb^{2+}
Add 1:1 HCl till complete precipitation, cool to a low temperature and filter[1]

Precipitate: AgCl white Hg_2Cl_2 white $PbCl_2$ white wash the filter with hot water several times

Residue: AgCl, Hg_2Cl_2 Add ammonia and filter		Filtrate: Pb^{2+} concentrate solution and cool again
Residue: Hg + Hg(NH_2)Cl Black Dissolve with conc. HNO_3 and dilute. Immerse Cu plate[2].	Filtrate: $[Ag(NH_3)_2]^+$ Colourless Acidify with HCl	

$PbCl_2$

The reprecipitated $PbCl_2$ filtered and dissolved with small amount of NH_4 acetate solution. Add $K_2Cr_2O_7$

| Hg | AgCl white | PbCrO_4 yellow |

Notes

1. It is often useful to dissolve the substance in HCl. In such cases the insoluble chlorides which result are filtered and can be worked further.
   The solution before filtering is cooled; thus Pb is precipitated as $PbCl_2$. Otherwise, a portion of this Pb always interferes in H_2S group. The filtrate is used for examination of cations if subsequent groups. If the substance is dissolved in HNO_3, the filtrate is evaporated and the residue in HCl is taken up.
2. It results in amalgam formation. The mercury adheres strongly to Cu-plate.

Qualitative inorganic analysis

Cation separation and identification —H₂S group

2n HCl solution with the ions:
Hg²⁺ Pb²⁺ Bi³⁺ Cu²⁺ Cd²⁺ As³⁺⁽⁵⁾⁺ Sb³⁺⁽⁵⁾⁺ Sn²⁺⁽⁴⁺⁾

2n HCl solution with the ions:
Hg^{2+} Pb^{2+} Bi^{3+} Cu^{2+} Cd^{2+} $As^{3+(5)+}$ $Sb^{3+(5)+}$ $Sn^{2+(4+)}$

Heat the solution to boiling. Pass H₂S into the cooled solution. Dilute and further pass H₂S till a complete precipitation occurs. Thereafter filter.[1]

Precipitate: HgS black PbS black Bi_2S_3 black brown CuS black CdS yellow $As_2S_{3(5)}$ yellow $Sb_2S_{3(5)}$ orange SnS_2 brown or yellow.
Add yellow $(NH_4)_2S$, warm upto 30–35°C and filter.[2]

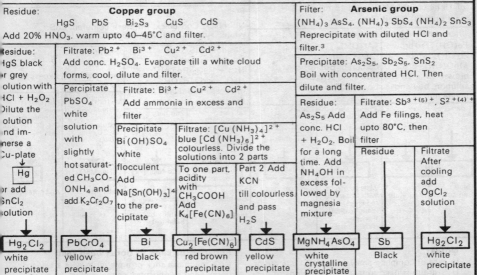

Residue: **Copper group**
        HgS   PbS Bi_2S_3 CuS   CdS
Add 20% HNO_3. warm upto 40–45°C and filter.

Filter: **Arsenic group**
$(NH_4)_3 AsS_4$. $(NH_4)_3 SbS_4$ $(NH_4)_2 SnS_3$
Reprecipitate with diluted HCl and filter.[3]

Precipitate: As_2S_5, Sb_2S_5, SnS_2
Boil with concentrated HCl. Then dilute and filter.

Residue: HgS black or grey olution with HCl + H₂O₂ Dilute the olution nd im- merse a Cu-plate ↓ [Hg] or add SnCl₂ solution	Filtrate: Pb²⁺ Bi³⁺ Cu²⁺ Cd²⁺ Add conc. H₂SO₄. Evaporate till a white cloud forms, cool, dilute and filter.				Residue: As₂S₅ Add conc. HCl + H₂O₂. Boil for a long time. Add NH₄OH in excess fol- lowed by magnesia mixture	Filtrate: Sb³⁺⁽⁵⁾⁺, S²⁺⁽⁴⁾⁺ Add Fe filings, heat upto 80°C, then filter
	Percipitate PbSO₄ white solution with slightly hot saturat- ed CH₃CO- ONH₄ and add K₂Cr₂O₇	Filtrate: Bi³⁺ Cu²⁺ Cd²⁺ Add ammonia in excess and filter				Residue	Filtrate After cooling add OgCl₂ solution
		Precipitate Bi(OH)SO₄ white flocculent Add Na[Sn(OH)₃]⁴ to the pre- cipitate	Filtrate: [Cu(NH₃)₄]²⁺ blue [Cd(NH₃)₆]²⁺ colourless. Divide the solutions into 2 parts				
			To one part, acidity with CH₃COOH Add K₄[Fe(CN)₆]	Part 2 Add KCN till colourless and pass H₂S		

Hg_2Cl_2	$PbCrO_4$	Bi	$Cu_2[Fe(CN)_6]$	CdS	$MgNH_4AsO_4$	Sb	Hg_2Cl_2
white precipitate	yellow precipitate	black	red brown precipitate	yellow precipitate	white crystalline precipitate	Black	white precipitate

Fluoroscence test

Test for Sn
In a small porcelain china dish of zinc granules, add spatula full of substance and 5 ml of 1:1 HCl.
Dip a test tube filled with cold water into this solution and hold a non-luminous flame. A bluish bright fringe appears around the test tube in case Sn is present.

Notes:

1. In case of complete analysis, the filtrate is necessary for the following groups after testing of ammonium sulphide group. Before further work, the H₂S is to be boiled off and the solution is to be concentrated to half the volume.

2. **Only the cations of Cu and As groups** to be tested; hence the separation with yellow ammonium sulphide left undone. In these cases the sulphides are used further.

3. The filterate is rejected.

4. The sodium stannite solution is prepared by treating SnCl₂ solution containing a drop of HCl with enough 2n NaOH so that the already precipitated Sn (OH)₂ is just redissolved on further addition.

Marsh's test
for testing As and Sb

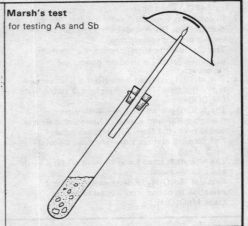

Qualitative inorganic analysis

Cation separation and identification—ammonium sulphide group

Oxidize with H_2O_2, the phosphate free[1] 1nHCl solution with the following ions:

$$Fe^{3+} \quad Al^{3+} \quad Cr^{3+} \quad Zr^{4+} \quad Ni^{2+} \quad Co^{2+} \quad Mn^{2+} \quad Zn^{2+}$$

Heat the solution to boiling, add NH_4OH in excess, further heat for a short time and filter.

Precipitate: ammonia group	Filtrate: Ammonium sulphide group

Precipitate: ammonia group

$Fe(OH)_3$ brown, $Al(OH)_3$ white, $Cr(OH)_3$ green, $Zr(OH)_4$ white. In 2n HCl solution, neutralize with solid soda. Pour a solution of a mixture of 15 ml of 30% NaOH and 15 ml. of 3% H_2O_2, heat upto 60°C and filter.

Filtrate: Ammonium sulphide group

$$Ni^{2+} \quad Co^{2+} \quad Mn^{2+} \quad Zn^{2+}$$

Add $(NH_4)_2S$ in excess, heat to boiling and filter[6]

Precipitate:

NiS black, CoS black. Mn flesh coloured, ZnS white: Add 2n HCl, warm to 50°C, and filter.

Precipitate:
$Fe(OH)_3$ red brown; $Zr(OH)_4$ white; $MnO(OH)_2$ black brown[2]. In a 2n HCl solution, boil and divide into parts

Filtrate
$[Al(OH)_4]^-$ colourless CrO_4^{2-} solution yellow coloured. Boil off H_2O_2. Add solid NH_4Cl in excess, boil and filter

Residue: NiS, CoS solution in conc. HCl + H_2O_2 Boil off H_2O_2; add NH_4OH in excess and divide the solution into parts

Filtrate Mn^{2+} Zn^{2+} Boil off H_2S. Add NaOH in excess and filter.

1 part Add NH_4SCN	2 part Add Na_2HPO_4 and boil	**Precipitate:** $Al(OH)_3$ white	**Filtrate:** yellow CrO_4^{2-} acidify with CH_3COOH, Add $BaCl_2$ and boil	1st part Add dimethyl glyoxime	2nd part Acidify with CH_3COOH and add α-nitroso-β-naphthol	**Precipitate:** $Mn(OH)_2$ white brown.[7] Melt in a mixture of Na_2CO_3 +KNO_3	**Filtrate:** $[Zn(OH)_3]$ colourless. Acidify with CH_3COOH and pass H_2S

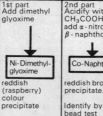

$Fe(SCN)_3$	$Zr_3(PO_4)_4$	$Al(OH)_3$	$BaCrO_4$	Ni-Dimethyl-glyoxime	Co-Naphthol	Na_2MnO_4	ZnS
blue red colour	white precipitate	white precipitate with Alizarin S red lake[4] or identification through cryolite test[5]	yellow precipitate	reddish (raspberry) colour precipitate	reddish brown precipitate. Identify by bead test	green melt	white precipitate

---

Notes:

1. In case of complete analysis, the HCl filterate of H_2S-group is used to test for the cations of ammonium sulphide group. The solution is to be tested for PO_4^{3-} before addition of ammonia, because by addition of NH_4OH the phosphates of alkaline earth group can be precipitated. The test for phosphate follows in HNO_3 and solution with ammonium molybdate. If PO_4^{3-} is present then the whole solution is treated with excess zirconium oxychloride and boiled for a long time. The phosphate is precipitated as $Zr(PO_4)_4$ and filtered.
2. In case of precipitation with NH_4OH, Mn can be precipitated. The Mn identification can be carried out with precipitate.
3. The Zr identification cannot be carried out when $ZrOCl_2$ is used for phosphate separation.
4. The $Al(OH)_3$ covered with alizarin S is washed out of the filter. The resulting lake is stable in dil. CH_3COOH.
5. The $Al(OH)_3$ is washed well with water, thereafter NaF and a drop of phenolphthalein is added on to the filter. $Al(OH)_3$ reacts with NaF forming cryolite simultaneously decolorizing of red colour.
6. The filterate is used for examining alkaline earth and alkaline metals.
7. First the $Mn(OH)_2$ undergoes atmospheric oxidation to become converted to brown black $MnO(OH)_2$.

Phosphorus salts or borax bead test

On a magnesia stick or a platinum wire a clear borax bead or phosphor salt is melted. After cooling, the pearl is moistened and small amount of substance is added. The bead shines with colour after remelting and making a total solution of substance in bead.

Colour of beads:

Oxidation beads		Reduction beads		Elements
Hot	cold	hot	cold	
blue	blue	blue	blue	Co
red-yel	green	green	green	Cr
violet	red-violet	colourless	colourless	Mn
yel-red	yel-colourless	greenish	greenish	Fe
green	blue-green-pale blue	colour-less	red brown in the presence of Sn	Cu
violet	red brown	colour-less or grey	colourless or grey	Ni

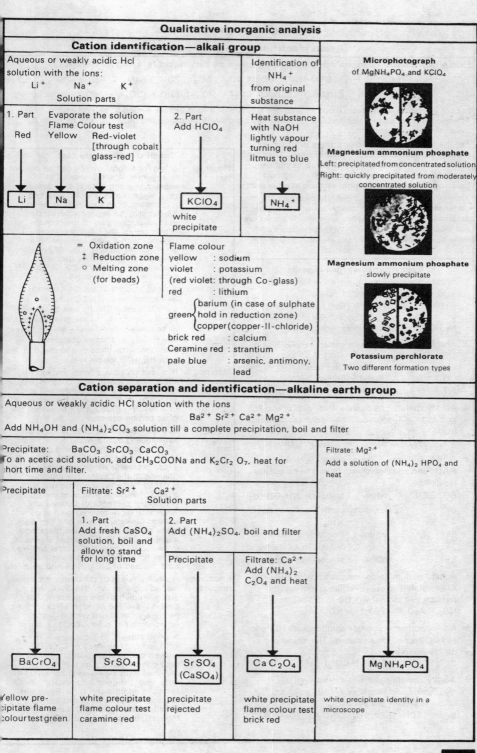

Qualitative inorganic analysis

Cation identification—alkali group

Aqueous or weakly acidic HCl solution with the ions:

Li$^+$ Na$^+$ K$^+$

Solution parts

Identification of NH$_4^+$ from original substance

Microphotograph
of MgNH$_4$PO$_4$ and KClO$_4$

1. Part Evaporate the solution
Flame Colour test

Red Yellow Red-violet [through cobalt glass-red]

↓ Li   ↓ Na   ↓ K

2. Part
Add HClO$_4$

↓ KClO$_4$

white precipitate

Heat substance with NaOH lightly vapour turning red litmus to blue

↓ NH$_4^+$

Magnesium ammonium phosphate
Left: precipitated from concentrated solution
Right: quickly precipitated from moderately concentrated solution

= Oxidation zone
‡ Reduction zone
○ Melting zone (for beads)

Flame colour
yellow : sodium
violet : potassium
(red violet : through Co-glass)
red : lithium
green { barium (in case of sulphate hold in reduction zone)
copper (copper-II-chloride)
brick red : calcium
Ceramine red : strontium
pale blue : arsenic, antimony, lead

Magnesium ammonium phosphate
slowly precipitate

Potassium perchlorate
Two different formation types

Cation separation and identification—alkaline earth group

Aqueous or weakly acidic HCl solution with the ions

Ba^{2+} Sr^{2+} Ca^{2+} Mg^{2+}

Add NH$_4$OH and (NH$_4$)$_2$CO$_3$ solution till a complete precipitation, boil and filter

Precipitate: BaCO$_3$ SrCO$_3$ CaCO$_3$
To an acetic acid solution, add CH$_3$COONa and K$_2$Cr$_2$O$_7$, heat for short time and filter.

Filtrate: Mg^{2+}
Add a solution of (NH$_4$)$_2$HPO$_4$ and heat

Precipitate

Filtrate: Sr^{2+} Ca^{2+}
Solution parts

1. Part
Add fresh CaSO$_4$ solution, boil and allow to stand for long time

2. Part
Add (NH$_4$)$_2$SO$_4$, boil and filter

Precipitate

Filtrate: Ca^{2+}
Add (NH$_4$)$_2$ C$_2$O$_4$ and heat

↓ BaCrO$_4$

↓ SrSO$_4$

↓ SrSO$_4$ (CaSO$_4$)

↓ CaC$_2$O$_4$

↓ MgNH$_4$PO$_4$

Yellow precipitate flame colour test green

white precipitate flame colour test caramine red

precipitate rejected

white precipitate flame colour test brick red

white precipitate identity in a microscope

Qualitative inorganic analysis

Anion identification—preliminary tests

From a large number of anions only the following anions are summarized here for their identification:

$$F^-, Cl^-, Br^-, I^-, S^{2-}, SO_3^{2-}, SO^{2-}, NO_2^-, NO_3^-, PO_4^{3-}, AsO_3^{3-}, AsO_4^{3-}$$

$$BO_3^{3-}, SON^-, CH_3COO^-, C_2O_4^{2-}, CO_3^{2-}, SiO_3^{2-}, (SiO_2), CrO_4^{2-}, MnO_4^-$$

In case of water soluble substances an aqueous solution or a soda extract is used to test for anions. Attention is paid, particularly, in case of preliminary tests since the cation reactions can also be given by closely existing anions.

For the identification of anions, there is no separation process as like that of cation identification. They are therefore to be found out by individual identification test, where care is to be taken expecially against breaking down of identification reactions by other anions.

Through preliminary tests a definite classification is undertaken; under certain conditions it is important to determine through preliminary tests which anions are not present. Preliminary tests cannot replace an individual identification. Suitability-wise one is not content with a single identification.

Behaviour towards reagents

1. Acidify the sodium carbonate extract with HCl acid. It forms precipitate in case of
   Al $(OH)_3$, Zn $(OH)_2$, Pb $(OH)_2$, Sn $(OH)_2$
   soluble in excess of HCl
   H_2SiO_3 insoluble in excess of HCl

2. Acidify sodium carbonate extract with HCl acid and add $BaCl_2$.
   The Barium salts results from
   SO_4^{2-} insoluble in conc. HCl
   SiF_6^{2-} F^- soluble in conc. HCl

3. Acidify sodium carbonate extract with acetic acid and add $CaCl_2$ solution. The calcium salts result from:
   SO_3^{2-}, PO_4^{3-}, BO_3^{3-}, F^-, $C_2O_4^{2-}$, (SO_4^{2-})

4. Acidify sodium carbonate extract with dil. HNO_3 and add $AgNO_3$ solution. The silver salts result from:
Cl^-, SCN^-	white	soluble in NH_4OH (dil.)
Br^-	pale yellow	partly soluble in conc. NH_3
I^-	yellow	insoluble in ammonia soluble in KCN solution
S^{2-}	black	soluble in conc. HNO_3

5. Acidify sodium carbonate extract with HCl. Add KI and starch solution. A blue colour results through free iodine with the ions:
   NO_2^-, AsO_4^{3-}, CrO_4^{2-}, (NO_3^-)

6. Acidify sodium carbonate solution with HCl and add dropwise iodine solution. The iodine is decolourized by:
   S^{2-}, SO_3^{2-}, AsO_3^{2-}, (SCN^-)

7. Acidify sodium carbonate extract with dil. H_2SO_4 and add dropwise $KMnO_4$ solution. The colour is discharged by:
   Br^-, I^-, SCN^-, S^{2-}, SO_3^{2-}, NO_2^-, $C_2O_4^{2-}$, AsO_3^{3-}

Specific preliminary test

Hepar test: A test for sulphur which can be present in any form. A small amount of substance along with soda is placed on a magnesia stick or Pt-wire and it is molten in an oxidizing flame to a bead and thereafter further heated in the reducing part of the flame. After cooling, the bead is pressed together with a drop of water on a silver (plate) coin. In case of presence of sulphur it develops a dark spot through formation of Ag_2S. Blind experiment must be done.

Creep test: A small amount of substance with 3–4 ml. of conc. H_2SO_4 is heated in a dry test tube. The formed gaseous HF creeps slowly on the glass walls in an upward direction by shaking the oily H_2SO_4 runs down the glass walls since the glass due to reaction with HF is no longer wet.

Etching test: In a lead crucible a small amount of substance is mixed with conc. H_2SO_4, the crucible is placed on a stand and slowly heated. In case of the presence of F it forms gaseous HF by which the glass can be itched.

Water droplet test: In a lead crucible a small amount of substance is mixed with (boiled) hot silicic acid and small amount of conc. H_2SO_4 and the crucible is closed with a lead protector which possesses an opening at the centre. The crucible is gently heated and a glass rod with a drop of water is held over the opening. Due to the presence of fluorine the water droplet is covered with silica jel. Instead of water droplet one can also place a piece of moistened black filter paper over the lid. The paper always is to be moist. The water droplet test also serves for the identification of silicic acid. Instead of SiO_2, NaF is to be mixed.

Rhodanide reaction: A small amount of substance is carefully melted together with $Na_2S_2O_3$ in a procelain dish by a small flame. The melt is dissolved in water (filtered it necessary), acidified with HCl and tested with $FeCl_3$ for SCN^-.

Chromyl chloride reaction: A small amount of substance is mixed with an equal quantity of $K_2Cr_2O_7$ in a given dry test tube and is covered with conc. H_2SO_4. The test tube is provided with a delivery tube whose other end is dipped into NaOH contained in another test tube. By carefully heating, the CrO_2Cl_2 is distilled: The already placed NaOH becomes yellow through the formation of CrO_4^{2-}.

Qualitative inorganic analysis

Anion identification—preliminary test

Hints about the anions	1. Substance + conc. H_2SO_4 heat → gas evolution colour/odour	2. Substance + $KHSO_4$ grind Note odour	3. Sodium carbonate extract + HCl + KI + starch ↑ Iodine—Blue colour	4. Sodium carbonate extract + HCl + iodine solution ↑ Decolourization	5. Sodium carbonate extract + H_2SO_4 + $KMnO_4$ solution ↑ Decolourization	6. Sodium carbonate extract + HNO_3 + $AgNO_3$ solution ↑ precipitate colour	6. Solubility in	7. Sodium concentrate extract + CH_3COOH + $CaCl_2$ solution ↑ white precipitate	8. Sodium carbonate extract + HCl + $BaCl_2$ solution ↑ white precipitate
CO_3^{2-}	CO_2								
BO_3^{3-}								$CO_3(BO_3)_2$	
CH_3COO^-	CH_3COOH^- pung smell	CH_3COOH							
PO_4^{3-}								$Ca_3(PO_4)_2$	
SiO_2									
F^-	HF (creep test)							CaF_2	BaF_2 soluble in conc. HCl.
Cl^-	HCl Pung odour					AgCl white	dil. NH_3 $(NH_4)_2CO_3$		
Br^-	Br_2 brown pung				Br^-	Ag Br pale yel	conc. NH_3 $Na_2S_2O_3$		
I^-	I_2 violet pung				I^-	AgI yellow	$(Na_2S_2O_3)$ KCN		
SCN	(HSCN) pung	(HSCN)			SCN^-	AgSCN white	dil. NH_3		
$C_2O_4^{2-}$	CO_2 CO				$C_2O_4^{2-}$			CaC_2O_4	
NO_2^-	NO_2 brown pung	NO_2	NO_2^-		NO_2^-				
NO_3^-	NO_2 brown pung	HNO_3	NO_3^- when str. acidic						
S^{2-}	H_2S rotten eggs	(H_2S)		S^{2-}	S^{2-}	Ag_2S (when 1:1 HNO_3 weakly acidic)			
SO_3^{2-}	SO_2 pungent	SO_2		SO_3^{2-}	SO_3^{2-}			$CaSO_3$	
SO_4^{2-}								$CaSO_4$ (when excess)	$BaSO_4$
AsO_3^{3-}				AsO_3^{3-}	AsO_3^{3-}			$Ca_3(AsO_3)_2$	
AsO_4^{3-}			AsO_4^{3-} (weak)					$Ca_3(AsO_4)_2$	
CrO_4^{2-}	CrO_2Cl_2 when Cl is present		CrO_4^{2-}						
MnO_4^-	Mn_2O_7 violet (explosive)		MnO_4^-						

Qualitative inorganic analysis

Anion identification—individual test (proof)

F^-	Cl^-	Br^- I^-	SCN^-	CH_3COO^-	$C_2O_4^{2-}$
Preliminary test: creep test. Note: In the presence of SiO_2, or BO_3^{3-} creep test can fail to serve because of the formation of SiF_4 or BF_3 Preliminary test Nos. 7 and 8 are positive. Identification through water droplet test by addition of SiO_2; see SiO_2 identification. F^- interferes in cation separation process. Eliminate by evaporation with conc. H_2SO_4 before cation identification. Attention: Insoluble alkaline earth sulphates can result.	Preliminary test No. 6 positive note colour and solubility of AgCl. Test, first of all for Br^-, I^- SCN^-. When no Br^-, I^- SCN^- are present, then sodium carbonate extract + HNO_3 + $AgNO_3$ $\downarrow$ AgCl white precipitate When Br^- and I^- are present sodium carbonate extract + CH_3COOH + $KMnO_4$. Boil Br_2 and I_2 are removed filter. Filtrate + HNO_3 + $AgNO_3$ $\downarrow$ AgCl. When SCN^- is present sodium carbonate extract + H_2SO_4 (weakly acidic) + H_2SO_3 (excess) + $CuSO_4$ $\downarrow$ CuSCN filter Filtrate H_2SO_3 boiled off. Add HNO_3 + $AgNO_3$ $\downarrow$ AgCl Cl^- does not interfere in cation separation process.	Preliminary tests 1, 5 & 6 positive In case of test No. 6, note colour and solubility Total identification with chlorine water sodium carbonate extract + HCl + chlorine water dropwise $\downarrow$ colour of chlorine layer Iodine violet ICl_3 (IO_3^-) colourless Bromine brown (brown colour can be overlapped) BrCl wine yellow When no white colour action occurs, I^- is not present. If violet colour disappears in case of further addition of chlorine water and the chloroform layer remains either colourless or pale green, then no Br^- is present. More chlorine water is to be added if the substance has a reducing effect. Bi^- and I^- do not interfere in cation separation process.	Preliminary test 5 & 6 positive Note: colour and solubility of AgSCN Test first of all for PO_4^{3-} $F^-dC_2O_4^{2-}$ When no F^-, PO_4^{3-} and $C_2O_4^{2-}$ are present sodium carbonate extract + HCl + $FeCl_3$ $\downarrow$ $Fe(SCN)_3$ red colour Note: I^- + $FeCl_3 \rightarrow I_2$ yellow brown colouration When F^-, PO_4^{3-} or $C_2O_4^{2-}$ is present sodium carbonate extract + CH_3COO + $CaCl_2$ filter Filtrate + HCl + $FeCl_3$ $\downarrow$ $Fe(SCN)_3$ red colour or Sodium carbonate extract + HCl + CO $(NO_3)_2$ + ether —amyl alcohol shake $\downarrow$ $CO(SCN)_2$ blue colour SCN^- interferes in cation separation process. Before cation identification, it is removed by boiling with conc. H_2SO_4 Attention: Insoluble alkaline earth sulphates can result.	Preliminary test No. 2 positive. However note also SCN^-, NO_2^-, NO_3^-, S^{2-}, SO_3^{2-} Substance Na_2CO_3 + AS_2O_3 heat $\rightarrow$ $\downarrow$ CH_3 CH_3 $\diagdown$ As-O-As $\diagup$ CH_3 CH_3 Repulsive odour of cacodyl oxide CH_3COO^- does not interfere in cation separation process.	Preliminary tests 5 and 7 positive. Test first of all for SO_3^{2-}, AsO_3^{3-} When no SO_3^{2-} AsO_3^{3-} are present sodium carbonate extract + CH_3COOH + $CaCl_2$ filter precipitate and dissolve in dil. H_2SO_4 + $KMnO_4$ $\downarrow$ decolourization shows presence of $C_2O_4^{2-}$ When As is present, sodium carbonate extract + HCl + pass + H_2S Filter Boil off H_2S. Neutralize with CH_3COOH + $CaCl$ precipitate. Filter and dissolve in dil. H_2SO_4 + $KMnO_4$ $\downarrow$ Decolourization When SO_3^{2-} present sodium carbonate extract + CH_3COOH + $CaCl_2$ precipitate. Filter and dissolve in dil H_2SO_4 Boil off SO_2 Add $KMnO_4$ $\downarrow$ decolourization $C_2O_4^{2-}$ interferes in cation separation process. Boil in acidic (HCl). analysis solution with H_2O_2 for a long time. Proceed thereafter with cation separation method.

CO_3^{2-}	BO_3^{3-}	PO_4^{3-}	SiO_2	S^{2-}	SO_3^{2-}	SO_4^{2-}
Preliminary test No. 1 positive	Preliminary test No. 7 positive	Preliminary test No. 7 positive	Preliminary test SiO_2 insoluble in acids	Preliminary tests No. 4, 5. and 6 positive. Note:	Preliminary tests 2, 4, 5 and 7 positive	Preliminary test No. 8 positive
Gas evolution with dil. HCl. Gas evolution with dil. CH_3COOH	Subs. + conc. H_2SO_4 + CH_3OH (careful), heat. Vapours ignited or brought into a non-luminous Bunsen flame	First test for SiO_2 and AsO_4^{3-} when present, make SiO_2 insoluble by boiling several times with HCl and filter. In strong HCl solution pass H_2S heat the solution As^- as is ppt sulphide filter	Identify by water droplet test subs. + CaF_2 + H_2SO_4 (conc.). Heat in a lead crucible	These preliminary tests can fail to work in case of insoluble sulphides	SO_3^{2-} identification in presence of CO_3^{2-} combined identification Subs. + dil. HCl. Heat	Sodium carbonate extract + dil HCl + $BaCl_2$
Presence of SO_3^{2-} must be tested first. If SO_3^{2-} is present substance + H_2O_2 heat. Then add dil. HCl. Gas evolution pass into $Ba(OH)_2$	$B(OCH_3)_3$ Green colour ation of flame	Oxidation of solution removes any reducing medium. From the filtrate of H_2S group test for PO_4^{3-} Boil off H_2S + HNO_3 + $(NH_4)_2 MoO_4$ warm	SiF_4 Turbidity of the water droplet held over the crucible	Subs. + HCl H_2S Lead acetate paper turned brown	$SO_2(CO_2)$ Pass into $Ba(OH_2)$ + HNO_3 + $KMnO_4$	$BaSO_4$ white fine powdered ppt. $BaSO_4$ is insoluble in in conc. HCl.
	Borate interferes in cation separation process since the alkaline earth borates interferes in ammonium sulphide group, before beginning cation separation process, distil. BO_3^- as		When F^- is present water droplet test is answered without addition of CaF_2	or in case of insoluble sulphides (only when SO_3^{2-} is not present) substance + Zn + HCl	Decolourization	SO_4^{2-} does not interfere in cation separation process. Alkaline earth sulphates result due to the presence of SO_4^{2-}
Ba CO_3 does not interfere in cation separation process		yellow ppt	When BO_3^{3-} is present, water droplet test fails to answer	H_2S	SO_3^{2-} does not interfere in cation identification. In principle along with SO_3^{2-}, SO_4^{2-} is present always.	
	$B(OCH_3)_3$ Attention! Insoluble alkaline earth sulphates can result	PO_4^{3-} interferes in cation saparation methods as alkaline earth phosphates in $(NH_4)_2 S$ group. After boiling off H_2S group, filter the H_2S group. Add $ZrOCl_2$ white ppt Filter Filtrate is used for further work	In case of the presence of $Br^- I^-$, Wo_2^-, NO_3^- the water droplet can be coloured. Be careful while rendering it turbid in these cases. $SiO_2 (SiO_3^2)$ interferes in cation separation process. Make it insoluble by boiling with acid and filter	or Neutralized sodium carbonate extract + $[(CdNH_3)_6]^{2+}$ ammoniacal col. salt solution CdS yellow		

Qualitative inorganic analysis					
Anion identification—individual tests					
NO_2^-	NO_3^-	AsO_3^{3-}	AsO_4^{3-}	CrO_4^{2-}	MnO_4^-
Preliminary test No. 1, 3 and 5 positive	Preliminary test No. 1 positive	Preliminary test No. 4 and 5 positive	Preliminary test No. 3 positive	Preliminary test No. 3 positive	Preliminary test No. 1 (careful
Test with Fe SO_4 as like NO_3^- (Br^- and I^- to be separated before hand). Instead of brown ring whole solution is brown.	Br^- and I^- must be proved first	Further tests. In cation separation process As must be identified.	Further tests. In cation separation process As must be identified.	Preliminary test No. 1 positive when simultaneously Cl^- present	explosive) and No. 3 positive
	When Br^- and I^- are present sodium carbonate extract $+ H_2SO_4$ $+ Ag_2SO_4$ ppt. Filter Filtrate $+$ $FeSO_4$ with conc. H_2SO_4 through sides	Preliminary tests 4 and 5 combined with As identification applies AsO_3^{3-}	Preliminary test No. 3 combined with As identification applies for AsO_4^{3-}	When CrO_4^{2-} is only present, sodium carbonate extract is coloured yellow. When CrO_4^{2-} is present, then during cation separation process Cr must be identified.	MnO_4^- is known from the violet colour of an aqueous solution or of sodium carbonate extract (under certain conditions formation of MnO_2 by boiling with sodium carbonate).
or Reduce with Devarda alloy. Note: NO_3^- also gives this test equally well		As^{3+} is separated as As_2S_3 in H_2S group. Hence no interference in case of further separation process.	or In the absence of SiO_2 and PO_4^{3-} sodium carbonate extract $+ HNO_3$ $+ (NH_4)_2 MoO_4$ Heat		
or Sodium carbonate extract $+$ HCl cool. Add sulphanilic acid $+$ α-naphthylamine	$Fe(NO)SO_4$ Brown ring. In case of presence of NO_2^- addition of $FeSO_4$, solution becomes brown		$\downarrow$ yellow ppt.	Preliminary test 3, yellow colouration of sodium carbonate extract and Cr identification applies for CrO_4^{2-}	Preliminary test No. 3 violet colour and Mn identification applies for MnO_4^-
$\downarrow$ red colour through the formation of an azodye, NO_2^- does not interfere in cation separation process. Along with NO_2^-, in principle NO_3^- will also exist.	when no Br^- and I^- are present, the addition of Ag_2SO_4 is not done.		As^{5+} lika As^{3+} is separated as sulphide. Hence no interference in case of further separation process.	CrO_4^{2-} does not interfere in cation separation methods	MnO_4^- does not interfere in cation separation process while by dissolving in HCl reduction occurs
	or (when no NO_2^- is present) sodium carbonate extract $+$ NaOH $+$ Devarda alloy Heat $\downarrow$ NH_3 odour litmus paper turned blue				
	NO_3^- does not interfere in cation separation process				

Note:

If a white flocculent precipitation is formed by acidifying sodium carbonate extract, it follows that the following compounds may be present:

$Al(OH)_3$, $Zn(OH)_2$, $Pb(OH)_2$, $Sn(OH)_2$ - soluble in dil. HCl.

H_2SiO_3 - insoluble in excess of HCl.

If the substance contains metals of As group besides S^{2-},

these can be precipitated by acidifying the corresponding sulphides.

Filter and for further work use sodium carbonate extract.

Qualitative organic analysis

Identification of elements

Element to find	Identification reaction	
C H	Organic substances, with a few exceptions, are detected by the fact they burn or char when heated on a platinum plate or metal spatula with a flame. The test for C and H follows simultaneously that a small amount of substance is mixed and heated with multiple quantities of freshly ignited CuO in a small test tube. Carbon is detected by the reaction of CO_2 thus formed producing turbidity when passed through $Ba(OH)_2$. The hydrogen present is converted into water that condenses at the colder parts of the test tube.	CO_2 H_2O $Ba(OH)_2$ Subs. + CuO Qualitative identification of C and H
N	The substance is dissolved by heating with a small piece of metallic potassium or sodium in a fusion tube and the hot tube is immersed in a dish containing cold water (careful — use protective goggles). The solution is filtered.	
N	The test tube breaks and the excess metal reacts with water. The solution is filtered, a drop of $FeSO_4$ and $FeCl_3$ is added and heated for a short time. Thereafter it is acidified with dil. HCl. A blue colour or blue precipitate shows the presence of N. Prussian blue.	
Halogen (Cl, Br, I)	Beilstein's test: A piece of copper wire is ignited in a non-luminous flame of bunsen burner so long till the flame is no longer coloured. Then a small amount of substance is added on the wire and again held at the flame. After burning off the organic substance, the presence of halogen is shown by green or bluish green flame colour.	
Halogen (Cl, Br, I)	In a test tube a small amount of substance is ignited with excess of pure CaO. After cooling, it is acidified with dil. HNO_3, filtered and as usual tested for halogens with $AgNO_3$. AgCl—white; AgBr—yellowish; AgI—yellow.	
S	Heper test: A drop of preserved (Sodium fusion extract) decomposed solution of the nitrogen identification is placed on a silver plate. A black spot shows S.	
S	A small portion of preserved solution of sodium fusion extract used for nitrogen identification, is acidified with dil. HCl, and a lead acetate paper is held over. The formation of black colour shows the presence of S.	

Melting and boiling points

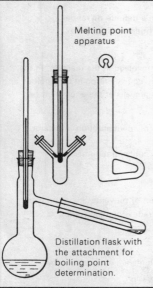

Melting point apparatus

Distillation flask with the attachment for boiling point determination.

Melting points determination

The melting point of a crystalline organic substance serves to check either the purity of an unknown substance or for its identification.

The melting point is determined in a melting point apparatus which is filled with pure concentrated sulphuric acid. The melting tube is so placed such that the lower end containing the substance touches the mercury bulb of the thermometer.

The melting point of a substance is the temperature at which the sample becomes a clear melt.

In case of impure substances the melting point is somewhat lower than of pure substances. To test for purity, frequently the mixed melting point is found out in which the pure substance is mixed thoroughly with the concerned substance and the melting point of the mixture determined. If both the melting points are equal, then the substance is pure or identical with the mixed substance (Exception - isomorphous substances.)

Boiling point determination

Like the melting point determinations, the determination of boiling point in case of liquid substances, serves to check the purity or identification of a given liquid.

For boiling point determination one needs a small distillation flask with thermometer.

The boiling point of a liquid is the temperature at which the vapour pressure of the liquid equals the atmospheric pressure therefore the boiling point is strongly dependent of the atmospheric pressure which is to be noted before carrying out the determination.

Quantitative inorganic analysis

Volumetric analysis

Volumetric methods are carried out for the reason being that they allow a faster work with greater accuracy. These are used only for such reactions in which one can distinguish definite final state through their faster reaction rate.

Neutralization analysis = Reaction between acids and bases

Redox analyses = Reaction between oxidation and reduction media

Precipitation analyses = Reactions by which solid compounds are precipitated.

All volumetric analytical methods are worked with normal solutions (n-solution) which contain the dissolved substance of one gram equivalent in one litre solvent.

The normality of a solution indicates as to how many gram equivalents are dissolved in one litre: 0.1 n, 0.2 n, 0.5 n, n, 2n, ...

In case of normal solutions accurate concentration data are necessary. With the help of the so-called standard titrimetric substances accurate volumetric solutions are prepared through weighing.

If no standard titrimetric substances are available, one must prepare the volumetric solutions either by weighing the corresponding substance or by diluting a concentrated solution. The concentrations of these volumetric solutions are to be checked, that is, these solutions are to be standardized with the help of standard primary substances or primary solutions and provide with the factor so ascertained.

With this factor, the given normality or the consumption of volumetric solution is corrected.

For detecting the end points in case of titrations an auxiliary agent is necessary in most of the cases. Besides electrical aids, for all fast volumetric analytical reactions, indicators are in use.

In principle, indicators are organic dyes which change their colour either in a definite p_H range (indicators for the neutralization reactions) or by a definite redoxpotential (indicators for redox analyses).

The choice of an indicator is made according to the substances which are involved in the reaction. In the neutralization analysis, the indicator shows the p_H range that prevails at the equivalent point by a colour change. In the redox analysis, the indicator shows the redox potential that prevails at the equivalent point.

The selection of suitable redox indicator is not so simple as that of neutralization indicator. This is because the redoxpotential changes with the p_H value and also other factors play an important role (colour changes of an indicator at different p_H, colour of salt solution, concentration of salt, etc).

In the following tables, the registered indicators allow also the measurement of p_H values by colorimetric means (indicators for the neutralization analysis) and the measurement of redoxpotential (redox indicators).

The notes given for the analysis, only gist is written. In a few cases these hints are enough, in other cases the search for the corresponding analysis in the original literature is made easier.

Volumetric analysis

1 val. (gram equivalent) of a substance is the quantity which corresponds to 1 gram hydrogen

$$1 \text{ val.} = \frac{1 \text{ Mole}}{\text{Valency}}$$

Valency = No. of replaceable H-atoms in acids

No. of replaceable OH-groups in bases

$$1 \text{ val.} = \frac{1 \text{ Mole}}{\text{No. of electrons transferred}}$$

Applies for oxidation and reduction media

For the preparation of normal solution the necessary weight (W) of a substance is given by:

$$W = \frac{1 \text{ mole}}{v} \times l \times n$$

m = molecular weight
v = valency or No. of electrons
l = litre of solution
n = normality of solution.

$$W = E \times l \times n$$

when $\frac{1 \text{ mole}}{v} = E$ (Equivalent weight in gram)

If a normal solution is to be prepared by weighing of a given % solution, then:

$$W = \frac{100}{p} \times E \times l \times n$$

where p = percentage of solution

Standard solutions are seldom prepared by weighing the initial concentrated solution. It is convenient to measure the volume [V] of these solutions. In these cases one must know their specific gravity or it can be ascertained through hydrometer. The volume of necessary initial solution can be calculated as follows:

$$V = \frac{E \times 100}{p \times \gamma} \times l \times n$$

γ = specific gravity of solution.

Calculation of factor (F) of standard solution.

$$F = \frac{\text{volume of consumed primary standard solution}}{\text{volume of given normal solution to standardize}}$$,

or

$$F = \frac{e}{a \times \text{volume of normal solution to be standardized}}$$

> l in case of approximately stronger, and
< l in case of approximately weaker n-solutions.

e = weighed portion of standard titrimetric substance in mg.

a = mg. of primary standard substance which corresponds to 1 ml. standard titrimetric solution of indicated normality.
The values can be taken from the table.

Application of factor (F)

$$F \times n = \text{Accurate normality of standardized solution}$$

$$F \times ml = \text{Volume in ml. used when the volumetric solution has the indicated normality.}$$

In case of titration the required weight is:

$$W = E n V F$$

n = a of the table when 0.1 n solutions are employed

W = weight in mg.
E = equivalent weight of the given substance in mg.
V = used up volume (ml.) of volumetric solution
F = factor of volumetric solution.

Volumetric analysis

Preparation of normal solutions. The given values are applicable to 0.1 n solution. For solutions of other normalities, multiply with the corresponding factor.

Substance	Contents g/l	Weighed in g or ml. conc. solutions diluted to 1000 ml.
H Cl	3.6465	10 ml. 35% solution
H_2SO_4	4.9041	3 ml. 96% solution
HNO_3	6.3016	8 ml. 60% solution
$H Cl O_4$	10.0465	28 ml. 30% solution
$H_2C_2O_4 \cdot 2H_2O$	6.3035	6.3035 g
$Ba(OH)_2 \cdot 8H_2O$	15.755	18 g
KOH	5.6108	6 g
NaOH	3.9999	4.5 g
Na_2CO_3	5.2992	5.2992 g
NH_4OH	3.5048	7.5 ml. 25% solution
$K Mn O_4$	3.1608	3.2 g
$K Br O_3$	2.78355	2.78355 g
$K_2Cr_2O_7$	4.9036	4.9036 g
$Ce(SO_4)_2 \cdot 4H_2O$	40.432	41 g
	12.691	23 g KI in 40 ml. water + 12.8 g I or 3.5668 g of KIO_3 + 37 g KI (+ HCl)
$Na_2S_2O_3 \cdot 5H_2O$	24.8194	25 g
As_2O_3	4.9455	4.9455 g As_2O_3 + 15 g $NaHCO_3$
$Ag NO_3$	16.9888	16.9888 g
Na Cl	5.8448	5.8448 g
NH_4SCN	7.6125	8 - 9 g
KSCN	9.719	10 - 12 g _according to moisture_
Complexon I Triplex I	19.11 (0.1m)	19.11 g
Complexon II Triplex II	29.224 (0.1m)	29.224 g
Complexon III Triplex III	37.21 (0.1m)	37.21 g

Volumetric analysis

Preparation of normal solutions. The given values are applicable to 0.1 n solution. For solutions of other normalities, multiply with the corresponding factor.

Stability of solution or sensitivity	Standardize with	Weighed e in mg	a mg = 1 ml 0.1 n solution	log a	D E log a
unlimited unlimited unlimited unlimited unlimited	Na_2CO_3	130.150	5.2992	72 425	27 575
sensitive to CO_2; glass sensitive to CO_2; glass sensitive to CO_2; glass unlimited sensitive to CO_2; glass	$H_2C_2O_4 \cdot 2H_2O$	160.180	6.3035	79 959	20 041
org. substances, light	$Na_2C_2O_4$	170.190	6.7002	82 608	17 392
unlimited					
unlimited	As_2O_3	120.140	4.9455	69 421	31 579
unlimited	$Na_2C_2O_4$	170.190	6.7002	82 608	17 392
org. substances, light I_2 escapes (vaporizes)					
sulphur bacteria O_2; air	KIO_3	90.110	3.5668	55 228	44 772
light	NaCl	140.160	5.8448	76 677	23 323
unlimited					
unlimited	$AgNO_3$	400.420	16.9888	23 016	76 984
unlimited					
unlimited in poly-ethylene flasks unlimited in poly-ethylene flasks unlimited in polyethylene flasks	$CaCO_3$	180.200	10 009	00 039	99 961

Volumetric analysis

Indicators for neutralization analysis and determination of pH

Substance	Chemical nomenclature	Turning point range pH	Colour change	Indicator solution
Cresol red (1. turning point)	o-Cresolsulphophthalein (Na-salt)	0.2–1.8	red-yellow	0.1% in water
Thymol blue (1. turning point)	Thymolsulphophthaline (Na-salt)	1.2–2.8	red-yellow	0.1% in water
Metanil yellow	m-Aminobenzenesulphonic acid azodiphenylamine (Na-salt)	1.2–2.8	red-yellow	0.1% in water
m-Cresol purple (1. turning point)	m-Cresolsulphophthalein	1.2–2.8	red-yellow	0.1% in alcohol
p-Xylenol blue (1. turning point)	1,4-Dimethyl-5-oxybenzene sulphophthalein	1.2–2.8	red-yellow	0.1% in alcohol
Tropolin 00	p-Benzenesulphonic acid azodiphenylamine (Na-salt)	1.3–3.2	red-yellow	0.1% in water
β-Dinitrophenol	2,6-Dinitrophenol	1.7–4.4	colourless yellow	0.1% in alcohol
α-Dinitrophenol	2,4-Dinitrophenol	2.6–4.6	colourless yellow	0.1% in 50% alcohol
Dimethyl yellow	Dimethylaminoazobenzene	2.9–4.0	red-yellow	0.1% in 20% alcohol
Methyl orange	Dimethylaminoazobenzene sulphonic acid (Na-salt)	3.1–4.4	red-yellow	0.1% in water
Bromophenol blue	Tetrabromophenol sulphopthlein	3.0–4.6	yel-red-violet	0.1% in alcohol
Bromocresol green	Tetrabromo-m-cresol sulphothalein (Na-salt)	3.8–5.4	yellow-blue	0.1% in water
Methyl red	Dimethylaminoazobenzene-o-Carbonic acid (Na-salt)	4.4–6.2	yellow-orange	0.1% in water
p-Nitrophenol	p-Nitrophenol	4.7–7.9	colourless yellow	0.1% in water
Chlorophenol red	Dichlorophenol sulphophthalein (Na-salt)	4.8–7.0	yellow-purple	0.1% in water
Bromocresol purple	Dibromo-o-cresol sulphophthalein (Na-salt)	5.2–6.8	yellow-purple	0.1% in water
Bromophenol red	Dibromophenol sulphophthalein (Na-salt)	5.2–6.8	yellow-purple	0.1% in water
Nitrazine yellow	Nitrazine yellow	6.0–7.0	yellow-violet	0.1% in water
Bromothymol blue	Dibromothymol sulphophthalein (Na-salt)	6.0–7.6	yellow-blue	0.1% in water
Phenol-red	Phenolsulphophthalein (Na-salt)	6.4–8.2	yellow-red	0.1% in water
m-Nitrophenol	m-Nitrophenol	6.6–8.6	colourless yellow	0.3% in water
Neutral red	Dimethylaminophenacine chloride (Na-salt)	6.8–8.0	bluishred-orange yellow	0.1% in water
Cresol red (2. turning point)		7.0–8.8	yellow-purple	
m-Cresol purple (2. turning point)		7.4–9.0	yellow-purple	
Thymol blue (2. turning point)		8.0–9.6	yellow-blue	
p-Xylenol blue (2. turning point)		8.0–9.6	yellow-blue	

Volumetric analysis — Indicator for neutralization analysis and determination of pH

Substance	Chemical nomenclature	Turning point pH range	Colour change	Indicator solution
Phenolphthalein	Di-p-dihydroxy diphenylphthalide	8.0–9.8	colourless-violet	0.1% in 70% alcohol
Thymolphthalein	Dithymolphthalide	9.3–10.5	colourless-blue	0.1% in 90% alcohol
Alizarin yellow G.G.	m-Nitrobenzene-azosalicyclic acid (Na-salt)	10.0–12.1	pale yellow-orange yellow	0.1% in water
Alizarin yellow R	p-Nitrobenzene-azosalicyclic acid (Na-salt)	10.0–12.1	pale yellow-orange red	0.1% in water
Alizarin yellow RS	p-Nitrobenzene-azosalicyclic acid-monosulphonic acid (Na-salt)	10.0–12.1	pale yellow-brownish red	0.1% in water
β-Naphthol violet	2-Naphthol-1-azo-p-nitrobenzene-3,6-disulphonic acid (Na-salt)	10.0–12.0	orange yellow-red violet	0.1% in water
Tropolin-o (Resorcinol yellow)	p-benzenesulphonic acid-azo-resorcinol	11.0–12.7	yellow-orange brown	0.1% in water

Volumetric analysis — Indicators for redox analysis and determination of redox potentials

Substances	Redox potential E_o (volt) 50% reduced 20°C	r_H values	Colour change reduced-oxidized	Indicator solution
Neutral red	−0.32	2–4.5	colourless-red	0.05% in 60% alcohol
Safranin T	−0.29	4–7.5	colourless-red	0.05% in water
Indigodisulphonate (k-salt)	−0.11	8.5–10.5	yellowish-blue	0.05% in water
Indigotrisulphonate (k-salt)	−0.07	9.5–12	yellowish-blue	0.05% in water
Indigo tetra sulphonate (k-salt)	−0.03 pH = 7	11.5–13.5	yellowish-blue	0.05% in water
Methylene blue	+0.01	13.5–15.5	colourless-blue	0.05% in water
Thionine	+0.06	15–17	colourless-violet	0.05% in 60% alcohol
Toluylene blue	+0.11	16–18	colourless-bluish violet	0.05% in 60% alcohol
Thymol indophenol	+0.18	17.5–20	colourless-blue[1]	0.02% in 60% alcohol
m-Cresol indophenol	+0.21	19–21.5	colourless-blue[2]	0.02% in 60% alcohol
2,6-Dichlorophenol	+0.23	20–22.5	colourless-blue	0.02% in water
Diphenylanine	+0.76	25–28	colourless-violet	1g in 100 ml. conc. H_2SO_4
Diphenyl benzidine	+0.76	25–28	colourless-violet	1g in 100 ml. conc. H_2SO_4
Diphenyl amine-p-sulphonic acid (Ba or Na salt)	+0.83 in 1 m H_2SO_4	27–29	colourless-violet	0.05% in water
N-Methyl diphenyl amine p-sulphonic acid (Na-salt)	+0.85	28–30	colourless-purple red	0.1% in water
Ferroin Tri-o-phenanthroline ferrous-sulphate	+1.12	39–41	red-bluish	0.025 m in water

[1] Intermediate reddish colour at pH 9:   [2] Intermediate reddish colour at pH 8.5.

Volumetric analysis

Weight	Titrate with 0.1 n solution of	Factor a mg/ml solution	log Factor	DE log Factor	Method
Ag	NaCl	10.788	03294	96706	Mohr's method in neutral solution with K_2CrO_4 as indicator.
	NH_4SCN	10.788	03294	96706	Volhard's method in acidic HCl solution with $Fe\,(NH_4)\,(SO_4)_2$ as indicator.
$AgNO_3$	NaCl	16.9888	23016	76984	As in case of Ag.
	NH_4SCN	16.9888	23016	76984	
Al	$KBrO_3$	0.2248	35170	64830	Precipitation in weakly acidified acetic acid solution with 8-hydroxy quinoline. After filtra-
Al_2O_3	$KBrO_3$	0.4247	62806	37194	tion, the precipitate is dissolved in warm dil. HCl. Before the titration one must add approx. 1 g KBr.
As	$KBrO_3$ KI_3 $Ce(SO_4)_2$	3.7455	57351	42649	With $KBrO_3$: Titration follows in strongly acidified HCl solution at about 60°C. Methyl red as indicator, is decolourized. With KI_3: Solution must contain an excess of $NaHCO_3$. Titration at room temperature.
As_2O_3	$KBrO_3$ KI_3 $Ce(SO_4)_2$	4.9455	69421	30579	With $Ce(SO_4)_2$: Titration in sulphuric acid solution by application of OsO_4 or ICl as catalyst.
As_2O_5	$KI_3(Na_2S_2O_3)$	5.7455	75933	24067	In strongly acid solution, I_2 Is made free from KI. Titration with NaS_2O_3.
$Ba(OH)_2$	HCl	8.5688	93392	06608	Methyl orange as indicator.
$BaCO_3$	HCl	9.8685	99425	00575	known excess of acid is added to drive off all CO_2 and back titrated with NaOH.
$BaCl_2$	$AgNO_3$	10.4137	01761	98239	Mohr's method
$BaCl_2 \cdot 2H_2O$	$AgNO_3$	12.2153	08691	91308	
Bi	$KBrO_3$	1.7417	24097	75903	As in case of Al with $KBrO_3$.
Br	$AgNO_3$	7.9916	90263	09737	Mohr's method
	$Na_2S_2O_3$	7.9916	90263	09737	Elemental Bromine liberates free I_2 from KI which is back titrated with $Na_2S_2O_3$.
CO_2	HCl	2.2005	34252	65748	From easily decomposable carbonates. Excess HCl is back titrated.
CH_3COOH	NaOH	6.0054	77854	22146	Indicator phenolphthalein.
Ca	$KMnO_4$ $Ce(SO_4)_2$	2.004	30190	69810	With $KMnO_4$: Ca is precipitated in ammoniacal solution as oxalate. The filtered CaC_2O_4 precipitata is dissolved in dil. H_2SO_4 and
CaO	$KMnO_4$ $Ce(SO_4)_2$	2.804	44778	55222	at 60°–80°C is titrated upto a pale permanent pink colour.
$Ca(OH)_2$	HCl	3.705	56879	43121	With $Ca(SO_4)_2$: Ca is likewise precipitated as oxalate. The filtered precipitate of CaC_2O_4 is
$CaCO_3$	HCl $KMnO_4$ $Ce(SO_4)_2$	5.0045	69936	30064	dissolved in conc. HCl, diluted and at 50°C titrated. Catalyst: Iodine monochloride Indicator: Ferroin
CaC_2O_4 H_2O	$KMnO_4$ $Ce(SO_4)_2$	7.306	86365	13635	With HCl: A known excess of HCl is added and is back titrated with NaOH. Indicator: Methyl orange.
$CaCl_2$	$AgNO_3$	5.5497	74427	25573	Mohr's method
$CaCl_2 \cdot 6H_2O$	$AgNO_3$	10.956	03959	96041	

Volumetric analysis

Required substance	Titrate with 0.1 n solution of	Factor a mg/ml solution	log Factor	DE log Factor	Method
Cd	$KBrO_3$	1.4051	14771	85229	Precipitation with 8-Hydroxy quinoline and the related titration.
Cl	$AgNO_3$ $Na_2S_2O_3$	3.5457	54970	45030	Mohr's method. Elemental chlorine liberates free I_2 from KI
Co	$KBrO_3$	0.7368	86732	13268	Precipitation with 8-Hydroxy quinoline and the connected titration.
Cr	$KMnO_4$ $Na_2S_2O_3$	1.7337	23897	76103	With $KMnO_4$: Cr^{6+} is reduced in sulphuric acid solution by $FeSO_4$ and the excess $FeSO_4$ is titrated with $KMnO_4$.
Cr_2O_3	$KMnO_4$ $Na_2S_2O_3$	2.5337	40376	59624	Control $FeSO_4$ solution through a blank experiment.
CrO_3	$KMnO_4$ $Na_2S_2O_3$	3.3334	52292	47708	With $Na_2S_2O_3$: Cr^{6+} liberates free I_2 from KI
$CrO_4{}^{2-}$	$KMnO_4$ $Na_2S_2O_3$	3.8670	58737	41263	in hydrochloric acid which is titrated at room temperature.
Cu	$Na_2S_2O_3$	6.354	80305	19695	Cu^{2+} forms Cu_2I_2 with KI in sulphuric acid solution thereby simultaneously liberating free I_2 that is titrated with NaS_2O_3.
CuO	$Na_2S_2O_3$	7.954	90058	09942	
$CuSO_4$	$Na_2S_2O_3$	15.961	20305	79695	
$CuSO_4 \cdot 5H_2O$	$Na_2S_2O_3$	24.969	39739	60261	
Fe	$KBrO_3$	0.4653	66777	33223	Precipitation with 8-Hydroxy quinoline and the related titration.
Fe	$KMnO_4$ $Ce(SO_4)_2$	5.585	74702	25298	With $KMnO_4$: Fe^{2+} is directly titrated in sulphuric acid solution and in HCl solution according to the method of Zimmormann-Reinhardt. If the solution contains Fe^{3+} before hand, then it is reduced at its boiling point with $SnCl_2$ in HCl solution. An excess of $SnCl_2$ is destroyed by addition of $HgCl_2$ after cooling.
FeO	$KMnO_4$ $Ce(SO_4)_2$	7.185	85643	14357	
Fe_2O_3	$KMnO_4$ $Ce(SO_4)_2$	7.985	90227	09773	
Fe_3O_4	$KMnO_4$ $Ce(SO_4)_2$	7.719	88753	11247	With $Ca(SO_4)_2$: $FeSO_4$ either in acidified HCl or H_2SO_4 solution can be directly titrated. Fe^{3+} can be reduced with $SnCl_2$ as in HCl solution.
$FeCO_3$	$KMnO_4$ $Ce(SO_4)_2$	11.586	06393	93607	Reduction with metal reductors is also possible Indicator—ferroin.
$FeSO_4$	$KMnO_4$ $Ce(SO_4)_2$	15.192	18160	81840	
$FeSO_4 \cdot 7H_2O$	$KMnO_4$ $Ce(SO_4)_2$	27.803	44408	55592	
$Fe_2(SO_4)_3$ $\cdot 9H_2O$	$KMnO_4$ $Ce(SO_4)_2$	28.102	44872	55128	
$FeSO_4(NH_4)_2$ $SO_4 \cdot 6H_2O$	$KMnO_4$ $Ce(SO_4)_2$	39.216	59346	40654	
$FeCl_3$	$KMnO_4$ $Na_2S_2O_3$	16.222	21011	78989	Direct titration, potentiometric indication.

Volumetric analysis

Required substance	Titrate with 0.1 n solution of	Factor a mg/ml solution	log Factor	DE log Factor	Method
H_3BO_3	NaOH	6.1844	79130	20870	In presence of Mannitol and Phenolphthalein as indicator
HBr	NaOH AgNO$_3$	8.0924	90808	09192	Methyl orange as indicator Mohr's method
HCOOH	NaOH KMnO$_4$	4.6027	66301	33699	Methyl orange as indicator
HCN	AgNO$_3$	2.7027	43180	56820	Volhard's method, in acidified HNO$_3$ solution
HCN	AgNO$_3$	5.4054	73283	26717	Liebig method in weakly alkaline solution till turbidity appears
$H_2C_2O_4$	NaOH KMnO$_4$ Ce(SO$_4$)$_2$	4.5018	65339	34661	With NaOH: Methyl orange as indicator
$H_2C_2O_4 \cdot 2 H_2O$	NaOH KMnO$_4$ Ce(SO$_4$)$_2$	6.3034	79958	20042	With KMnO$_4$ and Ce(SO$_4$)$_2$: Titration as in case of Ca
HCl	NaOH	3.6465	56188	43812	Indicator methyl orange or phenolphthalein
HClO	Na$_2$S$_2$O$_3$	2.6233	41885	58115	From KI, free I$_2$ is liberated and titrated with Na$_2$S$_2$O$_3$
HClO$_3$	Na$_2$S$_2$O$_3$	1.4078	14854	85146	
HClO$_4$	NaOH	10.0465	00201	99799	Indicator methyl orange
H_2CrO_4	KMnO$_4$ Na$_2$S$_2$O$_3$	3.934	59487	40513	As in case of Cr
HF	NaOH	2.0008	30121	69879	Indicator phenolphthalein. An excess of NaOH to a hot solution and is back titrated with HCl
HI	NaOH	12.792	10694	89306	Indicator methyl orange
	AgNO$_3$	12.792	10694	89306	Mohr's method
HIO$_3$	Na$_2$S$_2$O$_3$	2.932	46716	53284	In HCl acid solution free I$_2$ is liberated from KI and titrated with Na$_2$S$_2$O$_3$
HNO$_2$	KMnO$_4$	2.3508	37121	62879	Slow addition of nitrite to a warm H$_2$SO$_4$ acidified solution of KMnO$_4$
HNO$_3$	NaOH	6.3016	79945	20055	Indicator methyl orange
H_2O_2	KMnO$_4$ Na$_2$S$_2$O$_3$	1.7008	23065	76935	In sulphuric acid solution at low temperature. In sulphuric acid solution free I$_2$ is liberated from KI which is titrated with Na$_2$S$_2$O$_3$
H_3PO_4	NaOH	9.800	99125	00875	Methyl orange
	NaOH	4.900	69022	30978	Phenolphthalein
H_2S	KI$_3$	1.7041	23150	76850	An excess of I$_2$ solution added and the excess is back titrated with Na$_2$S$_2$O$_3$
	KBrO$_3$	0.4260	62944	37054	In HCl acid solution through an addition of excess of KBrO$_3$ and KBr, H$_2$S is oxidized to H$_2$SO$_4$
H_2SO_3	KI$_3$	4.1041	61322	38678	A known excess is added and excess is back titrated with Na$_2$S$_2$O$_3$

Volumetric analysis					
Required substance	Titrate with 0.1 n solution of	Factor a mg/ml solution	log Factor	DE log Factor	Method
H_2SO_4	NaOH	4.904	69055	30945	Methyl orange or phenolphthalein as indicator
$H_2S_2O_8$	$KMnO_4$	9.7074	98711	01289	Known excess of $FeSO_4$ added and the excess is back titrated with $KMnO_4$. Blank experiment with $FeSO_4$ solution to be done.
H_2SiF_6	NaOH	7.2055	85766	14234	An excess KCl reacts in presence of alcohol with formation of K_2SiF_6. The liberated free HCl is titrated against using Methyl red.
Hg	KI_3 NH_4SCN	10.0305	00132	99868	With KI_3: Hg_2^{2+} reacts with added I_2 and KI with formation of soluble complex HgI_4^{2-}.
HgO	KI_3 NH_4SCN	10.8305	03465	96535	The excess I_2 is back titrated. Hg^{2+} can be reduced before hand with NH_4 SCN—
$HgCl_2$	KI_3	13.576	13277	86723	Volhard method.
$HgClNH_2$	KI_3	12.605	10053	89947	
I	$Na_2S_2O_3$	12.691	10349	89651	Starch solution as indicator.
I_2O_5	$Na_2S_2O_3$	2.7818	44425	55575	In HCl solution, I_2 is liberated from KI
KOH	HCl	5.6108	74902	25098	Indicator methyl orange or phenolphthalein
$KAl(SO_4)_2$ $12 H_2O$	$KBrO_3$	3.9533	59696	40304	Precipitation with 8-hydroxy quinoline and the connected titration
KBr	$AgNO_3$	11.9012	07559	92441	Mohr or Volhard method.
$KBrO_3$	$Na_2S_2O_3$	2.7835	44460	55540	In HCl solution, free I_2 is liberated from KI.
KCN	$AgNO_3$	6.5119	81370	18630	See HCN
KCN	$AgNO_3$	13.0223	11473	88527	
KCNS	$AgNO_3$	9.719	98702	01238	Volhard method
K_2CO_3	HCl	6.9106	83942	16058	Indicator methyl orange
$KHCO_3$	HCl	10.012	00051	99949	
$K_2C_2O_4$	$KMnO_4$ $Ce(SO_4)_2$	8.3106	91963	08037	Titration as in case of Ca
KCl	$AgNO_3$	7.4557	87249	12751	Mohr method in neutral or Volhard method in nitric acid solution.
$KClO_3$	$Na_2S_2O_3$	2.0426	31018	68982	In HCl solution, free I_2 is liberated from KI.
K_2CrO_4	$KMnO_4$ $Na_2S_2O_3$	6.474	81115	18885	As in case of chromium
$K_2Cr_2O_7$	$KMnO_4$ $Na_2S_2O_3$	4.9036	69051	30949	
$K_4[Fe(CN)_6]$	$KMnO_4$ $Ce(SO_4)_2$	36.8360	56627	43373	With $KMnO_4$: The largely diluted solution is acidified with H_2SO_4 and then titrated.
KI	$AgNO_3$	16.601	22016	77984	Mohr or Volhard method
KIO_3	$Na_2S_2O_3$	3.5668	55228	44772	In HCl solution free I_2 is liberated from KI.

Volumetric analysis

Required substance	Titrate with 0.1 n solution of	Factor a mg/ml solution	log Factor	DE log Factor	Method
$KMnO_4$	$H_2C_2O_4$ $Na_2S_2O_3$	3.1608	49977	50023	As in case of Ca In an acidified solution free I_2 is liberated from KI.
KNO_2	$KMnO_4$	4.2554	62894	37106	As in case of HNO_2
KNO_3	H_2SO_4	10.1108	00477	99523	As in case of NO_3
K_2SO_3	KI_3	4.1041	61322	38678	As in case of H_2SO_4
$KHSO_3$	KI_3	6.0085	77877	22123	
Li_2CO_3	HCl	3.6946	56757	43243	Indicator methyl orange
Mg	$KBrO_3$	0.3040	66777	33223	
MgO	HCl	2.016	30449	69551	With HCl: Known excess of HCl added and the excess is back titrated with NaOH. Indicator methyl orange.
	$KBrO_3$	0.504	70243	29757	
$MgCO_3$	HCl $KBrO_3$	4.2165 1.0541	62495 02289	37505 97711	With $KBrO_3$: Mg is precipitated in ammoniacal solution in presence of NH_4Cl with 8-Hydroxy-quinoline. The precipitate is filtered and dissolved in warm dil. HCl. After addition of about 1g KBr, an excess bromate solution is added. The excess bromate can be titrated, after addition of KI, with $Na_2S_2O_3$.
$MgCl_2$	$KBrO_3$	1.1904	07568	92432	
$MgSO_4$	$KBrO_3$	1.5048	17749	82251	
$MgSO_4$ $7H_2O$	$KBrO_3$	3.081	48873	51127	
Mn	$KBrO_3$	0.6866	83672	16328	Precipitation with 8-hydroxy quinoline and the connected titration.
	$KMnO_4$	1.648	21696	78304	
	As_2O_3	1.3735	13783	86217	
MnO	$KMnO_4$	2.128	32797	67203	With $KMnO_4$: Volhard and Wolff method: Mn^{2+} is titrated in the presence of excess ZnO in boiling hot solution upto a permanent rose colour. End point is difficult to find due to the precipitation of MnO_2.
	As_2O_3	1.7735	24883	75117	
MnO_2	$KMnO_4$	2.608	41631	58369	With As_2O_3: Mn^{2+} is oxidized in nitric acid solution by $NH_4S_2O_8$ in the presence of $AgNO_3$ to Mn^{7+}. After decomposing the excess oxidation medium and precipitating Ag as AgCl, the $KMnO_4$ is titrated with As_2O_3.
	As_2O_3	2.1735	33716	66284	
	$Na_2S_2O_3$	4.3465	63814	36186	
$MnCO_3$	$KMnO_4$	3.4485	53763	46237	With $Na_2S_2O_3$: By decomposing with HCl in hot condition the chlorine evolved is led into an acidified solution of KI from which free I_2 is liberated.
$MnCl_2$ $4H_2O$	$KMnO_4$	5.938	77361	22639	
$MnSO_4$	$KMnO_4$	4.5308	65613	34387	
Mo	$Ce(SO_4)_2$	9.595	98204	01796	In strongly acidified solution with HCl, Mo^{6+} is reduced to Mo^{5+} by Hg. Filter and wash with Zn/HCl. Titrate in room temperature using Ferroin as indicator.
MoO_3	$Ce(SO_4)_2$	14.395	15821	84179	

Volumetric analysis

Required substance	Titrate with 0.1 n solution of	Factor a mg/ml solution	log Factor	DE log Factor	Method
N	H_2SO_4	1.4008	14638	85362	
5.55 N Gelatine	H_2SO_4	7.774	89064	10936	Nitrogen in ammonium compounds: with the help of strong NaOH, NH_3 is liberated from ammonium salts and is led into a measured excess of H_2SO_4. The excess H_2SO_4 left out is back titrated against NaOH. Indicator methyl orange.
6.25 N white of egg	H_2SO_4	8.755	94226	05774	
6.37 N Casein	H_2SO_4	8.923	95051	04949	
NH_3	H_2SO_4	1.7032	23126	76874	Nitrogen in organic compounds: By boiling with conc. H_2SO_4 in the presence of selenium reaction mixture, the organic substance is decomposed and nitrogen is bound as $(NH_4)_2$ SO_4. Further, work as described above for ammonium salts. (Refer also nitrogen determination by Kjeldahl's method.)
NH_4	H_2SO_4	1.8040	25624	74376	
NH_4OH	H_2SO_4	8.0048	90335	09665	
NH_4Cl	H_2SO_4	5.3496	72832	27168	
NH_4NO_3	H_2SO_4	3.5048	54466	45534	
$(NH_4)_2SO_4$	H_2SO_4	6.6070	82000	18000	
$NH_4Fe(SO_4)_2$ $12 H_2O$	H_2SO_4	48.221	68324	31676	
$(NH_4)_2Fe$ $(SO_4)_2 \cdot 6H_2O$	$KMnO_4$	39.216	59346	40654	Direct titration of Fe^{2+} in sulphuric acid solution with $KMnO_4$
NH_4CNS	$AgNO_3$	7.6125	88153	11847	Volhard meth.
NH_2OH	$KMnO_4$	1.6516	21791	78209	After addition : $NH_4(SO_4)_2$ solution in excess and dil H_2SO_4, it is boiled for 5 minutes and the formed Fe^{2+} is immediately titrated after dilution.
$NH_2OH \cdot HCl$	NaOH	38.994	59100	40900	Direct titration with NaOH. Indicator phenolphthalein.
N_2H_4	KI_3	8.012	90374	09626	Titration in alkaline bicarbonate solution. Starch solution as indicator. After addition of $NaHCO_3$, titrate immediately.
$N_2H_4 \cdot H_2SO_4$	KI_3	32.533	51232	48768	
N_2O_3	$KMnO_4$	1.9004	27884	72116	See HNO_2
N_2O_4	$KMnO_4$	4.6003	66283	33717	
NO_3	H_2SO_4	6.2008	79245	20755	Reduction of nitrates in alkaline solution with Devarda's alloy to NH_3. Further work as in case of ammonia distillation.
NaOH	HCl	3.9999	60205	39795	Indicator methyl orange or phenophthalein.
Na_2CO_3	HCl	5.2992	72425	27575	Indicator methyl orange. CO_2 is boiled down, and then titrated to the end point.
Na_2CO_3 $\cdot 2H_2O$	HCl	7.1013	85134	14866	
Na_2CO_3 $\cdot 10H_2O$	HCl	14.308	15557	84443	
$NaHCO_3$	HCl	8.4010	92433	07567	
$Na_2C_2O_4$	$KMnO_4$ $Ce(SO_4)_2$	6.7002	82608	17392	Titration as in case of Ca
NaBr	$AgNO_3$	10.2907	01245	98755	Mohr or Volhard method
NaCl	$AgNO_3$	5.8448	76677	23323	

Volumetric analysis

Required substance	Titrate with 0.1 n solution of	Factor a mg/ml solution	log Factor	DE log Factor	Method
$NaClO_3$	$Na_2S_2O_3$	1.7741	24898	75102	Free I_2 Is liberated from KI.
$NaNO_3$	$KMnO_4$	2.8335	45233	54767	$NaNO_3$ oxidizes in HCl acid solution on heating. Fe^{2+} to Fe^{3+}. The excess Fe^{2+} is back titrated.
	H_2SO_4	8.4999	92941	07059	As in case of NO_3
$Na_2Cr_2O_7$	$KMnO_4$ $Na_2S_2O_3$	4.3667	64015	35985	As in case of Cr
$Na_2B_4O_7$ 10 H_2O	HCl	19.071	28037	71963	Direct titration at room temperature. Methyl orange or methyl red as indicator.
Na_2SO_3	KI_3	6.3024	79951	20049	As in case of H_2SO_3
$NaHSO_3$	KI_3	5.2036	71630	28370	
$Na_2S_2O_3$	KI_3	15.8114	19897	80103	Direct titration at room temperature Starch solution as indicator.
$Na_2S_2O_3$ 5 H_2O	KI_3	24.8194	39479	60521	
Ni	$KBrO_3$	0.7336	86547	13453	Precipitation with 8-hydroxy quinoline and the connected titration.
NiO	$KBrO_3$	0.9336	97017	02983	
$NiSO_4$	$KBrO_3$	1.9345	28656	71344	
$NiSO_4$ 7 H_2O	$KBrO_3$	3.5108	54541	45459	
P	NaOH	0.1343	12808	87192	Precipitate as ammonium phosphomolybdate and dissolve in known excess of NaOH. The unused. excess is back titrated with HCl. Indicator: phenolphthalein.
PO_4	NaOH	9.498	97763	02237	Methyl orange
	NaOH	4.749	67660	32340	Phenolphthalein
P_2O_5	NaOH	7.098	85114	14886	Methyl orange
	NaOH	3.549	55011	44989	Phenolphthalein
PbO_2	$KMnO_4$ $Na_2S_2O_3$	11.96	07775	92225	With $Na_2S_2O_3$: as in case of MnO_2 With $KMnO_4$: The oxide is dissolved with HNO_3 after addition of known excess of standardized $H_2C_2O_4$. The unused excess of $H_2C_2O_4$ after acidification with H_2SO_4 is back titrated at 60°C.
Pb_3O_4	$KMnO_4$ $Na_2S_2O_3$	34.28	53506	46494	
S	KI_3	1.6033	20501	79499	H_2S is reacted with a known excess of I_2 solution. The unused excess is back titrated with $Na_2S_2O_3$.
	$KBrO_3$	0.4006	60296	39704	Refer H_2S
SO_2	KI_3	3.2033	50560	49440	As in case of H_2SO_3
SO_3	NaOH	4.0033	60242	39758	Titration as in H_2SO_4
SO_3^{2-}	KI_3	4.0033	60242	39758	As in case of H_2SO_3
SO_4^{2-}	NaOH	4.803	68154	31846	Titration as in H_2SO_4

Required substance	Titrate 0.1 n solution of	Factor a mg/ml solution	log Factor	DE. log Factor	Method
Sb	KI_3 $KBrO_3$	6.088	78447	21553	With $KBrO_3$: As in case of As.
Sb_2O_3	KI_3 $KBrO_3$	7.288	86261	13739	With KI_3: As in case of As. Solution must also contain in addition of tartaric acid so that no basic ammonium salts are precipitated.
Sn	KI_3 $KBrO_3$	5.935	77342	22658	With $KBrO_3$: As in case of As.
SnO	KI_3 $KBrO_3$	6.735	82834	17166	With KI_3: Sn^{2+} can be titrated at room temperature in HCl solution. Before the titration Sn^{4+} must be reduced, thereafter by employing CO_2 as protective
SnO_2	KI_3 $KBrO_3$	7.535	87708	12292	gas further work is carried out.
$SnCl_2$	KI_3 $KBrO_3$	9.4805	97683	02317	
Sr	$Ce(SO_4)_2$	4.3815	64162	35838	With $Ce(SO_4)_2$ precipitation as oxalate, thereafter titration as in case of Ca.
$SrCO_3$	$Ce(SO_4)_2$ HCl	7.382	86817	13183	With HCl: As in case of $CaCO_3$
Ti	$KBrO_3$	0.5988	77725	22275	Precipitation with 8 - hydroxyquinoline, thereafter titration.
TiO_2	$KBrO_3$	0.9988	99946	00054	
U	$KMnO_4$	11.907	07580	92420	With $KMnO_4$: A sulphuric acid solution of uranium is reduced to U^{4+} in a metal reductor and thereafter titrated at room temperature.
	$KBrO_3$	1.9845	29765	70235	
U_3O_8	$KMnO_4$	14.040	14738	85262	With $KBrO_3$: Precipitation with 8 - hydroxyquinoline, thereafter titration.
V	$KMnO_4$ $Ce(SO_4)_2$	5.095	70714	29286	With $KMnO_4$: V^{5+} oxidizes Fe^{2+} to Fe^{3+}. An unused excess of $FeSO_4$ is back titrated. Titration of $FeSO_4$ is to be first standardized.
V_2O_5	$KMnO_4$ $Ce(SO_4)_2$	9.095	95880	04120	With $Ce(SO_4)_2$: Reduction with $FeSO_4$ and connected potentiometric titration at 70°C.
Zn	$KBrO_3$	0.8173	91235	08765	
ZnO	$KBrO_3$	1.0173	00743	99257	With $KBrO_3$: As in case of Mg.
$ZnCO_3$	$KBrO_3$	1.5674	19518	80482	With KI_3: With the help of HCl, H_2S is liberated. Refer determination of H_2S
ZnS	$KBrO_3$	1.2181	08567	91433	
	KI_3	4.8725	68773	31227	
$ZnSO_4$	$KBrO_3$	2.0180	30493	69507	
$ZnSO_4$ · 7 H_2O	$KBrO_3$	3.5945	55564	44436	

Complexometric methods

Substance	Chemical name	Structural formula	Molecular weight	Solubility
Complexon I Titriplex I	Nitrilotriacetic acid	$HOOCH_2C-N \begin{cases} CH_2COOH \\ CH_2COOH \end{cases}$	191.14	Hard in water easily in alkali
Complexon II Titriplex II	Ethelenediamine tetra acetic acid	$\begin{cases} HOOCH_2C \\ HOOCH_2C \end{cases} N-CH_2-CH_2-N \begin{cases} CH_2COOH \\ CH_2COOH \end{cases}$	292.24	Hard in water easily in alkali
Complexon III Titriplex III	Disodium salt of ethylenediamine tetra acetic acid	$\begin{cases} HOOCH_2C \\ HOOCH_2C \end{cases} N-CH_2-CH_2-N \begin{cases} CH_2COONa \\ CH_2COONa \end{cases} +2H_2O$	372.25	easily in water

Complexon form with metal ions soluble but undissociated chelate complexes.
In volumetric analysis these can be used in two different ways.
1. By titration, H^+ ions are liberated which can be determined alkalimetrically. These reactions have no little significance.
2. The titration is carried out in a buffered solution (mostly with complexon III or Triplex III) whereby the end point of the reaction is shown by the so-called metal indicators. These metal indicators are substances with which the metal ions form complexes which have different colour than the pure indicator solutions. The indicator metal complexes are not so stable as the corresponding compounds with the complexons.

Towards the end of the titration, the indicator-metal complex is destroyed and this is accompanied with a change of colour.
These titrations can be carried out by different methods.

Direct titration : The metal ions are titrated directly in a buffed solution against a suitable indicator.

Substitution titration : When for direct titrations, no suitable indicators are available, or when the metal at that corresponding pH range is precipitated as compound then these types of titrations are carried out. In case of substitution titrations one adds Mg or Zn complexonate solution. Mg and Zn are bound by fewer solids, so that they interchange with the titrant metal and are titrated against corresponding connected indicators.

Back titration : When no substitute indicators for the direct titration are available, the complexon is added in known excess. The unused excess is then back titrated with standardized $ZnSO_4$ or $MgSO_4$.

Indirect titration: In case of indirect titrations often solid compounds are precipitated in which the metal ions of the precipitation medium or the cations of the precipitated compound is titrated.

Buffer mixture

pH = 1	n-HCl
pH = 2 — 4	p-chloroaniline and it is hydrochloride in solid form.
pH = 4 — 6.5	n-acetic acid and n-sodium acetate are mixed.
pH = 6.5 — 8	1 m ethanolamine and 1 n HCl are mixed when needed.
pH = 8 — 11	1 m NH_3 and 1 m NH_4Cl solutions are mixed when needed.
pH = 10	70 g of NH_4Cl and 570 ml. of NH_3 are dissolved and made upto 1 litre.
pH = 12 — 13	n-NaOH

Metallic indicators

Name	Indicator solution
Pyrocatechol violet	0.1 g in 100 ml. water
3, 3-dimethyl-naphthidine	1 g in 100 ml. glaciel acetic acid
Eriochrome-black T	1 g + 99 g NaCl are ground, solid form is employed saturated aqueous solution about 0.15% solution
Pthalein purple (metal phthalein)	0.1 g + 2 ml NH_3 (conc.) and dilute to 100 ml.
4-(2-pyridylazo) resorcinol	0.1% in water
Sulphosalicyclic acid	5% in water
Tiron	2% in water

Required substance	Factor to be applied for 0.01 m complexon	log Factor	DE log Factor	Remarks
Ag	2.1576	33397	66603	Indirect titration: Precipitate as AgCl and dissolve in $NH_3 + NH_4Cl$. After addition of $K_2[Ni(CN)_4]$ the free Ni liberated is titrated. Indicator murexide.
Al	0.2698	43104	56896	Back titration. Add known excess of complexon and back titrate the excess with standardized $ZnSO_4$. pH 7-8. Erio T as indicator.
Ba	1.3736	13786	86214	Direct titration at pH 11 with metalphthalein as indicator.
Bi	2.090	32015	67985	Direct titration at pH 2.5-3 with pyrocatechol violet as indicator. Colour changes from blue to yellow.
Br	1.5983	20366	79634	As in case of Ag. Precipitation as AgBr.
Ca CaO O_3	0.4008 0.5608 1.0009	60293 74881 00039	39707 25119 99961	Direct titration: Acid solution is neutralized with NaOH a small excess of NH_3 is added. Metalphthalein as indicator. Colour changes from red to grey.
°DH	0.5608	74881	25129	Direct titration at buffer pH 10. Erio T as indicator. Colour changes from red to blue.

Complexometric methods

Required substance	Factor to be employed for 0.01m complexon	log Factor	DE log Factor	Remarks
Cd	1.1241	05080	94920	Direct titration. Acid solution is neutralised with NaOH. Buffer p_H=10. Erio T as indicator. Colour changes from red to blue.
Ce	1.4018	14653	85347	Direct titration. Erio T as indicator. Colour changes from red to blue. Presence of ascorbic acid prevents oxidation. Standardize an acid solution with triethanol amine at p_H=7.
Cl⁻	0.7091	85073	14927	As in case of Ag.
CN⁻	1.0408	01735	98265	Indirect titration. Standard $NiSO_4$ solution is added in known excess, made ammonical and the excess is back titrated. Indicator Murexide.
CNS⁻	1.1617	06510	93490	As in case of Ag. Precipitation as AgCNS Colour changes from yellow to violet.
Co	0.5894	77041	22959	Back titration. Excess complexon left is back titrated with standardized $MgSO_4$ sol. Buffer p_H=20, Erio T as indicator. Colour changes from blue to red.
Cu	0.6354	80305	19695	Direct titration. Standardize with NH_3 at p_H=8. Murexide as indicator. Colour changes from yellow to violet.
F⁻	0.3800	57978	42022	Indirect titration. The neutral alkali fluoride solution is titrated with an excess of standardized $CaCl_2$ solution. After addition of alcohol and setting of the precipitate is standardized at p_H=10. Indicator Erio T.
Fe³⁺	0.5585	74702	25298	Direct titration. Fe must ba present as Fe^{3+}. p_H=2.5 Sulphosalicyclic acid as indicator. Titrate upto the disappearance of red colour.
Hg	2.0061	30235	69765	Substitution titration. Solution is treated with excess of Mg complexonate sol. Thereafter it is neutralised with NaOH. Buffer p_H=10, Erio T as indicator.
I⁻	2.5382	40452	59548	As in case of Ag. Precipitation as AgI. Due to the difficult solubility of $AgI-NH_3$ concentration is used.
In	1.1476	05979	94021	The solution is treated with tartaric acid and thereafter neutralised with NaCH. Buffer p_H=10. Titrate at boiling hot condition. Indicator Erio T. Colour changes from red to blue.
K	0.7820	89321	10649	Precipitation as $NaK_2[Co(NO_2)_6]H_2O$. The precipitate is dissolved in HCl and thereafter as in case of CO.
La	1.3892	14276	85724	As in case of Ce.
Mg MgO Mg CO₃	0.2432 0.4032 0.8433	38596 60552 92598	61404 39448 07402	Direct titration. First the acid sol is neutralised with NaCH. Buffer p_H=10, Erio T as indicator. Colour changes from red to blue.
Mn	0.5494	73989	26011	Substitution titration. Sol. is treated with ascorbic acid and is neutralised. Thereafter an excess of Mg Complexonate is added. Erio T as indicator.
Na	0.2299	36156	63844	Indirect titration. Precipitation of Na as $NaZn(UO_2)_3(CH_3COO)_9$. After dissolving in HCl an excess of $(NH_4)_2CO_3$ is added. Buffer p_H=10. Erio T as indicator (yellowish red - greenish blue).
Ni	0.5869	76856	23144	Direct titration. With Murexide as indicator the sol is neutralised with NH_3. Just before the end point, the titration sol is made strongly alkaline.
P PO₄ P₂O₅	0.3098 0.9498 0.7098	49101 97762 85111	50899 02238 14889	Indirect titration. Precipitation as $MgNH_4PO_4.6H_2O$. The precipitate is dissolved in HCl and the complexon sol is added in known excess. Thereafter it is neutralised with HaOH. Buffer pH = 10. The unused excess complexon sol is back titrated with $MgSO_4$ sol. Erio T as indicator. Colour changes from blue to red.
Pb	2.0721	31641	68359	Direct titration. Sol is treated with tartaric acid and neutralized with NaOH. Buffer pH = 10, Erio T as indicator.
Pd	1.067	02816	97184	As in case of Ag.
S SO₃²⁻ SO₄²⁻	0.3207 0.8007 0.9607	50604 90345 98257	49396 09655 01743	Indirect titration. SO_4 is precipitated as $BaSO_4$ with standardized known excess of $BaCl_2$ sol. The unused excess of $BaCl_2$ is back titrated as in case of Ba.
Sr	0.8763	94265	05735	As in case of Ba.
Th	2.3205	36559	63441	Indirect titration. The acid sol is standardized with NH_3 at p_H=2 and warmed to 40°C. Indicator Pyrocatechol violet. Colour changes from blue to yellow.
Tl	2.0439	31046	68954	Substitution titration. The weakly acidic sol is treated with excess of Mg complexonate. Then neutralised with NaOH. Buffer pH = 10. Erio T as indicator.
WO₄	2.4792	39431	60569	Indirect titration. Alkalitungstate is precipitated as calciumtungstate with $CaCl_2$ sol. The precipitate is decomposed with HCl and after separation of WO_3, the Ca is titrated.
Zn	0.6538	81544	18456	As in case Cd.
Zr	0.9122	96009	03991	The acid sol is treated with excess of complexon and is heated. Then standardized with NH_3 at p_H=6−7 and boiled for a short time. After cooling the p_H is brought to 5 with acetic acid and back titrated with $FeCl_3$ sol. Salicyclic acid as indicator.

Gravimetric analysis

In gravimetric analysis, in order to find out the constituents of solid compounds, they are to be isolated, converted and weighed. The conversions are subjected to stoichiometric principles. With the help of atomic or molecular weight, the known substances must not be related to the required substances unless one uses the calculation factor F for every calculation.

$$F = \frac{\text{Molecular weight of required substance}}{\text{Molecular weight of given substance}}$$

The application of this factor makes the calculation of analysis easier. It is:

$$\text{Required substance (g)} = [\text{weighed out portion}] \times F$$

or

$$\text{Required substance \%} = \frac{\text{Weighed out portion} \times F \times 100}{\text{Initial weight}}$$

By using logarithm the calculation is further simplified

$$\text{Required substance (g)} = \log \text{final weight} + \log F$$

or

$$\text{Required substance \%} = \log \text{final weight} + \log F + DE \log \text{initial weight}$$

The application of these calculations factors not only simplifies the calculation in case of gravimetric analytical work but they can also be used in case of other stoichiometric conversions. The proper choice of method is made through observations.

The described methods are written in the form of gist. In many cases hints are enough; in others the search for the corresponding methods is made easy, by referring the original literature.

Required	Given	Factor	log Factor	DE log Factor	Remarks
Ag	Ag Br	0.5745	75 925	24 075	**AgCl:** Ag is precipitated in weakly HNO_3 acid solution with NaCl solution of about 60° and allowed to stand in dark for a long time and filtered through A-1 filter crucible. A large excess of the precipitation medium is to be avoided. The crucible with its contents is dried in an oven at 130°–150°C and finally weighed. The precipitation of AgBr follows analogously to KBr.
	Ag Cl	0.7526	87 658	12 342	
	Ag_2S	0.8706	93 982	06 018	
$AgNO_3$	AgBr	0.9046	95 647	04 353	
	AgCl	1.1852	07 380	92 620	
Al	Al_2O_3	0.5291	72 357	27 643	
	$Al(Oxin)_3$	0.05871	76 868	23 132	
Al_2O_3	$Al(Oxin)_3$	0.1109	04 512	95 488	**Al_2O_3 :** Al is precipitated as Al (OH) from hydrochloric acid solution with NH_3. Thereafter it is boiled for a short time and filtered through a suitable Whatman filter. The wash water should contain one drop of NH_3 and some NH_4NO_3. The filter paper with its contents is incinerated in a porcelain crucible carefully and then ignited strongly. Finally weighed as Al_2O_3.
$Al_2(SO_4)_3$	Al_2O_3	3.3562	52 584	47 416	
	SO_3	1.4245	15 367	84 633	
$Al_2(SO_4)_3 \cdot 18\,H_2O$	Al_2O_3	6.5377	81 542	18 458	
	SO_3	2.7745	44 318	55 682	
As	As_2O_3	0.7574	87 930	12 070	
	As_2S_3	0.6090	78 460	21 540	**Al (Oxin)$_3$:** Al is precipitated in a weakly acidified acetic acid and buffered with sodium acetate solution at about 70°C. The precipitate after about 1 hour is filtered through a filter crucible and rinsed with cold water. Dry in an oven at 120°.
	As_2S_5	0.4831	68 400	31 600	
	$(NH_4MgAsO_4)_2 \cdot H_2O$	0.3937	59 516	40 484	
	$Mg_2As_2O_7$	0.4826	68 357	31 643	
As_2O_3	As	1.3200	12 070	87 930	
	As_2S_3	0.8041	90 530	09 470	For precipitation a 5% solution of 8-hydroxyquinoline in 12% acetic acid is employed.
	As_2S_5	0.6378	80 470	19 530	
	$(NH_4MgAsO_4)_2 \cdot H_2O$	0.5198	71 586	28 414	
	$Mg_2As_2O_7$	0.6372	80 427	19 573	

Gravimetric analytical methods

Required substance	Found as	Factor	log Factor	DE log Factor	Remarks
As_2O_5	As_2S_3	0.9342	97042	02958	$Mg_2As_2O_7$: The precipitation follows in ammonical solution with magnesia mixture as NH_4MgAsO_4. As must be present as As^{5+} As^{3+} can be oxidized in ammonical solution with H_2O_2. The precipitate filtered suitably through a filter crucible. After drying it is ignited in an electric furnace at 900°C weighed as: $Mg_2As_2O_7$. To weigh as (NH_4MgAsO_4). H_2O is unsuitable.
	As_2S_5	0.7410	86982	13018	
	$(NH_4MgAsO_4)_2 \cdot H_2O$	0.6039	78098	21902	
	$Mg_2As_2O_7$	0.7403	86939	13061	
AsO_3^{3-}	As_2S_3	0.9992	99965	00035	
	As_2S_5	0.7926	89905	10095	
	$(NH_4MgAsO_4)_2 \cdot H_2O$	0.6460	81021	18979	
	$Mg_2As_2O_7$	0.7918	89862	10138	
AsO_4^{3-}	As_2S_3	1.1293	05279	94721	
	As_2S_5	0.8958	95221	04779	
	$(NH_4MgAsO_4)_2 \cdot H_2O$	0.7301	86335	13665	
	$Mg_2As_2O_7$	0.8949	95176	04824	
Au	AuCN	0.8834	94617	05383	As_2S_3: As^{3+} can be precipitated as sulphide from a strongly acidified hydrochloric acid solution at boiling temperature by passing of H_2S gas. For qualitative determinations precipitation, with magnesia mixture is better.
	$AuCl_3$	0.6496	81249	18751	
	$KAu(CN)_2$	0.6840	83493	16507	
	$KAu(CN)_4 \cdot H_2O$	0.5503	74047	25953	
B	B_2O_3	0.3107	49240	50760	
	$Na_2B_4O_7 \cdot 10H_2O$	0.1135	05489	94511	
BO_2^-	B_2O_3	1.230	08982	91018	$BaCO_3$: Precipitation in ammonical solution in presence of NH_4Cl with $(NH_4)_2CO_3$. For quantitative determinations the precipitation of $BaSO_4$ is preferred.
BO_3^{3-}	B_2O_3	1.689	22770	77230	
$B_4O_7^{2-}$	B_2O_3	1.115	04723	95277	
Ba	$BaCO_3$	0.6960	84258	15742	
	$BaCl_2 \cdot 2H_2O$	0.5623	74993	25007	
	$BaCrO_4$	0.5421	73410	26590	
	$BaSO_4$	0.5885	76971	23029	$BaSO_4$: In dilute hydrochloric acid solution at boiling temperature the Ba as $BaSO_4$ is precipitated slowly with an equally hot dilute H_2SO_4. The solution is kept for a long time in hot condition. After first allowing to stand for an hour, it is filtered through a Whatman filter paper No. 41. The precipitate is washed with hot water and together with the filter incinerated in a porcelain crucible Ignition temperature $\approx$ 850°C.
$BaCO_3$	$BaSO_4$	0.8455	92712	07288	
$BaCl_2$	$BaSO_4$	0.8923	95049	04951	
$BaCl_2 \cdot 2H_2O$	$BaSO_4$	1.047	01978	98022	
BaF_2	$BaSO_4$	0.7512	87578	12422	
$Ba(NO_3)_2$	$BaSO_4$	1.120	04912	95088	
BaO	$BaCO_3$	0.7770	89043	10957	
	$BaCrO_4$	0.6053	78195	21805	$BaCrO_4$: The hydrochloric acid solution is just neutralized with NH_3 and buffered with ammonium acetate. Thereafter heated to boiling and Na is precipitated with $(NH_4)_2Cr_2O_7$. The precipitate is filtered through a filter crucible and after washing with a strongly diluted ammonium acetate solution, is ignited weakly. Weighed as: $BaCrO_4$.
	$Ba(OH)_2$	0.8949	95176	04824	
	$BaSO_4$	0.6570	81756	18244	
BaO_2	$BaSO_4$	0.7255	86066	13934	
$Ba(OH)_2$	$BaSO_4$	0.7342	86580	13410	
BaS	$BaSO_4$	0.7258	86083	13917	
Be	BeO	0.3603	55671	44329	
	$Be_2P_2O_7$	0.0939	97264	02736	
BeO	$Be_2P_2O_7$	0,2606	41593	58407	

Gravimetric analytical methods

Required substance	Found as	Factor	log Factor	DE log Factor	Remarks
Bi	Bi_2O_3	0.8970	95279	04721	Bi_2O_3: From a nearly neutral hydrochloric acid solution Bi in presence of sodium formate is precipitated as basic salt and filtered through a filter crucible. By moderately strong ignition Bi_2O_3 results.
	BiOCl	0.8024	90441	09559	
	$Bi(Oxin)_3$	0.3258	51297	48703	
	$Bi(Oxin)_3 \cdot H_2O$	0.3171	50096	49904	
	$BiPO_4$	0.6876	83731	16269	
	Bi_2S_3	0.8129	91005	08995	
					$BiPO_4$: From a hot nitric acid solution Bi is precipitated as $BiPO_4$ with the help of sodium phosphate. Filter through a filter crucible, first dried and finally ignited.
Br	$Ag\,Br$	0.4255	62894	37106	
	BrO_3	0.6248	79572	20428	
	KBr	0.6715	82702	17298	
	$Na\,Br$	0.7766	89018	10982	
C	CO_2	0.2729	43602	56398	$AgBr$: See AgCl.
	$CaCO_3$	0.1200	07919	92081	
	CaO	0.2142	33077	66923	CO_2: By burning of C-containing substances, by decomposition of carbonates with acids or by cracking from carbonates at high temperatures CO_2 results. It can be absorbed in a suitable absorption medium and weighed. Refer also elemental analysis and gas analysis.
	$BaCO_3$	0.0609	78430	21570	
CH_3O	$Ag\,I$	0.1322	12117	87883	
C_2H_5O	$Ag\,I$	0.1919	28310	71690	
CN	Ag	0.2412	38235	61765	
	$Ag\,CN$	0.1943	28851	71149	
CNS	$Ag\,CNS$	0.3502	54404	45596	
	$Cu\,CNS$	0.4776	67903	32097	CaO: $CaCO_3$ by igniting at a temperature above 900°C leaves CaO.
	$Fe(CNS)_3$	0.7543	87926	12074	
$CO(NH_2)_2$	N	2.1438	33117	66883	$AgSCN$: Rhodanide can be precipitated from a nitric acid solution with $AgNO_3$. The precipitation takes place under usual conditions as required for AgCl. Chloride should not be present.
	NH_3	1.7632	24629	75371	
CO_2	$Ca\,CO_3$	0.4397	64317	35683	
	$Ca\,O$	0.7848	89475	10525	
	$Mg\,O$	1.092	03804	96196	
	Na_2CO_3	0.4152	61828	38172	NH_3: By determination of N_2 according to Kjeldahl method.
	$Na_2CO_3 \cdot 10\,H_2O$	0.1538	18696	81304	
	$Na\,HCO_3$	0.5239	71923	28077	
CO_3^{2-}	CO_2	1.364	13467	86533	Na_2CO_3: Refer also CO_2 and CaO
$C_2O_4^{2-}$	CO_2	1.0000	0000	0000	$CaC_2O_4H_2O$: The hydrochloric acid solution is made weakly ammoniacal with NH_3 and heated to boiling. From the hot solution Ca is slowly precipitated with a similarly hot solution of ammonium oxalate and the precipitate is allowed to stand for a long time. After filtration through a sintered glass crucible, the precipitate is washed with a small amount of warm water and finally with acetone. The precipitate is then dried in an oven 110°C. Weighed as $CaC_2O_4 \cdot H_2O$.
	$Ca\,O$	1.570	19578	80422	
Ca	$CaCO_3$	0.4004	60254	39746	
	$CaC_2O_4 \cdot H_2O$	0.2743	43825	56175	
	CaF_2	0.5133	71039	28961	
	$Ca\,O$	0.7147	85412	14588	
	$CaSO_4$	0.2944	46894	53106	
CaC_2	$Ca\,O$	1.1430	05805	94195	
$Ca(CN)_2$	$Ca\,O$	1.4284	15486	84514	
	N	2.859	45622	54378	

Gravimetric analytical methods					
Required substance	Found as	Factor	log Factor	DE log Factor	Remarks
$CaCO_3$	CO_2	2.274	35684	64316	**$CaCO_3$**: The precipitation and filtration follows as like the final weighing as $CaC_2O_4 \cdot H_2O$. The filter crucible is then placed in a protective crucible after drying to be ignited at 450°–500°C Final weight as $CaCO_3$.
	$CaC_2O_4 \cdot H_2O$	0.6850	83571	16429	
	CaO	1.785	25158	74842	
	$CaSO_4$	0.7352	86638	13362	
$CaCl_2$	$CaCl_2 \cdot 6 H_2O$	0.5066	70468	29532	
	CaO	1.979	29647	70353	
	Cl	1.565	19455	80545	
CaF_2	CaO	1.392	14373	85627	**CaO**: Precipitation of Ca as $CaC_2O_4 \cdot H_2O$ as above. The precipitate is filtered through a Whatman filter paper No. 1 and washed with warm water containing $(NH_4)_2C_2O_4$. The precipitate with the filter is incinerated in a porcelain crucible and ignited at 1000°C. The left out CaO must be cooled in a covered crucible and brought to weighing. For quantitative determinations only smaller final weight is suited.
$Ca(NO_3)_2$	CaO	2.926	46630	53370	
CaO	Ca	1.399	14588	85412	
	CO_2	1.274	10525	89475	
	CaC_2	0.8750	94202	05798	
	$CaCN_2$	0.7000	84512.	15488	
	$CaCO_3$	0.5603	74842	25158	
	$CaC_2O_4 \cdot H_2O$	0.3838	58413	41587	
	CaF_2	0.7182	85627	14373	
	$[Ca_3(PO_4)_2]_3 Ca(OH)_2$	0.5549	74679	25321	
	$CaSO_4$	0.4119	61480	38520	
	$CaSO_4 \cdot 2 H_2O$	0.3257	51283	48717	
$Ca(OH)_2$	CaO	1.321	12101	87899	**$CaSO_4$**: The CaO obtained as above can be converted to $CaSO_4$ by boiling with H_2SO_4.
$Ca_3(PO_4)_2$	CaO	1.844	26570	73430	
	$Mg_2P_2O_7$	1.394	14411	85589	
$[Ca_3(PO_4)_2]_3$ $Ca(OH)_2$	CaO	1.792	25323	74677	
	$Mg_2P_2O_7$	1.504	17740	82260	
$Ca(HSO_3)_2$	CaO	3.6060	55703	44297	
$CaSO_4$	$BaSO_4$	0.5832	76585	23415	**$Cd(Oxin)_2$**: The precipitation with 8-hydroxy-quinoline follows similar to $Mg(Oxin)_2$
	CaO	2.428	38518	61482	
	SO_3	1.700	23056	76944	
$CaSO_4 \cdot 2 H_2O$	CaO	3.07	48715	51285	
Cd	CdO	0.8754	94220	05780	
	$Cd(Oxin)_2$	0.2805	44796	55204	
	$Cd(Oxin)_2 \cdot 1.5 H_2O$	0.2628	41961	58039	**$CdSO_4$**: The Cd is precipitated as sulphide from a dilute sulphuric acid solution. CdS is filtered and dissolved in 10% H_2SO_4. The solution is evaporated to drying in a porcelain crucible and finally ignited weakly. Weighed as $CdSO_4$.
	$Cd_2P_2O_7$	0.5638	75110	24890	
	CdS	0.7781	89101	10899	
	$CdSO_4$	0.5392	73175	26825	
CdO	$Cd_2P_2O_7$	0.6440	80890	19110	
	CdS	0.8888	94880	05120	
	$CdSO_4$	0.6160	78954	21046	
Ce	Ce_2O_3	0.8538	93134	06866	
	CeO_2	0.8141	91067	08933	
	$Ce_2(C_2O_4)_3$	0.5149	71170	28830	

Gravimetric analytical methods

Required substance	Found as	Factor	log Factor	DE log Factor	Remarks
Cl^-	Ag	0.3287	51676	48324	**$KClO_4$**: An aqueous sulphate free
	Ag Cl	0.2474	39334	60666	solution is evaporated with $HClO_4$
	Na Cl	0.6066	78293	21707	in an evaporating dish until white
ClO_3^-	Ag Cl	0.5822	76510	23490	clouds start to form. (Do not dry completely.) The residue containing
ClO_4^-	Ag Cl	0.6939	84128	15872	$HClO_4$ after cooling is titrated
	K Cl	1.3340	12515	87485	with alcohol and filtered through
	Na Cl	1.7017	23087	76913	a sintered glass crucible. After
	$KClO_4$	0.7178	85601	14399	being washed with the same mixtur (filtered liquid) it is rinsed with a
Co	$Co(\alpha\text{-Nitroso-}\beta\text{-naphtol})_3 \cdot 2H_2O$	0.0964	98404	01596	little pure absolute alcohol. The crucible with its contents is dried
	$Co(Oxin)_2 \cdot 2H_2O$	0.1538	18688	81312	in an oven at 120°C and weighed.
	Co_3O_4	0.7343	86584	13416	Weighed as $KClO_4$.
	$Co_2P_2O_7$	0.4039	60628	39372	
	$Co SO_4$	0.3802	58006	41994	**$Co(\alpha\text{-nitroso-}\beta\text{-naphthol})_3$**
CoO	Co	1.271	10430	89570	$\cdot 2H_2O$: Co is precipitated from an acetic acid solution with a similar
	$Co(Oxin)_2 \cdot 2H_2O$	0.1955	29118	70882	acetic acid solution of α-nitroso-
	$Co_2P_2O_7$	0.5136	71061	28939	β-naphthol under warm conditions
	$Co SO_4$	0.4835	68436	31564	and filtered through a filter crucible.
Cr	$Ba Cr O_4$	0.2053	31233	68767	The precipitate after rinsing with
	Cr_2O_3	0.6843	83522	16478	dilute acetic acid is dried in an
	$Cr PO_4$	0.3538	54880	45120	oven and finally weighed.
	$K_2Cr O_4$	0.2678	42782	57218	
	$K_2Cr_2O_7$	0.3536	54845	45155	**Co_3O_4**: By precipitating with
	$Pb Cr O_4$	0.1609	20659	79341	α-nitroso-β-naphthol the precipitat
Cr_2O_3	$Ba Cr O_4$	0.3000	47711	52289	is filtered and after washing is incinerated in α-porcelain crucible
	Cr	1.461	16478	83522	at a low temperature; thereafter
	$Pb Cr O_4$	0.2352	37137	62863	it is ignited strongly.
$Cr O_3$	$Ba Cr O_4$	0.3947	59628	40372	**$BaCrO_4$**: Refer precipitation of
	Cr_2O_3	1.316	11917	88083	Ba as $BaCrO_4$.
	$Pb Cr O_4$	0.3094	49054	50946	**Cr_2O_3**: Cr^{3+} like Al is precipitated
Cr_2O_7	$Ba Cr O_4$	0.4263	62970	37030	as hydroxide and by strong ignition
	Cr_2O_3	1.421	15259	84741	is converted to Cr_2O_3.
$Cr O_4^{2-}$	$Ba Cr O_4$	0.5479	66074	33926	**$CuSCN$**: To a neutral or weakly
	Cr_2O_7	1.526	18363	81637	acidic solution treated with an
	$Pb Cr O_4$	0.3589	55500	44500	excess of H_2SO_3, NH_4SCN is added
Cs	Cs_2SO_4	0.7345	86601	13399	dropwise with the help of a dropper to just enough excess. It is agitated
Cu	Cu SCN	0.5224	71801	28199	for a long time till the precipitate
	Cu O	0.7988	90246	09754	is white and finally settled down.
	Cu_2S	0.7985	90229	09771	The precipitate is filtered through
	Cu S	0.6646	82257	17743	a sintered glass crucible, washed
	$Cu(Oxin)_2$	0.1806	25669	74331	with cold water. Finally dried
	$Cu(C_{14}H_{11}O_2N)$ (Cupron)	0.2200	34241	65759	at 110°–120°C and weighed as $CuSCN$.

Gravimetric analytical methods

Required substance	Found as	Factor	log Factor	DE log Factor	Remarks
$CuCl_2$	CuO	1.690	22798	77202	**CuO:** Cu is precipitated from a hydrochloric acid solution as sulphide by passing of H_2S gas. After filtering through a Whatman filter No. 1, it is washed with water containing acidified H_2S. The filter is incinerated in a porcelain crucible and the residue is ignited strongly. Final weight: CuO
$CuFeS_2$	Cu_2S	2.306	36290	63710	
Cu_2O	CuO	0.8994	95396	04604	
CuO	Cu	1.252	09754	90246	
	$CuCNS$	0.6539	81554	18446	
	Cu_2S	0.9996	99983	00017	
$CuSO_4$	Cu	2.512	40002	59998	**Cu_2S**: By precipitation with H_2S, the resulting CuS is ignited with a bunsen burner in a rose crucible by simultaneously conducting H_2S gas through it. Thereafter it is cooled by circulating a current of hydrogen and weighed as Cu_2S.
	CuO	2.0066	30246	69754	
$CuSO_4 \cdot 5H_2O$	Cu	3.930	59437	40563	
	$CuCNS$	2.053	31235	68765	
	CuO	3.139	49680	50320	
	Cu_2S	3.138	49663	50337	
F	CaF_2	0.4867	68724	31276	**$Cu(Oxin)_2$**: The precipitation with 8-hydroxyquinoline follows as in case of precipitation of Mg.
	$CaSO_4$	0.2791	44579	55421	
	HF	0.9496	97754	02246	
	H_2SiF_6	0.7914	89839	10161	
	SiF_4	0.7301	86340	13660	
Fe	FeO	0.7773	89059	10941	**Cu (electrolytic):** From a strongly acidified nitric acid solution containing NH_4NO_3 the Cu is deposited on a platinum net cathode current strength 3–4 amp. at $\approx \Delta V$ in a stirred electrolyte. For decomposing of the NO_2^- urea is added. At the end of the electrolysis the net is rinsed with alcohol and carefully dried. Final weight: Cu
	Fe_2O_3	0.6994	84475	15525	
	Fe_3O_4	0.7236	88949	11051	
	$Fe(Oxin)_3$	0.1144	05835	94165	
$FeCl_2$	Fe	2.270	35596	64404	
	Fe_2O_3	1.5875	20071	79929	
	$FeCl_2 \cdot 4H_2O$	0.6376	80452	19548	
$FeCl_3$	Fe	2.905	46308	53692	
	Fe_2O_3	2.032	30783	69217	
	$FeCl_3 \cdot 6H_2O$	0.6001	77823	22177	
FeO	Fe	1.287	10941	89059	**Fe_2O_3:** The hydrochloric acid solution is treated with H_2O_2 for oxidation of iron and the excess is boiled off. From the boiling hot solution Fe is precipitated as $Fe(OH)_3$ with NH_3 and after settling for a short time, is filtered through a Whatman filter. The precipitate is washed with hot water and with its contents the filter placed in a porcelain crucible is incinerated; it is finally ignited with a bunsen burner. Final weight: Fe_2O_3.
	Fe_2O_3	0.8998	95416	04584	
Fe_2O_3	Fe	1.430	15525	84475	
	FeO	1.111	04584	85416	
	$FePO_4$	0.5294	72378	27622	
Fe_3O_4	Fe	1.3820	14051	85949	
	Fe_2O_3	0.9666	98525	01475	
FeS_2	Fe_2O_3	1.503	17680	82320	
$FeSO_4 \cdot 7H_2O$	Fe	4.978	69705	30295	
	Fe_2O_3	3.4818	54181	45819	
$Fe_2(SO_4)_3$	Fe	3.579	55390	44610	
	Fe_2O_3	2.504	39865	60135	

Gravimetric analytical methods

Required substance	Found as	Factor	log Factor	DE log Factor	Remarks
H	H_2O	0.1119	04884	95116	
HBO_2	B_2O_3	1.259	09994	90006	
H_3BO_3	B_2O_3	1.776	24944	75056	
HBr	AgBr	0.4309	63439	36561	**Ag Br**: Refer Br.
HCN	Ag	0.2505	39885	60115	
	AgCN	0.2018	30502	69498	
H_2CN_2	N	1.501	17626	82374	
HCO_3	CO_2	1.386	14190	85810	**CO_2**: Refer C.
H_2CO_3	CO_2	1.4094	14902	85098	
$H_2C_2O_4$	CO_2	1.023	00984	99016	
	CaO	1.606	20562	79438	**CaO**: As in case of Ca.
$H_2C_4H_4O_6$	$CaC_4H_4O_6 \cdot 4\,H_2O$	0.5768	76102	23898	
HCl	AgCl	0.2544	40552	59448	**AgCl**: Refer Ag.
	Cl	1.028	01218	98782	
$HClO_4$	AgCl	0.7009	84565	15435	
HF	CaF_2	0.5126	70974	29026	
	$CaSO_4$	0.2940	46829	53171	
	SiF_4	0.7690	88590	11410	
HI	AgI	0.5448	73626	26374	**AgI**: The precipitation follows as in case of AgCl.
HNO_3	NO	2.100	32221	67779	**$Mg_2P_2O_7$**: The hydrochloric acid solution is treated with about 5 g
	N_2O_5	1.167	06699	93301	NH_4Cl and 2 g $(NH_4)_2$ HPO_4 and boiled. To the boiling hot solution
H_3PO_2	Hg_2Cl_2	0.0699	84444	15556	NH_3 is added dropwise till a precipitate begins to fall. At that time
	$Mg_2P_2O_7$	0.5930	77305	22695	the solution is strongly stirred with a glass rod until the precipitate
H_3PO_3	Hg_2Cl_2	0.1737	23977	76023	becomes crystalline. Thereafter NH addition is continued till it
	$Mg_2P_2O_7$	0.7368	86733	13267	becomes red as phenophthalein. After allowing to stand for an hour
H_3PO_4	$Mg_2P_2O_7$	0.8805	94475	05525	the precipitate is filtered through a Whatman filter and washed with
	P_2O_5	1.3807	14011	85989	warm water containing NH_3- and NH_4NO_3. It is carefully incinerated
H_2S	$BaSO_4$	0.1460	16438	83562	in a porcelain crucible and is finally ignited in an electric oven at
	Ag_2S	0.1375	13838	86162	900° Final weight: $Mg_2P_2O_7$
H_2SO_3	$BaSO_4$	0.3516	54610	45390	**$BaSO_4$**: The sulphide sulphur is
	SO_2	1.2812	10762	89238	converted to sulphate by alkaline oxidation melt with Na_2CO_3-
H_2SO_4	$BaSO_4$	0.4202	62344	37656	Na_2O_2. After extraction of melt the sulphate is precipitated as $BaSO_4$ with $BaCl_2$. Refer also Ba.

Gravimetric analytical methods

Required substance	Found as	Factor	log Factor	DE log Factor	Remarks
H_2SiF_6	CaF_2	0,6150	78885	21115	
	$CaSO_4$	0,3527	54738	45262	SiO_2: Silicic acid is converted into insoluble SiO_2 by evaporation with HCl several times and finally at the end roasting at 130°C. The residue is taken up in dilute hot HCl and filtered through a Whatman filter No. 1. After washing with strongly diluted hot HCl and finally with hot water, it is incinerated and strongly ignited. Final weight: SiO_2. Initially the SiO_2 is not pure white, hence it can be evaporated with HF acid in the presence of H_2SO_4 in a platinum crucible, thereby it forms gaseous SiF_4. After evaporation, it is once again ignited and reweighed. Difference SiO_2. Si-alloys must be dissolved through oxidation before evaporation with HCl. SiO_2 containing minerals are dissolved by alkalimelt and there-after evaporated with HCl.
H_2SiO_3	SiO_2	1.300	11388	88612	
Hg	Hg_2Cl_2	0,8498	92932	07068	
	HgS	0,8622	93561	06439	
$HgCl_2$	HgS	1.167	06706	93294	
Hg_2Cl_2	HgS	1.0143	00628	99372	
I	Ag	1.176	07055	92945	
	AgI	0,5406	73285	26715	
	AgCl	0,8854	94713	05287	
IO_3	AgI	0,7450	87214	12786	
K	KCl	0,5244	71969	28031	
	$KClO_4$	0,2822	45055	54945	
	K_2O	0,8302	91916	08084	
	K_2SO_4	0,4487	65198	34802	
	K_2PtCl_6	0,1603	20486	79514	
KCN	HCN	2.409	38190	61810	
	AgCN	0,4863	68692	31308	
KCl	$KClO_4$	0,5381	73089	26911	$KClO_4$: Refer Cl.
	K_2O	1.583	19950	80050	
	K_2PtCl_6	0,3056	48515	51485	
$KHC_4H_4O_6$	$CaC_4H_4O_6 \cdot 4 H_2O$	0,7232	85925	14075	
K_2CO_3	CO_2	3.1404	49698	50302	
$KMnO_4$	Mn	2.877	45888	54112	
	MnO_2	1.8178	25955	74045	
	O_2	3.951	59671	40329	
KNO_3	N_2O_5	1.872	27232	72768	
K_2O	KCl	0,6317	80053	19947	
	$KClO_4$	0,3399	53139	46861	
	K_2PtCl_6	0,1931	28570	71430	
	K_2SO_4	0,5405	73282	26718	
$K_2O \cdot Al_2O_3 \cdot 6 SiO_2$	Al_2O_3	5.460	73719	26281	Al_2O_3: Refer Al.
	KCl	3.733	57210	42790	
	K_2O	5.910	77157	22843	
	K_2SO_4	3.194	50439	49561	
K_2SO_4	$BaSO_4$	0,7466	87307	12693	$BaSO_4$: Refer Ba.
	KCl	1.1687	06771	93229	
	K_2O	1.850	26718	73282	

Gravimetric analytical methods

Required substance	Found	Factor	log Factor	DE log Factor	Remarks
Li	$Li_2 CO_3$	0.1878	27380	72620	
	$Li Cl$	0.1637	21402	78598	
	$Li_2 O$	0.4645	66701	33299	
	$Li_3 PO_4$	0.1798	25479	74521	$Mg (Oxin)_2$: The hydrochloric
	$Li_2 SO_4$	0.1262	10121	89879	acid solution is treated with NH_4Cl
					and thereafter strongly made
$Li_2 O$	$LiCl$	0.3524	54701	45299	ammoniacal. After warming to
	$Li_3 PO_4$	0.3871	58778	41222	about 70°C, Mg is precipitated
	$Li_2 SO_4$	0.2718	43420	56580	with a 4% alcoholic solution of
					oxine. A larger excess of the pre-
Mg	$Mg O$	0.6032	78044	21956	cipitation medium is to be avoided.
	$Mg (Oxin)_2$	0.0778	89097	10903	After cooling it is filtered in a filter
	$Mg (Oxin)_2 \cdot 2 H_2O$	0.0698	84359	15641	crucible and washed with hot water
	$Mg_2 P_2 O_7$	0.2185	33948	66052	containing NH_3. Finally the crucible
	$NH_4 Mg PO_4 \cdot 6 H_2O$	0.0991	99602	00398	is dried at 130°C.
					Final weight: $Mg (Oxin)_2$.
$Mg CO_3$	$Mg CO_3 \cdot H_2O$	0.8240	91591	08409	
	$Mg O$	2.092	32046	67954	
	$Mg_2 P_2 O_7$	0.7577	87950	12050	$Mg_2P_2O_7$: Refer H_3PO_4.
$Mg Cl_2$	Cl	1.343	12804	87196	
	$Mg O$	2.362	37325	62675	
	$Mg_2 P_2 O_7$	0.8556	93229	06771	
MgO	CO_2	0.9161	96196	03804	CO_2: See under C.
	H_2O	2.238	34987	65013	SiO_2: Refer H_2SiO_3.
	Mg	1.658	21956	78044	$Mn_2P_2O_7$: The weakly acid
	$Mg_2 P_2 O_7$	0.3623	55904	44096	solution is treated with abundant
	$Mg SO_4$	0.3349	52494	47506	NH_4Cl and 5–10 ml. of cold
	$NH_4 Mg PO_4 \cdot 6 H_2O$	0.1643	21558	78442	saturated solution of sodium
					phosphate. The solution is carefully
$Mg(OH)_2$	$Mg O$	1.447	16045	83955	made ammoniacal and heated to
					boiling for a long time till the
$Mg SO_4$	$Mg O$	2.986	47506	52494	precipitate becomes crystalline.
	$Mg_2 P_2 O_7$	1.082	03410	96590	After filteration through a porcelain
					filter crucible and washed with
$Mg Si O_3$	$Mg O$	2.490	39625	60375	water containing NH_3, it is initially
	$Si O_2$	1.671	22297	77703	dried and ignited finally in an
					electric furnace.
$Mg_2 Si O_4$	$Mg O$	3.490	54286	45714	Final weight: $Mn_2P_2O_7$.
	$Si O_2$	2.342	36958	63042	$MnSO_4$: From a weakly ammoniacal
					solution containing. NH_4Cl Mn is
Mn	$Mn O$	0.7745	88900	11100	precipitated as MnS in cold freshly
	$Mn_3 O_4$	0.7203	85751	14249	prepared ammonium sulphide
	$Mn_2 P_2 O_7$	0.3871	58786	41214	solution. After filteration and
	$Mn S$	0.6314	80032	19948	washing with water containing
	$Mn SO_4$	0.3638	56088	43912	$(NH_4)_2 S$ it is heated with a bunsen
					burner to Mn_3SO_4. The so ob-
$Mn CO_3$	$Mn_3 O_4$	1.507	17813	82187	tained oxide in the porcelain
					crucible is dissolved in dil. H_2SO_4
					and small amount of H_2O_2. After
$Mn O$	Mn	1.291	11100	88900	evaporation of the excess H_2SO_4,
	$Mn_3 O_4$	0.9301	96851	03149	it is ignited in an electric furnace
	MnS	0.8153	91132	08868	at 450°–500°C.
					Final weight: $MnSO_4$.

equired	Given	Factor	log Factor	DE log Factor	Remarks
MnO_2	Mn	1.583	19933	80067	
$MnSO_4$	Mn	2.7487	43912	56088	
Mo	MoO_3	0.6665	82383	17617	MoO_3: The sulphuric acid solution
	MoS_2	0.5994	77770	22230	is saturated with H_2S in cold in a
	$PbMoO_4$	0.2613	41719	58281	small pressure bottle. Thereafter
					the closed bottle is suspended in
MoO_4	MoO_3	1.111	04578	95422	a water bath and heated so long
	$PbMoO_4$	0.4357	63914	36086	till the liberated MoS_3 has settled
					completely. MoS_3 thus obtained
N	NH_2	0.8742	94161	05839	is filtered through a filter crucible
	NH_3	0.8225	91512	08488	and washed first with dil. H_2SO_4
	NH_4	0.1428	15468	84532	and finally with alcohol. After
	NH_4Cl	0.2619	41806	58194	drying, the covered filter crucible is
	$(NH_4)_2PtCl_6$	0.0629	79834	20166	carefully heated till smell of SO_2
					is no longer detected. Finally the
NH_3	N	1.2159	08488	91512	cover is removed and weakly
	NH_4	0.9441	97502	02498	ignited.
	NH_4Cl	0.3184	50294	49706	Final weight MoO_3.
NH_4	N	1.2878	10986	89014	
	NH_3	1.059	02498	97502	
	NH_4Cl	0.3372	52792	47208	
NH_4Cl	N	3.8189	58194	41806	
	NH_3	3.1409	49706	50294	
	$(NH_4)_2PtCl_6$	0.2400	38028	61972	
NO_2	N_2O_3	1.210	08295	91705	
N_2O_3	NO_2	0.8261	91705	08295	
NO_3	NO	2.066	31521	68479	
	N_2O_5	1.148	05999	94001	
N_2O_5	KNO_3	0.5342	72793	27207	NaCl: In a hydrochloric acid
	$NaNO_3$	0.6354	80305	19695	solution free from $SO_4{}^{2-}$ Na can
	NO	1.800	25522	74478	be obtained as NaCl by evaporation
	NO_3	0.8710	94001	05999	and weakly igniting in a platinum
					crucible.
Na	Cl	0.6486	81186	18814	Na_2SO_4: Na is converted suitably
	NaCl	0.3934	59479	40521	to Na_2SO_4 by evaporation with
	$NaClO_4$	0.1878	27361	72639	H_2SO_4 in a platinum shell and
	$NaHCO_3$	0.2737	43723	56277	it is hardly volatile as NaCl. The
	Na_2CO_3	0.4339	63731	36269	determination is, however, possible
	Na_2O	0.7419	87033	12967	only when Na_2SO_4 results quantita-
	NaOH	0.5748	75951	24049	tively by evaporation with H_2SO_4.
	Na_2SO_4	0.3238	51015	48985	
	$NaMg(UO_2)_3 \cdot$				**Ag Br:** Refer AgCl
	$(CH_3COO)_9 \cdot 6\,H_2O$	0.01536	18634	81366	
	$Na\,Zn(UO_2)_3 \cdot$				**AgI:** See under AgCl
	$(CH_3COO)_9 \cdot 6\,H_2O$	0.01495	17457	82543	

Gravimetric analytical methods

Required	Given	Factor	log Factor	DE log Factor	Remarks
NaBr	AgBr	0.5480	73876	26124	
Na_2CO_3	CO_2	2.409	38172	61828	CO_2: See under C
	$NaHCO_3$	0.6309	79992	20008	
	NaOH	1.325	12220	87780	
	Na_2SO_4	0.7462	87285	12715	
NaCl	Cl	1.649	21707	78293	
	AgCl	0.4078	61041	38959	**AgCl**: Refer Ag
	$NaClO_4$	0.4774	67882	32118	
	Na_2SO_4	0.8229	91537	08463	
$NaHCO_3$	CO_2	1.909	28077	71923	
	Na_2CO_3	1.585	20008	79992	
Na I	Ag I	0.6385	80512	19488	
$NaNO_3$	N_2O_5	1.574	19692	80308	
Na_2O	CO_2	1.4086	14870	85130	
	N_2O_5	0.5739	75877	24123	
	NaCl	0.5303	72446	27554	
	Na_2SO_4	0.4364	63983	36017	
	SO_3	0.7743	88881	11119	
	SiO_2	1.032	01346	98654	
Na_3PO_4	PO_4	1.726	23708	76292	
	P_2O_5	2.310	36359	63641	
$Na_3PO_4 \cdot 12 H_2O$	P_2O_5	5.356	72883	27117	
$Na_2SO_3 \cdot 7 H_2O$	$BaSO_4$	1.080	03350	96650	
Na_2SO_4	$BaSO_4$	0.6086	78425	21575	
	$Na_2SO_4 \cdot 10 H_2O$	0.4409	64429	35571	
	NaCl	1.215	08463	91537	
	SO_3	1.774	24898	75102	
Ni	$Ni(C_2H_5N_4O_2)_2(Di)$	0.2250	35224	64776	$NiC_8H_{14}N_4O_4$ (**Gly.**): The dilute weakly acid solution is heated to boiling in a beaker. Thereafter for every 10 mg. of Ni, 6 ml. of 10% alcoholic solution of dimethyl glyoxime is added and the solution is made weakly ammoniacal. After settling down, the precipitate is filtered through a glass sintered crucible and washed with warm water. Finally the crucible with its contents is dried in an oven at $\approx$ 120°C and weighed.
	$NiC_8H_{14}N_4O_4$ (Gly.)	0.2032	30790	69210	
	$Ni(Oxin)_2 \cdot 2 H_2O$	0.1533	18546	81454	
	NiO	0.7858	89533	10467	
	$Ni_2P_2O_7$	0.4030	60530	39470	
	$NiSO_4$	0.3793	57900	42100	
NiO	Ni	1.273	10470	89530	
$NiSO_4 \cdot 7 H_2O$	NiO	3.760	57524	42476	
	$NiSO_4$	1.815	25885	74115	

Gravimetric analytical methods

Required	Given	Factor	log Factor	DE log Factor	Remarks
O	H_2O	0.8881	94847	05153	
O_2	$KMnO_4$	0.2531	40329	59671	
OCH_3	Ag I	0.1322	12114	87886	
OC_2H_5	Ag I	0.1919	28310	71690	
P	$Mg_2P_2O_7$	0.2783	44456	55544	$Mg_2P_2O_7$: See Mg.
	$(NH_4)_3PO_4 \cdot 12\,MoO_3$	0.01639	21458	78542	
	$(NH_4)_3PO_4 \cdot 14\,MoO_3$	0.01453	16212	83788	
	$P_2O_5 \cdot 24\,MoO_3$	0.01723	23612	76388	
	P_2O_5	0.4365	63990	36010	
PO_2	Hg_2Cl_2	0.0667	82408	17592	
	$Mg_2P_2O_7$	0.5659	75269	24731	
PO_3	Hg_2Cl_2	0.1673	22343	77657	
	$Mg_2P_2O_7$	0.7096	85101	14899	
P_2O_5	CaO	0.8438	92621	07379	$PbSO_4$: The neutral or acidic solution that contains Pb is boiled with excess dil. H_2SO_4 in a porcelain dish and finally evaporated. After cooling, it is diluted and an equal volume of alcohol is added and allowed to stand for setting the $PbSO_4$. It is filtered and washed with a 5% solution of H_2SO_4 in 50% alcohol. Towards the end rinsed with pure absolute alcohol. After drying the precipitate the crucible is protected with a spacious protective crucible and is heated over a bunsen flame at 450°C. Final weight: $PbSO_4$.
	$[Ca_3(PO_4)_2]_3 \cdot Ca(OH)_2$	0.4241	62722	37278	
	MgO	1.174	06963	93033	
	$Mg_2P_2O_7$	0.6377	80463	19537	
	$(NH_4)_3PO_4 \cdot 12\,MoO_3$	0.03755	57464	42536	
	$(NH_4)_3PO_4 \cdot 14\,MoO_3$	0.033428	52218	47782	
	Na_3PO_4	0.4329	63641	36359	
	PO_4	0.7473	87349	12651	
	$P_2O_5 \cdot 24\,MoO_3$	0.03947	59622	40378	
PO_4	$Mg_2P_2O_7$	0.8534	93111	06889	
	$(NH_4)_3PO_4 \cdot 12\,MoO_3$	0.05025	70113	29887	
	$(NH_4)_3PO_4 \cdot 14\,MoO_3$	0.04453	64867	35133	
	Na_3PO_4	0.5792	76292	23708	
	P_2O_5	1.338	12651	87349	
	$P_2O_5 \cdot 24\,MoO_3$	0.05281	72272	27728	
Pb	$PbCl_2$	0.7450	87218	12782	PbO_2: In strongly nitric acid solution which does not contain more than 0.1 g of Pb, that Pb is deposited out on a Pt. net as anode current strength 1–2 amps at 2–3 volt. In order that the entire Pb be deposited on the anode, the solution should contain some Cu which is deposited at the cathode during electrolysis. At the end of the electrolysis without switching off current, it is washed with water. The Pt. net with the deposited PbO_2 is dried for 2 hours in an oven at 180°C until all water is removed.
	$PbCrO_4$	0.6401	80625	19375	
	PbO	0.9283	96770	03230	
	PbO_2	0.8662	93763	06237	
	PbS	0.8660	93751	06249	
	$PbSO_4$	0.6832	83457	16543	
$Pb(C_2H_5)_4$	$PbSO_4$	1.067	02798	97202	
$PbCrO_4$	Cr_2O_3	4.252	62863	37137	
	PbO	1.448	16079	83921	

Gravimetric analytical methods

Required	Given	Factor	log Factor	DE log Factor	Remarks
PbO	$PbCl_2$	0.8026	90448	09552	
	$PbCrO_4$	0.6906	83921	16079	
	PbO_2	0.9331	96993	03007	
	PbS	0.9328	96981	03019	
	$PbSO_4$	0.7360	86687	13313	
PbS	$PbSO_4$	0.7890	89706	10294	
$PbSO_4$	PbO	1.359	13313	86687	
	$BaSO_4$	1.299	11367	88633	
Pd	$Pd(CN)_2$	0.6718	82719	17281	
	$Pd[C_{10}H_6O(NO)]_2$	0.2361	37302	62698	
Pt	$PtCl_4$	0.5790	76271	23729	
	$PtCl_6$	0.4784	67967	32033	
	$(NH_4)_2PtC_6$	0.4402	64366	35634	
Rb	$RbCl$	0.7068	86930	13070	
	Rb_2SO_4	0.6402	80634	19366	
S	$BaSO_4$	0.1374	13789	86211	
	$C_{12}H_{12}N_2SO_4$	0.1136	05629	94471	
	CuO	0.4031	60545	39455	
S_2O_3	$BaSO_4$	0.2402	38055	61945	**$BaSO_4$**: Refer Ba.
SO_2	$BaSO_4$	0.2745	43847	56153	
SO_3	$BaSO_4$	0.3430	53530	46470	
	CaO	1.428	15464	84536	
	MgO	1.986	29793	70207	
	Na_2SO_4	0.5636	75102	24898	
	NH_3	2.350	37108	62892	
	SO_4	0.8335	92088	07912	
SO_4	$BaSO_4$	0.4115	61442	38558	
	SO_3	1.200	07912	92088	
Sb	$SbCl_3$	0.5337	72732	27268	
	$SbCl_5$	0.4072	60977	39023	
	Sb_2O_4	0.7919	89866	10134	
	Sb_2S_3	0.7168	85542	14458	
	Sb_2S_5	0.6030	78032	21986	
Sb_2O_3	Sb	1.197	07814	92186	
	Sb_2O_4	0.9480	97680	02320	
	Sb_2S_3	0.8581	93355	06645	
	Sb_2S_5	0.7218	85846	14154	

Gravimetric analytical methods					
Required	Given	Factor	log Factor	DE log Factor	Remarks
Sb_2O_5	Sb	1.329	12337	87663	
Sb_2S_3	Sb Sb_2O_4	1.395 1.105	14458 04325	85542 95675	
Sb_2S_5	Sb	1.658	21968	78032	
Se	SeO_2	0.7116	85225	14775	
SeO_2	SeO_3	0.8740	94150	05850	
SeO_3	Se	1.608	20628	79372	
Se	SeO_3	0.6419	79372	20628	
SeO_2	Se	1.405	14775	85225	
Si	SiO_2	0.4675	66975	33025	**SiO_2**: See H_2SiO_3
SiF_6	CaF_2 $CaSO_4$ K_2SiF_6	0.6065 0.3478 0.6450	78290 54143 80957	21710 45857 19043	
SiO_3	SiO_2	1.266	10253	89747	
Si_2O_7	SiO_2	1.400	14595	85405	
SiO_4	SiO_2	1.533	18541	81459	
Sn	SnO_2	0.7877	89634	10366	
SnO_2	Sn	1.270	10366	89634	
Sr	$SrCO_3$ $SrC_2O_4 \cdot H_2O$ SrO $SrSO_4$	0.5935 0.4525 0.8456 0.4770	77344 65559 92716 67855	22656 34441 07284 32145	
$SrCO_3$	$Sr(NO_3)_2$ $Sr(OH)_2 \cdot 8H_2O$	0.6976 0.5555	84359 74469	15641 25531	
$Sr(OH)_2 \cdot 8H_2O$	$SrCO_3$ $Sr(NO_3)_2$	1.8002 1.256	25531 09890	74469 90110	
$SrSO_4$	$BaSO_4$	0.7870	89595	10405	
Ta	Ta_2O_5	0.8189	91327	08673	

Gravimetric analytical methods

Required	Given	Factor	log Factor	DE log Factor	Remarks
TeO_2	Te	1.251	09718	90282	
Th	$Th(NO_3)_4 \cdot 4\,H_2O$	0.4203	62353	37647	
	ThO_2	0.8788	94390	05610	
	$Th(Oxin)_4$	0.2870	45780	54220	
Ti	TiO_2	0.5995	77779	22221	
	$TiO(Oxin)_2$	0.1360	13354	86646	
	$Ti_3(PO_4)_4$	0.2744	43844	56156	
U	UO_2	0.8815	94524	05476	
	U_3O_8	0.8480	92840	07160	
	$(UO_2)_2 P_2O_7$	0.6668	82399	17601	
V	$AgVO_3$	0.2463	39153	60847	
	Ag_3VO_4	0.1162	06508	93492	
	V_2O_5	0.5602	74834	25166	
W	WO_3	0.7930	89929	10071	
Zn	$ZnNH_4PO_4$	0.3665	56405	43595	
	ZnO	0.8034	90492	09508	
	$Zn(Oxin)_2$	0.1849	26681	73319	
	$Zn_2P_2O_7$	0.4291	63259	36741	
	ZnS	0.6709	82668	17332	
	$ZnSO_4$	0.4050	60742	39258	
$ZnCO_3$	ZnO	1.541	18775	81225	
$ZnCl_2$	Cl	1.922	28373	71627	
	AgCl	0.4754	67707	32293	
	Zn	2.085	31902	68098	
	$ZnNH_4PO_4$	0.7640	88307	11693	
	ZnO	1.675	22394	77606	
ZnO	Zn	1.245	09508	90492	
	$Zn_2P_2O_7$	0.5341	72765	27235	
	ZnS	0.8352	92176	07824	
	$ZnSO_4$	0.5041	70250	29750	
ZnS	ZnO	1.197	07824	92176	
	$Zn_2P_2O_7$	0.6396	80589	19411	
	$BaSO_4$	0.4175	62061	37939	
$ZnSO_4$	ZnO	1.984	29750	70250	
$ZnSO_4\,7H_2O$	ZnO	3.533	54820	45180	
	ZnS	2.951	46997	53003	
Zr	ZrO_2	0.7403	86941	13059	
	ZrP_2O_7	0.3439	53657	46343	

$Zn_2P_2O_7$: To an acidic zinc salt solution NH_3 is added so that the solution reacts to show that it is only weakly acidic. Thereafter it is diluted to about 200 ml., heated to boiling and treated with an excess of ammonium acetate solution. The solution is kept at 90°C for a long time till the precipitate of $ZnNH_4PO_4$ becomes crystalline. Then it is filtered through a filter crucible and is first washed with hot 1% ammonium phosphate solution and at the end several times with cold water. After drying the precipitate, it is ignited in an electric over at $\approx$ 1000°C. Final weight: $Zn_2P_2O_7$

Zn: Zn can be deposited on a copperized Pt.-net as cathode by electrolysis of a strongly alkaline solution. The sulphuric acid solution is first neutralized, then an excess of about 5 g NaOH is added and the solution is diluted to 150 ml. The electrodes must stand the under voltage when they are dipped into solution. The electrolysis is carried out under normal temperature. Voltage 3-4 volt. After deposition of zinc further electrolysis is no longer continued, since the anode is attacked in other respects. After first complete washing, the current may be turned off. Finally, the electrode is rinsed with alcohol and dried for a short time.

Quantitative organic analytical methods

Required element	Determination of elements
	Method

<table>
<tr><td rowspan="1">C
H</td><td>

The organic substance burnt in a current of oxygen (volatile and easily inflammable substances are burnt in a current of air free from CO_2 and water vapour) and the combustion products— carbon dioxide and hydrogen are absorbed in suitable absorption media, oxygen or air is led through a gasometer and conducted with a constant velocity through the combustion tube. The combustion temperature depends upon the type of substance. The combustion tube is appropriately filled with granulated catalytic masses which essentially consist of heavy metal oxides. Halogen and sulphur are bound by silver wool or lead oxide respectively in the combustion tube, N_2 escapes as a gas.

Absorption medium for H_2O: $CaCl_2$ (fused)
                              P_2O_5
                              $Mg(ClO_4)_2$ (anhydrous)
        for CO_2:   Soda asbestos (ascarite)

Oxygen and combustion air are pretreated with their absorption medium.

Calculation:

$$\% C = \frac{mg\ CO_2 \times 0.2729 \times 100}{\text{Initial weight in mg}} \qquad \%H = \frac{mg\ H_2O \times 0.1119 \times 100}{\text{Initial weight in mg}}$$

</td></tr>
</table>

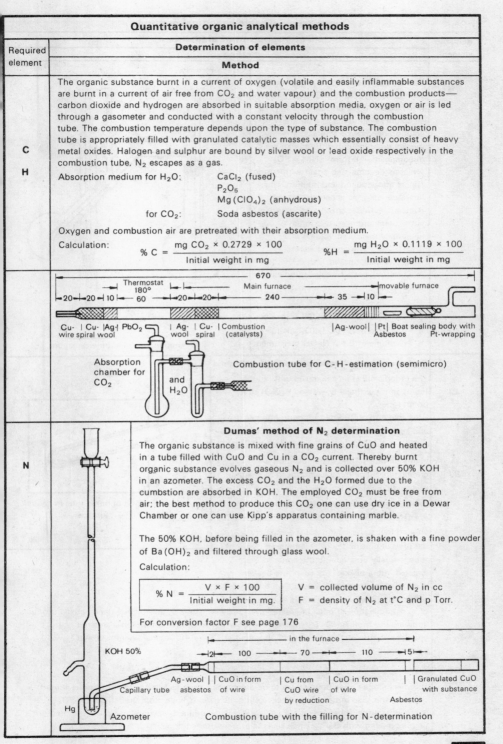

Dumas' method of N_2 determination

The organic substance is mixed with fine grains of CuO and heated in a tube filled with CuO and Cu in a CO_2 current. Thereby burnt organic substance evolves gaseous N_2 and is collected over 50% KOH in an azometer. The excess CO_2 and the H_2O formed due to the cumbstion are absorbed in KOH. The employed CO_2 must be free from air; the best method to produce this CO_2 one can use dry ice in a Dewar Chamber or one can use Kipp's apparatus containing marble.

The 50% KOH, before being filled in the azometer, is shaken with a fine powder of $Ba(OH)_2$ and filtered through glass wool.

Calculation:

$$\% N = \frac{V \times F \times 100}{\text{Initial weight in mg.}}$$

V = collected volume of N_2 in cc
F = density of N_2 at t°C and p Torr.

For conversion factor F see page 176

670

Thermostat 180° — Main furnace — movable furnace

20 — 20 — 10 — 60 — 20 — 20 — 240 — 35 — 10

Cu- | Cu- | Ag- | PbO$_2$ | Ag- | Cu- | Combustion | Ag-wool | Pt | Boat sealing body with
wire spiral wool   wool spiral (catalysts)   Asbestos   Pt-wrapping

Absorption chamber for CO_2

and H_2O

Combustion tube for C-H-estimation (semimicro)

KOH 50%

in the furnace

2 — 100 — 70 — 110 — 5

Ag-wool | CuO in form | Cu from | CuO in form | | Granulated CuO
Capillary tube asbestos of wire   CuO wire of wire   with substance
                                    by reduction   Asbestos

Hg   Azometer

Combustion tube with the filling for N-determination

	Quantitative organic analytical methods	
Required element	**Determination of elements**	
	method	

N

The organic substance is evaporated with conc. H_2SO_4 in a Kjeldahl flask, if necessary in the presence of selenium reaction mixture till the solution becomes colourless or is weakly green coloured. By the above decomposition it forms $(NH_4)_2SO_4$ which is decomposed again with strong NaOH after cooling and dilution. The liberated NH_3 gas is carried through distillation (under specific conditions) with steam and absorbed in an accurately measured volume of 0.1n H_2SO_4. The unused excess of H_2SO_4 is back titrated with 0.1n NaOH using methyl orange as indicator.

Calculation:

$$\% \text{ N} = \frac{(a-b) \times 1.4008 \times 100}{\text{Initial weight in mg}}$$

a = volume of 0.1 n H_2SO_4 taken
b = volume of 0.1 n NaOH consumed

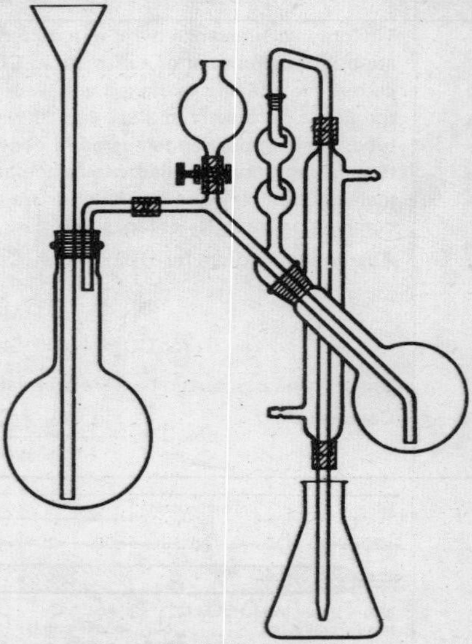

Ammonia distillation with steam

Cl

BJ

I

(S)

Carius method of determination of halogens

The organic substance is mixed with double the quantity of finely powdered $AgNO_3$ in a bomb tube.
After addition of 2–3 ml of fuming nitric acid and sealing of bombtube by melting it is heated for several hours at 250°–300°C in a carius furnace till the decomposition is completed. Since the tube is closed by melting and there is acid in the small tube which brings together the acid and substance, it reacts easily with nitric acid. After completely cooling by careful melting of the capillary the pressure inside the tube is rebased, the tube is finally opened and the contents are washed into a beaker. Above all the beaker is allowed to stand for 1 - 2 hours on a boiling water bath in case of Ag Br and AgI. In case of I_2 determination, it must be treated previously with H_2SO_3 so that any decomposed products combined to form $AgIO_3$ is reduced. The silver halide is filtered through a sintered glass crucible, washed with pure alcohol and dried at 130°–140°C. Final weight: AgCl
    AbBr
    AgI

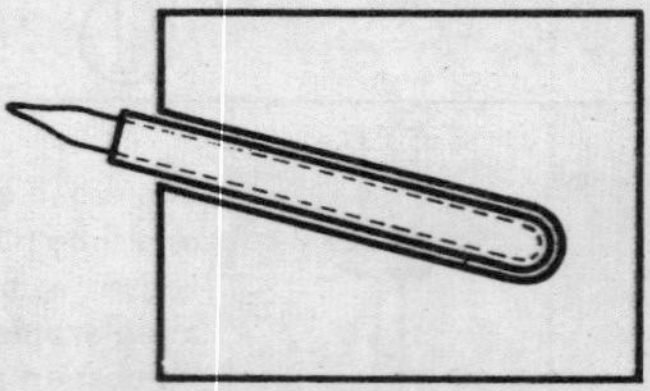

Position of bomb tube in carius furnace

Besides halogen, if sulphur is also to be determined, then first the decomposition is carried out as in the case of halogens. After filtering off silver halides, from the filtrate the SO_4^2 is precipitated with barium nitrate and $BaSO_4$ is determined gravimetrically. SO_4^2 should not be precipitated with $BaCl_2$, since silver is present in the filtrate.

	Quantitative organic methods
	Determination of elements
Required element	**Method**
S	**Carius method of determination of S** The determination follows exactly as in case of halogens in the bombtube instead of $AgNO_3$, 10 times the quantity of finely powdered anhydrous $BaCl_2$ is added. The resulting $BaSO_4$ is filtered and by gently igniting to red hot in a crucible, dried, allowed to cool and weighed.
Cl Br	**Determination of chlorine and bromine by combustion** The organic substance is burnt in a current of oxygen in a tube containing beads and platinum catalyst. The combustion products are absorbed in the portion of the tube which is filled with glass beads and moistened with 5% H_2O_2. Whereby HCl or HBr is evolved. After completing the combustion, the tube is washed and the halogen is determined volumetrically by K. Fajan's method with 0.02 n $AgNO_3$, employing an absorption indicator Indicator for Cl $^-$ = 0.01% solution of dichlorofluorescein in 60% alcohol for Br $^-$ = 0.01% aqueous solution of Ersin — sodium.
S	**Sulphur determination by combustion** The sulphur determination follows the same way as the determination of Cl and Br by combustion in a bead tube. The glass pearls are moistened with 5% H_2O_2. The combustion must be carried out slowly to make it possible for the complete absorption of SO_3 formed. After washing the tube the sulphate is determined as $BaSO_4$ by gravimetric analytical method.
Cl Br S	 Combustion tube for Cl and Br determination
I	**Determination of iodine by combustion** The substance is burnt in a current of O_2 with platinum catalyst and the liberated free I_2 is absorbed in an absorption tube which contains 10% sodium acetate in glacial acetic acid and a drop of bromine. Thereby the I_2 is oxidized to iodate. After destroying the excess bromine by a drop of conc. formic acid, solid KI is added in excess and the liberated I_2 is titrated with 0.1n $Na_2S_2O_3$. Using starch solution as indicator. Combustion tube for iodine determination

Quantitative organic methods
Molecular weight determination

To arrange the formula in organic chemistry one needs to know the molecular weight of the compound. Different methods are employed for determining molecular weight.

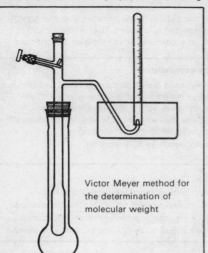

Victor Meyer method for the determination of molecular weight

1. By displacement of air: [For volatile compounds]
The substance is weighed in a small glass tube (Victor Meyer's tube) and after heating to a temperature which must lie well above the boiling point of the concerned substance, it is introduced through a device at the upper portion of the tube having the evaporating bulb. The substance immediately vapourizes whereby an equivalent amount of air quantity is displaced in the measuring tube. The vapourization temperature need not be measured since the air volume on cooling to the room temperature is noted. The noted volume of air is brought to standard conditions (NTP) after taking into consideration the aqueous tension, barometric pressure, and the temperature.
With the help of molar volume, the molecular weight can be calculated.

$$M = \frac{V_m}{V_o} \times s$$

M = molecular weight

V_m = molar volume = 22415 ml.

V_o = the noted air volume which has been reduced to NTP

s = initial weight in gram.

2. Depression in freezing point

The freezing point of a liquid is lowered by dissolving a substance in it. The depression in freezing point is dependent upon the concentration. It is constant in case of a given solvent, when an equal number of moles of different substances are dissolved in an equal quantity of solvent. Therefore the molecular weight of an unknown substance can be calculated.

First, one determines the freezing point of pure solvent and thereafter the freezing point of the prepared solution. With the help of the solvent constant the molecular weight is found out.

$$M = \frac{wK}{W\,\Delta T}$$

w = initial weight in gram
K = solvent constant (molal depression constant)
 = lowering of freezing point for 1 mole in 1000 grams of solvent
ΔT = lowering of freezing point in degrees.
W = weight of solvent (gram)

Solvent	Freezing point °C	Solvent constant K
Ethylene bromide	9.98	12.5
Formic acid	8.43	2.77
Benzene	5.4	5.1
Camphor	178	40
Cyclohexane	6.2	20.2
Glacial acetic acid	16.65	3.9
Water	0.0'	1.853

Molecular weight apparatus for the freezing point method

Frequently employed solvents are benzene or glacial acetic acid.

A simple molecular weight determination method, known as Rast method, depends on the melting point determination of known substance camphor and the substance mixture in comparison to the pure camphor. The accuracy of this method in many cases is adequate. The calculation is done as explained above.

Molecular weight determination

3. Elevation in boiling point

The boiling point of a liquid is raised by a substance dissolved in it. The elevation in boiling point is similarly dependent on the moles of dissolved substance in a given quantity of the liquid. For determination of molecular weight, first the boiling point of pure liquid and later the boiling point of the solution are determined. The calculation is the same as in the case of molecular weight determination by depression in freezing point, only the constant has another value (molal elevation constant).

Solvent	B.P. °C	Molal elevation constant K
Acetone	56.1	1.73
Ether	34.6	2.16
Alcohol	78.3	1.20
Aniline	184.4	3.69
Benzene	80.5	2.57
Chloroform	61.2	3.88

Molecular weight determinations by boiling point method does not have as much importance as the depression in freezing point method.

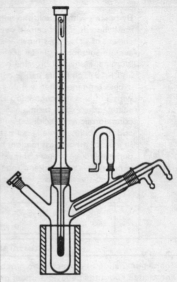

Molecular weight determination apparatus for elevation in boiling point

Water determination

1. Drying of the substance in an oven at 105°C. The water can be determined by this method only when the substance under the given conditions does not change and is not volatile at this temperature.

2. Xylene method: The substance is heated in a round bottomed flask with xylene saturated with water. Thereby the water is separated. It condenses together with the xylene distilling over in a set reflex condenser and collects in a measuring tube. The initial weight confirms to the water content.

3. By evolution of acetylene out of CaC_2 with the help of water and measuring the gaseous volume.
By evolution of ammonia out of Mg_3N_2 and volumetrically determination of ammonia with H_2SO_4.

4. By separation of water out of a substance with a current of dry gas under certain conditions at higher temperature and absorption of water in P_2O_5.

5. By titration with Karl-Fischer solution: By this method, very small quantities are determined. This method requires accurate check on the experimental conditions and frequent control of the employed solutions.

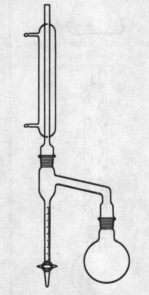

Water determination apparatus in xylene method

Quantitative organic methods

Methoxyl and ethoxyl estimation

Required	Method
CH$_3$O- C$_2$H$_5$O-	By boiling with HI acid the methoxyl in the form of methyliodide that is carried over with the aid of a current of CO$_2$ into a receiver containing 10% sodium acetate solution in glacial acetic acid and a drop of bromine. Elemental bromine converts CH$_3$I to CH$_3$Br and BrI which for its part is converted to HIO$_3$ by excess bromine. After washing into an Elenmeyer flask with a little water containing about 1.5 g of sodium acetate, the excess bromine is destroyed by a drop of conc. formic acid, acidified with H$_2$SO$_4$, solid KI added and the liberated free I$_2$ is titrated with 0.1 n Na$_2$S$_2$O$_3$. In case of ethoxyl estimation, first it is heated for 1 hour under reflux condenser, thereafter cold water circulation is stopped and the formed C$_2$H$_5$I distils over into the receiver.

Apparatus for methoxyl and ethoxyl determination

$$1 \text{ ml of } 0.1 \text{ n Na}_2\text{S}_2\text{O}_3 = 0.5171 \text{ mg CH}_3\text{O}$$
$$\text{or} = 0.7507 \text{ mg C}_2\text{H}_5\text{O}$$

Determination of characteristic number

The determination of characteristic numbers serves for the characterization of substances by which their knowledge about the behaviour under given conditions is important or by which the determination of accurate composition is not necessary.

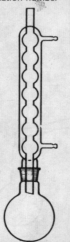

Baader flask for determination of saponification number

Round bottomed flask with reflux condenser for saponification

Neutralization number

By means of neutralization number the free acids contained in an oil or fat are ascertained. It is defined as how many mg KOH are required for neutralization of free acids contained in 1 g of substance.

The oil or fat is dissolved in a suitable solvent mixture mostly ether and alcohol and is titrated with 0.1 n or 0.5 n alcoholic potash using phenolphthalein as indicator upto a pale pink colour.

The initial weight increases towards the expected neutralization number.

If necessary, it is determined through a blank experiment whether the neutral solvent reacts.

$$NN = \frac{\text{ml } 0.5 \text{ n KOH} \times 28.054}{\text{weight in g}}$$

The factor 5.6108 is employed for the application of 0.1 n KOH.

Saponification number

By means of saponification number the free acids and the acids that are bound as esters contained in an oil or fat are determined. It is defined as how many mg KOH are necessary for the neutralization of acid and to saponify the esters in 1 g of substance.

The oil or fat, as in case of neutralization number, is dissolved in a suitable solvent and is heated under reflux in a saponification flask with 30-40 ml of 0.5 n alcoholic potash for 30 minutes. After cooling the unused excess of alkali is back titrated with 0.5 n HCl using phenolphthalein indicator. In case of dark coloured solutions alkali blue 6 B as indicator is better suited. In a similar manner, a blank experiment is carried out.

$$SN = \frac{(a - b) \times 28.054}{\text{Initial weight in g}}$$

a = volume of 0.5 n HCl in blank experiment

b = volume of 0.5 n HCl for the main experiment

Quantitative organic methods

Determination of characteristic numbers

Iodine number

By means of iodine number we ascertain whether the oil or fat contains any saturation or not. It is defined as how many g of halogen, calculated as iodine, that 100 g of the oil or fat can take up.

In case of iodine value according to Hanus method a solution of BrI in acetic acid is employed and in case of

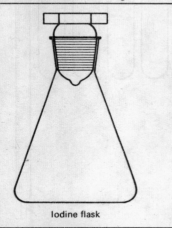

Iodine flask

iodine value according to the method of Kaufmann one uses a solution of bromine in water free methanol saturated with NaBr. The solutions are approximately 0.1 normal, and it must always be checked for accurate stability.

For determination of iodine value, 0.1-0.15 g of oil or fat weighed into an iodine flask and dissolved in 20 ml chloroform. Thereafter a known excess of iodine solution is added and allowed to stand in dark for a long time. At the end, a 10% solution of KI is added and the unused excess of iodine is back titrated with 0.1 n $Na_2S_2O_3$. Here also a blank experiment is necessary.

$$IN = \frac{(a-b) \times 1.2693}{\text{Initial weight in g}}$$

a = volume of 0.1 n $Na_2S_2O_3$ in blank experiment
b = volume of 0.1 n $Na_2S_2O_3$ for the main experiment

Reichert-Meissl and Polenski number

By means of Reichert-Meissl value, one can know about the water soluble and steam volatile acids contained in an oil or fat. These values are existing for low molecular weight fatty acids as for example, in butter and other foods.

The Reichert-Meissl value is defined as how many ml of 0.1 n alkali is necessary under definite conditions to neutralize the water soluble and steam volatile acids ascertained out of 5 g of fat.

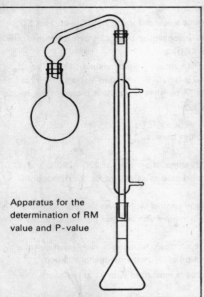

Apparatus for the determination of RM value and P-value

In case of determination, first the fat is saponified with 50% KOH, thereafter decomposed by water and dil. H_2SO_4 and 110 ml is distilled into a measuring flask. At the end from the distillate 100 ml is filtered through a dry filter paper in order to separate water insoluble acids and finally titrated with 0.1 alkali.

Example: consumed = 24.8 ml. 0.1 n alkali for 100 ml.
= 27.28 ml. alkali for 110 ml.
Blank experiment = 0.13 ml. of 0.1 n alkali.
RMN = 27.28 − 0.13 = 27.13

By means of Polenski number volatile but water insoluble acids are determined under the same conditions.

The determination of Polenski number always follows in the same apparatus as the RM value. The water insoluble acids hanging on to the reflux condenser and the measuring flask are dissolved with 90% alcohol and the solution is also poured through the filter in order to dissolve these acids. Thereafter it is titrated with 0.1 n alkali similarly.

Polenski number = consumed ml. of 0.1 n alkali

Quantitative organic methods
Gas analysis

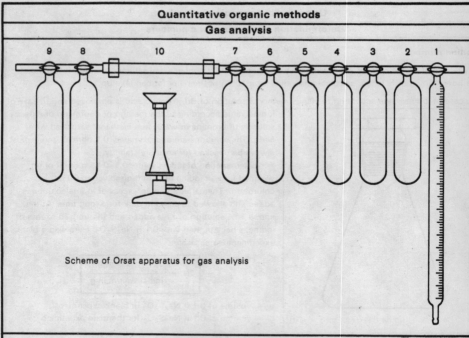

Scheme of Orsat apparatus for gas analysis

Arrangement of Orsat apparatus:

1. Gas burette: 100 ml. capacity divided into 1/10 ml. At the end of the gas burette a tube is connected with a levelling flask. An approximately 0.1 n H_2SO_4 with 5% NaCl serves as a sealing liquid.

2. – Gas absorption chambers.

	Absorption chamber for	Absorption medium
2.	CO_2	30% KOH
3.	Higher C-H substances (propylene)	Bromine water or fuming H_2SO_4 with 15% SO_3
4. + 5.	O_2	Alkali pyrogallol solution (5% pyrogallol in 30% KOH)
6. + 7.	CO	Neutral cuprous chloride solution 110 g Cu_2Cl_2 + 240 g NH_4Cl dissolved in 1 litre with airfree water. A polished Cu-plate is placed in the solution.
8.	CO_2 that is formed by combustion of CH_4.	30% KOH
9.	Blocking water for absorption of the gases in case of combustion of H_2.	As in gas burette

10. Tube of nonscaling steel, filled with CuO in wire form for combustion of H_2 at 270 – 290°C and methane at red heat. At either ends, cooling chambers are connected as safeguard for the compounds in tube.

The gas burette and absorption chambers are connected to one another with 3 ways stopcocks and at times through small pieces of tubes. The so arranged stopcock connection consists of capillary tubes in order to keep the volume as small as possible.

Before beginning the analysis, the stopcocks are flushed with nitrogen. Nitrogen is withdrawn out of a nitrogen cylinder or prepared through sucking air and absorption of O_2 in the pyrogallol solution.

After completion of the analysis copper in the combustion tube is reoxidized with air at red heat.

Quantitative organic methods

Gas analysis, volumetric analytical methods

In case of determination of individual gas compositions in a gaseous mixture, frequently volumetric analytical methods are employed.

Required gas	Method
O_2	By reaction with $Mn(OH)_2$ in alkaline solution, thereby an oxidation to $Mn(OH)_2$ results which is acidified with HCl and liberates free I_2 from KI. Iodine is titrated with $Na_2S_2O_3$.
H_2S	Passing of H_2S into cadmium acetate solution is bound as CdS. Thereafter it is decomposed with HCl in the presence of a measured quantity of I_2 solution whereby iodine is reduced. The unused excess is titrated with $Na_2S_2O_3$.
NH_3	Ammonia is absorbed in a known excess of 0.1 n H_2SO_4 in a receiver and the unused excess is titrated against 0.1 n NaOH using methyl orange as indicator.
HCN	HCN is absorbed in a suspension of nickel carbonate in soda solution, at end decomposed with H_2SO_4 and is distilled over into a receiver containing 0.1 n NaOH. The resulting NaCN is titrated with 0.1 n $AgNO_3$ by employing of KI as indicator.
CO	I_2O_5 reduces to I_2 at 130°C which is collected in a KI solution and is titrated with $Na_2S_2O_3$.

Gas analysis, physical method

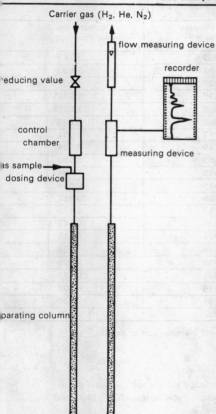

Carrier gas (H_2, He, N_2)

flow measuring device

recorder

reducing value

control chamber

measuring device

gas sample
dosing device

separating column

Measuring principle of gas chromotographs

Fractional condensation or distillation at low temperature.

Measurement of infrared absorption as rise in temperature pressure at constant volume.

Measurement of heat conductivity in comparison to air or with another similar gas.

Gas—Chromotographic method:

In a separating column, a gaseous mixture that is admitted through a current of carrier gas, is resolved into individual components either due to the basic differences in absorption at the surface active centres or due to the different solubility or affinity in the high boiling liquids. After flowing through the separating column the individual gas constituents are detected through thermal conductivity in comparison to that of the carrier gas by a measuring device (recorder).

Through gas chromotographic, practically all gases and liquids which are below and above 100°C can be detected. The time taken for the gas components upto the appearance in the measuring device gives information about the type and nature of gas. The pointer deflection of the instrument or the area defined by the curve is a measure of the concentration.

It is important to keep such factors as velocity of the carrier gas, temperature of the separating column, and a steady voltage at the hot wire of the thermal conductivity compartment, etc. a constant.

Carrier gas: H_2, He, N_2.
Fillings for the separation column: active charcoal silicagal or porous material impregnated with silicone oil or high boiling C-H substances.

Quantitative organic methods

Determination of viscosity of lubricating oils

Viscosity is the property of a liquid in which two adjacent oppositely moving layers set up a resistance.

It gives information about the action by the flow of liquids, in case of lubricating substances about the action of lubrication in ball bearings.

Dynamic viscosity: η or (V_d) for most liquids is only dependent on temperature and is constant for pressure.

Unit for dynamic viscosity: $\dfrac{1\ g}{cm \times s}$ = 1 poise (P)   Subunit 0.01 poise = 1 centipoise (cP)

In industry it is used as a derived unit = $\dfrac{kp \times s}{m^2}$

For inter conversion: $1\ \dfrac{kp \times s}{m^2}$ = 98.1 P or 1 P = 0.01020 $\dfrac{kp \times s}{m^2}$

In case of all viscosity measurements, attention must be paid to the accurate check on measuring temperature. In case of measuring the viscosity in physical mass units, a thermostat is necessary whose temperature fluctuation, especially in case of low temperatures, is not greater than $\pm\ 0.03°C$.

Density, dynamic, kinematic viscosities of pure water at different temperatures.

Temperature °C	Density g/cm³	Dynamic viscosity centipoise (cP)	Kinematic viscosity centistockes (cSt)
0	0.99984	1.792	1.792
5	0.99996	1.520	1.520
10	0.99970	1.307	1.307
15	0.99910	1.138	1.139
20	0.99820	1.002	1.0038
25	0.99705	0.890	0.893
30	0.99565	0.797	0.801
35	0.99403	0.719	0.724
40	0.99221	0.653	0.658
45	0.99022	0.598	0.604
50	0.98805	0.548	0.554
55	0.98570	0.505	0.512
60	0.98321	0.467	0.475
65	0.98057	0.434	0.443
70	0.97778	0.404	0.413
75	0.97486	0.378	0.388
80	0.97180	0.355	0.365
85	0.96862	0.334	0.345
90	0.96532	0.315	0.326
95	0.96189	0.298	0.310
100	0.95835	0.282	0.295

Kinematic viscosity: v (or V_k) is the quotient of dynamic viscosity and of density.

$$v = \frac{\eta}{\rho} = \frac{cm^2}{s} = 1\ \text{stokes (St)}$$

Subunit = 0.01 stokes = 1 centistokes (cSt)

The measurement of viscosity with absolute viscosimeters is too tedious and difficult. For this, the instrument constant is determined for the employed viscosimeter with the help of standard oils. The viscosity of pure water at 20°C serves as a standard value.

Pure water at 20°C has the dynamic viscosity = 1.002 cP

kinematic viscosity = 1.0038 cSt

Quantitative organic methods

Determination of viscosity of lubricating oils conversion table

Physical mass unit kinematic viscosity cSt	Engler-degree E	Conventional mass Redwood no. 1 viscosity (70°F) seconds (R)	Saybold universal viscosity (100°F) seconds (S)
2	1.1195	30.215	32.62
3	1.218	32.725	36.03
4	1.3075	35.33	39.14
5	1.394	37.94	42.35
6	1.4805	40.55	45.56
7	1.566	43.26	48.77
8	1.6535	46.07	52.09
9	1.743	48.93	55.50
10	1.834	51.80	58.91
12	2.023	58.02	66.04
14	2.222	64.50	73.57
16	2.435	71.32	81.30
18	2.646	78.31	89.44
20	2.876	85.64	97.77
22	3.11	93.17	106.4
24	3.35	100.7	115.0
26	3.59	108.4	123.7
28	3.83	116.2	132.5
30	4.08	124.1	141.3
35	4.71	143.75	163.7
40	5.35	163.7	186.3
45	5.995	183.8	209.1
50	6.64	203.9	232.1
55	7.30	224.0	255.2
60	7.95	244.2	278.3
100	13.20	406.1	463.5

Ist the kinematic viscosity $V_k > 60$ cSt, which gives $E = 0.1320 \times V_k$.

The valid customary units for the viscosity are:
Engler degree (°E) measured in Engler viscosimeter. This measured viscosity with the Engler viscosimeter may not be converted into kinematic viscosity on account of greater accuracy, in return the conversion of physical measuring value into Engler degree is possible.
Time of outflow in seconds in Redwood-apparatus (Redwood-seconds).
Time of outflow in seconds in Saybold-instrument (Saybold-seconds).

Quantitative organic methods
Determination of viscosity of lubricating oils

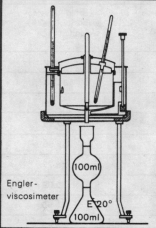

Engler-
viscosimeter

Measurement of viscosity with Engler viscosimeter

Unit of mass: Engler degree (°E)

Calculation of $E_t = \dfrac{\text{Time of outflow of 200 ml liquid at t°C}}{\text{Time of outflow of 200 ml distld. water at 20°C}}$

A calibrated thermometer is used for the measurement of temperature. Measuring temperatures are normally 20°, 50°, and 100°. The time of outflow of water is always measured at 20°C. It must lie between 50 to 52 seconds. The temperature of thermostat is to be kept constant during the measurement. The temperature of waterbath shall amount to:

$$\begin{array}{ll}20°C \text{ at} & 20°C \text{ measuring temperature}\\50.25°C \text{ at} & 50°C \text{ measuring temperature}\\101°C \text{ at} & 100°C \text{ measuring temperature}\end{array}$$

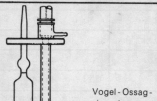
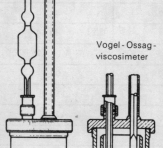

Vogel-Ossag-
viscosimeter

Measurement of viscosity with Vogel-Ossag-viscosimeter

Unit of mass: Centistokes (cSt)

Centipoise (cP)

Application range: Measurement of kinematic viscosity in range of 2 to 20000 cSt or the dynamic viscosity in the range of 2 to 100000 cP. Here a set of 5 capillaries with different diameters are necessary.

Characteristic number k or c of measuring capillars	Minimum flow time s
0.01	265
0.02	190
0.05	120
0.1	85
0.5	40
1.0	30

The flow time shall amount to as small as 30 seconds.

Calculation: Kinematic viscosity: V_κ v k $\times$ t_κ (cSt)

Dynamic viscosity: V_d η c $\times$ t_c (cP)

1st in case of the dynamic viscosity, the density d of liquid greater than 0.95 or smaller than 0.85 so that c is to be multiplied with the correction factor 0.196 (6-d).

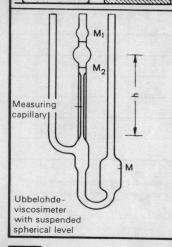

M_1

M_2

h

M

Measuring
capillary

Ubbelohde-
viscosimeter
with suspended
spherical level

Measurement of viscosity with Ubbelohde-viscosimeter

Unit of mass: Centistokes (cSt)

Range of application: Measurement of kinematic viscosity in the range of about 1 to 10000 (cSt).

Here a set of minimum 4 capillaries of different diameters is necessary.

Ubbelohde-Capillary	I	II	III	IV
Interior diameter of capillary mm	0.161 to 0.65	1.08 to 1.14	1.94 to 2.02	3.50 to 3.60
Application range (cSt)	1 to 10	4 to 100	40 to 1000	above 400
Limiting values minimum viscosity (cSt) shortest time of flow (s)	3 300	10 100	40 40	— 40
Characteristic number k	0.01	0.1	1	10
Calculation: Kinematic viscosity $V_\kappa = v = k \times t_\kappa$ (cSt)				

Quantitative organic methods
Determination of viscosity of lubricating oil

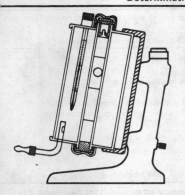

Measurement of viscosity with Höppler-viscosimeter

Unit of mass: Centipoise (cP)

Range of application: Measurement of dynamic viscosity in range of 0.6 to 250000 cP by employing 6 different spherical balls.
Time taken for a fall > 60s for ball no. 1
> 30s for ball no. 2 upto 6
Temperature range −20° to +120°C

Calculation: Dynamic viscosity $V_d = \eta = k \left((\rho_1 - \rho_2) \times t \right)$

K = constant (refer Table)

ρ_1 = density of metal or glass ball (see Table).

ρ_2 = density of the liquid to be examined in g/cm^3

t = full time for the ball in seconds.

Spherical balls for the falling ball viscosimeters with a measuring tube of 15.9 mm inner diameter

Ball size no.	Made out of	Density ρ g/cm^3	Diameter of the ball mm	Unspherical nature mm	Constant (approximate) K(cP × cm/g × s)	Measuring range cP
1	Instrumental glass 20	2.4	15.81 ± 0.01	± 0.0005	0.007	0.6 to 10
2	Instrumental glass 20	2.4	15.6 ± 0.05	± 0.0005	0.09	4 to 130
3	Ni-iron	8.1	15.6 ± 0.05	± 0.001	0.09	20 to 700
4	Ni-iron	8.1	15.2 ± 0.1	± 0.001	0.7	150 to 4800
5	Ni-iron or steel	7.7 to 8.1	14.0 ± 0.5	± 0.001	7	1500 to 45000
6	Steel	7.7 to 7.8	11.0 ± 1	± 0.002	35	> 7500

SAE - viscosity classification

for motor lubricating oil									for gearing oil								
SAE visco-sity class	Viscosity at								SAE visco-sity class	Viscosity at							
	−17.8°C (0°F)				98.9°C (210°F)					−17.8°C (0°F)				98.9°C (210°F)			
	least		best		least		best			least		best		least		best	
	cSt	E	cSt	E	cSt	E	cSt	E		cSt	E	cSt	E	cSt	E	cSt	E
5W	–	–	869	115					75	–	–	3257	430	4.18	1.33	–	–
10W	1303[1]	172[1]	2606	344	3.86	1.30	–	–	80	3257[1]	430[1]	21716	2867				
20W	2606[2]	344[2]	10423	1376					90					14.24	2.249	25.0[2]	3467[2]
20					5.73	1.46	9.62	1.80	140					25.0	3.467	42.7	5.707
30					9.62	1.80	12.94	2.12	250					42.7	5.707	–	–
40					12.92	2.12	16.77	2.52				53	62	72	78		
50					16.77	2.52	22.68	3.19									
				53	62	69	78										

(1) The least viscosity at −17.8° (0°F) drops as the limiting value when the viscosity at 98.9°C lies above 4.18 cSt (1.33E).

(2) The least viscosity falls as a limiting value when the viscosity at 98.9° lies above 5.73 cSt, i.e. (1.46E)

(1) The minimum viscosity at −17.8°C (0°F) falls to a limiting value when the viscosity at 98.9°C does not be below 6.66 cSt (1.536E).

(2) The best or maximum viscosity at 98.9°C (210°F) falls as a limiting value when the viscosity at −17.8°C does not lie below 162900 cSt (21500E).

The viscosity at −17.8° (0°F) is found out by extrapolation of primary measurements at temperatures which are at least 33.3°C apart. Here a suitable viscogram plot is used.

Quantitative organic methods

Determination of heat value of solid fuels

Heat of combustion (high calorific value) and heat value (low heat value) are a measure of heat quantities liberated by combustion of fuels.

Heat of combustion (high heat value) H_o	Calorific value (low heat value) H_u
Quantity of heat that is liberated by completely burning a unit mass of fuel (1 kg).	As in case of H_o
Temperature of fuel and the combustion products at 20°C.	As in case of H_o
In the available fuel, the formed water due to combustion exists in the liquid state after combustion at 20°C.	In the fuel available the formed water due to combustion exists in the vapour state after combustion at 20°C.
c and s burn totally without any residue of CO_2 and SO_2 which exist in gaseous state after combustion.	As in case of H_o
The N_2 present is not oxidized.	As in case of H_o

In a bomb calorimeter, 1 g of fuel is burnt in pure oxygen after electrical ignition, and the heat liberated is given to the water contained in that calorimeter vessel. The temperature rise of the calorimeter water is a measure of the heat quantity. The O_2 pressure in that bomb calorimeter is 30 atm.

The firmly powdered fuel is pressed into a tablet and bound to the ignition wire with a cotton thread.

As ignition wire, a thin platinum wire is used. Instead of platinum a nickeline wire can also be employed. In this case the wire is suitably pressed in with the table. In the bomb calorimeter 5 ml of water is added to the (dissolving) absorption of resulting sulphuric and nitric acids.

Initial weight and water equivalent are to be determined one after another so that the temperature rise in the Beckman thermometer amounts of 2 - 3°C. The bomb calorimeter must be covered with water completely. Through combustion of a substance with known heat of combustion (benzoic acid), the calorimeter is calibrated so that the water equivalent is determined.

Calculation:

$$H_o = \frac{W_w (t_m + C - t_o) - \Sigma B}{G}$$

W_w = Water equivalent of calorimeter in cal/°C
t_o = temperature at the initial reading of the main experiment
t_m = temperature at the final reading of main experiment
C = correction factor for the heat exchange between carolimeter and the surrounding.
ΣB = sum of correction factors for measured quantities of heat which do not correspond to the concept of explanations of heat of combustion. It therefore belongs to the heat of formation of nitric and sulphuric acids. This is determined by titration with 0.1 n $Ba(OH)_2$. Heat of combustion of admixed materials like cotton thread, nickeline wire or paraffin oil as admixtures are not easily combustible substances.
G = Initial weight in grams.

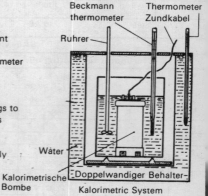

Beckmann thermometer — Thermometer Zundkabel
Ruhrer
Wäter
Kalorimetrische Bombe — Doppelwandiger Behalter
Kalorimetric System

Heat of formation for 1 ml of 0.1 n HNO_3 = 1.5 cal.
Heat of formation for 1 ml of 0.1 n H_2SO_4 = 3.6 cal.

For 1 mg nickeline wire, 0.65 cal and for 1 mg of cotton thread 4.0 cals are to be counted.

Calculation of water equivalent:

$$W_w = \frac{E \times i \times Bz}{t_m + C - t_o}$$

It signifies: E = heat of combustion of a standard substance in kcal/kg
i = initial weight of standard substance in gram
Bz = heat of combustion of admixed materials

The low heat value H_u of fuels are calculated from the heat of combustion (high calorific value H_o)

$$H_u = H_o - 5.85 \times W$$

W = by elemental analysis of resulting water in % (vapour of fuel + water out of the hydrogen compounds of fuel).

5.85 = latent heat of (vaporization) steam in kcal for 1.0 g f water at 20°C.

General Subject Index

In order to facilitate the search of information regarding concepts, definitions, rules, principles or methods for inorganic as well as organic compounds and their properties, the general index is arranged in five parts.

1. General subject index . . . .
2. Analytical methods . . . .
3. Names of inorganic compounds
   a) Names according to the rule of IUPAC nomenclature . . . .
   b) Trivial geological or minerological names . . . .
4. Names of organic compounds . . . .
5. Molecular formula of organic compounds . . . .

Before each individual section short explanations have been included.

Greek alphabets

Since Greek alphabets are frequently used in designating symbols for compositions, units of measures, particularly in case of names of chemical compounds, these alphabets are given before the general subject index.

A α Alpha (a)	Z ζ Zeta (z)	Λ λ Lambda (l)	Π π Pi (p)	Φ φ Phi (ph)
B β Beta (b)	H η Eta (e)	M μ Mu (m)	P ϱ Rho (r)	X χ Chi (ch)
Γ γ Gamma (g)	Θ ϑ Theta (th)	N ν Nu (n)	Σ σ Sigma (s)	Ψ ψ Psi (ps)
Δ δ Delta (d)	I ι Iota (i)	Ξ ξ Xi (x)	T τ Tau (t)	Ω ω Omega (o)
E ε Epsilon (e)	K κ Kappa (k)	O o Omicron	Υ υ Upsilon	

General Subject Index

Analytical methods

In the section of the book **"Methods"**, the analytical methods are arranged from an objective point of view. In this part of the index the usual heading of these methods are compiled in alphabetical order.

Inorganic compounds (nomenclature)

For arrangement of inorganic compounds in tables on pages 39–76 as well as for nomenclature see pp. 36–38. The clear and unequivocal method for the characterization of a chemical compound is its formula. The classification principles for the tables are shown in the initial chapter (see pp. 20–35), special properties of compounds, class and survey in pp. 34–35.

The classification of the (registered) compounds in the periodic system towards the beginning of tabular form as well as the colour scheme, the simultaneous universal property characteristics, simplifies not only the search of the desired information but also makes possible a methodical interpretation. Rules and examples for the search of inorganic compound according to their names.

1. **Binary compounds:** Arrangement according to the electronegative parts, the sequence of the tabular form after the "Gmelin number", e.g. NaCl—sodium *chloride* "chloride"-chloro compounds binary; Al^2O_3—Aluminium *oxide*. "Oxide"-oxygen compounds binary. Exception hydrogen compounds, e.g. sodium hydroxide, HCl, hydrogen *chloride*. Binary hydrogen compounds are all grouped together on p—39. F^2O better OF^2, oxygendi (fluoride) to be found under oxide.

2. **Ternary, quarternary and other compounds:** Apart from the few double compounds, generally these are complexes (oxy acids, and their salts, etc.) Arrangement is according to the central atom of the complexe sequence is as per the group of periodic system. Exception: According to the IUPAC nomenclature, the classical, trivial names are permitted; these are indicated below [e.g. H_2SO_4. Proper name Tetroxo sulphuric acid—commonly sulphuric acid]. For other trivial names as well as geological and mineralogical names, see p. 258.

Names of inorganic compounds

Trivial and geological or minerological names

...ial names which are permitted by IUPAC nomenclature, see p. 257 Geological or minerological names are common for all ...s (pp. 70–73), double oxides and for double sulphides (pp. 74–76) for which chemical composition is not either uniform or ...ometric (for explanation see pp. 70 and 74).

...72 (16)	Fahlerz (fahlore) 75 (8) and 17	Plagioclase 73 (13)
...ite 72 (11)	Faylite 70 (20)	Pollux 73 (5)
...75 (35)	Ferrosillite 72 (2)	Pollucite 73 (5)
...73 (73)	Fibrolite 72 (9)	Potash alum 75 (35)
...e 71 (17)	Fibrous serpentine 72 (16)	Potash feldspar 73 (8)
...jonite 76 (44)	Fluoraptatite 76 (41)	Polyhalite 76 (48)
...sulphonic acid 64 (39)	Forsterite 70 (12)	Prontite 75 (3)
...oole 72 (10)	Fulminic acid 44 and 62 (51)	Proustite 75 (3)
...e 73 (2)		Pyrargyrite 75 (9)
...42 (34)	Galenite 74 (5)	Pyrophantite 74 (17)
...site 71 (2)	Garnerite 72 (20)	
...te 71 (13)	Garnet 71 (9)	
...ite 73 (11)	Germanite 75 (22)	Quartz α 42 (22)
...ioclaus 73 (10)	Gesderfite 75 (16)	Quartz β 42 (23)
...ihylite 72 (17)	Glance cobalt 75 (5)	
...site 75 (21)	Glassrite 75 (30)	Red lead 74 (7)
...pyrites 75 (4)	Glauberite 75 (28)	Rhodonite 71 (25)
...: nickel glanze 75 (6)	Glaucophane 72 (13)	Rinnmann's green 74 (5)
...os 72 (16)	Grassubrite 71 (11)	Rutile 42 (32)
...inite 75 (26)	Gudmundite 75 (11)	
...72 (8)		Salfatarite 75 (29)
...71 (16)	Harmotome 73 (19)	Scheetite 60 (44)
...76 (39)	Hausmannite 74 (9)	Schlippe's salt 65 (42)
	Hemimorphite 71 (22)	Schoenite 75 (32)
...dite 71 (31)	Heulandite 73 (18)	Serpentine 72 (16)
...71 (29)	Hornblende 72 (10)	Siliceous calamine
...72 (26)	Hornblende, general 72 (12)	(hydrous zinc silicate) 71 (22)
...te 75 (26)	Hydrargylite 42 (18a)	Silicic acid 55 (53)
...te 54 (14)	Hydrocyanic acid 44 and 62 (21)	Sillimanite 72 (9)
...gerite 75 (7)	Hypersthene 72 (4)	Skladowskite 71 (19)
...te 75 (19)	Hydrazoic acid 62 (1)	Soda alum 75 (29)
...le 64 (18)		Soda feldspar 73 (7)
...e 75 (14)	Ilmenite 74 (18)	Soda microcline 73 (10)
...te 42 (33)	Ilvaite 74 (18)	Soda hornblende 72 (13)
...e 72 (3)		Soddyite 71 (18)
...nite 73 (15)	Jamesonite 75 (20)	Sphene 71 (6)
...idisulphide 51 (58)	Jahabsite 74 (11)	Spinell 74 (4)
...andun (see silicon	Jordanite 75 (1)	Spodumene 72 (5)
...ide) 53 (3)		Steatite 72 (15)
...te 74 (22)	Kainite 76 (43)	Stephanite 75 (10)
...acle 71 (15)	Kalinite 75 (35)	Stebnite 72 (15)
...e 71 (21)	Kasseler yellow 76 (22)	Sulphamide 64 (41)
...73 (1)	Kaolin 72 (18)	Syngenite 75 (34)
...a-Epidote 71 (7)		
...zite 73 (17)	Langebenite 75 (33)	Talc 72 (15)
...ersite 75 (15)	Lanmonitite 73 (12)	Takaedrite 75 (8)
...patite 76 (42)	Lead chamber crystal 64 (37)	Tellurium bismuth 75 (23)
...e 72 (28)	Leonite 75 (31)	Tetradymite 75 (25)
...oid 72 (21)	Lepidolite 72 (23)	Thalenite 71 (23)
...obalite α 42 (26)	Leucite 73 (4)	Thenard's blue 74 (6)
...obalite β 42 (27)	Leuna salt peter 76 (40)	Thiocyanic acid 44 and 62 (58)
...obalite (structure) 73	Lievrite 71 (8)	Thiophosgene 61 (58)
...-iron steel 74 (8)	Lithiamica 72 (23)	Thomsonite 73 (14)
...eberyl 74 (3)	Leonite 75 (5)	Thorite 70 (22)
...glance 75 (5)		Thortvetite 71 (23)
...pyrites 75 (16)	Magnetite 74 (12)	Tiechyhydrite 74 (31)
...r pyrites 75 (13)	(magnetiron ox)	Titanite 71 (6)
...ite 71 (30)	Malachite 76 (38)	Titanyl sulphate 57 (32)
...dun 42 (18a)	Manganese 75 (38)	Topas 71 (3)
...e 65 (7)	Margarite 72 (27)	Tourmaline 71 (33)
...hionite 76 (15)	Mendipite 76 (20)	Trevorite 74 (13)
...ite 75 (15)	Microdine 73 (9)	Tridynite (structure) 73
...: acid 44 and 62 (49)	Millon's base 66 (34)	Tridynite α 42 (24)
...e 71 (1)	Mohr's salt 75 (50)	Tridynite β 42 (25)
	Montmorilonite 72 (19)	Triple nitrite 76 (46)
...e 73 (6)	Muscovite 72 (25)	Triiphane 72 (5)
...ne 73 (16)		
...ire 42 (18a) and 54 (14)	Nessler's reagent 66 (37)	Ullamannite 75 (12)
...ite 71 (30)	Nepheline 73 (3)	Uranotile 71 (20)
...hogen 43 (32) and 44, 62 (22)	Neptunite 71 (27)	Uranyl sulphate 57 (58)
...de 72 (7)	Nickel glance 75 (12)	
...se 71 (28)	Nickel pyrites 75 (16)	Vanthoffite 75 (27)
...lane 44, 62 (60)		Vesuvianite 71 (32)
...n 71 (1)	Olivine 71 (4)	
...ite 76 (5)	Orthite 71 (17)	Water glass 70 (3–10)
	Orthoclase 73 (8)	White lead 76 (37)
...e 72 (16)	Ozone 43 (1)	Willemite 71 (22)
...e 73 (3)		Willmenite 70 (17)
...te 75 (2)	Perowskite 74 (16)	Wollastonite 71 (24)
...e 71 (7)	Petalite 73 (1)	
...te 72 (1)	Phenacite 70 (11)	Zeolite 73 (16)
	Phosgene 61 (1)	Zincspinel 74 (5)
	Pistalite 71 (7)	Zircon 70 (19)
		Zoisite 71 (5)

Names of organic substances (organic compounds)

For the arrangement of organic substances in tables see pp. 80—128 and for nomenclature see pp. 77—79, 86, 90 an
This arrangement permits a methodical interpretation of the tables. The important means of characterization of an organic com
are its structural formula and, secondly, the molecular formula (see pp. 267—272). For the nomenclature of an organic com
different possibilities that are overlapping and interdependent can arise.

1. **Trivial names:** Derived from naturally occurring substances; synthetic names; discoverer's names, and so on; e.g. butyr benzene, nicotin (Jean Nicot).

2. **Systematic names:** Named according to the rules of IUPAC nomenclature, e.g. 2-methylpentane.

3. **Parent substance names:** are half systematic and half trivial names, e.g. hydroxy acetone (parent substance acetone: sys propan-1-ol-2).

4. **Compound names** (fused): for example, naphthalene-2-acetic acid pyridinehydro chloride, etc.

5. **Subtractive or additive names:** for example dehydro benzene, ethylene bromide.

6. **Substitution names:** Based on the theory of Gerherd (1816—1856) and often similar to the systematic names, e.g., chloro methylbenzene.

7. **Radiofunctional names:** Based on the radical theory of Dumas (1800—1884) and Liebig (1803—1873). The basis of the was the dualistic conception of the structure of substances due to electrically opposite changed particles. (Berzelius 1759 and therefore the similarity with the names of inorganic compounds, e.g. methyl chloride and sodium chloride. Today these cc are partly held true due to the theory of electronegativity.

$$\left[\begin{array}{l}\text{Carbonium}\\\text{Carbanian}\end{array}\right]\begin{array}{l}\text{cation}\\\text{anion}\end{array}$$

8. **Replacement names or 'a' nomenclature:** For explanation and example see p. 86 heterochains.

9. **Hantzsch-Widman-names for heterocycles:** For explanation and examples see p. 90 heterorings.

The systematic names are given in the tables of this book. It is not possible to bring up all the names. In case of doubt, one can search the molecular formula index (pp. 267—272). Numbers, letters, and other symbols towards characteriza isomers (n, iso and others) are beyond the scope of alphabetical series.

Naphthionic acid 116 (5)
or 1-Napthol 107 (17)
or 2-Naphthol 107 (18)
Naphthol-8-amino-3,6-
  disulphonic acid 128 (1)
Naphthol-3,6-disulphonic acid
  127 (32)
Naphthol-6,8, disulphonic acid
  127 (33)
Naphthylamine 119 (16)
Naphthylamine 119 (17)
phthylmethylanine 100 (17)
Naphthylmethylanine 100 (16)
2-Naphthoquinone 111 (15)
4-Naphthoquinone 111 (16)
nphi-naphthoquinone 111 (17)
eooctane 80 (25)
eopentane 80 (8)
enol 106 (28)
cotine 104 (15)
cotinic acid 116 (19)
ctoninamide 122 (30)
cotinic acid diethylamide 114 (20)
cotyrin 104 (14)
nhydrin 117 (12)
Nitranilin 123 (8)
Nitranilin 123 (9) 131
Nitroanthraquinone 123 (7)
Nitrobenzaldehyde 123 (6)
trosobenzene 121 (14)
trobenzene 121 (21) 131
troethylene 121 (20)
troform 121 (18)
troglycerin 128 (25) 131
tromethane 121 (16) 131
Nitronaphthalene 121 (24)
Nitronaphthalene 121 (28)
trosodimethylaniline 121 (15)
Nitroso-N methyl urea 124 (12)
troso methylurethane 123 (3)
Nitrotoluene 121 (25) 131
Nonadecane 81 (15)
nadecanoic acid 115 (19)
Nonane 81 (5)
nanal 110 (11)
Nonanol 106 (15)
nanoic acid 115 (9)
nene-1 82 (12)

ctene-1 82 (11)
Octacosanol 106 (22)
Octadecane 18 (14)
ctadecanol 110 (12)
ctadecanoic acid 115 (18)
12-Octadecadienoic acid 115 (27)
7,9-Octadecatrienoic acid 115 (29)
s-Octadecene-1-ol 106 (31)
s-Octadecenoic acid 115 (25)
Octadecanol (stearylate) 106 (20)
-octane 80 (20)
Octane 80 (23) 81 (4) 131
ctanal 110 (10)
-Octanaphthalene 85 (5)
Octanol 106 (4)
ctanoic acid 115 (8)
ctyl-n-butyrate 112 (19)
ctyl-n-caproate 112 (20)
eyl alcohol 106 (31)
eic acid 115 (25)
ns-oleic acid 115 (26)
= R₂ = R₃ = Oleicester 112 (28)
enanthaldehyde 110 (9)
psopyrrole 101 (22)
rcin 109 (8)
rcinol 109 (8)
ornithine 124 (1)
rthoformic acid triethylester 87 (39)
rthoformic acid trimethylester 87 (38)
rthoacetic acid triethylester 87 (41)
rthocarbonic acid tetramethylester
  87 (42)
rthocarbonic acid tetrapropylester
  87 (43)
-Octahydronaphthalene 85 (5)
valene 84 (14)
xadiazole 94 (4)
2,5-Oxadiazole 94 (5)
3,5-Oxadiazole 94 (4)
xalacetic acid 112 (26)
xalacetic acid (ketoform) 118 (5)
xalaceticester diethylester 112 (26)
xalic acid 115 (31)
xalic acid diethylester 112 (24)

3,5-Oxa-4-methylheptane 87 (37)
Oxamide 122 (26)
Oxane 90 (6)
2,4-Oxapentane 87 (26)
Oxazole 94 (2)
1,2-Oxazole 94 (1)
1,3-Oxazole 94 (2)
Oxetane 90 (2)
Oxindole 113 (23)
Oxirane 90 (1)
2-oxobutane dicarboxylic acid 118 (5)
3-Oxobutyric acid 118 (3)
1-Oxocyclobutane 111 (7)
1-Oxocyclohexane 111 (9)
1-Oxocyclopentane 111 (8)
2-Oxoindoline (2-oxindole) 113 (23)
3-Oxoindoline (3-oxindole) 113 (22)
Oxole 90 (5)
Oxolane 90 (3)
2,4-Oxo-3-methylpentane 87 (36)
2-Oxooxetane 113 (8)
4-Oxopentanal 117 (2)
4-Oxopentanoic acid 118 (4)
2-Oxopropanol 117 (1)
2-Oxopropionic acid 118 (2)
4-Oxo-1,4-pyran 113 (10)
2-Oxotetrahydrofuran 113 (19)
2,2-Oxydiethanol 108 (6)

n-Palmitic acid 115 (16)
Parabonic acid 113 (19)
Parvulin 102 (17)
α-Parvulin 102 (16)
Pelargonaldehyde 110 (11)
n-Pelargonic acid 115 (9)
Pentacene 84 (8)
n-Pentacontane 81 (25)
n-Pentacosane 81 (21)
n-Pentadecane 81 (11)
Pentadecanoic acid 115 (15)
Pentadiene 1,3 82 (15)
Pentadiene 1,4 82 (16)
2-4, Pentadione 111 (2)
Pentaerythriotol 106 (40)
1,1,1,2,2, Pentafluoro-2
  chloroethane 126 (16)
Pentamethyl benzene 96 (20)
Pentamethylene 83 (5)
Pentamethylenediamine 119 (12)
Pentamethyleneoxide 90 (6)
Pentamethylenesulphide 91 (6)
Pentamethylenetetrazole 93 (24)
n-Pentane 80 (6), 81 (1) 131
Pentanal 110 (6)
2-Pentanone 110 (28)
3-Pentanone 110 (29)
Pentanoic acid 115 (5)
n-Pentanethiol 127 (6)
2,5,8,11,14-Pentaoxoipentadecane
  87 (35)
n-Pentatriacontane 81 (23)
Pentazole 92 (24)
n-Pentadecylcarboxylic acid 115 (15)
Pentene-1 82 (6)
Pentyne-1 82 (30)
Pentyne-2 82 (31)
Perbenzoic acid 116 (24)
Perylene 84 (12)
Δ¹,⁵-Phellandrene or α-phellandrene
  95 (19)
Δ⁽⁷⁾¹²-or β-phellandrene 95 (20)
Phene 83 (12)
Phenanthrene 84 (3)
Phenanthroquinone 111 (20)
3,4-Phenanthrenediol 107 (25)
Phenanthridine 93 (16)
4,5-Phenanthroline 93 (21)
Phenazine 93 (20)
Phenetol 99 (6)
Phenol 107 (7) 131
Phenothiazine 94 (9)
Phenoxazine 94 (6)
Phenoxybenzene 103 (1)
Phentiazine 93 (12) 94 (9)
Phenylacetic acid 116 (3)
Phenylacetylene 96 (25)
Phenylacetonitrile 122 (4)
trans-β-phenylacyl carboxylic
  acid 116 (4)
L-Phenylamine 124 (12)
N-Phenylaniline 103 (9)
Phenylazide 12 (4)
Phenylbenzene 98 (3)

N-Phenylbenzylamine 103 (10)
Phenylbromide 125 (43)
Phenylcarbinol 109 (9)
Phenylchloride 125 (22)
Phenylcyanide 122 (17)
Phenylcyclohexane 98 (2)
N-Phenyldiethylamine 100 (6)
1-Phenyl-4-dimethylamino-2,
  3-dimethylpyrazolone 114 (23)
Phenyldisulphide 103 (6)
Phenyldithiobenzene 103 (6)
m-Phenelenediamine 119 (19)
o-Phenelenediamine 119 (18)
p-Phenelenediamine 119 (20)
Phenylethane 96 (15)
Phenylethanol 109 (10)
Phenylether 103 (1)
β or 2-Phenylethylalcohol 109 (10)
β or 2-Phenylethylamine 120 (12)
Phenylethylene 96 (22)
Phenylfluoride 125 (11)
N-Phenylglycine 116 (11)
Phenylhydrazine 121 (3) 131
Phenylhydroxylamine 120 (2)
Phenylisocyanate 122 (3)
Phenylisothiocyanecite 127 (23)
Phenyliodide 126 (8)
Phenylmethane 96 (1)
2-Phenyl-2-methylpropane 96 (19)
Phenylphosphine 128 (12)
1-Phenylpropane 96 (2)
2-Phenylpropenal 110 (20)
1-Phenylpropene 96 (23)
2-Phenylpropene 96 (24)
3-Phenyl-2-propene-1-ol 109 (11)
2-Phenylpyridine 104 (1)
4-Phenylpyridine 104 (3)
3-Phenylpyridine 104 (2)
1-Phenylpyrazolone 104 (10)
2-Phenylquinoline 104 (6)
6-Phenylquinoline 104 (7)
8-Phenylquinoline 104 (8)
2-Phenylquinoline-4-carboxylic
  acid 116 (23)
p-Phenylquinonediimine 121 (7)
Phenylsulphide 103 (5)
Phenylthiobenzene 103 (5)
Phenylthiocyanate 127 (22)
Phenylvinylether 99 (8)
Phloroglucinol 107 (13)
Phloroglucinol triethylether 99 (21)
Phloroglucinol trimethylether 99 (20)
Phorone 110 (34)
Phosgene 126 (23)
Phosphaniline 128 (12)
1-Phospha-2,8,9-trioxaada
  mantane 94 (18)
p-Phosphanobenzoic acid 128 (14)
Phalhabazine 93 (8)
o-Phthalic acid 116 (9)
Phthalicanhydride 113 (16)
Pthalimide 113 (25)
Phyllopyrrole 101 (25)
Phytol 106 (32)
α or 2-Picoline 102 (7)
β or 3-Picoline 102 (8)
γ or 4-Picoline 102 (9)
Picolinic acid 116 (18)
Picric acid 123 (4)
Pimelic acid 115 (36)
Pinacoline 110 (32)
Pinacolone 32 (110)
d,l-Pinene 97 (19)
d,l-α-Pinene 97 (19)
Piperazine 92 (12)
2,5,-(1,4) Piperazinedione 108 (26)
Piperinic acid 116 (16)
Piperonal 110 (22)
α-2-Piperoline 102 (2)
β or 3-Piperoline 102 (3)
Piperidine 91 (18)
Piperidyne 91 (19)
Porphin 94 (10)
Potassiumbenzoate 118 (40)
Potassiumhydrogenoxalate 118 (32)
Potassiumhydrogentartrate 118 (37)
Potassiumoxalate 118 (31)
Potassiumtartrate 118 (39)
Prehnitene 96 (14)
Propyne-1 82 (28) 131
Propiolacetone 113 (8)
Propiolic acid 115 (30)
Propionaldehyde 110 (3)

Molecular formulae of organic compounds

Basic principles in arrangement:
1. Increasing number of C atoms (C_1-C_{70}).
2. Increasing number of H atoms or other atoms present.
   The molecular formulae of isomeric compounds are repeatedly entered, followed by molecular formula:
   (i) Number=Page number; (ii) Number in brackets= serial number in the arrangement on that page.

Sequence of atoms in molecular formulae follows according to "GMELIN-Number" of elements (exception C):

C	1 H	8 O	9 N	10 F	11 Cl	12 Br	13 I	15 S	16 Se	17 Te	19 B	21 Si	22 P	23 As	24 Sb	25 Bi	26 Li	27 Na	28 K	
																				(see also p. 77)

Peculiarities: $\{C_{\frac{1}{3}}\}_N$ Graphite 84 (15) $\{C_{\frac{1}{3}}\}_G$ Diamond 85 (8) N_5H 92 (24)

CHON 121 (8)
CHON 123 (10)
CHO$_6$N$_3$ 121 (18)
CHN 122 (6)
CHF$_3$ 125 (3)
CHFCl$_2$ 126 (11)
CHCl$_3$ 125 (14)
CHBr$_3$ 125 (35)
CHI$_3$ 126 (3)
CH$_2$O 110 (1)
CH$_2$O$_2$ 115 (1)
CH$_2$O$_4$N$_2$ 121 (17)
CH$_2$N$_2$ 121 (4)
CH$_2$N$_2$ 123 (13)
CH$_2$N$_4$ 92 (18)
CH$_2$F$_2$ 125 (2)
CH$_2$Cl$_2$ 125 (13)
CH$_2$Br$_2$ 125 (34)
CH$_2$I$_2$ 126 (2)
CH$_3$ON 121 (9)
CH$_3$ON 122 (24)
CH$_3$O$_2$N 121 (16)
CH$_3$F 125 (1)
CH$_3$Cl 125 (12)
CH$_3$Br 125 (33)
CH$_3$I 126 (1)
CH$_4$ 80 (1)
CH$_4$O 106 (1)
CH$_4$ON$_2$ 122 (27)
CH$_4$O$_3$P 128 (13)
CH$_4$O$_3$As 128 (17)
CH$_4$O$_4$S 128 (19)
CH$_4$N$_2$S 127 (24)
CH$_4$S 127 (1)
CH$_5$ON 120 (20)
CH$_5$ON 120 (21)
CH$_5$ON$_3$ 124 (16)
CH$_5$N 119 (1)
CH$_5$N$_3$ 123 (1)
CH$_5$N$_3$S 127 (25)
CH$_5$P 128 (10)
CH$_6$As 128 (15)
CH$_6$N$_2$ 121 (1)
CH$_6$B$_2$ 89 (29)
COCl$_2$ 126 (23)
CO$_6$N$_4$ 121 (19)
CF$_2$Cl$_2$ 126 (12)
CF$_3$Cl 126 (13)
CF$_3$I 126 (14)
CF$_4$ 125 (4)
CCl$_4$ 125 (15)
CBr$_4$ 125 (36)
CI$_4$ 126 (4)
CS$_2$ 127 (11)

C$_2$HOCl$_3$ 126 (21)
C$_2$HOBr$_3$ 126 (31)
C$_2$HO$_2$Cl$_3$ 126 (29)
C$_2$HCl$_3$ 125 (20)
C$_2$H$_2$ 82 (27)

C$_2$H$_2$O 111 (4)
C$_2$H$_2$ON$_2$ 94 (4)
C$_2$H$_2$ON$_2$ 94 (5)
C$_2$H$_2$O$_2$ 110 (16)
C$_2$H$_2$O$_2$Cl$_2$ 126 (28)
C$_2$H$_2$O$_3$ 118 (1)
C$_2$H$_2$O$_4$ 115 (31)
C$_2$H$_2$N$_4$ 92 (19)
C$_2$H$_2$N$_4$ 92 (20)
C$_2$H$_2$N$_4$ 92 (21)
C$_2$H$_2$N$_4$ 92 (22)
C$_2$H$_2$Cl$_2$ 125 (19)
C$_2$H$_2$Cl$_4$ 125 (17)
C$_2$H$_2$Br$_2$ 125 (40)
C$_2$H$_2$Br$_2$ 125 (41)
C$_2$H$_3$ON 122 (1)
C$_2$H$_3$ON 122 (2)
C$_2$H$_3$OCl 126 (22)
C$_2$H$_3$OBr$_3$ 126 (30)
C$_2$H$_3$O$_2$N 121 (20)
C$_2$H$_3$O$_2$Cl 126 (24)
C$_2$H$_3$O$_2$Cl 126 (27)
C$_2$H$_3$O$_2$Cl$_3$ 126 (20)
C$_2$H$_3$N 122 (7)
C$_2$H$_3$N 122 (8)
C$_2$H$_3$NS 127 (16)
C$_2$H$_3$NS 127 (17)
C$_2$H$_3$N$_3$ 92 (13)
C$_2$H$_3$N$_3$ 92 (14)
C$_2$H$_3$F 125 (8)
C$_2$H$_3$F$_3$ 125 (7)
C$_2$H$_3$Cl 125 (18)
C$_2$H$_3$Br 125 (39)
C$_2$H$_3$I 126 (6)
C$_2$H$_4$ 82 (1)
C$_2$H$_4$O 90 (1)
C$_2$H$_4$O 106 (24)
C$_2$H$_4$O 110 (2)
C$_2$H$_4$O$_2$ 112 (1)
C$_2$H$_4$O$_2$ 115 (2)
C$_2$H$_4$O$_2$ 117 (3)
C$_2$H$_4$O$_2$N$_2$ 122 (26)
C$_2$H$_4$O$_3$ 117 (15)
C$_2$H$_4$N$_4$ 92 (22)
C$_2$H$_4$N$_4$ 92 (23)
C$_2$H$_4$N$_{14}$ 104 (21)
C$_2$H$_4$F$_2$ 125 (6)
C$_2$H$_4$Br$_2$ 125 (38)
C$_2$H$_4$S 91 (1)
C$_2$H$_5$ON 121 (10)
C$_2$H$_5$ON 122 (25)
C$_2$H$_5$OCl 126 (19)
C$_2$H$_5$O$_2$N 123 (14)
C$_2$H$_5$O$_2$N 128 (26)
C$_2$H$_5$O$_2$N$_3$ 124 (17)
C$_2$H$_5$O$_2$N$_3$ 122 (28)
C$_2$H$_5$N 91 (13)
C$_2$H$_5$F 125 (5)
C$_2$H$_5$Cl 125 (16)
C$_2$H$_5$Br 125 (37)

C$_2$H$_5$I 126 (5)
C$_2$H$_6$ 80 (2)
C$_2$H$_6$O 87 (1)
C$_2$H$_6$O 106 (2)
C$_2$H$_6$O$_2$ 106 (33)
C$_2$H$_6$O$_2$S 127 (12)
C$_2$H$_6$O$_3$S 127 (13)
C$_2$H$_6$O$_3$S 128 (23)
C$_2$H$_6$O$_4$S 128 (21)
C$_2$H$_6$O$_4$S 128 (20)
C$_2$H$_6$S 88 (1)
C$_2$H$_6$S 127 (2)
C$_2$H$_6$S$_2$ 88 (10)
C$_2$H$_6$Se 88 (15)
C$_2$H$_6$Te 88 (17)
C$_2$H$_6$Be 89 (34)
C$_2$H$_6$Zn 89 (36)
C$_2$H$_6$Cd 89 (38)
C$_2$H$_6$Hg 89 (39)
(C$_2$H$_6$Sn) 89 (23)
C$_2$H$_7$ON 122 (19)
C$_2$H$_7$ON 120 (23)
C$_2$H$_7$ON 120 (22)
C$_2$H$_7$O$_3$NS 127 (27)
C$_2$H$_7$N 119 (2)
C$_2$H$_7$N 88 (19)
C$_2$H$_7$N$_3$ 88 (44)
C$_2$H$_7$P 128 (11)
C$_2$H$_7$P 89 (1)
C$_2$H$_7$As 89 (5)
C$_2$H$_7$As 128 (16)
C$_2$H$_8$N$_2$ 88 (40)
C$_2$H$_8$N$_2$ 121 (2)
C$_2$H$_8$N$_2$ 119 (9)
C$_2$H$_8$Si 89 (15)
C$_2$H$_{10}$B$_2$ 89 (30)
C$_2$F$_4$ 125 (9)
C$_2$F$_4$Cl$_2$ 126 (15)
C$_2$F$_6$Cl 126 (16)

C$_3$H$_2$O 110 (15)
C$_3$H$_2$O$_2$ 115 (30)
C$_3$H$_2$O$_3$N$_2$ 113 (19)
C$_3$H$_2$O$_5$ 118 (8a)
C$_3$H$_2$N$_2$ 122 (11)
C$_3$H$_3$ON 94 (1)
C$_3$H$_3$ON 94 (2)
C$_3$H$_3$O$_3$N$_3$ 108 (27)
C$_3$H$_3$N 122 (14)
C$_3$H$_3$NS 94 (7)
C$_3$H$_3$NS 94 (8)
C$_3$H$_3$N$_3$ 92 (15)
C$_3$H$_3$N$_3$ 92 (16)
C$_3$H$_3$N$_3$ 92 (17)
C$_3$H$_4$ 83 (2)
C$_3$H$_4$ 82 (26)
C$_3$H$_4$ 82 (28)
C$_3$H$_4$O 101 (2)
C$_3$H$_4$O 106 (26)
C$_3$H$_4$O 110 (13)

C$_3$H$_4$O$_2$ 113 (8)
C$_3$H$_4$O$_2$ 115 (21)
C$_3$H$_4$O$_2$ 117 (1)
C$_3$H$_4$O$_3$ 118 (2)
C$_3$H$_4$O$_4$ 115 (32)
C$_3$H$_4$O$_6$ 118 (8b)
C$_3$H$_4$N$_2$ 92 (7)
C$_3$H$_4$N$_2$ 92 (9)
C$_3$H$_4$N$_2$S 119 (32)
C$_3$H$_5$OCl 125 (32)
C$_3$H$_5$OBr 126 (32)
C$_3$H$_5$O$_2$Br 126 (33)
C$_3$H$_5$O$_4$ 117 (19)
C$_3$H$_5$O$_6$N$_3$ 128 (25)
C$_3$H$_5$N 122 (9)
C$_3$H$_5$N 122 (10)
C$_3$H$_5$NS 127 (18)
C$_3$H$_5$NS 127 (19)
C$_3$H$_5$F 125 (10)
C$_3$H$_5$Cl 125 (21)
C$_3$H$_5$Br 125 (42)
C$_3$H$_5$I 126 (7)
C$_3$H$_6$ 83 (1)
C$_3$H$_6$ 82 (3)
C$_3$H$_6$O 90 (2)
C$_3$H$_6$O 101 (1)
C$_3$H$_6$O 106 (25)
C$_3$H$_6$O 110 (3)
C$_3$H$_6$O 110 (26)
C$_3$H$_6$O$_2$ 92 (1)
C$_3$H$_6$O$_2$ 109 (21)
C$_3$H$_6$O$_2$ 112 (2)
C$_3$H$_6$O$_2$ 112 (3)
C$_3$H$_6$O$_2$ 115 (3)
C$_3$H$_6$O$_3$ 90 (10)
C$_3$H$_6$O$_3$ 117 (5)
C$_3$H$_6$O$_3$ 117 (10)
C$_3$H$_6$O$_3$ 117 (16)
C$_3$H$_6$O$_3$ 117 (17)
C$_3$H$_6$S 91 (2)
C$_3$H$_6$N$_2$ 92 (8)
C$_3$H$_6$S$_3$ 91 (10)
C$_3$H$_7$ON 121 (11)
C$_3$H$_7$O$_2$N 123 (16)
C$_3$H$_7$O$_2$N 123 (15)
C$_3$H$_7$O$_2$NS 127 (26)
C$_3$H$_7$O$_3$N 124 (7)
C$_3$H$_7$O$_5$NS 127 (28)
C$_3$H$_7$N 119 (8)
C$_3$H$_7$N 91 (14)
C$_3$H$_8$ 80 (3)
C$_3$H$_8$O 87 (2)
C$_3$H$_8$O 106 (3)
C$_3$H$_8$O 106 (4)
C$_3$H$_8$O$_2$ 87 (26)
C$_3$H$_8$O$_2$ 106 (34)
C$_3$H$_8$O$_2$ 106 (35)
C$_3$H$_8$O$_2$ 108 (1)
C$_3$H$_8$O$_3$ 106 (38)
C$_3$H$_8$S 88 (2)

268

C_3H_6S 127 (3)
C_3H_6S 127 (4)
C_3H_9ON 108 (10)
$C_3H_9O_3B$ 128 (32)
$C_3H_9O_4P$ 128 (28)
$C_3H_9O_6P$ 128 (30)
C_3H_9N 88 (20)
C_3H_9N 88 (34)
C_3H_9N 119 (3)
$C_3H_9N_3$ 91 (22)
C_3H_9B 89 (27)
C_3H_9P 89 (3)
C_3H_9As 89 (7)
C_3H_9Sb 89 (11)
C_3H_9Bi 89 (6)
C_3H_9Al 89 (32)
$C_3H_{10}N_2$ 119 (10)
C_3O_2 111 (6)

C_4H_2 82 (37)
$C_4H_2O_3$ 113 (12)
$C_4H_2O_4$ 115 (42)
$C_4H_2O_4N_2$ 113 (21)
C_4H_4 82 (39)
C_4H_4O 90 (5)
$C_4H_4O_2$ 114 (11)
$C_4H_4O_2N_2$ 108 (25)
$C_4H_4O_3$ 113 (11)
$C_4H_4O_3N_2$ 113 (20)
$C_4H_4O_4$ 113 (13)
$C_4H_4O_4$ 115 (40)
$C_4H_4O_4$ 115 (41)
$C_4H_4O_5$ 118 (5)
$C_4H_4O_6$ 118 (6)
$C_4H_4O_6$ 118 (7)
$C_4H_4N_2$ 92 (10)
$C_4H_4N_2$ 92 (11)
$C_4H_4N_2$ 92 (12)
$C_4H_4N_2$ 122 (12)
C_4H_4S 91 (5)
C_4H_5N 91 (17)
C_4H_5N 122 (15)
C_4H_5N 122 (16)
C_4H_5NS 119 (31)
C_4H_5NS 127 (20)
C_4H_5NS 127 (21)
C_4H_6 82 (29)
C_4H_6 83 (4)
C_4H_6 82 (14)
C_4H_6O 87 (24)
C_4H_6O 90 (4)
C_4H_6O 110 (14)
C_4H_6O 111 (5)
C_4H_6O 111 (7)
$C_4H_6O_2$ 111 (1)
$C_4H_6O_2$ 112 (21)
$C_4H_6O_2$ 112 (9)
$C_4H_6O_2$ 115 (22)
$C_4H_6O_2$ 115 (23)
$C_4H_6O_2$ 115 (24)
$C_4H_6O_2N_2$ 108 (26)
$C_4H_6O_2N_2$ 113 (18)
$C_4H_6O_3$ 113 (1)
$C_4H_6O_3$ 118 (3)
$C_4H_6O_4$ 113 (7)
$C_4H_6O_4$ 115 (33)
$C_4H_6O_5$ 117 (25)
$C_4H_6O_6$ 117 (26)
$C_4H_6O_8$ 118 (10)
C_4H_6S 91 (4)
C_4H_7O 101 (10)
$C_4H_7ON_3$ 123 (2)
$C_4H_7O_4N$ 124 (3)
C_4H_7N 91 (16)
C_4H_8 82 (3)

C_4H_8 82 (4)
C_4H_8 82 (5)
C_4H_8 83 (3)
C_4H_8 95 (1)
C_4H_8O 87 (21)
C_4H_8O 87 (22)
C_4H_8O 90 (3)
C_4H_8O 101 (3)
C_4H_8O 101 (4)
C_4H_8O 101 (5)
C_4H_8O 101 (9)
C_4H_8O 110 (4)
C_4H_8O 110 (5)
C_4H_8O 110 (27)
$C_4H_8O_2$ 92 (2)
$C_4H_8O_2$ 112 (4)
$C_4H_8O_2$ 112 (5)
$C_4H_8O_2$ 112 (6)
$C_4H_8O_2$ 115 (4)
$C_4H_8O_2$ 117 (4)
$C_4H_8O_2$ 117 (8)
$C_4H_8O_2N_2$ 121 (12)
$C_4H_8O_3$ 117 (18)
$C_4H_8O_3N_2$ 124 (5)
$C_4H_8O_3N_2$ 123 (3)
C_4H_8S 88 (8)
C_4H_8S 91 (3)
$C_4H_8S_2$ 91 (9)
C_4H_9ON 94 (3)
$C_4H_9O_2N_3$ 124 (18)
$C_4H_9O_3N$ 124 (8)
C_4H_9N 91 (15)
C_4H_{10} 80 (4)
C_4H_{10} 80 (5)
$C_4H_{10}O$ 87 (4)
$C_4H_{10}O$ 87 (3)
$C_4H_{10}O$ 106 (5)
$C_4H_{10}O$ 106 (6)
$C_4H_{10}O$ 106 (7)
$C_4H_{10}O$ 106 (8)
$C_4H_{10}ON$ 119 (21)
$C_4H_{10}O_2$ 87 (29)
$C_4H_{10}O_2$ 87 (36)
$C_4H_{10}O_2$ 106 (36)
$C_4H_{10}O_2$ 108 (2)
$C_4H_{10}O_2$ 87 (38)
$C_4H_{10}O_3$ 108 (6)
$C_4H_{10}O_3S$ 128 (24)
$C_4H_{10}O_4$ 106 (39)
$C_4H_{10}O_4S$ 128 (22)
$C_4H_{10}N_2$ 91 (21)
$C_4H_{10}S$ 127 (5)
$C_4H_{10}S$ 88 (3)
$C_4H_{10}S_2$ 88 (11)
$C_4H_{10}Se$ 88 (16)
$C_4H_{10}Te$ 88 (18)
$C_4H_{10}Be$ 89 (35)
$C_4H_{10}Zn$ 89 (37)
$C_4H_{10}Hg$ 89 (40)
$C_4H_{10}Sn$ 89 (24)
$C_4H_{11}ON$ 108 (11)
$C_4H_{11}ON$ 108 (13)
$C_4H_{11}O_2N$ 108 (15)
$C_4H_{11}N$ 88 (21)
$C_4H_{11}N$ 88 (22)
$C_4H_{11}N$ 88 (35)
$C_4H_{11}N$ 119 (4)
$C_4H_{11}P$ 89 (2)
$C_4H_{11}As$ 89 (6)
$C_4H_{12}N_2$ 119 (11)
$C_4H_{12}N_2$ 88 (41)
$C_4H_{12}As_2$ 89 (9)
$C_4H_{12}Si$ 89 (16)
$C_4H_{12}Ge$ 89 (19)
$C_4H_{12}Sn$ 89 (21)

$C_4H_{12}Pb$ 89 (25)
$C_4H_{13}N_3$ 119 (23)
$C_4H_{14}B_2$ 89 (31)

C_5H_4OS 110 (25)
$C_5H_4O_2$ 110 (23)
$C_5H_4O_2$ 113 (10)
$C_5H_4O_2N_4$ 108 (28)
$C_5H_4O_3$ 116 (12)
$C_5H_4O_3N_4$ 108 (29)
$C_5H_4N_4$ 93 (22)
C_5H_5ON 108 (19)
C_5H_5ON 108 (20)
C_5H_5ON 108 (21)
$C_5H_5ON_2$ 122 (23)
$C_5H_5O_2N$ 108 (22)
$C_5H_5O_2N$ 108 (23)
$C_5H_5O_3N$ 108 (24)
C_5H_5N 91 (20)
$C_5H_5N_5$ 119 (30)
C_5H_6 83 (7)
C_5H_6O 90 (8)
C_5H_6O 101 (6)
C_5H_6O 101 (7)
$C_5H_6ON_2$ 122 (22)
C_5H_6OS 109 (24)
$C_5H_6O_2$ 109 (23)
$C_5H_6O_2$ 114 (13)
$C_5H_6N_2$ 119 (16)
$C_5H_6O_5$ 118 (9)
$C_5H_6N_2$ 119 (25)
$C_5H_6N_2$ 119 (26)
$C_5H_6N_2$ 119 (27)
C_5H_6S 91 (8)
C_5H_6S 101 (13)
C_5H_6S 101 (14)
C_5H_7N 101 (20)
C_5H_8 82 (15)
C_5H_8 82 (16)
C_5H_8 82 (17)
C_5H_8 82 (30)
C_5H_8 82 (31)
C_5H_8 82 (32)
C_5H_8 83 (6)
C_5H_8O 90 (7)
C_5H_8O 111 (8)
$C_5H_8O_2$ 110 (24)
$C_5H_8O_2$ 111 (2)
$C_5H_8O_2$ 117 (2)
$C_5H_8O_3$ 118 (4)
$C_5H_8O_4$ 115 (34)
C_5H_8S 91 (7)
$C_5H_9O_2N$ 116 (17)
$C_5H_9O_4N$ 124 (4)
C_5H_9N 91 (19)
$C_5H_9N_2$ 120 (18)
C_5H_{10} 82 (6)
C_5H_{10} 82 (7)
C_5H_{10} 83 (5)
C_5H_{10} 95 (2)
C_5H_{10} 95 (3)
$C_5H_{10}O$ 87 (23)
$C_5H_{10}O$ 90 (6)
$C_5H_{10}O$ 101 (8)
$C_5H_{10}O$ 107 (1)
$C_5H_{10}O$ 110 (6)
$C_5H_{10}O$ 110 (7)
$C_5H_{10}O$ 110 (28)
$C_5H_{10}O$ 110 (29)
$C_5H_{10}O_2$ 109 (22)
$C_5H_{10}O_2$ 112 (7)
$C_5H_{10}O_2$ 112 (8)
$C_5H_{10}O_2$ 112 (9)
$C_5H_{10}O_2$ 112 (10)

$C_5H_{10}O_2$ 115 (5)
$C_5H_{10}O_3N_2$ 124 (6)
$C_5H_{10}S$ 91 (6)
$C_5H_{11}O_2N$ 123 (17)
$C_5H_{11}O_2N$ 128 (27)
$C_5H_{11}N$ 101 (19)
$C_5H_{11}N$ 91 (18)
C_5H_{12} 81 (1)
C_5H_{12} 80 (6)
C_5H_{12} 80 (7)
C_5H_{12} 80 (8)
$C_5H_{12}O$ 87 (5)
$C_5H_{12}O$ 87 (6)
$C_5H_{12}O$ 106 (9)
$C_5H_{12}O$ 106 (10)
$C_5H_{12}O$ 106 (11)
$C_5H_{12}O_2$ 87 (27)
$C_5H_{12}O_2$ 87 (30)
$C_5H_{12}O_2$ 106 (37)
$C_5H_{12}O_2$ 108 (3)
$C_5H_{12}O_2$ 108 (4)
$C_5H_{12}O_2N_2$ 124 (1)
$C_5H_{12}O_3$ 108 (7)
$C_5H_{12}O_4$ 106 (40)
$C_5H_{12}S$ 127 (6)
$C_5H_{13}O_2N$ 108 (16)
$C_5H_{13}N$ 88 (23)
$C_5H_{13}N$ 88 (24)
$C_5H_{13}N$ 88 (25)
$C_5H_{13}N$ 119 (5)
$C_5H_{14}N_2$ 119 (12)

$C_6H_3O_4N_2F$ 126 (18)
$C_6H_3O_6N_3$ 121 (23)
$C_6H_3O_7N_3$ 123 (4)
$C_6H_3O_8N_3$ 123 (5)
$C_6H_4O_2$ 111 (14)
$C_6H_4O_4N_2$ 121 (22)
$C_6H_4O_5$ 116 (15)
$C_6H_4N_4$ 93 (23)
$C_6H_4Cl_2$ 125 (23)
$C_6H_4Cl_2$ 125 (24)
$C_6H_4Br_2$ 125 (44)
C_6H_5ON 121 (14)
C_6H_5OCl 126 (25)
C_6H_5OI 126 (9)
$C_6H_5O_2I$ 126 (10)
$C_6H_5O_2$ 116 (18)
$C_6H_5O_2N$ 116 (19)
$C_6H_5O_2N$ 116 (20)
$C_6H_5O_2N$ 121 (21)
$C_6H_5N_3$ 93 (11)
$C_6H_5N_3$ 122 (4)
C_6H_5F 125 (11)
C_6H_5Cl 125 (22)
$C_6H_5Cl_3$ 125 (27)
C_6H_5Br 125 (43)
C_6H_5I 126 (8)
C_6H_6 82 (38)
C_6H_6 82 (40)
C_6H_6 83 (12)
C_6H_6 90 (5a)
C_6H_6 95 (3)
C_6H_6O 107 (7)
$C_6H_6ON_2$ 121 (5)
$C_6H_6ON_2$ 122 (30)
$C_6H_6O_2$ 107 (8)
$C_6H_6O_2$ 107 (9)
$C_6H_6O_2$ 107 (10)
$C_6H_6O_2N_2$ 123 (8)
$C_6H_6O_2N_2$ 123 (9)
$C_6H_6O_2S$ 127 (14)
$C_6H_6O_3$ 107 (11)
$C_6H_6O_3$ 107 (12)
$C_6H_6O_3$ 107 (13)

C$_6$H$_6$O$_3$S 127 (15)
C$_6$H$_6$O$_4$ 107 (14)
C$_6$H$_6$O$_4$ 107 (15)
C$_6$H$_6$O$_6$ 107 (16)
C$_6$H$_6$O$_6$ 116 (14)
C$_6$H$_6$N$_2$ 121 (6)
C$_6$H$_6$Cl$_6$ 125 (28)
C$_6$H$_6$S 127 (8)
C$_6$H$_7$ON 120 (24)
C$_6$H$_7$ON 122 (20)
C$_6$H$_7$O$_3$NS 127 (29)
C$_6$H$_7$O$_3$NS 127 (30)
C$_6$H$_7$N 102 (7)
C$_6$H$_7$N 102 (8)
C$_6$H$_7$N 102 (9)
C$_6$H$_7$N 119 (15)
C$_6$H$_7$P 128 (12)
C$_6$H$_8$ 83 (10)
C$_6$H$_8$ 83 (11)
C$_6$H$_8$O$_2$ 101 (9)
C$_6$H$_8$O$_2$N$_2$S 128 (4)
C$_6$H$_8$O$_3$As 128 (18)
C$_6$H$_8$O$_4$ 114 (14)
C$_6$H$_8$O$_6$ 116 (1)
C$_6$H$_8$O$_7$ 117 (20)
C$_6$H$_8$N$_2$ 102 (19)
C$_6$H$_3$N$_2$ 119 (18)
C$_6$H$_8$N$_2$ 119 (19)
C$_6$H$_8$N$_2$ 119 (20)
C$_6$H$_8$N$_2$ 121 (3)
C$_6$H$_8$N$_2$ 122 (13)
C$_6$H$_8$S 101 (15)
C$_6$H$_8$S 101 (16)
C$_6$H$_8$S 101 (17)
C$_6$H$_9$O$_2$N$_2$ 124 (14)
C$_6$H$_9$N 101 (21)
C$_6$H$_{10}$ 82 (18)
C$_6$H$_{10}$ 82 (19)
C$_6$H$_{10}$ 82 (20)
C$_6$H$_{10}$ 82 (33)
C$_6$H$_{10}$ 82 (34)
C$_6$H$_{10}$ 82 (35)
C$_6$H$_{10}$ 82 (36)
C$_6$H$_{10}$ 83 (9)
C$_6$H$_{10}$O 87 (25)
C$_6$H$_{10}$O 110 (33)
C$_6$H$_{10}$O 111 (9)
C$_6$H$_{10}$O$_2$ 111 (3)
C$_6$H$_{10}$O$_3$ 112 (22)
C$_6$H$_{10}$O$_3$ 113 (2)
C$_6$H$_{10}$O$_4$ 112 (24)
C$_6$H$_{10}$O$_4$ 115 (35)
C$_6$H$_{10}$N$_4$ 93 (24)
C$_6$H$_{10}$S 88 (9)
C$_6$H$_{10}$S$_3$ 88 (14)
C$_6$H$_{11}$ON 113 (17)
C$_6$H$_{11}$O$_2$N$_2$Br 126 (34)
C$_6$H$_{12}$ 82 (9)
C$_6$H$_{12}$ 83 (8)
C$_6$H$_{12}$ 95 (4)
C$_6$H$_{12}$O 107 (2)
C$_6$H$_{12}$O 110 (8)
C$_6$H$_{12}$O 110 (30)
C$_6$H$_{12}$O 110 (31)
C$_6$H$_{12}$O 110 (32)
C$_6$H$_{12}$O$_2$ 107 (3)
C$_6$H$_{12}$O$_2$ 107 (4)
C$_6$H$_{12}$O$_2$ 112 (11)
C$_6$H$_{12}$O$_2$ 115 (6)
C$_6$H$_{12}$O$_2$ 117 (9)
C$_6$H$_{12}$O$_3$N 94 (17)
C$_6$H$_{12}$O$_3$P 94 (18)
C$_6$H$_{12}$O$_6$ 107 (5)
C$_6$H$_{12}$O$_6$ 107 (6)
C$_6$H$_{12}$N$_2$ 88 (43)

C$_6$H$_{12}$N$_4$ 94 (16)
C$_6$H$_{13}$O$_2$N 123 (18)
C$_6$H$_{13}$O$_2$N 123 (19)
C$_6$H$_{13}$N 102 (1)
C$_6$H$_{13}$N 102 (2)
C$_6$H$_{13}$N 102 (3)
C$_6$H$_{13}$N 119 (14)
C$_6$H$_{14}$ 81 (2)
C$_6$H$_{14}$ 80 (9)
C$_6$H$_{14}$ 80 (10)
C$_6$H$_{14}$ 80 (11)
C$_6$H$_{14}$ 80 (12)
C$_6$H$_{14}$ 80 (13)
C$_6$H$_{14}$O 87 (7)
C$_6$H$_{14}$O 87 (8)
C$_6$H$_{14}$O 87 (9)
C$_6$H$_{14}$O 87 (10)
C$_6$H$_{14}$O 87 (11)
C$_6$H$_{14}$O 106 (12)
C$_6$H$_{14}$O$_2$ 87 (31)
C$_6$H$_{14}$O$_2$ 87 (37)
C$_6$H$_{14}$O$_2$ 108 (5)
C$_6$H$_{14}$O$_2$N$_2$ 124 (2)
C$_6$H$_{14}$O$_2$N$_4$ 124 (19)
C$_6$H$_{14}$O$_3$ 87 (32)
C$_6$H$_{14}$O$_3$ 108 (8)
C$_6$H$_{14}$O$_4$ 108 (9)
C$_6$H$_{14}$N$_2$ 88 (42)
C$_6$H$_{14}$S 88 (4)
C$_6$H$_{14}$S 88 (5)
C$_6$H$_{14}$S 127 (7)
C$_6$H$_{15}$ON 108 (12)
C$_6$H$_{15}$ON 108 (14)
C$_6$H$_{15}$O$_2$N 108 (17)
C$_6$H$_{15}$O$_2$N 119 (22)
C$_6$H$_{15}$O$_3$N 108 (18)
C$_6$H$_{15}$O$_4$P 128 (29)
C$_6$H$_{15}$N 88 (26)
C$_6$H$_{15}$N 88 (27)
C$_6$H$_{15}$N 88 (28)
C$_6$H$_{15}$N 88 (36)
C$_6$H$_{15}$N 119 (6)
C$_6$H$_{15}$B 89 (28)
C$_6$H$_{15}$P 89 (4)
C$_6$H$_{15}$As 89 (8)
C$_6$H$_{15}$Sb 89 (12)
C$_6$H$_{15}$Bi 89 (14)
C$_6$H$_{15}$AL 89 (33)
C$_6$H$_{16}$N$_3$ 119 (24)
C$_6$H$_{16}$N$_2$ 119 (13)
C$_6$H$_{18}$Si$_2$ 89 (18)

C$_7$H$_6$ON 122 (3)
C$_7$H$_6$OF 126 (17)
C$_7$H$_6$OCl 126 (26)
C$_7$H$_6$O$_3$N 123 (6)
C$_7$H$_6$O$_3$NS 128 (7)
C$_7$H$_6$O$_6$N$_3$ 121 (27)
C$_7$H$_6$N 122 (17)
C$_7$H$_6$NS 127 (22)
C$_7$H$_6$NS 127 (23)
C$_7$H$_6$N$_3$ 93 (12)
C$_7$H$_6$O 110 (17)
C$_7$H$_6$O$_2$ 116 (2)
C$_7$H$_6$O$_2$ 117 (6)
C$_7$H$_6$O$_3$ 116 (24)
C$_7$H$_6$O$_3$ 117 (23)
C$_7$H$_6$O$_4$ 117 (24)
C$_7$H$_6$O$_4$N$_2$ 121 (26)
C$_7$H$_6$O$_6$ 117 (27)
C$_7$H$_6$O$_6$P 128 (14)
C$_7$H$_7$N$_2$ 93 (6)
C$_7$H$_7$Cl$_2$ 125 (26)
C$_7$H$_7$ON 122 (29)
C$_7$H$_7$ON 122 (31)

C$_7$H$_7$ON 122 (32)
C$_7$H$_7$O$_2$N 121 (25)
C$_7$H$_7$O$_2$N 124 (9)
C$_7$H$_7$O$_2$N 124 (10)
C$_7$H$_7$O$_3$N 124 (11)
C$_7$H$_7$O$_4$NS 128 (6)
C$_7$H$_7$N$_3$ 122 (5)
C$_7$H$_7$Cl 125 (25)
C$_7$H$_8$ 83 (16)
C$_7$H$_8$ 96 (1)
C$_7$H$_8$ 99 (1)
C$_7$H$_8$O 109 (1)
C$_7$H$_8$O 109 (2)
C$_7$H$_8$O 109 (3)
C$_7$H$_8$O 109 (9)
C$_7$H$_8$O$_2$ 109 (8)
C$_7$H$_8$O$_2$ 109 (12)
C$_7$H$_8$N$_2$ 121 (13)
C$_7$H$_8$S 127 (9)
C$_7$H$_8$S 127 (10)
C$_7$H$_9$ON 120 (25)
C$_7$H$_9$N 100 (3)
C$_7$H$_9$N 102 (10)
C$_7$H$_9$N 102 (11)
C$_7$H$_9$N 102 (12)
C$_7$H$_9$N 120 (1)
C$_7$H$_9$N 120 (2)
C$_7$H$_9$N 120 (3)
C$_7$H$_9$N 120 (11)
C$_7$H$_{10}$ 83 (15)
C$_7$H$_{10}$O$_2$ 102 (27)
C$_7$H$_{10}$N$_2$ 120 (10)
C$_7$H$_{10}$S 101 (18)
C$_7$H$_{11}$N 101 (22)
C$_7$H$_{12}$ 82 (21)
C$_7$H$_{12}$ 83 (14)
C$_7$H$_{12}$ 95 (13)
C$_7$H$_{12}$O 111 (10)
C$_7$H$_{12}$O$_3$ 112 (23)
C$_7$H$_{12}$O$_4$ 112 (25)
C$_7$H$_{12}$O$_4$ 115 (36)
C$_7$H$_{12}$O$_6$ 117 (28)
C$_7$H$_{13}$N 93 (19)
C$_7$H$_{14}$ 82 (10)
C$_7$H$_{14}$ 83 (13)
C$_7$H$_{14}$ 95 (7)
C$_7$H$_{14}$O 110 (9)
C$_7$H$_{14}$O$_2$ 102 (26)
C$_7$H$_{14}$O$_2$ 112 (12)
C$_7$H$_{14}$O$_2$ 112 (13)
C$_7$H$_{14}$O$_2$ 112 (14)
C$_7$H$_{14}$O$_2$ 115 (7)
C$_7$H$_{14}$O$_3$ 94 (12)
C$_7$H$_{14}$N$_3$ 94 (15)
C$_7$H$_{15}$N 100 (1)
C$_7$H$_{15}$N 102 (4)
C$_7$H$_{15}$N 102 (5)
C$_7$H$_{16}$ 81 (3)
C$_7$H$_{16}$ 80 (14)
C$_7$H$_{16}$ 80 (15)
C$_7$H$_{16}$ 80 (16)
C$_7$H$_{16}$ 80 (17)
C$_7$H$_{16}$ 80 (18)
C$_7$H$_{16}$ 80 (19)
C$_7$H$_{16}$ 80 (20)
C$_7$H$_{16}$ 80 (21)
C$_7$H$_{16}$ 80 (22)
C$_7$H$_{16}$O 87 (12)
C$_7$H$_{16}$O 87 (13)
C$_7$H$_{16}$O 106 (13)
C$_7$H$_{16}$O$_2$ 87 (28)
C$_7$H$_{16}$O$_3$ 87 (39)

C$_8$H$_4$ONK 113 (25a)
C$_8$H$_4$O$_3$ 113 (16)

C$_8$H$_5$O$_2$N 113 (25)
C$_8$H$_5$O$_2$N 113 (24)
C$_8$H$_6$ 96 (25)
C$_8$H$_6$O 90 (12)
C$_8$H$_6$O$_3$ 110 (22)
C$_8$H$_6$O$_4$ 116 (9)
C$_8$H$_6$N$_2$ 93 (7)
C$_8$H$_6$N$_2$ 93 (8)
C$_8$H$_6$N$_2$ 93 (9)
C$_8$H$_6$N$_2$ 93 (10)
C$_8$H$_6$S 91 (12)
C$_8$H$_7$ON 123 (11)
C$_8$H$_7$ON 123 (12)
C$_8$H$_7$ON 113 (23)
C$_8$H$_7$ON 113 (22)
C$_8$H$_7$O$_2$N 117 (13)
C$_8$H$_7$N 91 (24)
C$_8$H$_7$N 122 (18)
C$_8$H$_8$ 83 (22)
C$_8$H$_8$ 96 (22)
C$_8$H$_8$O 114 (4)
C$_8$H$_8$O 90 (11)
C$_8$H$_8$O 99 (8)
C$_8$H$_8$O 110 (18)
C$_8$H$_8$O 110 (19)
C$_8$H$_8$O$_2$ 110 (21)
C$_8$H$_8$O$_2$ 116 (3)
C$_8$H$_8$O$_3$ 117 (7)
C$_8$H$_8$O$_3$ 117 (21)
C$_8$H$_8$S 91 (11)
C$_8$H$_9$ON 114 (10)
C$_8$H$_9$O$_2$N 116 (11)
C$_8$H$_9$N 91 (23)
C$_8$H$_{10}$ 83 (21)
C$_8$H$_{10}$ 96 (2)
C$_8$H$_{10}$ 96 (3)
C$_8$H$_{10}$ 96 (4)
C$_8$H$_{10}$ 96 (5)
C$_8$H$_{10}$ 96 (6)
C$_8$H$_{10}$O 99 (2)
C$_8$H$_{10}$O 99 (3)
C$_8$H$_{10}$O 99 (4)
C$_8$H$_{10}$O 99 (5)
C$_8$H$_{10}$O 99 (6)
C$_8$H$_{10}$O 109 (10)
C$_8$H$_{10}$ON$_2$ 121 (15)
C$_8$H$_{10}$O$_2$ 99 (13)
C$_8$H$_{10}$O$_2$ 99 (14)
C$_8$H$_{10}$O$_2$ 99 (15)
C$_8$H$_{10}$O$_2$ 116 (8)
C$_8$H$_{10}$O$_4$N$_2$ 100 (25)
C$_8$H$_{10}$O$_3$N 122 (21)
C$_8$H$_{11}$N 100 (4)
C$_8$H$_{11}$N 100 (5)
C$_8$H$_{11}$N 100 (8)
C$_8$H$_{11}$N 100 (9)
C$_8$H$_{11}$N 100 (10)
C$_8$H$_{11}$N 102 (13)
C$_8$H$_{11}$N 102 (14)
C$_8$H$_{11}$N 102 (15)
C$_8$H$_{11}$N 120 (4)
C$_8$H$_{11}$N 120 (5)
C$_8$H$_{11}$N 120 (6)
C$_8$H$_{11}$N 120 (7)
C$_8$H$_{11}$N 120 (8)
C$_8$H$_{11}$N 120 (9)
C$_8$H$_{11}$N 120 (12)
C$_8$H$_{12}$ 83 (19)
C$_8$H$_{12}$ 83 (20)
C$_8$H$_{12}$O$_2$ 114 (12)
C$_8$H$_{12}$O$_3$N$_2$ 114 (17)
C$_8$H$_{12}$O$_4$ 116 (7)
C$_8$H$_{12}$O$_6$ 112 (26)
C$_8$H$_{12}$N$_2$ 100 (24)
C$_8$H$_{12}$N$_2$ 120 (17)

$H_{13}ON$ 114 (18)
$H_{13}N$ 101 (23)
$H_{13}N$ 101 (24)
H_{14} 83 (18)
$H_{14}O$ 110 (35)
$H_{14}O$ 111 (11)
$H_{14}O_3$ 113 (3)
$H_{14}O_4$ 104 (18)
$H_{14}O_4$ 115 (37)
H_{16} 82 (11)
H_{16} 83 (17)
H_{16} 95 (8)
H_{16} 95 (9)
$H_{16}O$ 110 (10)
$H_{16}O_2$ 115 (8)
$H_{16}O_2N_2S_2$ 104 (22)
$H_{16}N_2$ 94 (14)
$H_{17}N$ 102 (6)
$H_{17}N$ 100 (2)
H_{18} 81 (4)
H_{18} 80 (23)
H_{18} 80 (24)
H_{18} 80 (25)
$H_{18}O$ 87 (14)
$H_{18}O$ 87 (15)
$H_{18}O$ 106 (14)
$H_{18}O_3$ 87 (33)
$H_{18}O_3$ 87 (41)
$H_{18}O_4$ 87 (34)
$H_{18}S$ 88 (6)
$H_{18}S_2$ 88 (12)
$H_{19}N$ 88 (29)
$H_{19}N$ 88 (30)
$H_{20}O_4Si$ 128 (31)
$H_{20}Si$ 89 (17)
$H_{20}As_2$ 89 (10)
$H_{20}Ge$ 89 (20)
$H_{20}Sn$ 89 (22)
$H_{20}Pb$ 89 (26)

$_9H_6O_2$ 113 (14)
$_9H_6O_2$ 113 (15)
$_9H_6O_4$ 117 (12)
$_9H_7N$ 93 (1)
$_9H_7N$ 93 (2)
$_9H_8$ 90 (12a)
$_9H_8$ 86 (2)
$_9H_8O$ 110 (20)
$_9H_8O_2$ 116 (4)
$_9H_8O_4$ 118 (11)
$C_9H_8N_2$ 119 (28)
$C_9H_8N_2$ 119 (29)
$C_9H_9O_2N_3S_2$ 128 (8)
$C_9H_9O_3$ 117 (22)
$C_9H_9O_3Ni_2$ 126 (38)
C_9H_9N 102 (20)
C_9H_{10} 96 (23)
C_9H_{10} 96 (24)
C_9H_{10} 90 (11a)
C_9H_{11} 86 (1)
C_9H_{11} 90 (11a)
$C_9H_{11}O_2N$ 124 (12)
$C_9H_{11}O_3N$ 124 (13)
$C_9H_{11}N$ 93 (3)
C_9H_{12} 96 (6)
C_9H_{12} 96 (7)
C_9H_{12} 96 (8)
C_9H_{12} 96 (9)
C_9H_{12} 96 (10)

C_9H_{12} 96 (11)
C_9H_{12} 96 (12)
C_9H_{12} 96 (13)
$C_9H_{12}O$ 99 (7)
$C_9H_{12}O_2$ 109 (17)
$C_9H_{12}O_3$ 99 (19)
$C_9H_{12}O_3$ 99 (20)
$C_9H_{13}N$ 100 (7)
$C_9H_{13}N$ 100 (11)
$C_9H_{13}N$ 100 (12)
$C_9H_{13}N$ 100 (13)
$C_9H_{13}N$ 102 (16)
$C_9H_{13}N$ 102 (17)
$C_9H_{14}O$ 110 (34)
$C_9H_{14}O_6$ 116 (13)
$C_9H_{15}N$ 101 (25)
C_9H_{16} 82 (22)
C_9H_{16} 82 (23)
$C_9H_{16}O_4$ 115 (38)
$C_9H_{17}ON(H_2O)$ 114 (15)
$C_9H_{17}N$ 93 (4)
$C_9H_{17}N$ 93 (5)
C_9H_{18} 82 (12)
C_9H_{18} 95 (10)
$C_9H_{18}O$ 106 (27)
$C_9H_{18}O$ 110 (11)
$C_9H_{18}O_2$ 112 (15)
$C_9H_{18}O_2$ 115 (9)
$C_9H_{18}S$ 94 (13)
C_9H_{20} 81 (5)
$C_9H_{20}O$ 106 (15)
$C_9H_{20}O_4$ 87 (42)
$C_9H_{21}N$ 88 (37)

$C_{10}H_6O_2$ 111 (15)
$C_{10}H_6O_2$ 111 (16)
$C_{10}H_6O_2$ 111 (17)
$C_{10}H_6O_8N_2S$ 128 (2)
$C_{10}H_7O_2N$ 116 (21)
$C_{10}H_7O_2N$ 121 (24)
$C_{10}H_7O_2N$ 121 (28)
$C_{10}H_8$ 86 (8)
$C_{10}H_8$ 84 (1)
$C_{10}H_8O$ 107 (17)
$C_{10}H_8O$ 107 (18)
$C_{10}H_8O_2$ 107 (22)
$C_{10}H_8O_2$ 107 (21)
$C_{10}H_8O_2$ 107 (20)
$C_{10}H_8O_2$ 107 (19)
$C_{10}H_8O_7S_2$ 127 (33)
$C_{10}H_8O_7S_2$ 127 (32)
$C_{10}H_8O_8S_2$ 127 (34)
$C_{10}H_8N_2$ 104 (16)
$C_{10}H_9O_3NS$ 127 (31)
$C_{10}H_9O_7NS_2$ 128 (1)
$C_{10}H_9N$ 102 (21)
$C_{10}H_9N$ 102 (22)
$C_{10}H_9N$ 102 (23)
$C_{10}H_9N$ 119 (16)
$C_{10}H_9N$ 119 (17)
$C_{10}H_{10}$ 85 (2)
$C_{10}H_{10}$ 85 (1)
$C_{10}H_{10}O_2$ 101 (12)
$C_{10}H_{10}O_2$ 101 (11)
$C_{10}H_{10}O_2N_4S$ 128 (9)
$C_{10}H_{10}N_2$ 104 (14)
$C_{10}H_{11}O_2N$ 104 (19)
$C_{10}H_{12}$ 85 (3)
$C_{10}H_{12}O$ 99 (23)
$C_{10}H_{12}O$ 99 (22)
$C_{10}H_{12}O_2$ 109 (18)
$C_{10}H_{12}O_2$ 109 (19)
$C_{10}H_{12}O_3$ 109 (20)
$C_{10}H_{13}N$ 102 (25)

$C_{10}H_{14}$ 85 (4)
$C_{10}H_{14}$ 96 (14)
$C_{10}H_{14}$ 96 (15)
$C_{10}H_{14}$ 96 (16)
$C_{10}H_{14}$ 96 (17)
$C_{10}H_{14}$ 96 (18)
$C_{10}H_{14}$ 96 (19)
$C_{10}H_{14}O$ 109 (4)
$C_{10}H_{14}O$ 109 (5)
$C_{10}H_{14}O$ 114 (2)
$C_{10}H_{14}ON_2$ 114 (20)
$C_{10}H_{14}O_2$ 99 (16)
$C_{10}H_{14}O_2$ 99 (17)
$C_{10}H_{14}O_2$ 99 (18)
$C_{10}H_{14}N_2$ 104 (15)
$C_{10}H_{15}N$ 100 (6)
$C_{10}H_{16}$ 85 (5)
$C_{10}H_{16}$ 85 (7)
$C_{10}H_{16}$ 95 (16)
$C_{10}H_{16}$ 95 (17)
$C_{10}H_{16}$ 95 (18)
$C_{10}H_{16}$ 95 (19)
$C_{10}H_{16}$ 95 (20)
$C_{10}H_{16}$ 95 (21)
$C_{10}H_{16}$ 97 (16)
$C_{10}H_{16}$ 97 (18)
$C_{10}H_{16}$ 97 (19)
$C_{10}H_{16}$ 97 (20)
$C_{10}H_{16}O$ 114 (6)
$C_{10}H_{16}O_4$ 116 (6)
$C_{10}H_{16}N$ 100 (23)
$C_{10}H_{17}Cl$ 125 (29)
$C_{10}H_{18}$ 82 (24)
$C_{10}H_{18}$ 82 (25)
$C_{10}H_{18}$ 85 (6)
$C_{10}H_{18}$ 95 (14)
$C_{10}H_{18}$ 95 (15)
$C_{10}H_{18}$ 97 (15)
$C_{10}H_{18}$ 97 (17)
$C_{10}H_{18}O$ 106 (28)
$C_{10}H_{18}O$ 106 (29)
$C_{10}H_{18}O$ 106 (30)
$C_{10}H_{18}O$ 109 (14)
$C_{10}H_{18}O$ 109 (15)
$C_{10}H_{18}O$ 114 (1)
$C_{10}H_{18}O_3$ 113 (4)
$C_{10}H_{18}O_4$ 115 (39)
$C_{10}H_{20}$ 82 (13)
$C_{10}H_{20}$ 95 (11)
$C_{10}H_{20}$ 95 (12)
$C_{10}H_{20}O$ 109 (13)
$C_{10}H_{20}O_2$ 112 (16)
$C_{10}H_{20}O_2$ 112 (17)
$C_{10}H_{20}O_2$ 115 (10)
$C_{10}H_{22}$ 81 (6)
$C_{10}H_{22}O$ 87 (16)
$C_{10}H_{22}O$ 87 (17)
$C_{10}H_{22}O$ 106 (16)
$C_{10}H_{22}O_3$ 87 (40)
$C_{10}H_{22}O_5$ 87 (35)

$C_{10}H_{22}S_2$ 88 (13)
$C_{10}H_{23}N$ 88 (31)
$C_{10}H_{23}N$ 88 (32)

$C_{11}H_7O_4N$ 116 (22)
$C_{11}H_8O_2$ 116 (5)
$C_{11}H_9N$ 104 (1)
$C_{11}H_9N$ 104 (2)
$C_{11}H_9N$ 104 (3)
$C_{11}H_{10}$ 97 (1)
$C_{11}H_{10}$ 97 (2)
$C_{11}H_{10}O$ 99 (10)
$C_{11}H_{10}O$ 99 (9)
$C_{11}H_{10}ON$ 104 (12)

$C_{11}H_{10}NS$ 104 (13)
$C_{11}H_{11}N$ 100 (16)
$C_{11}H_{11}N$ 100 (17)
$C_{11}H_{11}N$ 102 (24)
$C_{11}H_{12}ON$ 114 (22)
$C_{11}H_{12}O_2N_2$ 124 (15)
$C_{11}H_{14}O_2$ 99 (24)
$C_{11}H_{14}O_2$ 99 (25)
$C_{11}H_{16}$ 96 (20)
$C_{11}H_{17}N$ 100 (14)
$C_{11}H_{17}N$ 100 (15)
$C_{11}H_{22}O_2$ 112 (11)
$C_{11}H_{22}O_2$ 115 (18)
$C_{11}H_{24}$ 81 (7)

$C_{12}H_6O_{12}$ 116 (10)
$C_{12}H_8$ 86 (5)
$C_{12}H_8O$ 92 (4)
$C_{12}H_8N_2$ 93 (20)
$C_{12}H_8N_2$ 93 (21)
$C_{12}H_8S_2$ 92 (6)
$C_{12}H_9O_4$ 116 (16)
$C_{12}H_9ON$ 94 (6)
$C_{12}H_9N$ 93 (13)
$C_{12}H_9NS$ 94 (9)
$C_{12}H_{10}$ 98 (3)
$C_{12}H_{10}$ 86 (4)
$C_{12}H_{10}O$ 103 (1)
$C_{12}H_{10}ON$ 124 (20)
$C_{12}H_{10}ON_2$ 124 (21)
$C_{12}H_{10}O_3N_2S$ 128 (3)
$C_{12}H_{10}O_4$ 118 (12)
$C_{12}H_{10}N$ 103 (25)
$C_{12}H_{10}N_2$ 121 (7)
$C_{12}H_{10}N_2$ 103 (14)
$C_{12}H_{10}S$ 103 (5)
$C_{12}H_{10}S_2$ 103 (6)
$C_{12}H_{11}N$ 103 (9)
$C_{12}H_{11}N$ 104 (5)
$C_{12}H_{11}N$ 104 (4)
$C_{12}H_{11}N_2$ 120 (19)
$C_{12}H_{11}N_3$ 103 (20)
$C_{12}H_{12}$ 97 (3)
$C_{12}H_{12}$ 97 (4)
$C_{12}H_{12}$ 97 (5)
$C_{12}H_{12}$ 97 (6)
$C_{12}H_{12}$ 97 (7)
$C_{12}H_{12}$ 97 (6)
$C_{12}H_{12}$ 97 (7)
$C_{12}H_{12}$ 97 (8)
$C_{12}H_{12}$ 97 (9)
$C_{12}H_{12}$ 97 (10)
$C_{12}H_{12}$ 97 (11)
$C_{12}H_{12}$ 97 (12)
$C_{12}H_{12}$ 97 (13)
$C_{12}H_{12}$ 97 (14)
$C_{12}H_{12}N_2$ 103 (13)
$C_{12}H_{13}O_2N_6S$ 128 (5)
$C_{12}H_{13}N$ 100 (18)
$C_{12}H_{13}N$ 100 (19)
$C_{12}H_{13}N$ 100 (20)
$C_{12}H_{13}N$ 100 (21)
$C_{12}H_{14}N_2$ 120 (13)
$C_{12}H_{16}$ 98 (2)
$C_{12}H_{16}O_7Br_2$ 126 (36)
$C_{12}H_{18}$ 96 (21)
$C_{12}H_{18}O_3$ 99 (21)
$C_{12}H_{22}$ 98 (1)
$C_{12}H_{22}O_3$ 113 (5)
$C_{12}H_{24}O_2$ 112 (19)
$C_{12}H_{24}O_2$ 115 (12)
$C_{12}H_{26}$ 81 (8)
$C_{12}H_{26}O$ 87 (18)
$C_{12}H_{26}O$ 106 (17)
$C_{12}H_{27}N$ 119 (7)

C$_{12}$H$_{27}$N 88 (38)

C$_{13}$H$_2$O 110 (36)
C$_{13}$H$_9$N 93 (14)
C$_{13}$H$_9$N 93 (16)
C$_{13}$H$_9$N 93 (17)
C$_{13}$H$_9$N 93 (18)
C$_{13}$H$_{10}$ 86 (3)
C$_{13}$H$_{10}$O 92 (5)
C$_{13}$H$_{10}$O 114 (7)
C$_{13}$H$_{11}$N 93 (15)
C$_{13}$H$_{12}$ 98 (4)
C$_{13}$H$_{12}$O 99 (11)
C$_{13}$H$_{12}$O 99 (12)
C$_{13}$H$_{13}$N$_2$ 103 (18)
C$_{13}$H$_{13}$N 103 (10)
C$_{13}$H$_{13}$N 103 (22)
C$_{13}$H$_{17}$ON$_2$ 114 (23)
C$_{13}$H$_{20}$O 110 (36)
C$_{13}$H$_{20}$O 114 (3)
C$_{13}$H$_{26}$O$_2$ 115 (13)
C$_{13}$H$_{28}$ 81 (9)

C$_{14}$H$_7$O$_4$N 123 (7)
C$_{14}$H$_8$O$_2$ 111 (18)
C$_{14}$H$_8$O$_2$ 111 (20)
C$_{14}$H$_9$Cl 5 125 (31)
C$_{14}$H$_{10}$ 84 (2)
C$_{14}$H$_{10}$ 84 (3)
C$_{14}$H$_{10}$ 98 (9)
C$_{14}$H$_{10}$O 107 (23)
C$_{14}$H$_{10}$O 111 (19)
C$_{14}$H$_{10}$O$_2$ 107 (24)
C$_{14}$H$_{10}$O$_2$ 107 (25)
C$_{14}$H$_{10}$O$_2$ 114 (8)
C$_{14}$H$_{10}$O$_3$ 117 (14)
C$_{14}$H$_{10}$O$_4$ 114 (21)
C$_{14}$H$_{10}$N$_{10}$ 104 (20)
C$_{14}$H$_{12}$ 86 (9)
C$_{14}$H$_{12}$ 98 (7)
C$_{14}$H$_{12}$ 98 (8)
C$_{14}$H$_{12}$O$_2$ 117 (11)
C$_{14}$H$_{12}$O$_3$ 117 (30)

C$_{14}$H$_{12}$N$_2$ 103 (19)
C$_{14}$H$_{13}$N$_2$ 104 (11)
C$_{14}$H$_{14}$ 98 (5)
C$_{14}$H$_{14}$ 98 (6)
C$_{14}$H$_{14}$O 103 (2)
C$_{14}$H$_{14}$O$_2$ 103 (3)
C$_{14}$H$_{14}$N$_2$ 120 (16)
C$_{14}$H$_{14}$S 103 (7)
C$_{14}$H$_{14}$S$_2$ 103 (8)
C$_{14}$H$_{16}$ 98 (17)
C$_{14}$H$_{16}$N 103 (11)
C$_{14}$H$_{16}$O$_2$N$_2$ 120 (15)
C$_{14}$H$_{16}$N$_2$ 103 (12)
C$_{14}$H$_{16}$N$_2$ 120 (14)
C$_{14}$H$_{17}$N 100 (22)
C$_{14}$H$_{19}$O$_9$Br 126 (36)
C$_{14}$H$_{28}$O$_2$ 112 (20)
C$_{14}$H$_{28}$O$_2$ 115 (14)
C$_{14}$H$_{30}$ 81 (10)
C$_{14}$H$_{30}$O 106 (18)
C$_{14}$H$_{30}$O 87 (19)
C$_{14}$H$_{31}$N 88 (33)

C$_{15}$H$_{11}$N 104 (6)
C$_{15}$H$_{11}$N 104 (7)
C$_{15}$H$_{11}$N 104 (8)

C$_{15}$H$_{11}$O$_4$NI$_4$ 126 (39)
C$_{15}$H$_{16}$O 103 (4)
C$_{15}$H$_{30}$O$_2$ 115 (15)
C$_{15}$H$_{32}$ 81 (11)
C$_{15}$H$_{33}$N 88 (39)

C$_{16}$H$_{10}$ 84 (7)
C$_{16}$H$_{10}$ 86 (6)
C$_{16}$H$_{11}$O$_2$N 116 (23)
C$_{16}$H$_{13}$N 104 (9)
C$_{16}$H$_{30}$O 114 (5)
C$_{16}$H$_{32}$O$_2$ 115 (16)
C$_{16}$H$_{34}$ 81 (12)
C$_{16}$H$_{34}$O 106 (19)

C$_{17}$H$_{10}$O 111 (21)
C$_{17}$H$_{30}$O 111 (13)

C$_{17}$H$_{32}$O 111 (12)
C$_{17}$H$_{34}$O$_2$ 115 (17)
C$_{17}$H$_{36}$ 81 (13)

C$_{18}$H$_{12}$ 84 (4)
C$_{18}$H$_{12}$ 84 (5)
C$_{18}$H$_{12}$ 84 (6)
C$_{18}$H$_{12}$N$_2$ 104 (17)
C$_{18}$H$_{14}$ 98 (10)
C$_{18}$H$_{14}$ 98 (11)
C$_{18}$H$_{15}$N 103 (23)
C$_{18}$H$_{16}$ 97 (22)
C$_{18}$H$_{16}$ 97 (21)
C$_{18}$H$_{30}$O$_2$ 115 (28)
C$_{18}$H$_{30}$O$_2$ 115 (29)
C$_{18}$H$_{32}$O$_2$ 115 (27)
C$_{18}$H$_{34}$O$_2$ 115 (25)
C$_{18}$H$_{34}$O$_2$ 115 (26)
C$_{18}$H$_{36}$O 106 (31)
C$_{18}$H$_{36}$O 110 (12)
C$_{18}$H$_{36}$O$_2$ 115 (18)
C$_{18}$H$_{38}$ 81 (14)
C$_{18}$H$_{38}$O 106 (20)

C$_{19}$H$_{14}$O 114 (9)
C$_{19}$H$_{15}$Cl 125 (30)
C$_{19}$H$_{16}$ 98 (12)
C$_{19}$H$_{16}$ 98 (13)
C$_{19}$H$_{16}$O 109 (16)
C$_{19}$H$_{38}$O$_2$ 115 (19)
C$_{19}$H$_{40}$ 81 (15)

C$_{20}$H$_{12}$ 84 (10)
C$_{20}$H$_{12}$ 84 (12)
C$_{20}$H$_{14}$ 86 (7)
C$_{20}$H$_{14}$ 84 (8)
C$_{20}$H$_{14}$ 98 (18)
C$_{20}$H$_{14}$ 98 (19)
C$_{20}$H$_{14}$ 98 (20)
C$_{20}$H$_{14}$N$_2$ 103 (16)
C$_{20}$H$_{14}$N$_2$ 103 (17)
C$_{20}$H$_{14}$N$_4$ 94 (10)
C$_{20}$H$_{15}$N$_3$ 103 (21)

C$_{20}$H$_{16}$ 86 (10)
C$_{20}$H$_{16}$N$_4$ 94 (11)
C$_{20}$H$_{40}$O 106 (32)
C$_{20}$H$_{40}$O$_2$ 115 (20)
C$_{20}$H$_{42}$ 81 (16)
C$_{21}$H$_{44}$ 81 (17)
C$_{22}$H$_{14}$ 84 (9)
C$_{22}$H$_{46}$ 81 (18)
C$_{23}$H$_{48}$ 81 (19)
C$_{24}$H$_{12}$ 84 (13)
C$_{24}$H$_{16}$ 84 (11)
C$_{24}$H$_{18}$ 98 (14)
C$_{24}$H$_{18}$N$_2$ 103 (15)
C$_{24}$H$_{20}$N$_2$ 103 (24)
C$_{24}$H$_{50}$ 81 (20)
C$_{24}$H$_{50}$O 87 (20)
C$_{25}$H$_{20}$ 98 (15)
C$_{25}$H$_{52}$ 81 (21)
C$_{26}$H$_{54}$O 106 (21)
C$_{28}$H$_{58}$O 106 (22)

C$_{30}$H$_{62}$ 81 (22)
C$_{30}$H$_{62}$O 106 (23)
C$_{32}$H$_{12}$ 84 (14)
C$_{34}$H$_{10}$O$_2$ 111 (22)
C$_{34}$H$_{16}$O$_2$ 111 (23)
C$_{35}$H$_{72}$ 81 (23)
C$_{36}$H$_{70}$O$_3$ 113 (6)
C$_{38}$H$_{30}$ 98 (16)
C$_{40}$H$_{56}$ 98 (21)
C$_{40}$H$_{56}$ 98 (22)
C$_{40}$H$_{82}$ 81 (24)
C$_{50}$H$_{102}$ 81 (25)
C$_{51}$H$_{98}$O$_6$ 112 (27)
C$_{54}$H$_{104}$O$_6$ 112 (28)
C$_{54}$H$_{110}$O$_6$ 112 (29)
C$_{60}$H$_{122}$ 81 (26)
C$_{62}$H$_{126}$ 81 (27)
C$_{64}$H$_{130}$ 81 (28)
C$_{70}$H$_{142}$ 81 (29)